# Autophagy and Inflammation in Neuropsychological Disorders

Tahmineh Mokhtari
*Hubei University of Medicine, China*

Kadir Uludag
*Shanghai Jiao Tong University, China*

IGI Global
Scientific Publishing
Publishing Tomorrow's Research Today

Vice President of Editorial Melissa Wagner
Director of Acquisitions Mikaela Felty
Director of Book Development Jocelynn Hessler
Production Manager Mike Brehm
Cover Design Jose Rosado

Published in the United States of America by
IGI Global Scientific Publishing
701 East Chocolate Avenue
Hershey, PA, 17033, USA
Tel: 717-533-8845
Fax: 717-533-7115
Website: https://www.igi-global.com E-mail: cust@igi-global.com



Library of Congress Cataloging-in-Publication Data

Names: Mokhtari, Tahmineh, 1986- editor http://id.loc.gov/rwo/agents/n2025014281 | Uludağ, Kadir, 1994- editor http://id.loc.gov/rwo/agents/n2025003700

Title: Autophagy and inflammation in neuropsychological disorders / edited by: Tahmineh Mokhtari, Kadir Uludag.

Description: Hershey PA : IGI Global Scientific Publishing, [2026] | Includes bibliographical references. | Summary: "The primary objective of the book is to provide a comprehensive and up-to-date exploration of the interplay between autophagy and inflammation in the context of neuropsychological disorders. We welcome not only books directly focused on autophagy and inflammation but also topics associated with inflammation to enhance the breadth of our book and enhance our comprehension of the relationship between autophagy and inflammation"-- Provided by publisher.

Identifiers: LCCN 2025003466 (print) | LCCN 2025003467 (ebook) | ISBN 9798369359082 hardcover | ISBN 9798369359099 ebook

Subjects: LCSH: Nervous system--Diseases--Treatment--Case studies | Inflammation--Immunological aspects--Case studies | Cell death--Case studies | LCGFT: Case studies. https://id.loc.gov/authorities/genreForms/gf2017026140

Classification: LCC RC349.8 .A65 2025 (print) | LCC RC349.8 (ebook) | DDC 616.8--dc23/eng/20250411

LC record available at https://lccn.loc.gov/2025003466

LC ebook record available at https://lccn.loc.gov/2025003467

British Cataloguing in Publication Data

A Cataloguing in Publication record for this book is available from the British Library.

# Table of Contents

**Preface** ........................................ xi

**Chapter 1**
Autophagy and Its Role in Neurodegenerative Diseases ........................................ 1
*Fatih Kara, Faculty of Medicine, Kafkas University, Turkey*
*Abdulsamet Mavzer, Faculty of Medicine, Kafkas University, Turkey*
*İrem Karalik, Faculty of Medicine, Kafkas University, Turkey*
*Ayşe S. Baran, Faculty of Medicine, Kafkas University, Turkey*

**Chapter 2**
Autophagy-Based Treatment Approaches in Neuropsychological Diseases ...... 27
*Şebnem Yıldırımcan Kadıçeşme, Kafkas University, Turkey*

**Chapter 3**
Modulating Neuroinflammation and Autophagy in Psychiatric Disorders Through Mechanisms and Therapies: Therapeutic Strategies and Pathways ...... 61
*Reyed M. Reyed, Genetic Engineering and Biotechnology Research Institute (GEBRI), City of Scientific Research and Technological Applications (SRTA-City), Alexandria, Egypt*
*Pranav Kumar Prabhakar, Department of Biotechnology, School of Engineering and Technology, Nagaland University, Kohima, India*

**Chapter 4**
Autophagy Pathways in Major Depressive Disorder ........................................ 125
*Golnaz Shafiei, Anatomical Sciences Research Center, Institute for Basic Sciences, Kashan University of Medical Sciences, Kashan, Iran*
*Maryam Saadat, Department of Anatomical Sciences, School of Medicine, Semnan University of Medical Sciences, Semnan, Iran*
*Shima Farajpour, Department of Anatomy, School of Medicine, Tehran University of Medical Sciences, Tehran, Iran*
*Shima Mohammadi, Department of Neuroscience, School of Medicine, Shahroud University of Medical Sciences, Shahroud, Iran*
*Nasrin Dashti, Medical Laboratory Sciences, Allied Medical Sciences, Tehran University of Medical Sciences, Tehran, Iran*
*Tahmineh Mokhtari, Department of Entomology and Nematology, University of California, Davis, USA*

**Chapter 5**
Autophagy in Spinal Cord Injury: A Double-Edged Sword in Preclinical Studies of Synthetic and Natural Compounds .................................................. 163
*Tahmineh Mokhtari, Department of Entomology and Nematology, University of California, Davis, USA*
*Farzaneh Majidi, Department of Anatomy, Tehran University of Medical Sciences, Tehran, Iran*
*Fatemeh Sheikhbahaei, Department of Anatomical Sciences, Kerman University of Medical Sciences, Kerman, Iran*
*Parvaneh Behi Shahreza, Department of Anatomy, Tehran University of Medical Sciences, Tehran, Iran*
*Mahnaz Poorhassan, Department of Anatomical Sciences and Cognitive Neuroscience, Tehran Medical Sciences, Islamic Azad University, Tehran, Iran*

**Chapter 6**
Autophagy and Inflammation in Neuropsychological Disorders: Autophagy in the Neurological-Visual Disease - Mechanisms and Therapeutic Potential .. 199
*Fatemeh Azimi, Iran University of Medical Sciences, Tehran, Iran*

**Chapter 7**
Polycystic Ovary Syndrome PCOS and Depression Role of Neuroinflammation ........ 229
*Mohammad Reza Ramezanpour, Pediatric Urology and Regenerative Medicine Research Center, Gene, Cell, and Tissue Research Institute, Children's Medical Center, Tehran University of Medical Sciences, Tehran, Iran*
*Omid Bahrami, Pediatric Urology and Regenerative Medicine Research Center, Gene, Cell, and Tissue Research Institute, Children's Medical Center, Tehran University of Medical Sciences, Tehran, Iran*
*Muhammad Parsa Pashazadeh, Pediatric Urology and Regenerative Medicine Research Center, Gene, Cell, and Tissue Research Institute, Children's Medical Center, Tehran University of Medical Sciences, Tehran, Iran*
*Rojin Sarallah, Islamic Azad University Medical Branch of Tehran, Tehran, Iran*
*Reza Nikfar, Pediatric Urology and Regenerative Medicine Research Center, Gene, Cell, and Tissue Research Institute, Children's Medical Center, Tehran University of Medical Sciences, Tehran, Iran*
*Masoumeh Majidi Zolbin, Iranian Tissue Bank and Research Center, Tehran University of Medical Sciences, Tehran, Iran*

**Chapter 8**
Unveiling Nano-Specific Toxicity: Empowering Medical Literacy and Exploring Its Connection to Autophagy ........ 263
*Kadir Uludag, Shanghai Jiao Tong University, China*
*Amjad Islam Amjad, School Education Department, Government of Punjab, Kasur, Pakistan*

**Compilation of References** ........ 281

**About the Contributors** ........ 383

**Index** ........ 389

# Detailed Table of Contents

**Preface** ........................................................................................................ xi

**Chapter 1**
Autophagy and Its Role in Neurodegenerative Diseases ...................................... 1
*Fatih Kara, Faculty of Medicine, Kafkas University, Turkey*
*Abdulsamet Mavzer, Faculty of Medicine, Kafkas University, Turkey*
*İrem Karalik, Faculty of Medicine, Kafkas University, Turkey*
*Ayşe S. Baran, Faculty of Medicine, Kafkas University, Turkey*

Autophagy is an intrinsic cellular process whereby cells degrade their own constituents, enabling them to acquire vital components during periods of nutrient deprivation and to eliminate damaged organelles and misfolded proteins. Neurodegenerative diseases are predominantly characterized by the accumulation of protein aggregates and tend to be more common with advancing age. Given that autophagy becomes less effective with age and is crucial for maintaining cellular homeostasis, its potential link to neurodegenerative disorders raises significant questions. Moreover, since neurons are post-mitotic and possess limited regenerative capacity, they are particularly reliant on autophagic mechanisms to sustain their viability. Understanding the role of autophagy in neurodegenerative diseases is critical not only for elucidating the pathogenesis of these diseases but also for developing effective therapeutic interventions. This chapter elucidates the mechanisms underlying autophagy in the context of recent findings and underscores its role in the progression of neurodegenerative diseases.

**Chapter 2**
Autophagy-Based Treatment Approaches in Neuropsychological Diseases ...... 27
*Şebnem Yıldırımcan Kadıçeşme, Kafkas University, Turkey*

Autophagy is an extremely important molecular pathway for the maintenance of cellular and organismal balance. According to recent research, cognitive diseases like dementia, Alzheimer's disease, Parkinson's disease, Huntington's disease, attention-deficit hyperactivity disorder, schizophrenia, and bipolar disorder are all influenced by defective autophagy. Neuropsychiatric disorders are extremely important because they affect the brain function, emotions, and mood of patients. Treatment of neuropsychiatric disorders is a very difficult process, and some problems are encountered in response to treatment. Therefore, new treatment approaches are needed. Given its significance in the pathophysiology of neuropsychological diseases, the autophagy pathway appears to be a promising target for the development of novel therapeutic approaches for these conditions. New research should be focused on explaining the relationship between this pathway and neuropsychological disorders. The data obtained from these studies will pave the way for the development of new therapeutic agents.

**Chapter 3**
Modulating Neuroinflammation and Autophagy in Psychiatric Disorders Through Mechanisms and Therapies: Therapeutic Strategies and Pathways ...... 61
*Reyed M. Reyed, Genetic Engineering and Biotechnology Research Institute (GEBRI), City of Scientific Research and Technological Applications (SRTA-City), Alexandria, Egypt*
*Pranav Kumar Prabhakar, Department of Biotechnology, School of Engineering and Technology, Nagaland University, Kohima, India*

The chapter presents the relationship between autophagy and neuroinflammation in major depression, bipolar disorder, and schizophrenia. It emphasizes that oxidative stress coupled with mitochondrial dysfunction can activate mTOR, AMPK, and FOXO1, thereby influencing synaptic function and behavior. The influence of microglia, HMGB1, and Toll-like receptors is discussed. The chapter explores how altered gut microbiota and nutrition affect autophagy by regulating short-chain fatty acid levels and inflammatory signals. The effects of lifestyle modifications, such as time-restricted feeding and the use of key phytochemicals (curcumin, resveratrol, hydroxytyrosol, thymoquinone, saffron), on autophagy and inflammation levels are investigated. Summary tables and case studies support the results presented. Neuronal organoids and 3D-printed brain models are introduced as innovative tools to explore disease pathways for future clinical initiatives.

**Chapter 4**
Autophagy Pathways in Major Depressive Disorder ....................................... 125
*Golnaz Shafiei, Anatomical Sciences Research Center, Institute for Basic Sciences, Kashan University of Medical Sciences, Kashan, Iran*
*Maryam Saadat, Department of Anatomical Sciences, School of Medicine, Semnan University of Medical Sciences, Semnan, Iran*
*Shima Farajpour, Department of Anatomy, School of Medicine, Tehran University of Medical Sciences, Tehran, Iran*
*Shima Mohammadi, Department of Neuroscience, School of Medicine, Shahroud University of Medical Sciences, Shahroud, Iran*
*Nasrin Dashti, Medical Laboratory Sciences, Allied Medical Sciences, Tehran University of Medical Sciences, Tehran, Iran*
*Tahmineh Mokhtari, Department of Entomology and Nematology, University of California, Davis, USA*

Major depressive disorder (MDD) is a chronic condition with high prevalence, especially in individuals with physical illnesses. This chapter explores autophagy's involvement in MDD, connecting it to oxidative stress, increased pro-inflammatory cytokine levels, and impaired mitochondrial function. Dysregulated autophagy contributes to MDD pathophysiology, offering insights into novel therapeutic strategies. The authors review preclinical and clinical evidence, the effects of conventional antidepressants on autophagy, and the potential of herbal products in modulating these pathways. Understanding autophagy's role in MDD underscores the need for further research to develop targeted treatments.

**Chapter 5**
Autophagy in Spinal Cord Injury: A Double-Edged Sword in Preclinical Studies of Synthetic and Natural Compounds ................................................. 163
*Tahmineh Mokhtari, Department of Entomology and Nematology, University of California, Davis, USA*
*Farzaneh Majidi, Department of Anatomy, Tehran University of Medical Sciences, Tehran, Iran*
*Fatemeh Sheikhbahaei, Department of Anatomical Sciences, Kerman University of Medical Sciences, Kerman, Iran*
*Parvaneh Behi Shahreza, Department of Anatomy, Tehran University of Medical Sciences, Tehran, Iran*
*Mahnaz Poorhassan, Department of Anatomical Sciences and Cognitive Neuroscience, Tehran Medical Sciences, Islamic Azad University, Tehran, Iran*

Autophagy is a crucial cellular process involved in spinal cord injury (SCI) pathophysiology. It can protect neural tissues by removing damaged cellular components, thereby promoting recovery. However, excessive autophagy may lead to neuronal apoptosis, inflammation, and disrupted cellular balance. This chapter explores the balance between autophagic activity and cellular health, emphasizing the need for precise regulation to harness autophagy's therapeutic potential. Recent studies highlight how autophagy interacts with mitochondrial dysfunction, reactive oxygen species, and inflammatory responses in SCI. Moreover, it discusses how natural compounds and nutraceuticals can modulate autophagy, offering promising strategies for managing consequences of SCI. Understanding autophagy is vital for advancing therapeutic interventions in SCI.

**Chapter 6**
Autophagy and Inflammation in Neuropsychological Disorders: Autophagy in the Neurological-Visual Disease - Mechanisms and Therapeutic Potential .. 199
*Fatemeh Azimi, Iran University of Medical Sciences, Tehran, Iran*

Vision is a highly complex neural process that begins in the retina and is transmitted to the brain through the optic nerve, relying on the precise function of highly specialized neural and glial cells. It is of great importance for the integrity of vision to maintain the health of these cells. Autophagy, a lysosome-dependent degradation pathway, plays a crucial role in maintaining cellular homeostasis by eliminating misfolded proteins and damaged organelles. Dysregulation of autophagy has been recognized as a pivotal factor in the pathogenesis of multiple neurological-visual diseases, such as glaucoma, age-related macular degeneration (AMD), and inherited retinal dystrophies (IRDs). This chapter studies how autophagy is essentially governed at the molecular level during physiological/pathophysiological processes, particularly in the retina and optic nerve. Finally, emerging therapeutic approaches that aim to regulate autophagy are discussed to offer potential avenues to slow disease progression and preserve vision in neurodegenerative eye disorders.

**Chapter 7**
Polycystic Ovary Syndrome PCOS and Depression Role of Neuroinflammation ........ 229
*Mohammad Reza Ramezanpour, Pediatric Urology and Regenerative Medicine Research Center, Gene, Cell, and Tissue Research Institute, Children's Medical Center, Tehran University of Medical Sciences, Tehran, Iran*
*Omid Bahrami, Pediatric Urology and Regenerative Medicine Research Center, Gene, Cell, and Tissue Research Institute, Children's Medical Center, Tehran University of Medical Sciences, Tehran, Iran*
*Muhammad Parsa Pashazadeh, Pediatric Urology and Regenerative Medicine Research Center, Gene, Cell, and Tissue Research Institute, Children's Medical Center, Tehran University of Medical Sciences, Tehran, Iran*
*Rojin Sarallah, Islamic Azad University Medical Branch of Tehran, Tehran, Iran*
*Reza Nikfar, Pediatric Urology and Regenerative Medicine Research Center, Gene, Cell, and Tissue Research Institute, Children's Medical Center, Tehran University of Medical Sciences, Tehran, Iran*
*Masoumeh Majidi Zolbin, Iranian Tissue Bank and Research Center, Tehran University of Medical Sciences, Tehran, Iran*

Polycystic ovary syndrome (PCOS) is a multidimensional condition with metabolic, hormonal, and neuropsychiatric effects. Depression is common among PCOS patients, and emerging research shows neuroinflammation as a probable reason. Chronic low-grade inflammation, insulin resistance, hyperandrogenism, and gut dysbiosis all contribute to neuroinflammatory processes, which result in hypothalamic dysfunction, neurotransmitter imbalances, and hypothalamic-pituitary-adrenal (HPA) axis dysregulation. Abnormal microglial activation and oxidative stress may further damage neuroplasticity, interfering with proper mood regulation. This chapter investigates how neuroinflammatory pathways influence the relationship between PCOS and depression, focusing on their effects on brain function and mood regulation. This chapter addresses neuroinflammation as a mechanism contributing to psychiatric symptoms in PCOS by identifying key inflammatory drivers, as well as prospective treatment targets.

**Chapter 8**
Unveiling Nano-Specific Toxicity: Empowering Medical Literacy and Exploring Its Connection to Autophagy ............................................................ 263
*Kadir Uludag, Shanghai Jiao Tong University, China*
*Amjad Islam Amjad, School Education Department, Government of Punjab, Kasur, Pakistan*

This chapter explores the misconception surrounding nano-specific toxicity and emphasises the critical role of improved medical literacy in addressing this issue. By delving into the realm of nanotechnology and its potential health implications, this study highlights the importance of accurate information dissemination and education to dispel unfounded fears. Through enhanced medical literacy, stakeholders can navigate the complexities of nano-toxicity concerns more effectively, paving the way for informed decision-making and fostering a deeper understanding of nanomaterials' true impact on health and well-being. Furthermore, this chapter delves into the fascinating relationship between nano-specific toxicity and autophagy, unveiling insights that could have profound implications for both medicine and nanotechnology. The authors have also discussed the potential connection between fasting and medical understanding of autophagy.

**Compilation of References** ............................................................ 281

**About the Contributors** ............................................................ 383

**Index** ............................................................ 389

# Preface

In recent years, the intersection of autophagy, inflammation, and neuropsychological disorders has become a burgeoning area of research, shedding light on the intricate molecular pathways that underlie a range of debilitating conditions. *Autophagy and Inflammation in Neuropsychological Disorders* seeks to offer an in-depth exploration of these phenomena and their complex relationship with neuropsychiatric diseases, such as Alzheimer's disease, schizophrenia, and mood disorders.

This book brings together the work of leading experts in the fields of neuroscience, biology, and clinical research, offering a comprehensive overview of the role of autophagy and inflammation in brain health. The content is organized into several key sections:

This work is not only for researchers and clinicians but for anyone invested in the future of neuropsychology, providing a roadmap to understanding the molecular basis of disease and the potential pathways to improved treatments.

# Chapter 1
# Autophagy and Its Role in Neurodegenerative Diseases

**Fatih Kara**
https://orcid.org/0000-0002-9729-5380
*Faculty of Medicine, Kafkas University, Turkey*

**Abdulsamet Mavzer**
https://orcid.org/0009-0004-2599-6497
*Faculty of Medicine, Kafkas University, Turkey*

**İrem Karalik**
https://orcid.org/0009-0005-9766-9257
*Faculty of Medicine, Kafkas University, Turkey*

**Ayşe S. Baran**
https://orcid.org/0009-0005-3151-4420
*Faculty of Medicine, Kafkas University, Turkey*

## ABSTRACT

*Autophagy is an intrinsic cellular process whereby cells degrade their own constituents, enabling them to acquire vital components during periods of nutrient deprivation and to eliminate damaged organelles and misfolded proteins. Neurodegenerative diseases are predominantly characterized by the accumulation of protein aggregates and tend to be more common with advancing age. Given that autophagy becomes less effective with age and is crucial for maintaining cellular homeostasis, its potential link to neurodegenerative disorders raises significant questions. Moreover, since neurons are post-mitotic and possess limited regenerative capacity, they are particularly reliant on autophagic mechanisms to sustain*

DOI: 10.4018/979-8-3693-5908-2.ch001

*their viability. Understanding the role of autophagy in neurodegenerative diseases is critical not only for elucidating the pathogenesis of these diseases but also for developing effective therapeutic interventions. This chapter elucidates the mechanisms underlying autophagy in the context of recent findings and underscores its role in the progression of neurodegenerative diseases.*

## AUTOPHAGY

Autophagy is a process in which cells break down their organelles and proteins for recycling (J. Liu et al., 2024; S. Liu et al., 2023). Autophagy is a physiological mechanism aimed at maintaining homeostasis in cells (Gómez-Virgilio et al., 2022; Pavlinov et al., 2020). Autophagy provides crucial contributions to cells by balancing energy sources in response to nutrient stress, removing misfolded or aggregated proteins, clearing damaged organelles, and eliminating intracellular pathogens (Bai et al., 2020; Morimoto et al., 2023). During periods of starvation, autophagy is triggered to support cell survival by supplying essential nutrients through non-selective autophagy (Almannai et al., 2022; Iriondo et al., 2023; Yu et al., 2023). Additionally, it serves a critical housekeeping function by selectively degrading misfolded or aggregated proteins, damaged or unnecessary organelles, and intracellular pathogens through selective autophagy (Almannai et al., 2022; Iriondo et al., 2023). However, disruption of autophagy or excessive autophagic activity can lead to impaired cellular functions, which may be associated with diseases (Gómez-Virgilio et al., 2022; S. Liu et al., 2023). Additionally, the agents targeting the control and regulation of autophagy may contribute to the treatment of diseases (S. Liu et al., 2023).

## AUTOPHAGY TYPES

There are three types of autophagy: macroautophagy, microautophagy, and chaperone-mediated autophagy (CMA) (L. Chen et al., 2024; Morimoto et al., 2023) (Figure 1). Macroautophagy is the predominant mechanism of autophagy (Tutas et al., 2025). In macroautophagy, the phagophore, an isolation membrane, engulfs and sequesters cellular material (i.e. misfolded and aggregated proteins, damaged or unnecessary organelles, or microorganisms), leading to the formation of a double-membrane autophagosome (Demeter et al., 2020; Gómez-Virgilio et al., 2022). The autophagosome fuses with a lysosome, forming a single-membrane autolysosome, where the outer membrane merges with the lysosomal membrane, allowing the degradation of the inner membrane and its enclosed cargo (Holzer et al., 2024). Macroautophagy is capable of degrading cellular cargo that is larger than a single lysosome (Holzer et al., 2024). It has been shown that the activity of macroautophagy during nutrient scarcity peaks approximately 4 to 6 hours after the start of fasting (Gómez-Virgilio et al., 2022). In microautophagy, cellular material is directly engulfed into the lysosome via invaginations in the membrane (Gómez-Virgilio et al., 2022). In CMA, proteins containing the KFERQ pentapeptide motif or a sequence biochemically similar to KFERQ are recognized by Hsc70 and directed to the lysosome, where they are subsequently internalized through LAMP2A, a protein located in the lysosomal membrane that acts as a channel for protein translocation (L. Chen et al., 2024; Gómez-Virgilio et al., 2022; Kacal et al., 2021). KFERQ-like motifs include, in addition to glutamine, alkaline, acidic, and hydrophobic residues (Gómez-Virgilio et al., 2022; Schnebert et al., 2022). Such sequences may be inherent in the protein's primary structure or generated through post-translational modifications (Gómez-Virgilio et al., 2022). Some chaperones, such as Hsp90, Hsp40, Hop, Hip, and BAG-1, assist the functions of Hsc70 (Gómez-Virgilio et al., 2022; Schnebert et al., 2022). CMA serves as a primary defense mechanism against protein aggregation by facilitating the degradation of proteins that have undergone damage or partial unfolding, thereby contributing to protein quality control (Gómez-Virgilio et al., 2022; Hormazabal et al., 2022). CMA plays a role in the breakdown of proteins associated with neurodegenerative diseases, including α-synuclein (α-syn) and tau (Y.-T. Wang & Lu, 2022). With aging, there is a decline in protein quality control within neurons, and dysregulated CMA may play a role in neurodegenerative diseases (Gómez-Virgilio et al., 2022; Y.-T. Wang & Lu, 2022). Starvation upregulates the activity of CMA (Krause et al., 2023). After at least 8 hours of fasting, CMA provides significant levels of free amino acids to the cells, and CMA activity peaks 12 to 20 hours after the onset of fasting (Gómez-Virgilio et al., 2022).

*Figure 1. Types of autophagy*

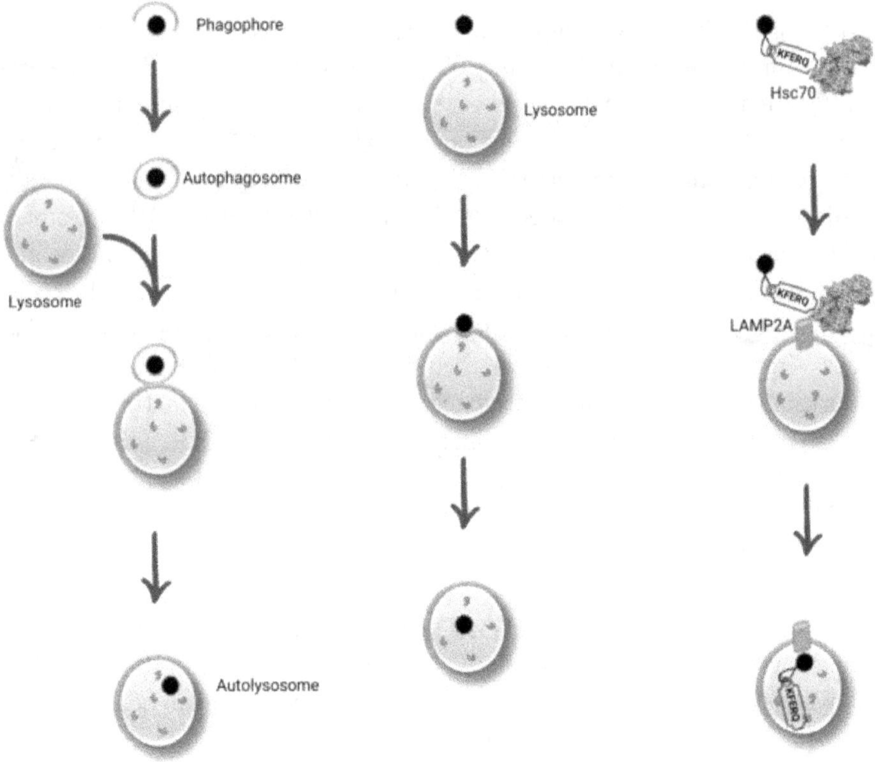


## Macroautophagy Process

The major mammalian proteins involved in the process of macroautophagy, the stages at which they function, their roles, and their yeast counterparts are presented in Table 1.

*Table 1. The major proteins involved in the process of macroautophagy*

| Mammalian Protein | Macroautophagy Stage | Function | Yeast Counterpart[1] |
|---|---|---|---|
| ULK1, ULK2 | Induction | Catalytic subunit of ULK complex.<br><br>Under nutrient-rich conditions, mTORC1 phosphorylates ULK1/2 and ATG13, resulting in their inactivation. Upon nutrient deprivation, this inhibitory phosphorylation is lifted, allowing ULK1/2 to become activated. Once activated, ULK1/2 exerts its protein kinase activity to initiate autophagy. | Atg1 |
| ATG2A, ATG2B | Elongation | Lipid transfer protein.<br><br>ATG2A and ATG2B support autophagosome membrane expansion by mediating lipid transport.<br><br>For simplicity, ATG2A and ATG2B will be referred to as ATG2 henceforth. | Atg2 |
| ATG3 | Elongation | E2-like enzyme.<br><br>ATG3 is involved in the conjugation reaction with PE to convert LC3-I to LC3-II. | Atg3 |
| ATG4A, ATG4B, ATG4C, ATG4D | Elongation | Cysteine protease.<br><br>ATG4A, ATG4B, ATG4C, and ATG4D[2] are involved in processing the mammalian orthologs of yeast Atg8, namely the LC3 and GABARAP subfamilies.<br><br>For simplicity, the ATG4 family will be referred to as ATG4 henceforth. | Atg4 |
| ATG5 | Elongation | Component of the ATG12-ATG5-ATG16L1 complex.<br><br>The ATG12-ATG5-ATG16L1 complex, acting as an E3-like enzyme, is involved in the conjugation reaction with PE to convert LC3-I to LC3-II. | Atg5 |
| Beclin 1 | Nucleation | Scaffold protein in the PI3KC3-C1 complex.<br><br>The PI3KC3-C1 complex converts PI to PI3P. | Atg6 |
| ATG7 | Elongation | E1-like enzyme.<br><br>ATG7 directs LC3-I and ATG12 to their E2-like enzymes, ATG3 and ATG10, respectively. | Atg7 |

continued on following page

*Table 1. Continued*

| Mammalian Protein | Macroautophagy Stage | Function | Yeast Counterpart[1] |
|---|---|---|---|
| MAP1LC3 (LC3), GABARAP[3] | Elongation | Ubiquitin-like modifiers.<br><br>The formation of LC3-II triggers the elongation of the autophagosomal membrane.<br><br>For simplicity, the mammalian orthologs of yeast Atg8 will be referred to as LC3 henceforth. | Atg8 |
| ATG9A, ATG9B[4] | Nucleation | Transmembrane protein with lipid scramblase activity.<br><br>ATG9A/B is capable of transferring phospholipids between the inner and outer leaflets of membranes. ATG2A/B takes phospholipids from the endoplasmic reticulum and delivers them to the outer layer of the expanding phagophore, where ATG9A/B redistributes them across the phagophore membranes, facilitating its expansion.<br><br>For simplicity, ATG9A and ATG9B will be referred to as ATG9 henceforth. | Atg9 |
| ATG10 | Elongation | E2-like enzyme.<br><br>ATG10 facilitates the binding of ATG5 to ATG12. | Atg10 |
| RB1CC1 | Induction | Regulatory scaffold protein in the ULK complex.<br><br>RB1CC1 is essential for supporting the kinase activity of ULK. The presence of RB1CC1 influences the phosphorylation state of ULK. | Atg11[5], Atg 17[6] |
| ATG12 | Elongation | Ubiquitin-like protein and component of the ATG12-ATG5-ATG16L1 complex.<br><br>The ATG12-ATG5-ATG16L1 complex, acting as an E3-like enzyme, is involved in the conjugation reaction with PE to convert LC3-I to LC3-II. | Atg12 |
| ATG13 | Induction | Regulatory scaffold protein in the ULK complex.<br><br>ATG13 regulates autophagy initiation by influencing the enzymatic activity, subcellular localization, and stability of ULK. | Atg13 |
| AG14L, ATG14S[7] | Nucleation | Component of the PI3KC3-C1 complex.<br><br>The PI3KC3-C1 complex converts PI to PI3P.<br><br>For simplicity, ATG14L and ATG14S will be referred to as ATG14 henceforth. | Atg14 |

continued on following page

*Table 1. Continued*

| Mammalian Protein | Macroautophagy Stage | Function | Yeast Counterpart[1] |
|---|---|---|---|
| Lysosomal lipases[8] | Degradation | In yeast, Atg15 functions mainly as a lipase, mediating the breakdown of autophagic bodies. In mammals, this function is carried out by lysosomal lipases. | Atg15 |
| ATG16L1, ATG16L2[9] | Elongation | ATG16L1 serves as a scaffold protein in the ATG12-ATG5-ATG16L1 complex.<br><br>The ATG12-ATG5-ATG16L1 complex, acting as an E3-like enzyme, is involved in the conjugation reaction with PE to convert LC3-I to LC3-II. | Atg16 |
| WIPI1, WIPI2, WIPI3, WIPI4[10] | Nucleation | PI3P-binding proteins.<br><br>WIPIs bind to PI3P and function as its effectors during autophagosome formation. They mediate the recruitment of the ATG12-ATG5-ATG16L1 complex. | Atg18 |

[1]ATG101, a stabilizing protein, is present in the ULK complex in mammals but absent in yeast, whereas Atg29 and Atg31 are components of the Atg1 complex in yeast but have no homologs in mammals (Holzer et al., 2024).

[2]The ATG4 family exhibits substrate specificity. For instance, ATG4B demonstrates the broadest specificity, capable of processing all proteins within the LC3/GABARAP subfamilies, while ATG4A shows a more specific preference for the GABARAP subfamily (Morimoto et al., 2023).

[3]The LC3/GABARAP protein family is categorized into two subfamilies: the LC3 subfamily, which includes LC3A, LC3B, and LC3C, and the GABARAP subfamily, consisting of GABARAP, GABARAPL1, and GABARAPL2 (Iriondo et al., 2023).

[4]Although ATG9B can functionally replace ATG9A, only ATG9A is extensively expressed (Holzer et al., 2024).

[5]Another proposed mammalian counterpart of Atg11 is the Huntingtin protein (Zientara-Rytter & Subramani, 2020).

[6]RB1CC1 bears a closer resemblance to Atg11 than to Atg17 (Zientara-Rytter & Subramani, 2020).

[7]While the yeast Atg14 gene produces a single protein, the mammalian ATG14 gene generates two splice isoforms: a long form (ATG14L) and a short form (ATG14S). Both yeast Atg14 and mammalian ATG14L contain a cysteine repeat domain in their N-terminal regions—a feature that is absent in ATG14S (González-Rodríguez, Delorme-Axford, et al., 2022). This repeat domain is not essential for ATG14 activity in the PI3KC3-C1 complex (González-Rodríguez, Klionsky, et al., 2022). While both isoforms can facilitate phagophore formation, they have opposing roles in regulating autophagosome-lysosome fusion: ATG14L promotes autolysosome formation, whereas ATG14S inhibits it (González-Rodríguez, Klionsky, et al., 2022).

[8]Atg15 is a lipase involved in the degradation of membranes transported through autophagy (Hirata et al., 2021; Kagohashi et al., 2023; Watanabe et al., 2023). Orthologs of Atg15 have been found solely in fungi (Hirata et al., 2021). The intricacy of lysosomal hydrolases presents a challenge in studying the interactions involved in lysosomal membrane breakdown (Kagohashi et al., 2023).

[9]While ATG16L1 is well characterized, the functions of ATG16L2 remain largely unknown. It has been proposed that ATG16L2 plays a promotive role in autophagy regulation (D. Wang et al., 2022).

[10]In yeast, the PROPPIN family comprises three proteins—Atg18, Atg21, and Hsv2—whereas mammals express four homologs known as WIPIs: WIPI1, WIPI2, WIPI3 (WDR45B), and WIPI4 (WDR45) (Almannai et al., 2022). WIPI1 supports WIPI2 in its function; however autophagosome formation can still occur in its absence (Sporbeck et al., 2023). In contrast, cells lacking WIPI2 are unable to generate autophagosomes (Fracchiolla et al., 2020). WIPI2 attaches to newly generated PI3P and, by engaging with ATG16L1, directs the ATG12-ATG5-ATG16L1 complex to the phagophore (Almannai et al., 2022; Fracchiolla et al., 2020). WIPI3 and WIPI4 have auxiliary roles in autophagy and support WIPI2's PI3P effector functions (Yamaguchi et al., 2020). Beyond canonical autophagy, growing evidence has revealed the presence of an alternative autophagy

pathway, which plays a role in processes such as mitochondrial clearance in erythrocytes and the defense of intestinal epithelial cells against bacterial infections (Gong et al., 2024). Alternative autophagy utilizes the same initiation machinery as canonical autophagy, including the ULK complex and PI3KC3-C1 (Gong et al., 2024). However, unlike canonical autophagy, it operates independently of WIPI2 (Gong et al., 2024). Instead, it relies on WIPI3, which detects the PI3P signal from the PI3KC3-C1 complex (Gong et al., 2024). Additionally, WIPI3 and WIPI4 form complexes with ATG2 and contribute to the ATG2-mediated autophagic process (Almannai et al., 2022; Bayly-Jones et al., 2024; Zhu et al., 2024).

The stages of the macroautophagy process and the key proteins involved in these stages are presented in Figure 2.

*Figure 2. The macroautophagy process*

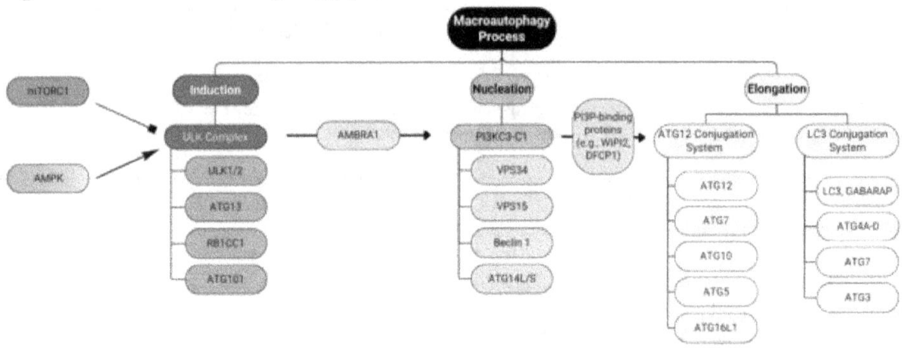


## INDUCTION

During the induction and nucleation stages of macroautophagy (hereafter referred to simply as autophagy), two complexes play key roles: the ULK complex and the PI3KC3-C1 complex (Boukhalfa et al., 2020). The former exhibits protein kinase activity and catalyzes the phosphorylation of proteins, whereas the latter exhibits lipid kinase activity and phosphorylates PI to generate PI3P. The ULK complex not only phosphorylates downstream factors but also provides a scaffolding function (Holzer et al., 2024). The ULK complex consists of ULK1 (or ULK2), ATG13, RB1CC1, and ATG101 (Demeter et al., 2020). Under nutrient-rich conditions, mTORC1 inhibits autophagy by phosphorylating ULK1/2 and ATG13 (Almannai et al., 2022; Bhattacharya et al., 2024; Demeter et al., 2020; Gómez-Virgilio et al., 2022; Son et al., 2021). Under starvation conditions, mTORC1 is inhibited, whereas AMPK is activated (Almannai et al., 2022; Gómez-Virgilio et al., 2022). Activated AMPK may directly phosphorylate and inhibit mTORC1 activity, leading to the activation of ULK1/2 (Gómez-Virgilio et al., 2022; Klionsky et al., 2021). Once dephosphorylated and released from mTORC1, ULK1/2 initiates the autophagy process (Demeter et al., 2020; Gómez-Virgilio et al., 2022). Next, the ULK complex phosphorylates AMBRA1, facilitating the recruitment of the PI3KC3-C1 complex (Gómez-Virgilio et al., 2022). AMBRA1 serves as a regulatory bridge between the ULK complex and the PI3KC3-C1 complex (Gómez-Virgilio et al., 2022). Moreover, ATG9 is recruited by the ULK complex and acts as a lipid scramblase (Son et al., 2021).

## NUCLEATION

The PI3KC3-C1 complex (also known as the PtdIns3K nucleation complex and the VPS34 complex) consists of VPS34, VPS15, Beclin 1, and ATG14 (Gómez-Virgilio et al., 2022; Pavlinov et al., 2020). Under nutrient-starvation conditions, vesicles carrying ATG9 proteins show increased co-localization with autophagosomal markers (Holzer et al., 2024). ATG2 removes phospholipids from the endoplasmic reticulum and incorporates them into the outer membrane of the enlarging phagophore, where ATG9 rearranges them across the bilayer, promoting its growth (Holzer et al., 2024). During autophagosome biogenesis, ATG9 vesicles act as a structural foundation for assembling the autophagy machinery (Holzer et al., 2024). The PI3KC3-C1 complex, when recruited to ATG9 vesicles abundant in PI, can generate PI3P (Holzer et al., 2024). VPS34 is a lipid kinase that catalyzes the phosphorylation of PI (Pavlinov et al., 2020; Tran et al., 2021). AMBRA1 facilitates the catalytic activity of VPS34 (Klionsky et al., 2021). PI3P is essential for autophagosome formation; therefore, blocking VPS34 activity impairs PI3P synthesis and ultimately inhibits autophagy (González-Rodríguez, Delorme-Axford, et al., 2022; Pavlinov et al., 2020; Sporbeck et al., 2023). PI3P attracts PI3P-binding effector proteins—like DFCP1 (also termed ZFYVE1) and WIPI2—at autophagosomal membrane formation sites; these proteins subsequently promote autophagosome formation by recruiting further components of the autophagy machinery (Boukhalfa et al., 2020; González-Rodríguez, Klionsky, et al., 2022; Wei et al., 2024; Zachari et al., 2020). WIPI proteins are characterized by a β-propeller structure composed of seven blades (Almannai et al., 2022). WIPI2 serves as a bridge between PI3P synthesis by the PI3KC3-C1 complex and LC3 conjugation by recruiting the ATG12-ATG5-ATG16L1 complex (Fracchiolla et al., 2020; Son et al., 2021).

## ELONGATION

There are two ubiquitin-like conjugation systems involved in autophagosome elongation (L. Chen et al., 2024; Iriondo et al., 2023). The first system is composed of the proteins ATG7, ATG10, ATG12, ATG5, and ATG16L1, while the second system consists of the LC3/GABARAP protein family members (hereafter referred to simply as LC3) along with the proteins ATG4, ATG7, and ATG3 (Gómez-Virgilio et al., 2022; Iriondo et al., 2023). The second conjugation system also requires the first conjugation system that functions as an E3-like enzyme (González-Rodríguez, Delorme-Axford, et al., 2022).

## ATG12 CONJUGATION SYSTEM

First, the E1-like enzyme ATG7 activates ubiquitin-like ATG12 (Iriondo et al., 2023; J. Liu et al., 2024). Next, the activated ATG12 conjugates with ATG5 through the E2-like enzyme ATG10 (L. Chen et al., 2024; Iriondo et al., 2023). Finally, the ATG12-ATG5 assembles with ATG16L1 (Iriondo et al., 2023; S. Zhang et al., 2022). ATG12–ATG5-ATG16L1 complex functions as an E3-like enzyme (Iriondo et al., 2023).

## LC3 CONJUGATION SYSTEM

First, ATG4 processes the proLC3 to generate LC3-I (Gómez-Virgilio et al., 2022; Iriondo et al., 2023). Next, the E1-like enzyme ATG7 and the E2-like enzyme ATG3, together with the E3-like enzyme ATG12-ATG5-ATG16L1 complex, catalyze the conjugation of LC3-I to PE, forming LC3-II (L. Chen et al., 2024; Iriondo et al., 2023; Ye et al., 2021; S. Zhang et al., 2022) (Figure 3). The covalent conjugation of protein LC3-I to the amino group of PE lipids (also termed lipidation) anchors LC3-II to the autophagosomal membrane (Gómez-Virgilio et al., 2022; Iriondo et al., 2023; Ye et al., 2021). After anchoring to the membrane, LC3-II plays key roles in expansion, closure, and fusion processes (Iriondo et al., 2023; Morimoto et al., 2023; Ye et al., 2021). Normally, LC3 is almost evenly distributed within the cell, but upon autophagy induction, it clusters in the autophagosome (Gómez-Virgilio et al., 2022).

*Figure 3. Conjugation systems responsible for elongation*

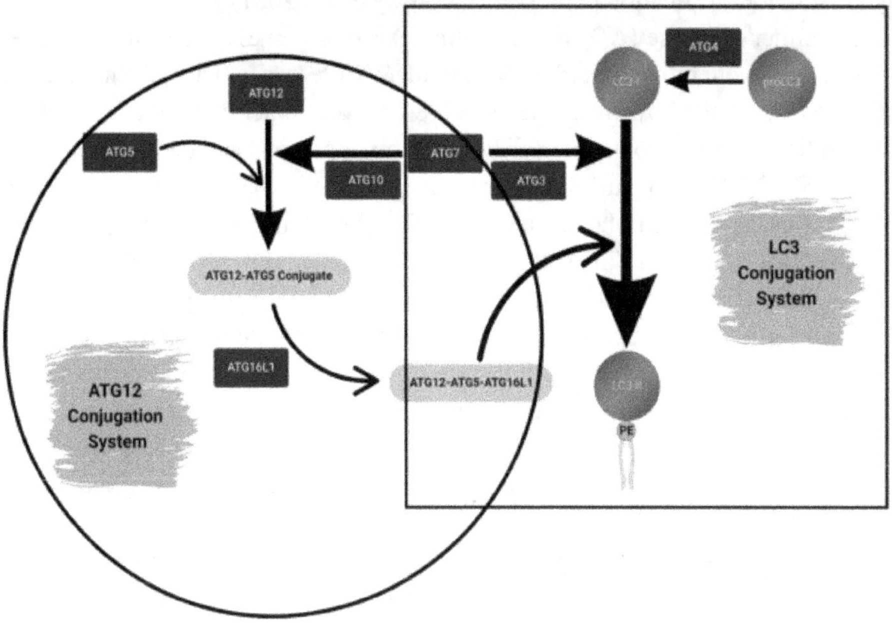


## NEURODEGENERATIVE DISEASES

Fully matured and differentiated cells that can no longer divide through mitosis are called post-mitotic (von Zglinicki et al., 2021). Such cells largely depend on autophagy to clear damaged organelles or misfolded and aggregated proteins (Gómez-Virgilio et al., 2022). As post-mitotic cells, neurons are highly sensitive to disruptions in the autophagic process (Gómez-Virgilio et al., 2022; Klionsky et al., 2021; Tutas et al., 2025). Autophagic efficiency declines as the organism ages, and impaired autophagy increases susceptibility to a range of age-related diseases, including neurodegenerative disorders (Gómez-Virgilio et al., 2022; Lei & Klionsky, 2023). The buildup of proteins prone to aggregation is a common feature of many neurodegenerative diseases (Chopra et al., 2022; Klionsky et al., 2021; Shu et al., 2023). However, autophagy's neuroprotective roles are believed to extend beyond its established functions in protein homeostasis and organelle turnover. For example, studies have shown that the ATG machinery is involved in various neural processes, including synaptic transmission, the breakdown of synaptic cargo and vesicles, and myelination/demyelination processes (Klionsky et al., 2021).

Alzheimer's disease (AD) is the leading cause of dementia and the most prevalent neurodegenerative disorder (Sheppard & Coleman, 2020; Z. Zhang et al., 2021). AD is marked by impairments in memory and other cognitive abilities due to synaptic loss and atrophy primarily in the hippocampus and cerebral cortex (S. Chen et al., 2020; Sheppard & Coleman, 2020). AD is characterized by the presence of amyloid-β (Aβ) plaques and neurofibrillary tau tangles (Caballero et al., 2021; Le Bras, 2022; Z. Zhang et al., 2021). Aβ plaques are protein clumps called beta-amyloid between nerve cells, while neurofibrillary tau tangles are bundles of filaments largely made up of a protein called tau within neurons (Le Bras, 2022; Sheppard & Coleman, 2020). Both intracellular MAPT/tau tangles and extracellular Aβ plaques are protein aggregates, and autophagy plays a key role in clearing tau and Aβ (Klionsky et al., 2021; Sheppard & Coleman, 2020). There is evidence suggesting that neuronal autophagy fails before the accumulation of amyloid plaques and neurofibrillary tau tangles (Carosi et al., 2021; Le Bras, 2022; X.-W. Zhang et al., 2023). Moreover, it is proposed that boosting autophagy could help alleviate AD-related damage and potentially restore memory function (Bai et al., 2020; J. Chen et al., 2021; Fernandes et al., 2025).

Parkinson's disease (PD) is characterized by the gradual degeneration of dopaminergic neurons within the substantia nigra and the formation of Lewy bodies in various regions of the central nervous system, including the substantia nigra, hippocampus, and cerebral cortex. (Milán-Tomás et al., 2021; Themistokleous et al., 2023; Zhou et al., 2023). The α-synuclein (α-syn) protein, encoded by the SNCA gene, can form aggregates that are the major components of Lewy bodies (Hou et al., 2020; Minchev et al., 2022; Zhou et al., 2023). PD has been proposed to represent a manifestation of Lewy body disease (Borghammer et al., 2024). However, some researchers have suggested that the term "diseases with Lewy bodies" would be more appropriate than "Lewy body disease" (Menšíková et al., 2022). Parkinson's disease primarily manifests through notable disruptions in motor and autonomic functions, while cognitive and psychological impairments may also occur (Klionsky et al., 2021; Lizama & Chu, 2021; Lu et al., 2020). Lewy body disorders, such as PD, PD dementia, and dementia with Lewy bodies, are the second most common cause of primary neurodegenerative dementia after AD (Cohen et al., 2025). Autophagy has been reported to play a role in the degradation of α-syn, and its activation decreases the accumulation of α-syn and protects dopaminergic neurons (Klionsky et al., 2021; Nechushtai et al., 2023). Additionally, it has been observed that anomalies in this protein (such as overexpression and mutations) are toxic to autophagy (Klionsky et al., 2021; Tu et al., 2021).

Amyotrophic lateral sclerosis (ALS) is a neurodegenerative disease that primarily affects the motor system, which controls voluntary muscle movements, ultimately resulting in loss of motor function (Masrori & Van Damme, 2020). Non-motor symptoms, such as cognitive and behavioral changes, are present in up to 50% of cases (Feldman et al., 2022; Masrori & Van Damme, 2020). Evidence suggests that protein aggregation and disrupted autophagy play important roles in the development of ALS (Chua et al., 2022; Feldman et al., 2022; Masrori & Van Damme, 2020). Moreover, in addition to autophagic dysfunction, disruption of the endolysosomal system is also considered a critical factor in ALS pathology (Todd et al., 2023).

Huntington's chorea, also known as Huntington's disease (HD), is one of the trinucleotide repeat disorders, in which abnormal expansions of CAG repeats in the associated gene result in proteins with extended polyglutamine tracts (Arning & Nguyen, 2022; Belgrad & Khvorova, 2024). Expansion of the CAG repeat may occur throughout an individual's lifetime, a phenomenon known as somatic expansion (Bunting et al., 2022; Scahill et al., 2025). The causative mutation associated with HD occurs in the huntingtin gene (HTT) on chromosome 4, leading to the translation of an elongated polyglutamine tract in the huntingtin protein (Htt) (Arning & Nguyen, 2022; Chouksey & Pandey, 2020). Consequently, the mutant protein (mHtt) gradually damages neurons (Arning & Nguyen, 2022). HD is a progressive, fatal neurodegenerative condition characterized by a spectrum of motor abnormalities such as chorea and loss of coordination, along with cognitive and psychiatric disturbances (Arning & Nguyen, 2022; Bunting et al., 2022). The differences in age of onset and progression of HD are attributed to the number of CAG repeats and other genetic and environmental factors (Arning & Nguyen, 2022; Bunting et al., 2022). In healthy individuals, the number of glutamine repeats in the relevant protein region is typically fewer than 36, usually ranging between 16 and 20 (Arning & Nguyen, 2022; Findlay Black et al., 2024). When there are 36-39 repeats, a late-onset form of HD with slower symptom progression may be observed (Arning & Nguyen, 2022; Ruiz de Sabando et al., 2024). However, if the number of CAG repeats exceeds 39, the mutation is inevitably associated with disease symptoms, and when the repeat number is much higher, early-onset (juvenile) HD may occur (Arning & Nguyen, 2022; Ruiz de Sabando et al., 2024). There are other similar diseases caused by expanded CAG repeats in different genes (Bunting et al., 2022). All these disorders, including HD and several forms of spinocerebellar ataxias, are collectively referred to as polyglutamine diseases (Bunting et al., 2022). Variations in CAG repeat length that do not reach the threshold for HD might still be linked to psychiatric conditions, including depression and anxiety (Ruiz de Sabando et al., 2024; Vater et al., 2024). It has been shown that disruptions in the autophagic process can lead to elevated levels of polyglutamine-expanded proteins (Bunting et al., 2022; Lin et al., 2024). Inducing autophagy has the potential to reduce mHtt levels and alleviate the progression of HD (Bunting et al., 2022; Zhao et al., 2024).

## CONCLUSION AND RECOMMENDATIONS

Protein aggregates in neurodegenerative diseases represent a striking finding that is frequently implicated in the underlying pathogenesis of these disorders. Owing to its ability to clear such aggregates, autophagy has emerged as a key research focus—not only for understanding the development of these conditions but also for identifying potential therapeutic strategies. Given the inherently limited regenerative capacity of neural tissue through cellular division, robust autophagic mechanisms are especially critical. In this light, the age-related decline in autophagy may partly explain why neurodegenerative diseases predominantly affect older individuals. Today, there is an urgent demand for effective treatments for these conditions. Emerging evidence indicates that lifestyle modifications and therapeutic interventions which enhance autophagic activity could play an important role in both preventing and managing neurodegenerative disorders. Overall, it seems that, from a health perspective, occasional fasting might offer benefits over a continuously fed state.

## REFERENCES


Almannai, M., Marafi, D., & El-Hattab, A. W. (2022). WIPI proteins: Biological functions and related syndromes. *Frontiers in Molecular Neuroscience, 15*, 1011918. Advance online publication. DOI: 10.3389/fnmol.2022.1011918 PMID: 36157071

Arning, L., & Nguyen, H. P. (2022). Huntington disease update: New insights into the role of repeat instability in disease pathogenesis. *Medizinische Genetik : Mitteilungsblatt des Berufsverbandes Medizinische Genetik e.V, 33*(4), 293–300. DOI: 10.1515/medgen-2021-2101 PMID: 38835439

Bai, B., Wang, X., Li, Y., Chen, P.-C., Yu, K., Dey, K. K., Yarbro, J. M., Han, X., Lutz, B. M., Rao, S., Jiao, Y., Sifford, J. M., Han, J., Wang, M., Tan, H., Shaw, T. I., Cho, J.-H., Zhou, S., Wang, H., & Peng, J. (2020). Deep Multilayer Brain Proteomics Identifies Molecular Networks in Alzheimer's Disease Progression. *Neuron, 106*(4), 700. DOI: 10.1016/j.neuron.2020.04.031 PMID: 32437656

Bayly-Jones, C., Lupton, C. J., D'Andrea, L., Chang, Y.-G., Jones, G. D., Steele, J. R., Venugopal, H., Schittenhelm, R. B., Halls, M. L., & Ellisdon, A. M. (2024). Structure of the human TSC:WIPI3 lysosomal recruitment complex. *Science Advances, 10*(47), eadr5807. Advance online publication. DOI: 10.1126/sciadv.adr5807 PMID: 39565846

Belgrad, J., & Khvorova, A. (2024). More than 185 CAG repeats: A point of no return in Huntington's disease biology. *Brain, 147*(5), 1601–1603. DOI: 10.1093/brain/awae105 PMID: 38588052

Bhattacharya, A., Torggler, R., Reiter, W., Romanov, N., Licheva, M., Ciftci, A., Mari, M., Kolb, L., Kaiser, D., Reggiori, F., Ammerer, G., Hollenstein, D. M., & Kraft, C. (2024). Decoding the function of Atg13 phosphorylation reveals a role of Atg11 in bulk autophagy initiation. *EMBO Reports, 25*(2), 813–831. DOI: 10.1038/s44319-023-00055-9 PMID: 38233718

Borghammer, P., Okkels, N., & Weintraub, D. (2024). Parkinson's Disease and Dementia with Lewy Bodies: One and the Same. *Journal of Parkinson's Disease, 14*(3), 383–397. DOI: 10.3233/JPD-240002 PMID: 38640172

Boukhalfa, A., Nascimbeni, A. C., Ramel, D., Dupont, N., Hirsch, E., Gayral, S., Laffargue, M., Codogno, P., & Morel, E. (2020). PI3KC2α-dependent and VPS34-independent generation of PI3P controls primary cilium-mediated autophagy in response to shear stress. *Nature Communications, 11*(1), 294. DOI: 10.1038/s41467-019-14086-1 PMID: 31941925

Bunting, E. L., Hamilton, J., & Tabrizi, S. J. (2022). Polyglutamine diseases. *Current Opinion in Neurobiology*, *72*, 39–47. DOI: 10.1016/j.conb.2021.07.001 PMID: 34488036

Caballero, B., Bourdenx, M., Luengo, E., Diaz, A., Sohn, P. D., Chen, X., Wang, C., Juste, Y. R., Wegmann, S., Patel, B., Young, Z. T., Kuo, S. Y., Rodriguez-Navarro, J. A., Shao, H., Lopez, M. G., Karch, C. M., Goate, A. M., Gestwicki, J. E., Hyman, B. T., & Cuervo, A. M. (2021). Acetylated tau inhibits chaperone-mediated autophagy and promotes tau pathology propagation in mice. *Nature Communications*, *12*(1), 2238. DOI: 10.1038/s41467-021-22501-9 PMID: 33854069

Carosi, J. M., Denton, D., Kumar, S., & Sargeant, T. J. (2021). Retromer dysfunction at the nexus of tauopathies. *Cell Death and Differentiation*, *28*(3), 884–899. DOI: 10.1038/s41418-020-00727-2 PMID: 33473181

Chen, J., He, H.-J., Ye, Q., Feng, F., Wang, W.-W., Gu, Y., Han, R., & Xie, C. (2021). Defective Autophagy and Mitophagy in Alzheimer's Disease: Mechanisms and Translational Implications. *Molecular Neurobiology*, *58*(10), 5289–5302. DOI: 10.1007/s12035-021-02487-7 PMID: 34279771

Chen, L., Gao, T., Zhou, P., Xia, W., Yao, H., Xu, S., & Xu, J. (2024). Recent advances of vacuolar protein-sorting 34 inhibitors targeting autophagy. *Bioorganic Chemistry*, *143*, 107039. DOI: 10.1016/j.bioorg.2023.107039 PMID: 38134519

Chen, S., Zhou, Q., Ni, Y., & Le, W. (2020). Autophagy and Alzheimer's Disease. *Advances in Experimental Medicine and Biology*, *1207*, 3–19. DOI: 10.1007/978-981-15-4272-5_1 PMID: 32671736

Chopra, G., Shabir, S., Yousuf, S., Kauts, S., Bhat, S. A., Mir, A. H., & Singh, M. P. (2022). Proteinopathies: Deciphering Physiology and Mechanisms to Develop Effective Therapies for Neurodegenerative Diseases. *Molecular Neurobiology*, *59*(12), 7513–7540. DOI: 10.1007/s12035-022-03042-8 PMID: 36205914

Chouksey, A., & Pandey, S. (2020). Phenotypic variability in huntington's disease. *Annals of Indian Academy of Neurology*, *23*(2), 153. DOI: 10.4103/aian.AIAN_543_19 PMID: 32189853

Chua, J. P., De Calbiac, H., Kabashi, E., & Barmada, S. J. (2022). Autophagy and ALS: Mechanistic insights and therapeutic implications. *Autophagy*, *18*(2), 254–282. DOI: 10.1080/15548627.2021.1926656 PMID: 34057020

Cohen, J. S., Phillips, J., Das, S. R., Olm, C. A., Radhakrishnan, H., Rhodes, E., Cousins, K. A. Q., Xie, S. X., Nasrallah, I. M., Yushkevich, P. A., Wolk, D. A., Lee, E. B., Weintraub, D., Irwin, D. J., & McMillan, C. T. (2025). Posterior hippocampal sparing in Lewy body disorders with Alzheimer's copathology: An in vivo MRI study. *NeuroImage. Clinical*, *45*, 103714. DOI: 10.1016/j.nicl.2024.103714 PMID: 39675237

Demeter, A., Romero-Mulero, M. C., Csabai, L., Ölbei, M., Sudhakar, P., Haerty, W., & Korcsmáros, T. (2020). ULK1 and ULK2 are less redundant than previously thought: Computational analysis uncovers distinct regulation and functions of these autophagy induction proteins. *Scientific Reports*, *10*(1), 10940. DOI: 10.1038/s41598-020-67780-2 PMID: 32616830

Feldman, E. L., Goutman, S. A., Petri, S., Mazzini, L., Savelieff, M. G., Shaw, P. J., & Sobue, G. (2022). Amyotrophic lateral sclerosis. *Lancet*, *400*(10360), 1363–1380. DOI: 10.1016/S0140-6736(22)01272-7 PMID: 36116464

Fernandes, S. M., Mayer, J., Nilsson, P., & Shimozawa, M. (2025). How close is autophagy-targeting therapy for Alzheimer's disease to clinical use? A summary of autophagy modulators in clinical studies. *Frontiers in Cell and Developmental Biology*, *12*, 1520949. Advance online publication. DOI: 10.3389/fcell.2024.1520949 PMID: 39845082

Findlay Black, H., Kay, C., Dawson, J., Bortnick, S., Javier, K., Xia, Q., Chau, C. H., Leavitt, T., Arning, L., Nguyen, H. P., & Hayden, M. R. (2024). Ascertainment of uninterrupted CAG repeat length and disease-modifying variants in fragment-based genetic testing for Huntington Disease. *Genetics in Medicine Open*, *2*, 101882. DOI: 10.1016/j.gimo.2024.101882 PMID: 39669608

Fracchiolla, D., Chang, C., Hurley, J. H., & Martens, S. (2020). A PI3K-WIPI2 positive feedback loop allosterically activates LC3 lipidation in autophagy. *The Journal of Cell Biology*, *219*(7), e201912098. Advance online publication. DOI: 10.1083/jcb.201912098 PMID: 32437499

Gómez-Virgilio, L., Silva-Lucero, M.-D.-C., Flores-Morelos, D.-S., Gallardo-Nieto, J., Lopez-Toledo, G., Abarca-Fernandez, A.-M., Zacapala-Gómez, A.-E., Luna-Muñoz, J., Montiel-Sosa, F., Soto-Rojas, L. O., Pacheco-Herrero, M., & Cardenas-Aguayo, M.-D.-C. (2022). Autophagy: A Key Regulator of Homeostasis and Disease: An Overview of Molecular Mechanisms and Modulators. *Cells*, *11*(15), 2262. Advance online publication. DOI: 10.3390/cells11152262 PMID: 35892559

Gong, X., Wang, Y., Zhou, Y., & Pan, L. (2024). Structure of the WIPI3/ATG16L1 Complex Reveals the Molecular Basis for the Recruitment of the ATG12~ATG5-ATG16L1 Complex by WIPI3. *Cells, 13*(24), 2113. DOI: 10.3390/cells13242113 PMID: 39768203

González-Rodríguez, P., Delorme-Axford, E., Bernard, A., Keane, L., Stratoulias, V., Grabert, K., Engskog-Vlachos, P., Füllgrabe, J., Klionsky, D. J., & Joseph, B. (2022). SETD2 transcriptional control of ATG14L/S isoforms regulates autophagosome–lysosome fusion. *Cell Death & Disease, 13*(11), 953. DOI: 10.1038/s41419-022-05381-9 PMID: 36371383

González-Rodríguez, P., Klionsky, D. J., & Joseph, B. (2022). Autophagy regulation by RNA alternative splicing and implications in human diseases. *Nature Communications, 13*(1), 2735. DOI: 10.1038/s41467-022-30433-1 PMID: 35585060

Hirata, E., Shirai, K., Kawaoka, T., Sato, K., Kodama, F., & Suzuki, K. (2021). Atg15 in Saccharomyces cerevisiae consists of two functionally distinct domains. *Molecular Biology of the Cell, 32*(8), 645–663. DOI: 10.1091/mbc.E20-07-0500 PMID: 33625870

Holzer, E., Martens, S., & Tulli, S. (2024). The Role of ATG9 Vesicles in Autophagosome Biogenesis. *Journal of Molecular Biology, 436*(15), 168489. DOI: 10.1016/j.jmb.2024.168489 PMID: 38342428

Hormazabal, J., Saavedra, F., Espinoza-Arratia, C., Martinez, N. W., Cruces, T., Alfaro, I. E., & Loyola, A. (2022). Chaperone mediated autophagy contributes to the newly synthesized histones H3 and H4 quality control. *Nucleic Acids Research, 50*(4), 1875–1887. DOI: 10.1093/nar/gkab1296 PMID: 35037039

Hou, X., Watzlawik, J. O., Fiesel, F. C., & Springer, W. (2020). Autophagy in Parkinson's Disease. *Journal of Molecular Biology, 432*(8), 2651–2672. DOI: 10.1016/j.jmb.2020.01.037 PMID: 32061929

Iriondo, M. N., Etxaniz, A., Varela, Y. R., Ballesteros, U., Lázaro, M., Valle, M., Fracchiolla, D., Martens, S., Montes, L. R., Goñi, F. M., & Alonso, A. (2023). Effect of ATG12–ATG5-ATG16L1 autophagy E3-like complex on the ability of LC3/GABARAP proteins to induce vesicle tethering and fusion. *Cellular and Molecular Life Sciences : CMLS, 80*(2), 56. DOI: 10.1007/s00018-023-04704-z PMID: 36729310

Kacal, M., Zhang, B., Hao, Y., Norberg, E., & Vakifahmetoglu-Norberg, H. (2021). Quantitative proteomic analysis of temporal lysosomal proteome and the impact of the KFERQ-like motif and LAMP2A in lysosomal targeting. *Autophagy, 17*(11), 3865–3874. DOI: 10.1080/15548627.2021.1876343 PMID: 33446043

Kagohashi, Y., Sasaki, M., May, A. I., Kawamata, T., & Ohsumi, Y. (2023). The mechanism of Atg15-mediated membrane disruption in autophagy. *The Journal of Cell Biology*, *222*(12), e202306120. Advance online publication. DOI: 10.1083/jcb.202306120 PMID: 37917025

Klionsky, D. J., Petroni, G., Amaravadi, R. K., Baehrecke, E. H., Ballabio, A., Boya, P., Bravo-San Pedro, J. M., Cadwell, K., Cecconi, F., Choi, A. M. K., Choi, M. E., Chu, C. T., Codogno, P., Colombo, M. I., Cuervo, A. M., Deretic, V., Dikic, I., Elazar, Z., Eskelinen, E., & Pietrocola, F. (2021). Autophagy in major human diseases. *The EMBO Journal*, *40*(19), e108863. Advance online publication. DOI: 10.15252/embj.2021108863 PMID: 34459017

Krause, G. J., Kirchner, P., Stiller, B., Morozova, K., Diaz, A., Chen, K.-H., Krogan, N. J., Agullo-Pascual, E., Clement, C. C., Lindenau, K., Swaney, D. L., Dilipkumar, S., Bravo-Cordero, J. J., Santambrogio, L., & Cuervo, A. M. (2023). Molecular determinants of the crosstalk between endosomal microautophagy and chaperone-mediated autophagy. *Cell Reports*, *42*(12), 113529. DOI: 10.1016/j.celrep.2023.113529 PMID: 38060380

Le Bras, A. (2022). New insights into the origin of amyloid plaques. *Lab Animal*, *51*(7), 187–187. DOI: 10.1038/s41684-022-01007-x

Lei, Y., & Klionsky, D. J. (2023). Transcriptional regulation of autophagy and its implications in human disease. *Cell Death and Differentiation*, *30*(6), 1416–1429. DOI: 10.1038/s41418-023-01162-9 PMID: 37045910

Lin, T.-H., Chen, W.-L., Hsu, S.-F., Chen, I.-C., Lin, C.-H., Chang, K.-H., Wu, Y.-R., Chen, Y.-R., Yao, C.-F., Lin, W., Lee-Chen, G.-J., & Chen, C.-M. (2024). Small Molecules Inducing Autophagic Degradation of Expanded Polyglutamine Protein through Interaction with Both Mutant ATXN3 and LC3. *International Journal of Molecular Sciences*, *25*(19), 10707. DOI: 10.3390/ijms251910707 PMID: 39409036

Liu, J., Xiao, Y., Cao, L., Lu, S., Zhang, S., Yang, R., Wang, Y., Zhang, N., Yu, Y., Wang, X., Guo, W., Wang, Z., Xu, H., Xing, C., Song, X., & Cao, L. (2024). Insights on E1-like enzyme ATG7: Functional regulation and relationships with aging-related diseases. *Communications Biology*, *7*(1), 382. DOI: 10.1038/s42003-024-06080-1 PMID: 38553562

Liu, S., Yao, S., Yang, H., Liu, S., & Wang, Y. (2023). Autophagy: Regulator of cell death. *Cell Death & Disease*, *14*(10), 648. DOI: 10.1038/s41419-023-06154-8 PMID: 37794028

Lizama, B. N., & Chu, C. T. (2021). Neuronal autophagy and mitophagy in Parkinson's disease. *Molecular Aspects of Medicine*, *82*, 100972. DOI: 10.1016/j.mam.2021.100972 PMID: 34130867

Lu, J., Wu, M., & Yue, Z. (2020). Autophagy and Parkinson's Disease. *Advances in Experimental Medicine and Biology*, *1207*, 21–51. DOI: 10.1007/978-981-15-4272-5_2 PMID: 32671737

Masrori, P., & Van Damme, P. (2020). Amyotrophic lateral sclerosis: A clinical review. *European Journal of Neurology*, *27*(10), 1918–1929. DOI: 10.1111/ene.14393 PMID: 32526057

Menšíková, K., Matěj, R., Colosimo, C., Rosales, R., Tučková, L., Ehrmann, J., Hraboš, D., Kolaříková, K., Vodička, R., Vrtěl, R., Procházka, M., Nevrlý, M., Kaiserová, M., Kurčová, S., Otruba, P., & Kaňovský, P. (2022). Lewy body disease or diseases with Lewy bodies? *NPJ Parkinson's Disease*, *8*(1), 3. DOI: 10.1038/s41531-021-00273-9 PMID: 35013341

Milán-Tomás, Á., Fernández-Matarrubia, M., & Rodríguez-Oroz, M. C. (2021). Lewy Body Dementias: A Coin with Two Sides? *Behavioral Sciences (Basel, Switzerland)*, *11*(7), 94. DOI: 10.3390/bs11070094 PMID: 34206456

Minchev, D., Kazakova, M., & Sarafian, V. (2022). Neuroinflammation and Autophagy in Parkinson's Disease—Novel Perspectives. *International Journal of Molecular Sciences*, *23*(23), 14997. DOI: 10.3390/ijms232314997 PMID: 36499325

Morimoto, M., Bhambhani, V., Gazzaz, N., Davids, M., Sathiyaseelan, P., Macnamara, E. F., Lange, J., Lehman, A., Zerfas, P. M., Murphy, J. L., Acosta, M. T., Wang, C., Alderman, E., Adam, M., Alvarez, R. L., Alvey, J., Amendola, L., Andrews, A., Ashley, E. A., & Malicdan, M. C. V. (2023). Bi-allelic ATG4D variants are associated with a neurodevelopmental disorder characterized by speech and motor impairment. *NPJ Genomic Medicine*, *8*(1), 4. DOI: 10.1038/s41525-022-00343-8 PMID: 36765070

Nechushtai, L., Frenkel, D., & Pinkas-Kramarski, R. (2023). Autophagy in Parkinson's Disease. *Biomolecules*, *13*(10), 1435. DOI: 10.3390/biom13101435 PMID: 37892117

Pavlinov, I., Salkovski, M., & Aldrich, L. N. (2020). Beclin 1–ATG14L Protein–Protein Interaction Inhibitor Selectively Inhibits Autophagy through Disruption of VPS34 Complex I. *Journal of the American Chemical Society*, *142*(18), 8174–8182. DOI: 10.1021/jacs.9b12705 PMID: 32320221

Ruiz de Sabando, A., Ciosi, M., Galbete, A., Cumming, S. A., Álvarez, V., Martinez-Descals, A., Mila, M., Trujillo-Tiebas, M. J., López-Sendón, J. L., Fenollar-Cortés, M., Legarda, I., Bernal Noguera, S., Millán, J. M., Durán-Herrera, C., Ruiz-Martínez, J., Ruiz Onandi, R., Monckton, D. G., & Ramos-Arroyo, M. A. (2024). Somatic CAG repeat instability in intermediate alleles of the HTT gene and its potential association with a clinical phenotype. *European Journal of Human Genetics : EJHG, 32*(7), 770–778. DOI: 10.1038/s41431-024-01546-6 PMID: 38433266

Scahill, R. I., Farag, M., Murphy, M. J., Hobbs, N. Z., Leocadi, M., Langley, C., Knights, H., Ciosi, M., Fayer, K., Nakajima, M., Thackeray, O., Gobom, J., Rönnholm, J., Weiner, S., Hassan, Y. R., Ponraj, N. K. P., Estevez-Fraga, C., Parker, C. S., Malone, I. B., & Tabrizi, S. J. (2025). Somatic CAG repeat expansion in blood associates with biomarkers of neurodegeneration in Huntington's disease decades before clinical motor diagnosis. *Nature Medicine, 31*(3), 807–818. DOI: 10.1038/s41591-024-03424-6 PMID: 39825149

Schnebert, S., Goguet, M., Vélez, E. J., Depincé, A., Beaumatin, F., Herpin, A., & Seiliez, I. (2022). Diving into the Evolutionary History of HSC70-Linked Selective Autophagy Pathways: Endosomal Microautophagy and Chaperone-Mediated Autophagy. *Cells, 11*(12), 1945. DOI: 10.3390/cells11121945 PMID: 35741074

Sheppard, O., & Coleman, M. (2020). Alzheimer's Disease: Etiology, Neuropathology and Pathogenesis. In *Alzheimer's Disease: Drug Discovery* (pp. 1–22). Exon Publications. DOI: 10.36255/exonpublications.alzheimersdisease.2020.ch1

Shu, F., Xiao, H., Li, Q.-N., Ren, X.-S., Liu, Z.-G., Hu, B.-W., Wang, H.-S., Wang, H., & Jiang, G.-M. (2023). Epigenetic and post-translational modifications in autophagy: Biological functions and therapeutic targets. *Signal Transduction and Targeted Therapy, 8*(1), 32. DOI: 10.1038/s41392-022-01300-8 PMID: 36646695

Son, S. M., Park, S. J., Fernandez-Estevez, M., & Rubinsztein, D. C. (2021). Autophagy regulation by acetylation—Implications for neurodegenerative diseases. *Experimental & Molecular Medicine, 53*(1), 30–41. DOI: 10.1038/s12276-021-00556-4 PMID: 33483607

Sporbeck, K., Haas, M. L., Pastor-Maldonado, C. J., Schüssele, D. S., Hunter, C., Takacs, Z., Diogo de Oliveira, A. L., Franz-Wachtel, M., Charsou, C., Pfisterer, S. G., Gubas, A., Haller, P. K., Knorr, R. L., Kaulich, M., Macek, B., Eskelinen, E.-L., Simonsen, A., & Proikas-Cezanne, T. (2023). The ABL-MYC axis controls WIPI1-enhanced autophagy in lifespan extension. *Communications Biology, 6*(1), 872. DOI: 10.1038/s42003-023-05236-9 PMID: 37620393

Themistokleous, C., Bagnoli, E., Parulekar, R., & Muqit, M. M. K. (2023). Role of Autophagy Pathway in Parkinson's Disease and Related Genetic Neurological Disorders. *Journal of Molecular Biology*, *435*(12), 168144. DOI: 10.1016/j.jmb.2023.168144 PMID: 37182812

Todd, T. W., Shao, W., Zhang, Y., & Petrucelli, L. (2023). The endolysosomal pathway and ALS/FTD. *Trends in Neurosciences*, *46*(12), 1025–1041. DOI: 10.1016/j.tins.2023.09.004 PMID: 37827960

Tran, S., Fairlie, W. D., & Lee, E. F. (2021). BECLIN1: Protein Structure, Function and Regulation. *Cells*, *10*(6), 1522. DOI: 10.3390/cells10061522 PMID: 34204202

Tu, H., Yuan, B., Hou, X., Zhang, X., Pei, C., Ma, Y., Yang, Y., Fan, Y., Qin, Z., Liu, C., & Hu, L. (2021). α-synuclein suppresses microglial autophagy and promotes neurodegeneration in a mouse model of Parkinson's disease. *Aging Cell*, *20*(12), e13522. Advance online publication. DOI: 10.1111/acel.13522 PMID: 34811872

Tutas, J., Tolve, M., Özer-Yildiz, E., Ickert, L., Klein, I., Silverman, Q., Liebsch, F., Dethloff, F., Giavalisco, P., Endepols, H., Georgomanolis, T., Neumaier, B., Drzezga, A., Schwarz, G., Thorens, B., Gatto, G., Frezza, C., & Kononenko, N. L. (2025). Autophagy regulator ATG5 preserves cerebellar function by safeguarding its glycolytic activity. *Nature Metabolism*, *7*(2), 297–320. DOI: 10.1038/s42255-024-01196-4 PMID: 39815080

Vater, M., Rost, N., Eckstein, G., Sauer, S., Tontsch, A., Erhardt, A., Lucae, S., Brückl, T., Klopstock, T., Sämann, P. G., & Binder, E. B. (2024). Huntingtin CAG repeat size variations below the Huntington's disease threshold: Associations with depression, anxiety and basal ganglia structure. *European Journal of Human Genetics : EJHG*. Advance online publication. DOI: 10.1038/s41431-024-01737-1 PMID: 39572770

von Zglinicki, T., Wan, T., & Miwa, S. (2021). Senescence in Post-Mitotic Cells: A Driver of Aging? *Antioxidants & Redox Signaling*, *34*(4), 308–323. DOI: 10.1089/ars.2020.8048 PMID: 32164429

Wang, D., Yuan, T., Liu, J., Wen, Z., Shen, Y., Tang, J., Wang, Z., & Wu, X. (2022). ATG16L2 inhibits NLRP3 inflammasome activation through promoting ATG5-12-16L1 complex assembly and autophagy. *European Journal of Immunology*, *52*(8), 1321–1334. DOI: 10.1002/eji.202149764 PMID: 35426127

Wang, Y.-T., & Lu, J.-H. (2022). Chaperone-Mediated Autophagy in Neurodegenerative Diseases: Molecular Mechanisms and Pharmacological Opportunities. *Cells*, *11*(14), 2250. DOI: 10.3390/cells11142250 PMID: 35883693

Watanabe, Y., Iwasaki, Y., Sasaki, K., Motono, C., Imai, K., & Suzuki, K. (2023). Atg15 is a vacuolar phospholipase that disintegrates organelle membranes. *Cell Reports*, *42*(12), 113567. DOI: 10.1016/j.celrep.2023.113567 PMID: 38118441

Wei, B., Fu, Y., Li, X., Chen, F., Zhang, Y., Chen, H., Tong, M., Li, L., Pan, Y., Zhang, S., Chen, S., Liu, X., & Zhong, Q. (2024). ANKFY1 bridges ATG2A-mediated lipid transfer from endosomes to phagophores. *Cell Discovery*, *10*(1), 43. DOI: 10.1038/s41421-024-00659-y PMID: 38622126

Yamaguchi, H., Honda, S., Torii, S., Shimizu, K., Katoh, K., Miyake, K., Miyake, N., Fujikake, N., Sakurai, H. T., Arakawa, S., & Shimizu, S. (2020). Wipi3 is essential for alternative autophagy and its loss causes neurodegeneration. *Nature Communications*, *11*(1), 5311. DOI: 10.1038/s41467-020-18892-w PMID: 33082312

Ye, Y., Tyndall, E. R., Bui, V., Tang, Z., Shen, Y., Jiang, X., Flanagan, J. M., Wang, H.-G., & Tian, F. (2021). An N-terminal conserved region in human Atg3 couples membrane curvature sensitivity to conjugase activity during autophagy. *Nature Communications*, *12*(1), 374. DOI: 10.1038/s41467-020-20607-0 PMID: 33446636

Yu, Y. S., Kim, H., Kim, K., & Baek, S. H. (2023). Epigenetic regulation of autophagy by histone-modifying enzymes under nutrient stress. *Cell Death and Differentiation*, *30*(6), 1430–1436. DOI: 10.1038/s41418-023-01154-9 PMID: 36997734

Zachari, M., Longo, M., & Ganley, I. G. (2020). Aberrant autophagosome formation occurs upon small molecule inhibition of ULK1 kinase activity. *Life Science Alliance*, *3*(12), e202000815. DOI: 10.26508/lsa.202000815 PMID: 33109685

Zhang, S., Yazaki, E., Sakamoto, H., Yamamoto, H., & Mizushima, N. (2022). Evolutionary diversification of the autophagy-related ubiquitin-like conjugation systems. *Autophagy*, *18*(12), 2969–2984. DOI: 10.1080/15548627.2022.2059168 PMID: 35427200

Zhang, X.-W., Zhu, X.-X., Tang, D.-S., & Lu, J.-H. (2023). Targeting autophagy in Alzheimer's disease: Animal models and mechanisms. *Zoological Research*, *44*(6), 1132–1145. DOI: 10.24272/j.issn.2095-8137.2023.294 PMID: 37963840

Zhang, Z., Yang, X., Song, Y.-Q., & Tu, J. (2021). Autophagy in Alzheimer's disease pathogenesis: Therapeutic potential and future perspectives. *Ageing Research Reviews*, *72*, 101464. DOI: 10.1016/j.arr.2021.101464 PMID: 34551326

Zhao, D. Y., Bäuerlein, F. J. B., Saha, I., Hartl, F. U., Baumeister, W., & Wilfling, F. (2024). Autophagy preferentially degrades non-fibrillar polyQ aggregates. *Molecular Cell*, *84*(10), 1980–1994.e8. DOI: 10.1016/j.molcel.2024.04.018 PMID: 38759629

Zhou, Z. D., Yi, L. X., Wang, D. Q., Lim, T. M., & Tan, E. K. (2023). Role of dopamine in the pathophysiology of Parkinson's disease. *Translational Neurodegeneration*, *12*(1), 44. DOI: 10.1186/s40035-023-00378-6 PMID: 37718439

Zhu, Y., Fujimaki, M., Snape, L., Lopez, A., Fleming, A., & Rubinsztein, D. C. (2024). Loss of WIPI4 in neurodegeneration causes autophagy-independent ferroptosis. *Nature Cell Biology*, *26*(4), 542–551. DOI: 10.1038/s41556-024-01373-3 PMID: 38454050

Zientara-Rytter, K., & Subramani, S. (2020). Mechanistic Insights into the Role of Atg11 in Selective Autophagy. *Journal of Molecular Biology*, *432*(1), 104–122. DOI: 10.1016/j.jmb.2019.06.017 PMID: 31238043

## KEY TERMS AND DEFINITIONS

**Amphisome:** When an endosome fuses with an autophagosome, the resulting structure is called an amphisome. An amphisome is a single-membrane structure in which the inner autophagosome membrane is degraded.

**Autolysosome:** When an autophagosome or an amphisome fuses with a lysosome, the resulting structure is called an autolysosome.

**Autophagolysosome:** When an LC3-associated phagosome or an autophagosome that has engulfed a phagosome fuses with a lysosome, the resulting structure is called an autophagolysosome, which forms during xenophagy.

**Autophagosome:** An autophagosome is a double-membrane structure formed during macroautophagy.

**Autophagy:** Autophagy is an intracellular process in which cells break down their organelles and proteins for recycling.

**Endocytosis:** Endocytosis is a process by which cells take up large particles such as bacteria (phagocytosis/eating) and fluids or various molecules (pinocytosis/drinking) from the surrounding medium.

**Endosome:** An endosome is a single-membrane structure formed during endocytosis.

**Macroautophagy:** Macroautophagy is a subtype of autophagy that involves the use of autophagosomes.

**Phagocytosis:** Phagocytosis is a subtype of endocytosis by which cells take up large particles such as bacteria or cell debris.

**Phagolysosome:** When a phagosome fuses with a lysosome, the resulting structure is called a phagolysosome.

**Phagosome:** A phagosome is a single-membrane structure formed during phagocytosis.

## ENDNOTE

1 For a more detailed explanation, please refer to the following source: Klionsky, D. J., Eskelinen, E. L., & Deretic, V. (2014). Autophagosomes, phagosomes, autolysosomes, phagolysosomes, autophagolysosomes... wait, I'm confused. Autophagy, 10(4), 549–551. https://doi.org/10.4161/auto.28448

Chapter 2

# Autophagy-Based Treatment Approaches in Neuropsychological Diseases

**Şebnem Yıldırımcan Kadıçeşme**
https://orcid.org/0000-0002-8908-9464
*Kafkas University, Turkey*

## ABSTRACT

*Autophagy is an extremely important molecular pathway for the maintenance of cellular and organismal balance. According to recent research, cognitive diseases like dementia, Alzheimer's disease, Parkinson's disease, Huntington's disease, attention-deficit hyperactivity disorder, schizophrenia, and bipolar disorder are all influenced by defective autophagy. Neuropsychiatric disorders are extremely important because they affect the brain function, emotions, and mood of patients. Treatment of neuropsychiatric disorders is a very difficult process, and some problems are encountered in response to treatment. Therefore, new treatment approaches are needed. Given its significance in the pathophysiology of neuropsychological diseases, the autophagy pathway appears to be a promising target for the development of novel therapeutic approaches for these conditions. New research should be focused on explaining the relationship between this pathway and neuropsychological disorders. The data obtained from these studies will pave the way for the development of new therapeutic agents.*

DOI: 10.4018/979-8-3693-5908-2.ch002

## INTRODUCTION

Autophagy is a physiological process that eliminates infections, damaged macromolecules, long-lived proteins, misfolded or aggregated proteins, and malfunctioning organelles (Klionsky et al., 2021). Autophagy maintains intracellular homeostasis by breaking down and recycling proteins and organelles (Liu et al., 2023). Autophagy occurs through three different mechanisms: microautophagy, macroautophagy, and chaperone-mediated autophagy; however, the most prominent mechanism is macroautophagy (Liu et al., 2023).

Numerous neuropsychiatric and neurodegenerative disorders have been linked to abnormalities in the autophagy system. The onset and progression of neurodegenerative disorders are significantly influenced by precipitate-forming protein aggregates in nerve cells. The ubiquitin-proteasome system and autophagy do not remove these disease-associated aggregates from the cell. Therefore, disruptions in these pathways trigger synapse loss, nerve cell abnormalities, and cell death. For this reason, creating autophagy-based treatment plans for various illnesses requires an understanding of the molecular mechanisms underlying autophagy (Li et al., 2024). An overview of the connection between neuropsychological diseases and the autophagy pathway, as well as autophagy-based therapy options, is intended to be given to readers in this section.

## BACKGROUND

### Autophagy

Long-lived proteins, malfunctioning organelles, cytosolic fragments, damaged macromolecules, and pathogens are all degraded by autophagy, which uses proteolysis to maintain cellular and organismal homeostasis (Yamamoto, 2023). Based on membrane dynamics and cargo delivery mechanisms, three categories can be used to classify autophagy: chaperone-mediated autophagy, microautophagy, and macroautophagy (Figure 1). In macroautophagy, a structure with two membranes known as the autophagosome is formed as a result of membrane elongation (Yamamoto, 2023). The precursor of the autophagosome is the cup-shaped, double-membrane phagophore. The phagophore isolates cytoplasmic material and closes to form the autophagosome (Menzies et al., 2017). The autophagosome then fuses with lysosomes, and the cytoplasmic contents are degraded. Macroautophagy is regulated by 'autophagy-related proteins' or Atg proteins for short (Vargas et al; 2023; Yamamoto, 2023).

In microautophagy, intraluminal vesicles containing cytosolic materials are formed by membrane internalization, and the cytosolic contents are then degraded in lysosomes. Similar to macroautophagy, microautophagy can be selective or non-selective in its cargo intake and destruction (Wang et al., 2023). The majority of research on microautophagy has concentrated on alterations involving endosomes, vacuoles, or lysosomes. Accordingly, three types of microautophagy have been defined based on the morphological changes of lysosomes in both yeast and mammalian cells: type 1 is characterized by vacuolar or lysosomal protrusion, Type 2 by vacuolar or lysosomal invagination, and Type 3 by endosomal invagination (Oku & Sakai, 2018).

In recent years, depending on the mode of transfer of microautophagic cargoes to lysosomes and late endosomes (endolysosomes), microautophagy has been divided into two types: fission type and fusion type. In macroautophagy, lysosomes fuse with autophagosomes or amphisomes (union of autophagosomes and endosomes) (Zhao et al., 2021). The core autophagy complex and soluble N-ethylmaleimide-sensitive factor attachment protein receptor (SNARE) complexes are necessary for fusion-type microautophagy, just like macroautophagy. The fission type, however, operates independently of the core autophagy complex, with endosomal sorting complexes required for transport (ESCRTs) playing a role in membrane division or budding (Wang et al., 2023).

To date, various types of macroautophagy have been identified, such as mitophagy (mitochondria), lysophagy (lysosomes), reticulophagy (endoplasmic reticulum), aggrephagy (protein aggregates), and xenophagy (microorganisms). For microautophagy, selective types of microautophagy with different roles regulated by various molecular mechanisms (micromitophagy, endosomal microautophagy, micronucleophagy, etc.) have been identified (Kirkin & Rogov, 2019).

CMA does not require membrane deformation, in contrast to macroautophagy and microautophagy. In this kind of autophagy, the lysosomal membrane and the cytosolic molecular chaperone HSC70 (or HSPA8) facilitate the direct transport of particular cytosolic proteins with a peptide sequence to lysosomes. This pathway does not require vesicular trafficking (Yamamoto, 2023). CMA is responsible for regulating physiological processes such as metabolism, aging, immunity, and the cell cycle. Furthermore, the development of diseases like cancers, atherosclerosis, and neurodegenerative disorders is influenced by CMA dysfunction (Yao & Shen, 2023). Because of their involvement in the pathophysiology of infectious, metabolic, cancer, and neurodegenerative disorders, all three forms of autophagy have drawn a lot of interest (Klionsky et al., 2021).

*Figure 1. The main pathways of autophagy*

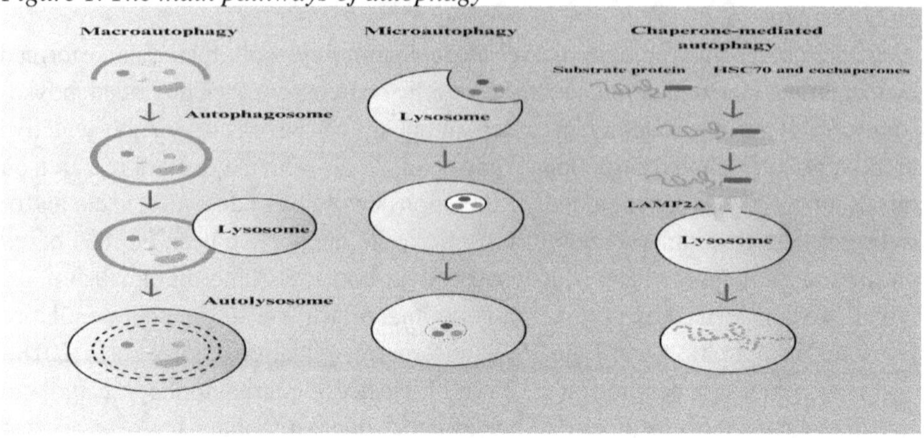


## Molecular Mechanisms of Macroautophagy

Macrophagy, which is mostly induced by hunger or stress, is the main mechanism observed in autophagy. In mammals, during the initiation phase of macroautophagy, the ULK1 (or ULK2) and FIP200 (or RB1CC1), ATG13, and ATG101 come together near the endoplasmic reticulum (ER) to form the ULK complex. It has been demonstrated that the liquid–liquid phase separation (LLPS) of FIP200 for the specification of ER autophagosome starting sites is triggered by calcium changes in the ER membrane (Zheng et al., 2022).

In yeast, the pre-autophagosomal structure (PAS), known as the homologous Atg1 complex, is formed by the assembly of the Atg1, Atg13, and Atg17–Atg29–Atg31 subcomplexes (Yamamoto, 2023). In PAS, it is a biomolecular condensate formed by LLPS (Fujioka et al., 2020). It has been suggested that hunger-induced activation of the ULK/Atg1 complex is necessary for autophagosome formation. The ULK complex assembles by cargo-driven means in addition to famine. SQSTM1 (also called p62), a ubiquitin-binding macroautophagy adapter, recognizes autophagic cargo that has been ubiquitin-labeled. p62 forms a liquid-like biomolecular condensate with ubiquitinated cargoes and interacts with FIP200 in the ULK complex. As a result, the p62 condensate's ubiquitinated cargo proteins degrade preferentially (Turco et al., 2021). After the formation of the ULK complex, class III phosphatidylinositol 3-kinase complex I and ATG9 vesicles are activated (Sawa-Makarska et al., 2020). In the ULK complex, ATG9 vesicles frequently engage with the ATG13/ATG101 subcomplex and go to the autophagosome formation zone (Ren et al; 2023; Nguyen et al; 2023).

The phagophore membrane then spreads out. ATG9 is a unique ATG protein that is anchored in the phagophore membrane and forms a homotrimeric complex. ATG9 functions as a scramblase by permitting the interchange of phospholipids between the phagophore membrane's outer and inner leaflets when it interacts with the lipid transfer protein ATG2 (Maeda et al., 2020). According to the most recent membrane elongation model, ATG2 bridges the gap between the two membranes by moving phospholipids from the ER membrane's outer leaflet to the phagophore membrane's outer leaflet (Valverde et al., 2019). The transfer of phospholipids produced from the ER to the inner leaflet of the phagophore membrane through the ATG9 scramblase complex causes the phagophore membrane to elongate.

Membrane elongation is also influenced by the ATG12 and ATG8 conjugation systems. The E1-like enzyme ATG7 and the E2-like enzyme ATG3 conjugate ubiquitin-like ATG8 family proteins (LC3 and GABARAP in mammals) with phosphatidylethanolamine (PE) to generate ATG8-PE (LC3-PE and GABARAP-PE in mammals). The ubiquitin-like protein ATG12 and its substrate ATG5 are conjugated by ATG7 and the E2-like enzyme ATG10. The ATG12–ATG5–ATG16L1 complex, which has E3-like activity in the LC3/GABARAP conjugation reaction, is created when the ATG12–ATG5 conjugate subsequently joins forces with ATG16L1. It is believed that phagophore membrane-localized LC3 and GABARAP contribute to the production of autophagosomes.

ATG8 family members are also involved in cargo recognition in selective macroautophagy. When misfolded proteins are tagged with ubiquitin in ubiquitin-dependent selective macroautophagy, they are identified by receptors that include ubiquitin-binding domains, including SQSTM1, NBR1, OPTN, NDP52, TAX1BP1, and TOLLIP. These receptors have LC3-interacting region (LIR) motifs that identify the autophagosome-associated protein LC3, which controls the pathway between ubiquitinated cargo and the autophagosome (Menzies et al., 2017). ATG8 family proteins in phagophores recognize these receptors containing the ATG8-interacting motif (LIR), and selective degradation occurs.

LC3 recognizes a number of cargo-resident receptors with LIR motifs to be present in autophagosomes during ubiquitin-independent selective macroautophagy. These comprise CALCOCO1 for macroGolgiphagy, NIX (also known as BNIP3L), BNIP3 FKBP8, FUNDC1, TRIM5, and BCL2L13 for macromitophagy, TEX264, CCPG1 FAM134A/B/C (also known as RETREG2/1/3), RTN3L, SEC62, and ATL3 for macroER-phagy, and TAX1BP1 for macroferritinophagy (Yamamoto, 2023). A double-membraned autophagosome is created when the phagophore membrane shuts after elongation. The separation of membranes during closure is regulated by the endosomal sorting complex required for transport (ESCRT) complex (Zhou et al., 2019). SNARE complexes (YKT6–SNAP29–STX7 and STX17–SNAP29–

VAMP7/8) mediate the fusion of lysosomes and the closed autophagosome. The materials within the resulting autolysosome are then degraded.

## The Role of Autophagy in the Nervous System

Neurons are specialized, polarized cells composed of three main structures: the axon, soma, and dendrites (Liénard et al., 2024). In these structures, all steps from synthesis to degradation of neuronal proteins are tightly controlled. Autophagy occurs in all of these neuronal compartments (Valencia et al., 2021). Neurons are extremely vulnerable to metabolic diseases, energy shortages, inflammation, neurotoxicity, traumas, and neurodegeneration. In these circumstances, autophagy is variedly activated. Autophagy happens at a basal level in the majority of cells. Nonetheless, under normal circumstances, neurons display persistently active autophagy, which sets them apart from other non-neuronal cells. (Valencia et al., 2021). It is believed that neurons continuously produce autophagosomes under normal conditions, unlike other cell types.

Neurons have a structure-specific regulation of autophagy. At the axon's distal end, autophagosome biogenesis takes place. The motor proteins dynein and kinesin (retrograde transport) assist formed autophagosomes in moving along microtubules toward the soma after they have undergone bidirectional movement. Autophagic vesicles (AVs) combine with endosomes and/or lysosomes during this transport process to create amphisomes and/or autolysosomes. These AVs mature further near the soma and exhibit bidirectional movement. Biogenesis also occurs in somato-dendritic structures. Dendritic autophagosome biogenesis is regulated by synaptic activity. Activation of the α-amino-3-hydroxy-5-methyl-4-isoxazolepropionic acid receptor (AMPAR)/N-methyl-D-aspartate receptor (NMDAR) or depolarization brought on by potassium chloride (KCl) increases the density of AVs in dendrites. After depolarization, lysosomes originating from the soma accumulate in dendrites. In autophagy occurring in the soma, AVs are less mobile than axons or dendrites. Autolysosomes are primarily degraded in this region (Liénard et al., 2024).

Protein aggregation, damaged organelles, and neural homeostasis can all be protected by moderate autophagy. By recycling fatty, amino, and nucleic acids, it also maintains energy balance. As a result, mild to moderate autophagy helps neurons survive and is essential for preserving the central nervous system's (CNS) homeostasis. Nevertheless, excessive or persistent autophagy can result in the buildup of autophagosomes in the cytoplasm, the breakdown of vital components, and possibly even the induction of autophagic cell death or other mechanisms of cell death (Li et al., 2024).

Autophagy plays important roles in axonal homeostasis, synapse development, and dendritic spine formation. Neuronal autophagy loss can result in abnormalities in cargo-specific axonal transport, which can cause axonal swelling and dystrophy. Near synapses, autophagosome biogenesis takes place, and autophagy is crucial for the formation of synapses and neural plasticity. According to studies, autophagy controls axonal ER calcium release in hippocampus neurons to regulate presynaptic neurotransmission, while rapamycin-induced autophagy quickly suppresses neurotransmitter release in dopaminergic neurons (Hernandez et al; 2012; Kurashige et al; 2021). All of these findings point to the critical role autophagy plays in neuronal function (Valencia et al., 2021).

Neuronal autophagy has a role in neuronal survival, aging, development, and death. Since neurons are long-lived post-mitotic cells, they are extremely dependent on degradation mechanisms because they are prone to collecting damaged organelles or misfolded proteins (Liénard et al., 2024). Furthermore, damage accumulates in neurons as a result of the fast movement of synaptic proteins in response to repeated electrical stimulations. Thus, autophagy's ability to regulate protein quality is essential for neuronal function. When neurons' autophagy is disrupted, aberrant proteins build up and increasing neuronal malfunction occurs, resulting in neurodegeneration. Studies show that the loss of autophagic regulatory genes in mice leads to abnormal behaviors, emphasizing the importance of autophagy in neuronal function (Valencia et al., 2021).

## The Role of Autophagy in Neuropsychological Disorders

In recent years, autophagy has become a focal point, especially in neurological diseases. neurodegenerative diseases are disorders related to advancing age. As neurons age, autophagy declines and damaged organelles and misfolded proteins tend to accumulate (Djajadikerta et al., 2020). The build-up of misfolded proteins causes neurodegenerative disorders. These misfolded proteins have a propensity to group together to form tangles or plaques that are rich in β-sheets and hinder cellular processes. The ubiquitin-proteasome system and chaperone-mediated autophagy are two of the cell's protein quality control systems that are interfered with when such aggregates build up. These mechanisms digest and recycle damaged proteins and organelles. Toxic aggregates build up when these systems malfunction or are compromised, further upsetting cellular functions and causing neuronal damage and death (Sevriev et al., 2024). Therefore, autophagy is considered a crucial mechanism that could alleviate the onset and progression of neurological degenerative lesions in neurodegenerative diseases (Li et al., 2024).

The accumulations observed in neurodegenerative diseases can consist of different proteins in different diseases and can be seen in different types of nerve cells and different intracellular organelles. Point mutations, amino acid repeats, or structures prone to polymerization (e.g., beta plaques) may cause aggregation and deposition. These abnormalities can affect many pathways within the nervous system, leading to synapse loss and cell death. (Kocaturk & Gozuacik, 2017).

After research on the essential proteins ATG5 and ATG7 associated with autophagy, studies on the role of autophagy in the nervous system have gained momentum (Kocaturk & Gozuacik, 2017). Studies have demonstrated that when ATG5 or ATG7 proteins are knocked out in mouse neurons, the brain's baseline autophagy level dramatically drops and ubiquitinated proteins, p62, and undegraded protein aggregates build up (Komatsu et al., 2007). Mice lacking Atg7 exhibited neuronal loss in the cerebellar cortex and brain, while Atg5 knockout mice developed progressive motor dysfunction (Hara et al; 2006; Komatsu et al; 2007). These studies highlight the essential role of autophagy in maintaining axonal balance and show that defects in autophagy lead to axonal degeneration (Komatsu et al., 2007).

In addition, autophagy regulates the level of proteins that tend to form aggregates inside the cell that cause neurodegenerative diseases (Berger et al., 2006). Autophagy and the ubiquitin-proteasome system are the primary pathways responsible for eliminating these toxic accumulations within cells (Kocaturk & Gozuacik, 2017). Therefore, defective autophagy causes the formation and progression of neurodegenerative diseases as it cannot clear the accumulated proteins.

Several studies have emphasized the importance of autophagy in preventing neurodegeneration (Cragnaz et al; 2021; Fang et al; 2019). Impaired autophagy is linked to various neurological developmental disorders and neurodegenerative diseases. Due to aberrant mTOR pathway, a negative regulator of autophagy, and decreased autophagy during neural development, patients with autism spectrum disorders exhibit damaged neuronal structures (Winden et al., 2018). The genetic form of autism spectrum condition known as fragile X syndrome is caused by a mutation in the FMR1 (fragile X messenger ribonucleoprotein 1) gene. A study found that whilst autophagy activation restored synaptic and cognitive impairments in Fmr1-knockout (KO) mice, decreased autophagy caused increased mTOR pathway activity in hippocampus neurons (Yan et al., 2018). Furthermore, a number of neurodegenerative illnesses, including Alzheimer's, Parkinson's, Amyotrophic lateral sclerosis, Huntington's, and Lafora disease, have been linked to impaired autophagy (Zatyka et al., 2020) .

In mice, the absence of Atg proteins, which are necessary for the production of autophagosomes, results in neurodegeneration, behavioral abnormalities, and neuronal death. A characteristic of neurodegenerative diseases is the aberrant buildup of misfolded proteins and damaged organelles. Therefore, one possible therapeutic approach for the treatment of neurodegenerative illnesses is to promote autophagy by blocking the mTOR pathway. Although autophagy is required to preserve neuronal function, there are several situations in which it is not absolutely necessary. Overautophagy during the neonatal period can lead to hypoxic-ischemic brain damage, while inhibiting autophagy can alleviate brain injury. Therefore, one possible treatment strategy for prenatal asphyxia could be autophagy inhibition (Corti et al., 2020). Significant remodeling occurs in the immature brain in comparison to the adult brain, including the growth of neuronal cells and the creation and destruction of synapses. Furthermore, mature and immature brains exhibit different patterns of autophagy-related protein expression. Indeed, there is more lysosomal activity and increased expression of autophagosome markers such LC3 and beclin-1 in young brains. These findings underscore the critical importance of maintaining autophagy balance for proper neural activity. Therefore, comprehending neuronal autophagy and how it is regulated is crucial for creating novel treatment approaches for a range of brain illnesses (Valencia et al., 2021).

## Neuropsychological Disorders Associated With Impaired Autophagy

### Alzheimer's Disease

Alzheimer's Disease (AD) is the most commonly observed neurodegenerative disorder, characterized by intracellular neurofibrillary tangles formed by aggregated hyperphosphorylated tau protein and amyloid-beta (Aβ) plaques, which are the cleaved products of amyloid precursor proteins (APPs) (Guo et al., 2018). Tau protein hyperphosphorylation in AD results in the development of neurofibrillary tangles. AD causes irreversible, progressive, and extremely severe deterioration of cognitive and emotional brain function. Several investigations have demonstrated the connection between AD pathophysiology and defective autophagy (Litwiniuk et al., 2023).

Under normal conditions, autophagosomes are rarely found in the brain. However, in AD brains, autophagosomal vesicles have been observed in dystrophic neurites. Another study demonstrated an increase in autophagic vacuoles in regions rich in presenilin 1 (PS1) (Guo et al., 2018). PS1 is a catalytic subunit of the gamma-secretase complex, which plays a role in the formation of amyloid-beta (Aβ) from amyloid-beta precursor protein (APP) (Zhang et al., 2014). β-secretase first cleaves

APP to create the β-C-terminal fragment (βCTF), which PS1 subsequently cleaves to produce Aβ. It is suggested that mutations in PS1, which inhibit the cleavage of APP, lead to the onset of AD. According to recent research, PS1 also lowers Aβ levels by encouraging autophagic βCTF breakdown (Bustos et al., 2017). PS1 also contributes to the fusion of lysosomes and autophagosomes. The phosphorylation of PS1 at Ser367 in mouse brains prevented the fusion of lysosomes and autophagosomes. Therefore, inhibition of autophagy may reduce βCTF degradation and lead to Aβ accumulation in the brain (Bustos et al., 2017).

Moreover, PS1 is involved in the lysosomal turnover of autophagic substrates (Lee et al; 2010). Another study showed that PS1 regulates lysosomal acidification, which helps maintain calcium ($Ca^{2+}$) balance. Acidification loss prevents lysosome-autophagosome fusion and causes lysosome malfunction (Lee et al; 2010). Autophagy-related proteins are strongly linked to AD, according to genome-wide association studies (GWAS). For example, it has been demonstrated that AD brains exhibit transcriptional repression of Beclin 1, which is in charge of autophagosome production (Pickford et al., 2008).

## Parkinson's Disease

The loss of dopaminergic neurons causes Parkinson's disease (PD), the second most prevalent neurodegenerative illness, to produce insufficient amounts of dopamine in the brain. While it can have genetic origins, PD is most often triggered by exposure to neurotoxic agents (chemicals). Symptoms include tremors in the hands, arms, legs, chin, and face; stiffness in the arms, legs, and trunk; slowness of movement; and poor balance and coordination. PD, which is more common in men than in women, usually starts around the age of 60 (Ben-Shlomo et al., 2024).

Neurons in PD patients' post-mortem brain samples have been found to have malfunctioning lysosomes and autophagosome accumulation. Lewy bodies are mostly made up of misfolded and aggregated α-synuclein. The increase in α-synuclein level when the lysosome is dysfunctional suggests a relationship between α-synuclein degradation and autophagy.

It has been demonstrated in vivo that the leucine-rich repeat kinase 2 (LRRK2) mutation LRRK2 G2019S, linked to autosomal-dominant PD, intensifies age-related reduction of autophagic flux (Saha et al., 2015). The VPS35 D620N mutation, which is also associated with autosomal-dominant PD, destabilisez the WASH complex, causing defects in autophagosome formation and leading to abnormal trafficking of ATG9.

Lysosomal-related genes that contribute to the etiology of PD have been found by genome-wide association studies (GWAS). For example, the lysosomal ATPase-related protein ATP13A2 is mutant in autosomal recessively inherited PD patients.

When ATP13A2 is downregulated, α-synuclein builds up and dopaminergic neurons' lysosomal degradation is decreased. (Usenovic et al., 2012). Defects in α-synuclein aggregation and the autophagosome-lysosome pathway are caused by autosomal recessive mutations in GBA, the gene encoding the lysosomal hydrolase Glucocerebrosidase (Abeliovich & Gitler, 2016).

Mutations in the Vacuolar Protein Sorting 13 Homolog C (VPS13C) gene, which is responsible for the regulation of mitochondrial function through the endolysosomal pathway in neurons, have been shown to result in mitochondrial malfunction and lysosomal dysfunction and are connected to autosomal recessive Parkinson's disease (Monfrini et al; 2022; Schröder et al; 2024).

## Huntington's Disease

An autosomal dominant neurodegenerative disease, Huntington's disease (HD) is typified by profound abnormalities in cognition, motor function, and mental health. It is the most common polyglutamine (polyQ) disorder. The HTT gene, responsible for HD, contains CAG repeat trinucleotides in its first exon (Lepinay & Cicchetti, 2023). These CAG repeats cause polyglutamine (PQ) expansions and accumulation of mutant HTT protein aggregates (Lepinay & Cicchetti, 2023). HTT plays a role in the regulation of autophagosome transport. Studies have shown that HTT depletion causes defective autophagosome transport (Wong & Holzbaur, 2014). Furthermore, the association of the V471A polymorphism in the Atg7 gene with the early-onset form of HD has been demonstrated (Metzger et al., 2010).

Studies with cell lines and animal models have shown defects in the loading of cargo into autophagosomes in HD (Martinez-Vicente et al ; 2010). These investigations have demonstrated the link between HD neurodegeneration and the autophagy process.

## Amyotrophic Lateral Sclerosis

A motor neuron disease that develops in adulthood, amyotrophic lateral sclerosis (ALS) is typified by the degeneration of both upper and lower motor neurons in the brain and spinal cord (Cozzi & Ferrari, 2022). These patients show decreases in muscle strength over the years. About 10% of ALS cases are familial, although the majority (~90%) are sporadic. Genetic, epigenetic, and environmental factors play a role in the development of ALS (Cozzi & Ferrari, 2022). It has symptoms such as impaired speech, difficulty swallowing, muscle cramps, and difficulty walking (Riva et al., 2024).

According to recent data, autophagy failure is a major factor in the etiology of ALS (Cozzi & Ferrari, 2022). The disorder causes harmful protein aggregates to accumulate, which reduces autophagy-mediated breakdown. The fact that multiple genes associated with ALS code for proteins engaged in different phases of the autophagic pathway emphasizes the significance of autophagy in the onset and development of the illness (Cozzi & Ferrari, 2022).

Dysfunction in autophagy-related proteins leads to neurotoxicity and disrupts proteostasis in ALS (Chua et al., 2021). In ALS, mutations are observed in genes encoding the basic components of the autophagy pathway. The *TARDBP* gene, which codes for TDP-43, regulates the processing of mRNA for many proteins involved in neuronal survival and neurodegenerative diseases. Additionally, it regulates the expression of the ATG7 gene, which is involved in the autophagic process (Bose et al., 2011). Approximately 98% of all ALS cases show neuronal and glial inclusions of TDP-43 and relocalisation from the cytoplasm to the nucleus. TARDBP mutations are found in approximately 3% of familial cases and less than 1% of sporadic cases (Cai & Ganesan, 2022).

SQSTM1/p62 and optineurin (OPTN) are autophagy receptors that interact with LC3 to facilitate the segregation of autophagic cargo within autophagosomes. It has been demonstrated that ALS is caused by mutations in the LC3-binding region of SQSTM1/p62 (L341V) (Goode et al., 2016). OPTN mutations result in decreased autophagy, protein clearance abnormalities, and compromised mitophagy (Zhao et al., 2023). Additionally, other autophagy-related genes, such as ALS2, C9ORF72, CHMP2B, TBK1, UBQLN2, and VCP, have been implicated in ALS pathology (Chua et al; 2021; Cozzi & Ferrai, 2022). According to these findings, autophagy is crucial for preserving cellular homeostasis and halting the formation of harmful aggregates in ALS.

## Schizophrenia

Schizophrenia (SCZ) is a heterogeneous, polygenic psychiatric disorder with a frequency of 0.5-1%. Significant disturbances in thinking, feeling, and behavior are observed in SCZ. The patient lives in his or her own world, detached from interpersonal relationships and reality, and the illness usually begins in adolescence (Farsi & Sheng, 2023). Recent research indicates that the pathophysiology of schizophrenia may involve autophagy. For example, a study found that the post-mortem hippocampus samples of patients with schizophrenia had 40% lower levels of Beclin 1 mRNA than controls (Wagner et al., 2015). Furthermore, reduced Beclin-1 mRNA and protein levels have been observed in the hippocampus of schizophrenia mouse models (Wagner et al., 2014).

In another study, autophagy-related genes that may be risk factors for schizophrenia were investigated. At the end of the study, genes such as VAMP7, PTEN, WIPI2, PARP1, DNAJB9, SH3GLB1, ATF4, EIF4G1, EGFR, and CDKN1A were found to be associated with schizophrenia risk (Tan et al., 2024). According to these results, autophagy failure may have a role in the onset and progression of schizophrenia.

## Bipolar Disorder

Bipolar disorder (BD) is a mood disorder characterized by depressive episodes as well as hypomanic and/or manic episodes. Serotonin, norepinephrine, dopamine, and neuropeptides play a role in its etiology. It has been suggested that dysfunctional autophagy resulting in impaired mitochondrial function is involved in the pathophysiology (Lam et al; 2023; Toker & Agam, 2015).

Studies have indicated that circadian dysregulation, one of the fundamental mechanisms underlying BD, is associated with dysfunctional autophagy. Additionally, defects related to mitochondrial quality control, neurotransmission, and ion channels have also been linked to disturbances in the autophagy pathway (Yang et al., 2023). The volume and amount of autophagic vacuoles in the rat heart have been shown to vary with day and night, light and darkness (Ma et al., 2011). In various mouse tissues (liver, heart, skeletal muscle, etc.), it has been shown that the autophagic flux peaks in the morning and decreases in the dark, based on the degradation rate of LC3-II.

Autophagy activation, phagophore maturation, and autophagosome formation are the three stages of autophagy that can be controlled by circadian rhythms. The circadian clock controls autophagy through several mechanisms at each stage. This lends credence to the idea that autophagy could result in BD-specific symptoms (Yang et al., 2023).

Further research has found decreased levels of LC3 and p62, autophagy-related proteins, at both gene and protein levels in the peripheral blood mononuclear cells (PBMCs) of BD patients. On the other hand, decreased protein and mRNA levels of LC3 and p62 were found in PBMCs of patients with BD, and a negative correlation was found between autophagic proteins and symptoms of depression and mania (Scaini et al., 2019). Studies suggest that impaired autophagy may be related to the pathophysiology of BD.

## Autophagy-Based Treatment Approaches for Neuropsychological Disorders

Abnormal autophagy characteristics are observed in psychiatric and neurodegenerative diseases. Autophagy-based treatments are therefore a viable choice for the management or prevention of certain illnesses (Park et al., 2020) . In the nervous system, autophagy preserves cellular homeostasis by controlling inflammation and apoptosis, removing damaged organelles, and breaking down misfolded proteins. Beclin 1, TP53, AMPK-m-TOR, endoplasmic reticulum stress, and other signaling pathways, which play a role in the regulation of autophagy, are good therapeutic targets for neurological diseases (Li et al., 2024). In this regard, boosting autophagy has been suggested as a treatment strategy to reduce neuropathology and neurodegeneration, which has led to research into autophagy-enhancing substances (Khalifeh et al., 2021).

According to recent research, upregulating autophagy in AD, HD, PD, and ALS improves the prognosis of the disease by lowering the buildup of intracellular aggregate-prone proteins (Park et al., 2020). However, excessive autophagy can be harmful for maintaining cellular homeostasis. Therefore, caution must be exercised when targeting autophagy in order to cure neurodegenerative illnesses.

In autophagy-based therapeutic approaches, autophagy is activated by various signals, both mTOR-dependent and mTOR-independent. mTOR-dependent autophagy inducers inhibit mTORC1 activity directly, leading to the activation of the ULK1 complex. mTOR-independent agents upregulate autophagy activity through various intracellular signaling cascades or lysosomal biogenesis (Guo et al., 2018). It has been demonstrated that rapamycin inhibits mTORC1 in AD, reducing amyloid plaques and fibrillary tangles in the brain and improving cognitive deficits (Davoody et al., 2024). A phytochemical named Arctigenin can trigger autophagy and inhibit Aβ production by means of the suppression of the AKT/mTOR pathway (Zhao et al., 2024).

In an AD mouse model, the quinolone derivative GTM-1 was shown to induce mTOR-independent autophagy, reducing Aβ pathology and correcting cognitive dysfunction (Uddin et al., 2018). The tyrosine kinase inhibitor Nilotinib was found to improve autophagic flux in an AD model, reducing Aβ levels, leading to neuroprotective effects, and improving memory function in mice (Nobili et al., 2023). Another study shown that the natural substance curcumin suppresses the PI3K/Akt/mTOR pathway, boosting autophagic flux and the removal of cellular aggregates while also increasing the expression levels of autophagy-related proteins such ATG5, ATG16L1, and Beclin-1. Furthermore, it was demonstrated that curcumin provides protection against disease progression and development when taken before AD symptoms appear (Azzini et al., 2024).

Carbamazepine (CBZ) was primarily developed for epilepsy treatment. Recent studies have shown that CBZ-induced autophagy provides protection against memory impairment and increased Aβ content in the brain in an AD mouse model. Clinical studies have shown that lithium can trigger autophagy independently of mTOR and improve AD (Uddin et al., 2018). A meta-analysis of preclinical studies has also shown that lithium has neuroprotective effects. Lithium treatment has been shown to reduce Aβ and tau levels and improve cognitive behavior in AD animal models (Singulani et al., 2024).

There is potential for treating AD with nicotinamide mononucleotide (NMN), a precursor of nicotinamide adenine dinucleotide (NAD+). A study on AD mice demonstrated that NMN supplementation could alleviate neuronal damage by restoring NAD+ levels. In vitro studies using Aβ-induced PC12 cells showed that autophagy inhibitors (chloroquine, bafilomycin A1) blocked the clearance of p-tau and antioxidant stress effects of NMN. As a result, it was observed that NMN triggers p-tau clearance by activating the autophagy pathway and thus has a curative effect on AD (Ma et al., 2024).

Resveratrol is a natural polyphenolic compound found in particularly high amounts in grape seeds and pericarp. It has a wide range of pharmacological effects (Zhao et al., 2024). Clinical trials to determine its efficacy in AD treatment are in phase II. Studies on transgenic AD mice have shown that resveratrol increases AMPK activity, reduces oxidative stress, and consequently decreases tau pathology (Dhapola et al., 2024).

Experimental PD models have studied several drugs that trigger autophagy through the mTOR signaling pathway, including isorinkophilin, resveratrol, rapamycin, PREP inhibitor (KYP-2047), and metformin. mTOR-independent drugs include nilotinib, trehalose, sodium valproate, CBZ, latrepirdin, lithium, and spermidine (Sanchez-Mirasierra et al., 2022). Additionally, medications that impact lysosome function, like ambroxol, isophagomine, and acidic nanoparticles, are employed. Ambroxol, for example, acts as a chaperone for the lysosomal enzyme glucocerebrosidase (GCase), which plays a role in lysosomal autophagy. The drug Ambroxol increases GCase activity in patients with mutations in the GBA gene, which encodes glucocerebrosidase (GCase), and modulates phosphorylated α-synuclein levels. The results of the study showed that the drug is safe (Öksüz et al., 2022).

In treatment strategies for PD, targeting α-synuclein is extremely popular. The goal of these treatments is to prevent α-synuclein aggregates, reduce α-synuclein production, and increase its degradation. The use of rapamycin, an inhibitor of the mTOR pathway, has improved motor function and decreased α-synuclein aggregation (Öksüz et al., 2022).

Another approach to PD treatment is targeting mitophagy. Signaling pathways involved in mitochondrial quality control are triggered by PINK1/Parkin when damage to the mitochondria occurs. The most frequent causes of autosomal recessive and early-onset Parkinson's disease (PD) (before the age of 45) are loss-of-function mutations in PINK1 and PRKN (Ge et al., 2020). Additionally, defective mitophagy has been associated with sporadic PD. Mitophagy pathways controlled by PINK1/Parkin and FUNDC1 with BNIP3L can be targeted for PD treatment (Masaldan et al., 2022). Examples of mitophagy-targeted treatments include niclosamide and its analogues and the PINK1 activator KTP (kinetin triphosphate) and its analogues (Nechushtai et al., 2023).

Leucine-rich repeat (LRR) and kinase domains are found in LRRK2, which is linked to lysosomal activity and the neuronal autophagy process. Mutations in the *LRRK2* gene have been linked to PD. Therefore, specific LRRK2 kinase inhibitors (LRRK2-IN1, GSK2578215A, GNE7915, DNL201, etc.) have been developed for PD treatment (Taymans et al., 2023).

Another autophagy-based approach for PD treatment involves miRNAs. miRNAs regulate autophagy and show neuroprotective effects. Baicalein is a natural compound that affects miRNA levels. In a study with rats, it was demonstrated that this compound triggered mitochondrial autophagy through miR-30b-5p and the SIRT1/AMPK/mTOR pathway, showing neuroprotective effects (Chen et al., 2022).

Recently, an approach called AUTOphagy-TArgeting Chimera (AUTOTAC) has been developed, which induces degradation of targeted proteins via the macroautophagy/autophagy-lysosome pathway (ALP). This therapeutic approach aims to eliminate disease-related proteins. In this technology, AUTOTAC carries ligands that bind to the target and to the autophagic receptor. This triggers macroautophagy, resulting in the degradation of the target proteins. In PD, untreated α-synuclein aggregates cause cellular toxicity and promote disease progression. A group of researchers synthesized the AUTOTAC chemical (ATC161), which binds to α-synuclein aggregates and the autophagy receptor (p62/SQSTM1). The study showed that oral administration of ATC161 in a PD mouse model induced degradation of α-synuclein aggregates, improved muscle strength and motor activity, and alleviated glial inflammatory responses (Lee et al ; 2023; Lee et al; 2024).

Activating cellular breakdown processes like autophagy is one of numerous treatment approaches intended to lower mutant huntingtin levels in HD. Increasing evidence suggests that accumulation of mHTT aggregates and autophagic and/or lysosomal dysfunction are the major pathogenic mechanisms underlying HD. In this situation, increasing autophagy may be a useful therapeutic approach to get rid of protein aggregates and enhance cellular performance.

The first identified mTOR-dependent autophagy inducer is rapamycin. It inhibits mTORC1 kinase activity through allosteric binding. By triggering autophagy, rapamycin prevents some of the functional disorders caused by mutant huntingtin. In a study with Drosophila, it was found that rapamycin increased the expression of *Atg1* (a gene responsible for autophagosome formation) and inhibited TOR. The data showed that rapamycin prevented the aggregation of mutant huntingtin in the flies' brains and locomotor performance loss (Roth et al., 2023).

Calpains are proteolytic enzymes activated by calcium that participate in essential cellular processes such as cell proliferation, apoptosis, signaling pathways, neurogenesis, and synaptic plasticity. Overactivation of the calpain system has been linked to HD. Calpastatin (CAST) is an endogenous inhibitor of calpains and inhibits calpains by activating autophagic pathways. Overexpression of CAST decreased calpain activation and calpain-mediated HTT breakdown and aggregation, according to a study conducted in transgenic mice. Moreover, an increase in autophagosome levels and improvement in motor function were observed in the mice (Incebacak Eltemur et al; 2022; Kim et al; 2021).

Ubiquilins are ubiquitin-binding proteins associated with the aggregation of misfolded proteins and protein degradation. It has been shown that ubiquilins control autophagic flux by controlling the mTOR pathway and lysosomal acidification. In a mouse model, it was shown that high expression of Ubiquilin-2 (UBQLN2) reduced pathogenic HTT levels (Gerson et al., 2020).

Transcription factor EB (TFEB), the major transcriptional regulator of the autophagy pathway, regulates the expression of genes important for lysosomal biogenesis, lysosomal functions, and macroautophagy. Many neurodegenerative illnesses, including HD, are influenced by TFEB. Due to its ability to regulate lysosomal activity, which is involved in the removal of defective proteins accumulated in neurodegenerative diseases, TFEB is an important candidate for the treatment of HD. A study in a mouse model showed that inducing TFEB activity stimulated autophagy and lysosomal activity, leading to a reduction in mHTT levels (Lee et al., 2024). However, another study showed that triggering TFEB activation could increase the formation of mHTT aggregates, potentially worsening the disease (Yang et al., 2023).

Berberine, a natural alkaloid obtained from various plants, has been shown to trigger autophagy through the AMPK/mTOR pathway in HD mice, reducing mutant HTT aggregation and motor dysfunction (Sunhe et al; 2024; Tian et al; 2023). Studies have shown that small molecules with affinity for dopamine (DA) D2 and D3 receptors (D2R/D3R) can regulate autophagy. It has been demonstrated that pramipexole (a D2R/D3R agonist), used in the treatment of PD, can activate autophagy (Dong et al., 2023). Death of the medium-sized spiny neurons in the striatum, where high expression of D2R and D3R is observed, is observed in HD.

Therefore, the effect of pramipexole (PPX) on HD treatment has been investigated. The results showed that PPX application led to autophagy activation in mice, which was associated with a reduction in mHTT levels and a protective effect in striatal neurons (Luis-Ravelo et al., 2018).

One naturally occurring disaccharide that promotes autophagy is trehalose. According to studies, trehalose can increase autophagic activity, which will improve cellular homeostasis and increase the survival of neuronal cells (Sevriev et al., 2024). Trehalose triggers autophagy by inhibiting the mTOR pathway. The mTOR pathway plays a crucial regulatory role in mammalian metabolism and physiology. In terms of brain function, it is responsible for the proliferation and differentiation of neural progenitor cells, the formation and maintenance of synaptic connections, and the regulation of complex cycles such as circadian rhythms. Disruption of signaling in this pathway has been linked to neurodegenerative and neuropsychiatric diseases.

AMP-activated protein kinase (AMPK) has regulatory effects in various metabolic pathways. AMPK activates autophagy by phosphorylating Unc-51 Like Autophagy Activating Kinase 1 (ULK1). In addition, AMPK can also initiate the autophagy process by activating ULK1 through inhibition of mTORC1. It has been observed that trehalose activates AMPK, which inhibits mTOR and initiates autophagy activation. Trehalose dramatically raised SIRT1 levels and AMPK and ULK1 phosphorylation, according to a study conducted on the NSC34 cell line. The data showed that this agent increased motor neuron survival by triggering autophagy and inhibiting apoptosis (Li et al., 2021). By boosting the activity of proteins like TSC1 (hamartin) and TSC2 (tuberin), which are in charge of mTORC1 suppression, trehalose can also block mTORC1 activation and induce autophagy. Trehalose can also trigger autophagy independent of the mTOR pathway. According to a study by Del Bello et al., trehalose initiates autophagy without the assistance of mTOR suppression or AMPK activation (Del Bello et al., 2022). Through a variety of ways, trehalose can trigger autophagy by activating TFEB, the primary transcriptional regulator of the autophagy-lysosomal pathway (Jeong et al., 2021). Trehalose can boost autophagic clearance in AD, which lowers the buildup of hyperphosphorylated tau tangles and amyloid-beta plaques. In PD, it can alleviate α-synuclein aggregation, which plays a significant role in disease progression by improving lysosomal function. In HD, it can prevent the aggregation of huntingtin protein (Sevriev et al., 2024).

ALS is characterized by the loss of nuclear splicing activity and the formation of cytoplasmic TDP-43 aggregates, a product of *TARDBP*. Since TDP-43 pathology is frequently observed in ALS, therapeutic approaches aimed at improving TDP-43 pathology (such as cytoplasmic mislocalization, expression levels, post-translational modifications, division, aggregation, and stress granule recruitment) are gaining attention. Considering the relationship between *TARDBP* pathology and defects in the autophagy pathway, *TARDBP* is a promising target for autophagy-based therapies in ALS (Chua et al., 2021). In human cell studies, rapamycin has been shown to prevent the loss of TDP-43 splicing activity and the formation of aggregates (Casiraghi et al., 2024).

Preclinical studies have demonstrated that ibudilast, a phosphodiesterase inhibitor with anti-inflammatory effects, induces autophagy and lysosomal biogenesis through the mTORC1-TFEB pathway. Ibudilast increases the number of autolysosomes by inhibiting mTORC1 activity and also enhances TFEB nuclear translocation. Research has shown that this drug significantly contributes to the clearance of TDP-43 and superoxide dismutase 1 (SOD1) protein aggregates, which are commonly seen in ALS. Therefore, ibudilast, an autophagy enhancer, can be applied as a therapeutic approach for various neurodegenerative diseases (Angelopoulou et al; 2022; Chen et al; 2020).

In ALS models involving *TARDBP*, inducing autophagy improves the disease course; however, there are uncertainties in triggering autophagy in the presence of *SOD1* mutations. Inducing autophagy in SOD1-linked ALS may delay the onset of the disease and improve survival. However, in the SOD1-G93A mouse model, a study with rapamycin showed that motor neuron degeneration and disease progression accelerated, with neurotoxicity increased due to excessive mitophagy (Chua et al., 2021).

Sirtuin 1 (SIRT1), a histone deacetylase class enzyme, regulates chromatin remodeling and gene transcription by deacetylating histone and non-histone proteins. It also regulates various signaling pathways, including oxidative stress, mitochondrial function, and autophagy. SIRT1 plays a role in autophagy control and the deacetylation of autophagy-related Atg proteins, thereby enhancing autophagy. Its role in the formation of neurodegenerative diseases has been demonstrated. Preclinical studies have shown that SIRT1 activators increase autophagic flow, reduce oxidative stress, and improve neuronal survival. The role of SIRT1 in the treatment of AD, HD, PD, and ALS has been well studied (Thapa et al., 2024). In a study with Baicalein, an active flavonoid, it was shown that this flavonoid activates SIRT1 and triggers mitophagy via the AMPK/mTOR pathway. In a study on rats with defective mitophagy in PD, induction of the SIRT1/AMPK pathway showed neuroprotective effects (Chen et al., 2022). SIRT1 activators, such as SRT1720, resveratrol, and SRT2104, have been shown to reduce the symptoms of neurodegenerative diseases.

The pathophysiology of SCZ has been connected to abnormalities in the autophagy process. One of the most effective, most commonly used antipsychotic drugs in the treatment of disorders such as SCZ and BD is olanzapine. However, treatment with olanzapine has been shown to cause changes in brain structure. It is particularly associated with aging in the brain. Mitophagy is intimately associated with aging. Studies have shown that this drug disrupts mitochondrial function, leading to abnormal mitophagy and triggering aging. The negative effects of mitochondrial damage and defective mitophagy caused by olanzapine were reduced by urolithin A, a mitophagy inducer. Therefore, urolithin A may be a good therapeutic approach to reverse the adverse effects of olanzapine treatment (Chen et al., 2023).

Activity-dependent neuroprotective protein (ADNP) is essential for brain development and cognitive function in mammals. Dysregulation of ADNP expression has been shown in AD, PD, and SCZ. One of the most important pathways related to ADNP is autophagy. The 8-amino-acid sequence NAPVSIPQ (NAP) of ADNP has been shown to interact with microtubules, protect microtubules, and trigger autophagy (Sragovich et al., 2017). The deregulation of ADNP, which can bind to MAP1LC3B/LC3B (microtubule-associated protein 1 light chain 3 β), leads to decreased autophagy. The neuropeptide-like drug NAP (davunetid, CP201) has been shown to restore the ADNP-LC3B interaction. Phase II studies with NAP have shown that it preserves brain function and improves daily living activities. In cell lines, adding NAP to clozapine (CLZ), a drug used in SCZ treatment, prevented CLZ's effects on blocking autolysosome formation and reducing neuronal cell viability, thereby increasing the autophagic process.

Studies have shown that medications used in SCZ, including CLZ, inhibit autophagy by pulling BECN1 into microtubules. NAP increases BECN1 expression. For effective treatment of schizophrenia, efficient autophagy is required. Therefore, combining drugs that can correct autophagy with those that might disrupt it can be highly beneficial. In this context, NAP could be a good option for autophagy-based treatment of schizophrenia (Merenlender-Wagner et al., 2014).

Defective autophagy and impaired mitochondrial function are thought to play a role in the pathophysiology of BD. In psychiatric disorders, inflammation increases, and oxidative stress occurs. Correcting autophagy may be effective in BD, SCZ, and depression. Recent studies have shown that psychoactive drugs such as tricyclic antidepressants, antipsychotics, lithium, and selective serotonin reuptake inhibitors (SSRIs) strengthen autophagy in neurons. Most of the antidepressants and mood stabilizers used in the treatment of psychiatric disorders activate autophagy. Autophagy may play a major role in the mechanism of action of these drugs (Mangan, 2015).

Studies suggest that increased autophagy is linked to better treatment responses. A study with patients using various antidepressants showed that this group had higher Beclin-1 and LC3 levels compared to untreated patients, supporting the idea that treatment responses improve with autophagy (Alcocer-Gómez et al., 2017). Other studies found that high Beclin-1 levels were connected with poor response to antidepressant treatment and symptom remission (Pierone et al; 2020; Scaini et al; 2019). These conflicting results suggest that not all psychiatric patients have defects in the autophagy pathway. However, if a defect exists, antidepressant treatment may increase autophagy in peripheral cells.

Lithium, the primary drug used in BD treatment, increases the lifespan and patients' quality of life and is particularly effective in treating acute manic episodes. It is suggested that the therapeutic effects of lithium in neurodegenerative and neuropsychiatric diseases are significantly influenced by the autophagy pathway (Puglisi-Allegra et al., 2023). In BD, Akt signaling modulation, which contributes to the control of autophagy, has been shown to be effective in lithium treatment (Khayachi et al; 2023; Ou et al; 2024). Various pharmacological studies have proven the role of autophagy in BD. These drugs regulate autophagy upwards through different mechanisms. Lithium triggers autophagy by inhibiting glycogen synthase kinase-3 (GSK3β) and inositol monophosphatase (IMPase), which regulate the autophagy pathway. By lowering the levels of inositol and inositol-1,4,5-trisphosphate (IP3), the antiepileptic medication CBZ activates AMPK. Activated AMPK inhibits mTORC1 phosphorylation and phosphorylates ULK1, which in turn induces autophagy. Like CBZ, valproate (VPA) reduces IP3 levels. Inhibiting histone deacetylases (HDAC) also results in increased histone acetylation and a disruption of histone binding to nuclear DNA. This increases DNA's sensitivity to transcription and induces autophagy by inhibiting mTORC1 signaling. VPA inhibits HDAC5 activity, thereby blocking the mTORC1 signal. Additionally, melatonin downregulates the PI3K/AKT/mTOR pathway to induce autophagy (Yang et al., 2023).

## CONCLUSION

Although the autophagy pathway is a good therapeutic target in the treatment of neuropsychological disorders, more research is needed. Knowing the molecular causes of illnesses and selecting the most appropriate pathways and genes for therapeutic targets is extremely important. Although autophagy is a good target for neurodegenerative diseases, the treatment may have some negative outcomes. When considering autophagy as a therapeutic target, care must be taken not to disrupt cellular homeostasis. In addition, when effective drugs are applied through autophagy, certain defects in autophagy regulation may occur. Or, the prolonged usage of these medications can cause severe adverse effects and toxicity. For example, since mTOR-dependent downstream pathways are also affected in the induction of autophagy through mTOR inhibitors, long-term exposure to the drug can lead to multi-systemic toxicities. Therefore, drugs that can achieve more specific targeting should be designed to selectively modulate autophagy. In autophagy-based treatment strategies, the focus has been on mTOR-independent autophagy inducers to reduce off-target effects.

Elderly people are more likely to suffer from neurodegenerative illnesses and strokes, and most studies on how autophagy affects neurological disorders are based on studies on cells and young animals. These differences in age and species in the current research pose a problem for the clinical applicability of autophagy-based therapies. Furthermore, in order to improve medication efficacy and decrease adverse effects, it is crucial to consider how central nervous system medications will penetrate the blood-brain barrier and target the neurological system.

The increase in data regarding autophagy and related mechanisms is of great importance for elucidating the pathogenesis of neuropsychological disorders and developing new treatment strategies.

## REFERENCES


Abeliovich, A., & Gitler, A. D. (2016). Defects in trafficking bridge Parkinson's disease pathology and genetics. *Nature*, *539*(7628), 207–216. DOI: 10.1038/nature20414 PMID: 27830778

Alcocer-Gómez, E., Casas-Barquero, N., Williams, M. R., Romero-Guillena, S. L., Cañadas-Lozano, D., Bullón, P., Sánchez-Alcazar, J. A., Navarro-Pando, J. M., & Cordero, M. D. (2017). Antidepressants induce autophagy dependent-NLRP3-inflammasome inhibition in Major depressive disorder. *Pharmacological Research*, *121*, 114–121. DOI: 10.1016/j.phrs.2017.04.028 PMID: 28465217

Angelopoulou, E., Pyrgelis, E. S., & Piperi, C. (2022). Emerging Potential of the Phosphodiesterase (PDE) Inhibitor Ibudilast for Neurodegenerative Diseases: An Update on Preclinical and Clinical Evidence. *Molecules (Basel, Switzerland)*, *27*(23), 8448. Advance online publication. DOI: 10.3390/molecules27238448 PMID: 36500540

Azzini, E., Peña-Corona, S. I., Hernández-Parra, H., Chandran, D., Saleena, L. A. K., Sawikr, Y., Peluso, I., Dhumal, S., Kumar, M., Leyva-Gómez, G., Martorell, M., Sharifi-Rad, J., & Calina, D. (2024). Neuroprotective and anti-inflammatory effects of curcumin in Alzheimer's disease: Targeting neuroinflammation strategies. *Phytotherapy Research*, *38*(6), 3169–3189. DOI: 10.1002/ptr.8200 PMID: 38616356

Ben-Shlomo, Y., Darweesh, S., Llibre-Guerra, J., Marras, C., San Luciano, M., & Tanner, C. (2024). The epidemiology of Parkinson's disease. *Lancet*, *403*(10423), 283–292. DOI: 10.1016/S0140-6736(23)01419-8 PMID: 38245248

Berger, Z., Ravikumar, B., Menzies, F. M., Oroz, L. G., Underwood, B. R., Pangalos, M. N., Schmitt, I., Wullner, U., Evert, B. O., O'Kane, C. J., & Rubinsztein, D. C. (2006). Rapamycin alleviates toxicity of different aggregate-prone proteins. *Human Molecular Genetics*, *15*(3), 433–442. DOI: 10.1093/hmg/ddi458 PMID: 16368705

Bose, J. K., Huang, C. C., & Shen, C. K. J. (2011). Regulation of autophagy by neuropathological protein TDP-43. *The Journal of Biological Chemistry*, *286*(52), 44441–44448. DOI: 10.1074/jbc.M111.237115 PMID: 22052911

Bustos, V., Pulina, M. V., Bispo, A., Lam, A., Flajolet, M., Gorelick, F. S., & Greengard, P. (2017). Phosphorylated Presenilin 1 decreases β-amyloid by facilitating autophagosome-lysosome fusion. *Proceedings of the National Academy of Sciences of the United States of America*, *114*(27), 7148–7153. DOI: 10.1073/pnas.1705240114 PMID: 28533369

Bustos, V., Pulina, M. V., Kelahmetoglu, Y., Sinha, S. C., Gorelick, F. S., Flajolet, M., & Greengard, P. (2017). Bidirectional regulation of Aβ levels by Presenilin 1. *Proceedings of the National Academy of Sciences of the United States of America*, *114*(27), 7142–7147. DOI: 10.1073/pnas.1705235114 PMID: 28533411

Cai, Q., & Ganesan, D. (2022). Regulation of neuronal autophagy and the implications in neurodegenerative diseases. *Neurobiology of Disease*, *162*(105582), 1–20. DOI: 10.1016/j.nbd.2021.105582 PMID: 34890791

Casiraghi, V., Sorce, M. N., Santangelo, S., Invernizzi, S., Bossolasco, P., Lattuada, C., Battaglia, C., Venturin, M., Silani, V., Colombrita, C., & Ratti, A. (2024). Modeling of TDP-43 proteinopathy by chronic oxidative stress identifies rapamycin as beneficial in ALS patient-derived 2D and 3D iPSC models. *Experimental Neurology*, *383*(115057), 1–17. DOI: 10.1016/j.expneurol.2024.115057 PMID: 39536963

Chen, M., Peng, L., Gong, P., Zheng, X., Sun, T., Zhang, X., & Huo, J. (2022). Baicalein Induces Mitochondrial Autophagy to Prevent Parkinson's Disease in Rats via miR-30b and the SIRT1/AMPK/mTOR Pathway. *Frontiers in Neurology*, *12*(646817), 1–11. DOI: 10.3389/fneur.2021.646817 PMID: 35237220

Chen, X., Wang, Z., Zheng, P., Dongol, A., Xie, Y., Ge, X., Zheng, M., Dang, X., Seyhan, Z. B., Nagaratnam, N., Yu, Y., & Huang, X. F. (2023). Impaired mitophagosome–lysosome fusion mediates olanzapine-induced aging. *Aging Cell*, *22*(11), 1–19. DOI: 10.1111/acel.14003 PMID: 37828862

Chen, Y., Wang, H., Ying, Z., & Gao, Q. (2020). Ibudilast enhances the clearance of SOD1 and TDP-43 aggregates through TFEB-mediated autophagy and lysosomal biogenesis: The new molecular mechanism of ibudilast and its implication for neuroprotective therapy. *Biochemical and Biophysical Research Communications*, *526*(1), 231–238. DOI: 10.1016/j.bbrc.2020.03.051 PMID: 32204915

Chua, J. P., De Calbiac, H., Kabashi, E., & Barmada, S. J. (2021). Autophagy and ALS: Mechanistic insights and therapeutic implications. *Autophagy*, *18*(2), 254–282. DOI: 10.1080/15548627.2021.1926656 PMID: 34057020

Corti, O., Blomgren, K., Poletti, A., & Beart, P. M. (2020). Autophagy in neurodegeneration: New insights underpinning therapy for neurological diseases. *Journal of Neurochemistry*, *154*(4), 354–371. DOI: 10.1111/jnc.15002 PMID: 32149395

Cozzi, M., & Ferrari, V. (2022). Autophagy Dysfunction in ALS: From Transport to Protein Degradation. *Journal of Molecular Neuroscience*, *72*(7), 1456–1481. DOI: 10.1007/s12031-022-02029-3 PMID: 35708843

Davoody, S., Asgari Taei, A., Khodabakhsh, P., & Dargahi, L. (2024). mTOR signaling and Alzheimer's disease: What we know and where we are? *CNS Neuroscience & Therapeutics*, *30*(4), 1–17. DOI: 10.1111/cns.14463 PMID: 37721413

Del Bello, B., Gamberucci, A., Marcolongo, P., & Maellaro, E. (2022). The autophagy inducer trehalose stimulates macropinocytosis in NF1-deficient glioblastoma cells. *Cancer Cell International*, *22*(1), 1–17. DOI: 10.1186/s12935-022-02652-5 PMID: 35864494

Dhapola, R., Beura, S. K., Sharma, P., Singh, S. K., & HariKrishnaReddy, D. (2024). Oxidative stress in Alzheimer's disease: Current knowledge of signaling pathways and therapeutics. *Molecular Biology Reports*, *51*(1), 1–18. DOI: 10.1007/s11033-023-09021-z PMID: 38165499

Djajadikerta, A., Keshri, S., Pavel, M., Prestil, R., Ryan, L., & Rubinsztein, D. C. (2020). Autophagy Induction as a Therapeutic Strategy for Neurodegenerative Diseases. *Journal of Molecular Biology*, *432*(8), 2799–2821. DOI: 10.1016/j.jmb.2019.12.035 PMID: 31887286

Farsi, Z., & Sheng, M. (2023). Molecular mechanisms of schizophrenia: Insights from human genetics. *Current Opinion in Neurobiology*, *81*, 102731. DOI: 10.1016/j.conb.2023.102731 PMID: 37245257

Fujioka, Y., Alam, J. M., Noshiro, D., Mouri, K., Ando, T., Okada, Y., May, A. I., Knorr, R. L., Suzuki, K., Ohsumi, Y., & Noda, N. N. (2020). Phase separation organizes the site of autophagosome formation. *Nature*, *578*(7794), 301–305. DOI: 10.1038/s41586-020-1977-6 PMID: 32025038

Ge, P., Dawson, V. L., & Dawson, T. M. (2020). PINK1 and Parkin mitochondrial quality control: A source of regional vulnerability in Parkinson's disease. *Molecular Neurodegeneration*, *15*(1), 1–18. DOI: 10.1186/s13024-020-00367-7 PMID: 32169097

Gerson, J. E., Safren, N., Fischer, S., Patel, R., Crowley, E. V., Welday, J. P., Windle, A. K., Barmada, S., Paulson, H. L., & Sharkey, L. M. (2020). Ubiquilin-2 differentially regulates polyglutamine disease proteins. *Human Molecular Genetics*, *29*(15), 2596–2610. DOI: 10.1093/hmg/ddaa152 PMID: 32681165

Goode, A., Butler, K., Long, J., Cavey, J., Scott, D., Shaw, B., Sollenberger, J., Gell, C., Johansen, T., Oldham, N. J., Searle, M. S., & Layfield, R. (2016). Defective recognition of LC3B by mutant SQSTM1/p62 implicates impairment of autophagy as a pathogenic mechanism in ALS-FTLD. *Autophagy*, *12*(7), 1094–1104. DOI: 10.1080/15548627.2016.1170257 PMID: 27158844

Guo, F., Liu, X., Cai, H., & Le, W. (2018). Autophagy in neurodegenerative diseases: Pathogenesis and therapy. *Brain Pathology (Zurich, Switzerland)*, *28*(1), 3–13. DOI: 10.1111/bpa.12545 PMID: 28703923

Incebacak Eltemur, R. D., Nguyen, H. P., & Weber, J. J. (2022). Calpain-mediated proteolysis as driver and modulator of polyglutamine toxicity. *Frontiers in Molecular Neuroscience*, *15*, 1–17. DOI: 10.3389/fnmol.2022.1020104 PMID: 36385755

Jeong, S. J., Stitham, J., Evans, T. D., Zhang, X., Rodriguez-Velez, A., Yeh, Y. S., Tao, J., Takabatake, K., Epelman, S., Lodhi, I. J., Schilling, J. D., DeBosch, B. J., Diwan, A., & Razani, B. (2021). Trehalose causes low-grade lysosomal stress to activate TFEB and the autophagy-lysosome biogenesis response. *Autophagy*, *17*(11), 3740–3752. DOI: 10.1080/15548627.2021.1896906 PMID: 33706671

Khalifeh, M., Barreto, G., & Sahebkar, A. (2021). Therapeutic potential of trehalose in neurodegenerative diseases: The knowns and unknowns. *Neural Regeneration Research*, *16*(10), 2026–2027. DOI: 10.4103/1673-5374.308085 PMID: 33642389

Khayachi, A., Abuzgaya, M., Liu, Y., Jiao, C., Dejgaard, K., Schorova, L., Kamesh, A., He, Q., Cousineau, Y., Pietrantonio, A., Farhangdoost, N., Castonguay, C.-E., Chaumette, B., Alda, M., Rouleau, G. A., & Milnerwood, A. J. (2023). Molecular signatures of hyperexcitability and lithium responsiveness in bipolar disorder patient neurons provide alternative therapeutic strategies. *BioRxiv*, 1-30. DOI: 10.1101/2023.07.21.550088

Kim, A., Lalonde, K., Truesdell, A., Welter, P. G., Brocardo, P. S., Rosenstock, T. R., & Gil-mohapel, J. (2021). New avenues for the treatment of huntington's disease. *International Journal of Molecular Sciences*, *22*(16), 1–50. DOI: 10.3390/ijms22168363 PMID: 34445070

Kirkin, V., & Rogov, V. V. (2019). A Diversity of Selective Autophagy Receptors Determines the Specificity of the Autophagy Pathway. *Molecular Cell*, *76*(2), 268–285. DOI: 10.1016/j.molcel.2019.09.005 PMID: 31585693

Klionsky, D. J., Petroni, G., Amaravadi, R. K., Baehrecke, E. H., Ballabio, A., Boya, P., Bravo-San Pedro, J. M., Cadwell, K., Cecconi, F., Choi, A. M. K., Choi, M. E., Chu, C. T., Codogno, P., Colombo, M. I., Cuervo, A. M., Deretic, V., Dikic, I., Elazar, Z., Eskelinen, E., & Pietrocola, F. (2021). Autophagy in major human diseases. *The EMBO Journal*, *40*(19), 1–64. DOI: 10.15252/embj.2021108863 PMID: 34459017

Kocaturk, N., & Gozuacik, D. (2017). Otofaji ve Nörodejeneratif Hastalıklar. Turkiye Klinikleri J Pharmacol-Special Topics, 5(1):11-20. https://www.researchgate.net/publication/316168897

Komatsu, M., Qing, J. W., Holstein, G. R., Friedrich, V. L., Iwata, J. I., Kominami, E., Chait, B. T., Tanaka, K., & Yue, Z. (2007). Essential role for autophagy protein Atg7 in the maintenance of axonal homeostasis and the prevention of axonal degeneration. *Proceedings of the National Academy of Sciences of the United States of America*, *104*(36), 14489–14494. DOI: 10.1073/pnas.0701311104 PMID: 17726112

Komatsu, M., Waguri, S., Koike, M., Sou, Y., Ueno, T., Hara, T., Mizushima, N., Iwata, J., Ezaki, J., Murata, S., Hamazaki, J., Nishito, Y., Iemura, S., Natsume, T., Yanagawa, T., Uwayama, J., Warabi, E., Yoshida, H., Ishii, T., & Tanaka, K. (2007). Homeostatic Levels of p62 Control Cytoplasmic Inclusion Body Formation in Autophagy-Deficient Mice. *Cell*, *131*(6), 1149–1163. DOI: 10.1016/j.cell.2007.10.035 PMID: 18083104

Lam, X. J., Xu, B., Yeo, P. L., Cheah, P. S., & Ling, K. H. (2023). Mitochondria dysfunction and bipolar disorder: From pathology to therapy. *IBRO Neuroscience Reports*, *14*, 407–418. DOI: 10.1016/j.ibneur.2023.04.002 PMID: 37388495

Lee, J., Sung, K. W., Bae, E. J., Yoon, D., Kim, D., Lee, J. S., Park, D., Park, D. Y., Mun, S. R., Kwon, S. C., Kim, H. Y., Min, J.-O., Lee, S.-J., Suh, Y. H., & Kwon, Y. T. (2023). Targeted degradation of α-synuclein aggregates in Parkinson's disease using the AUTOTAC technology. *Molecular Neurodegeneration*, *18*(1), 1–21. DOI: 10.1186/s13024-023-00630-7 PMID: 36597122

Lee, J., Yoon, D., Sung, K. W., Bae, E. J., Park, D. H., Suh, Y. H., & Kwon, Y. T. (2024). Targeted degradation of SNCA/α-synuclein aggregates in neurodegeneration using the AUTOTAC chemical platform. *Autophagy*, *20*(2), 463–465. DOI: 10.1080/15548627.2023.2274711 PMID: 37915239

Lee, S. W., Oh, Y. M., Victor, M. B., Yang, Y., Chen, S., Strunilin, I., Dahiya, S., Dolle, R. E., Pak, S. C., Silverman, G. A., Perlmutter, D. H., & Yoo, A. S. (2024). Longitudinal modeling of human neuronal aging reveals the contribution of the RCAN1–TFEB pathway to Huntington's disease neurodegeneration. *Nature Aging*, *4*(1), 95–109. DOI: 10.1038/s43587-023-00538-3 PMID: 38066314

Lepinay, E., & Cicchetti, F. (2023). Tau: A biomarker of Huntington's disease. *Molecular Psychiatry*, *28*(10), 4070–4083. DOI: 10.1038/s41380-023-02230-9 PMID: 37749233

Li, B., Li, P., Weng, R., Wu, Z., Qin, B., Fang, J., Wang, Y., Qiu, S., Yang, J., & Gu, L. (2021). Trehalose protects motorneuron after brachial plexus root avulsion by activating autophagy and inhibiting apoptosis mediated by the AMPK signaling pathway. *Gene*, *768*(145307), 1–7. DOI: 10.1016/j.gene.2020.145307 PMID: 33197516

Li, Y. Y., Qin, Z. H., & Sheng, R. (2024). The Multiple Roles of Autophagy in Neural Function and Diseases. *Neuroscience Bulletin*, *40*(3), 363–382. DOI: 10.1007/s12264-023-01120-y PMID: 37856037

Liénard, C., Pintart, A., & Bomont, P. (2024). Neuronal Autophagy: Regulations and Implications in Health and Disease. *Cells*, *13*(1), 1–33. DOI: 10.3390/cells13010103 PMID: 38201307

Litwiniuk, A., Juszczak, G. R., Stankiewicz, A. M., & Urbańska, K. (2023). The role of glial autophagy in Alzheimer's disease. *Molecular Psychiatry*, *28*(11), 4528–4539. DOI: 10.1038/s41380-023-02242-5 PMID: 37679471

Liu, S. Z., Yao, S. J., Yang, H., Liu, S. J., & Wang, Y. J. (2023). Autophagy: Regulator of cell death. *Cell Death & Disease*, *14*(10), 1–13. DOI: 10.1038/s41419-023-06154-8 PMID: 37794028

Luis-Ravelo, D., Estévez-Silva, H., Barroso-Chinea, P., Afonso-Oramas, D., Salas-Hernández, J., Rodríguez-Núñez, J., Acevedo-Arozena, A., Marcellino, D., & González-Hernández, T. (2018). Pramipexole reduces soluble mutant huntingtin and protects striatal neurons through dopamine D3 receptors in a genetic model of Huntington's disease. *Experimental Neurology*, *299*, 137–147. DOI: 10.1016/j.expneurol.2017.10.019 PMID: 29056363

Ma, D., Panda, S., & Lin, J. D. (2011). Temporal orchestration of circadian autophagy rhythm by C/EBPβ. *The EMBO Journal*, *30*(22), 4642–4651. DOI: 10.1038/emboj.2011.322 PMID: 21897364

Ma, R. Y., Li, L., Yang, H., Zou, B., Ma, R. X., Zhang, Y., Wu, M. M., Chen, P., Yao, Y. & Li, J. (2024). Therapeutic effect of nicotinamide mononucleotide on Alzheimer's disease through activating autophagy and anti-oxidative stress. *Biomedicine and Pharmacotherapy, 178*(July). DOI: 10.1016/j.biopha.2024.117199

Maeda, S., Yamamoto, H., Kinch, L. N., Garza, C. M., Takahashi, S., Otomo, C., Grishin, N. V., Forli, S., Mizushima, N., & Otomo, T. (2020). Structure, lipid scrambling activity and role in autophagosome formation of ATG9A. *Nature Structural & Molecular Biology*, *27*(12), 1194–1201. DOI: 10.1038/s41594-020-00520-2 PMID: 33106659

Mangan, D. (2015). *Autophagy activation as a mechanism of action of psychoactive drugs, revised and expanded version*. The Winnower., DOI: 10.15200/winn.143268.86120

Martinez–Vicente, M., Talloczy, Z., Wong, E., Tang, G., Koga, H., Kaushik, S., de Vries, R., Arias, E., Harris, S., Sulzer, D., & Cuervo, A. M. (2010). Cargo recognition failure is responsible for inefficient autophagy in Huntington's disease m. *Physiology & Behavior, 13*(1), 567–576. DOI: 10.1038/nn.2528.CARGO PMID: 20383138

Masaldan, S., Callegari, S., & Dewson, G. (2022). Therapeutic targeting of mitophagy in Parkinson's disease. *Biochemical Society Transactions, 50*(2), 783–797. DOI: 10.1042/BST20211107 PMID: 35311891

Menzies, F. M., Fleming, A., Caricasole, A., Bento, C. F., Andrews, S. P., Ashkenazi, A., Füllgrabe, J., Jackson, A., Jimenez Sanchez, M., Karabiyik, C., Licitra, F., Lopez Ramirez, A., Pavel, M., Puri, C., Renna, M., Ricketts, T., Schlotawa, L., Vicinanza, M., Won, H., & Rubinsztein, D. C. (2017). Autophagy and Neurodegeneration: Pathogenic Mechanisms and Therapeutic Opportunities. *Neuron, 93*(5), 1015–1034. DOI: 10.1016/j.neuron.2017.01.022 PMID: 28279350

Merenlender-Wagner, A., Malishkevich, A., Shemer, Z., Udawela, M., Gibbons, A., Scarr, E., Dean, B., Levine, J., Agam, G., & Gozes, I. (2015). Autophagy has a key role in the pathophysiology of schizophrenia. *Molecular Psychiatry, 20*(1), 126–132. DOI: 10.1038/mp.2013.174 PMID: 24365867

Merenlender-Wagner, A., Shemer, Z., Touloumi, O., Lagoudaki, R., Giladi, E., Andrieux, A., Grigoriadis, N. C., & Gozes, I. (2014). New horizons in schizophrenia treatment: Autophagy protection is coupled with behavioral improvements in a mouse model of schizophrenia. *Autophagy, 10*(12), 2324–2332. DOI: 10.4161/15548627.2014.984274 PMID: 25484074

Metzger, S., Saukko, M., Van Che, H., Tong, L., Puder, Y., Riess, O., & Nguyen, H. P. (2010). Age at onset in Huntington's disease is modified by the autophagy pathway: Implication of the V471A polymorphism in Atg7. *Human Genetics, 128*(4), 453–459. DOI: 10.1007/s00439-010-0873-9 PMID: 20697744

Nechushtai, L., Frenkel, D., & Pinkas-kramarski, R. (2023). Autophagy in Parkinson ' s Disease. *Biomolecules, 13*(10), 1–16. DOI: 10.3390/biom13101435 PMID: 37892117

Nobili, A., D'Amelio, M., & Viscomi, M. T. (2023). Nilotinib: From animal-based studies to clinical investigation in Alzheimer's disease patients. *Neural Regeneration Research, 18*(4), 803–804. DOI: 10.4103/1673-5374.350700 PMID: 36204843

Öksüz, N., Öztürk, Ş., & Doğu, O. (2022). Future Prospects in Parkinson ' s Disease Diagnosis and Treatment. *Nöro Psikiyatri Arşivi, 59*, 36–41. DOI: 10.29399/npa.28169 PMID: 36578989

Oku, M., & Sakai, Y. (2018). Three Distinct Types of Microautophagy Based on Membrane Dynamics and Molecular Machineries. *BioEssays*, *40*(6), 1–6. DOI: 10.1002/bies.201800008 PMID: 29708272

Ou, A. H., Rosenthal, S. B., Adli, M., Akiyama, K., Akula, N., Alda, M., Amare, A. T., Ardau, R., Arias, B., Aubry, J. M., Backlund, L., Bauer, M., Baune, B. T., Bellivier, F., Benabarre, A., Bengesser, S., Bhattacharjee, A. K., Biernacka, J. M., Cervantes, P., & Kelsoe, J. R. (2024). Lithium response in bipolar disorder is associated with focal adhesion and PI3K-Akt networks: A multi-omics replication study. *Translational Psychiatry*, *14*(1), 109. Advance online publication. DOI: 10.1038/s41398-024-02811-4 PMID: 38395906

Park, H., Kang, J. H., & Lee, S. (2020). Autophagy in neurodegenerative diseases: A hunter for aggregates. *International Journal of Molecular Sciences*, *21*(9), 1–24. DOI: 10.3390/ijms21093369 PMID: 32397599

Pickford, F., Masliah, E., Britschgi, M., Lucin, K., Narasimhan, R., Jaeger, P. A., Small, S., Spencer, B., Rockenstein, E., Levine, B., & Wyss-Coray, T. (2008). The autophagy-related protein beclin 1 shows reduced expression in early Alzheimer disease and regulates amyloid β accumulation in mice. *The Journal of Clinical Investigation*, *118*(6), 2190–2199. DOI: 10.1172/JCI33585 PMID: 18497889

Pierone, B. C., Pereira, C. A., Garcez, M. L., & Kaster, M. P. (2020). Stress and signaling pathways regulating autophagy: From behavioral models to psychiatric disorders. *Experimental Neurology*, *334*(113485), 1–13. DOI: 10.1016/j.expneurol.2020.113485 PMID: 32987001

Puglisi-Allegra, S., Lazzeri, G., Busceti, C. L., Giorgi, F. S., Biagioni, F., & Fornai, F. (2023). Lithium engages autophagy for neuroprotection and neuroplasticity: Translational evidence for therapy. *Neuroscience and Biobehavioral Reviews*, *148*(105148), 1–19. DOI: 10.1016/j.neubiorev.2023.105148 PMID: 36996994

Riva, N., Domi, T., Pozzi, L., Lunetta, C., Schito, P., Spinelli, E. G., Cabras, S., Matteoni, E., Consonni, M., Bella, E. D., Agosta, F., Filippi, M., Calvo, A., & Quattrini, A. (2024). Update on recent advances in amyotrophic lateral sclerosis. *Journal of Neurology*, *271*(7), 4693–4723. DOI: 10.1007/s00415-024-12435-9 PMID: 38802624

Roth, J. R., de Moraes, R. C. M., Xu, B. P., Crawley, S. R., Khan, M. A., & Melkani, G. C. (2023). Rapamycin reduces neuronal mutant huntingtin aggregation and ameliorates locomotor performance in Drosophila. *Frontiers in Aging Neuroscience*, *15*, 1–11. DOI: 10.3389/fnagi.2023.1223911 PMID: 37823007

Saha, S., Ash, P. E. A., Gowda, V., Liu, L., Shirihai, O., & Wolozin, B. (2015). Mutations in LRRK2 potentiate age-related impairment of autophagic flux. *Molecular Neurodegeneration, 10*(1), 1–13. DOI: 10.1186/s13024-015-0022-y PMID: 26159606

Sanchez-Mirasierra, I., Ghimire, S., Hernandez-Diaz, S., & Soukup, S. F. (2022). Targeting Macroautophagy as a Therapeutic Opportunity to Treat Parkinson's Disease. *Frontiers in Cell and Developmental Biology, 10*, 1–12. DOI: 10.3389/fcell.2022.921314 PMID: 35874822

Sawa-Makarska, J., Baumann, V., Coudevylle, N., von Bülow, S., Nogellova, V., Abert, C., Schuschnig, M., Graef, M., Hummer, G., & Martens, S. (2020). Reconstitution of autophagosome nucleation defines Atg9 vesicles as seeds for membrane formation. *Science, 369*(6508), 1–23. DOI: 10.1126/science.aaz7714 PMID: 32883836

Scaini, G., Barichello, T., Fries, G. R., Kennon, E. A., Andrews, T., Nix, B. R., Zunta-Soares, G., Valvassori, S. S., Soares, J. C., & Quevedo, J. (2019). TSPO upregulation in bipolar disorder and concomitant downregulation of mitophagic proteins and NLRP3 inflammasome activation. *Neuropsychopharmacology : Official Publication of the American College of Neuropsychopharmacology, 44*(7), 1291–1299. DOI: 10.1038/s41386-018-0293-4 PMID: 30575805

Sevriev, B., Dimitrova, S., Kehayova, G., & Dragomanova, S. (2024). Trehalose : Neuroprotective Effects and Mechanisms — An Updated Review. *NeuroSci, 5*(4), 429–444. DOI: 10.3390/neurosci5040032 PMID: 39484301

Singulani, M. P., Ferreira, A. F. F., Figueroa, P. S., Cuyul-Vásquez, I., Talib, L. L., Britto, L. R., & Forlenza, O. V. (2024). Lithium and disease modification: A systematic review and meta-analysis in Alzheimer's and Parkinson's disease. *Ageing Research Reviews, 95*, 1–20. DOI: 10.1016/j.arr.2024.102231 PMID: 38364914

Sragovich, S., Merenlender-Wagner, A., & Gozes, I. (2017). ADNP Plays a Key Role in Autophagy: From Autism to Schizophrenia and Alzheimer's Disease. *BioEssays, 39*(11), 1–10. DOI: 10.1002/bies.201700054 PMID: 28940660

Sunhe, Y.-X., Zhang, Y.-H., Fu, R.-J., Xu, D.-Q., & Tang, Y.-P. (2024). Neuroprotective effect and preparation methods of berberine. *Frontiers in Pharmacology, 15*, 1–23. DOI: 10.3389/fphar.2024.1429050 PMID: 39309003

Tan, Y., Zhu, J., & Hashimoto, K. (2024). Autophagy-related gene model as a novel risk factor for schizophrenia. *Translational Psychiatry, 14*(1), 1–11. DOI: 10.1038/s41398-024-02767-5 PMID: 38351068

Taymans, J. M., Fell, M., Greenamyre, T., Hirst, W. D., Mamais, A., Padmanabhan, S., Peter, I., Rideout, H., & Thaler, A. (2023). Perspective on the current state of the LRRK2 field. *NPJ Parkinson's Disease*, *9*(1), 1–9. DOI: 10.1038/s41531-023-00544-7 PMID: 37393318

Thapa, R., Moglad, E., Afzal, M., Gupta, G., Bhat, A. A., Hassan almalki, W., Kazmi, I., Alzarea, S. I., pant, K., Singh, T. G., Singh, S. K., & Ali, H. (2024). The role of sirtuin 1 in ageing and neurodegenerative disease: A molecular perspective. *Ageing Research Reviews*, *102*(102545), 1–14. DOI: 10.1016/j.arr.2024.102545 PMID: 39423873

Tian, E., Sharma, G., & Dai, C. (2023). Neuroprotective Properties of Berberine: Molecular Mechanisms and Clinical Implications. *Antioxidants*, *12*(10), 1–17. DOI: 10.3390/antiox12101883 PMID: 37891961

Toker, L., & Agam, G. (2015). Mitochondrial dysfunction in psychiatric morbidity: Current evidence and therapeutic prospects. *Neuropsychiatric Disease and Treatment*, *11*, 2441–2447. DOI: 10.2147/NDT.S70346 PMID: 26442764

Turco, E., Savova, A., Gere, F., Ferrari, L., Romanov, J., Schuschnig, M., & Martens, S. (2021). Reconstitution defines the roles of p62, NBR1 and TAX1BP1 in ubiquitin condensate formation and autophagy initiation. *Nature Communications*, *12*(1), 1–16. DOI: 10.1038/s41467-021-25572-w PMID: 34471133

Uddin, M. S., Stachowiak, A., Al Mamun, A., Tzvetkov, N. T., Takeda, S., Atanasov, A. G., Bergantin, L. B., Abdel-Daim, M. M., & Stankiewicz, A. M. (2018). Autophagy and Alzheimer's disease: From molecular mechanisms to therapeutic implications. *Frontiers in Aging Neuroscience*, *10*, 1–18. DOI: 10.3389/fnagi.2018.00004 PMID: 29441009

Usenovic, M., Tresse, E., Mazzulli, J. R., Taylor, J. P., & Krainc, D. (2012). Deficiency of ATP13A2 leads to lysosomal dysfunction, α-synuclein accumulation, and neurotoxicity. *The Journal of Neuroscience : The Official Journal of the Society for Neuroscience*, *32*(12), 4240–4246. DOI: 10.1523/JNEUROSCI.5575-11.2012 PMID: 22442086

Valencia, M., Kim, S. R., Jang, Y., & Lee, S. H. (2021). Neuronal autophagy: Characteristic features and roles in neuronal pathophysiology. *Biomolecules & Therapeutics*, *29*(6), 605–614. DOI: 10.4062/biomolther.2021.012 PMID: 33875624

Valverde, D. P., Yu, S., Boggavarapu, V., Kumar, N., Lees, J. A., Walz, T., Reinisch, K. M., & Melia, T. J. (2019). ATG2 transports lipids to promote autophagosome biogenesis. *The Journal of Cell Biology*, *218*(6), 1787–1798. DOI: 10.1083/jcb.201811139 PMID: 30952800

Wang, L., Klionsky, D. J., & Shen, H. M. (2023). The emerging mechanisms and functions of microautophagy. *Nature Reviews. Molecular Cell Biology*, *24*(3), 186–203. DOI: 10.1038/s41580-022-00529-z PMID: 36097284

Winden, K. D., Ebrahimi-Fakhari, D., & Sahin, M. (2018). Abnormal mTOR Activation in Autism. *Annual Review of Neuroscience*, *41*(1), 1–23. DOI: 10.1146/annurev-neuro-080317-061747 PMID: 29490194

Wong, Y. C., & Holzbaur, E. L. F. (2014). The regulation of autophagosome dynamics by huntingtin and HAP1 is disrupted by expression of mutant huntingtin, leading to defective cargo degradation. *The Journal of Neuroscience : The Official Journal of the Society for Neuroscience*, *34*(4), 1293–1305. DOI: 10.1523/JNEUROSCI.1870-13.2014 PMID: 24453320

Yamamoto, H. (2023). Molecular mechanisms of macroautophagy, microautophagy, and chaperone-mediated autophagy. *J Nippon Med Sch.*, *1302*(2003), 91–102. .DOI: 10.1272/jnms.JNMS.2024_91-102

Yan, J., Porch, M. W., Court-Vazquez, B., Bennett, M. V. L., & Zukin, R. S. (2018). Activation of autophagy rescues synaptic and cognitive deficits in fragile X mice. *Proceedings of the National Academy of Sciences of the United States of America*, *115*(41), 9707–9716. DOI: 10.1073/pnas.1808247115 PMID: 30242133

Yang, J., Xu, H., Zhang, C., Yang, X., Cai, W., & Chen, X. (2023). A prion-like domain of TFEB mediates the co-aggregation of TFEB and mHTT. *Autophagy*, *19*(2), 544–550. DOI: 10.1080/15548627.2022.2083857 PMID: 35635192

Yang, Y., Yuan, R., Lu, Y., Zhu, C., Zhang, C., Lue, H., & Zhang, X. (2023). The engagement of autophagy in maniac disease. *CNS Neuroscience & Therapeutics*, *29*(12), 3684–3692. DOI: 10.1111/cns.14353 PMID: 37438945

Yao, R., & Shen, J. (2023). Chaperone-mediated autophagy: Molecular mechanisms, biological functions, and diseases. *MedComm*, *4*(5), 1–27. DOI: 10.1002/mco2.347 PMID: 37655052

Zatyka, M., Sarkar, S., & Barrett, T. (2020). Autophagy in Rare (NonLysosomal) Neurodegenerative Diseases. *Journal of Molecular Biology*, *432*(8), 2735–2753. DOI: 10.1016/j.jmb.2020.02.012 PMID: 32087199

Zhang, X., Li, Y., Xu, H., & Zhang, Y. W. (2014). The γ-secretase complex: From structure to function. *Frontiers in Cellular Neuroscience*, *8*, 1–10. DOI: 10.3389/fncel.2014.00427 PMID: 25565961

Zhao, S., Chen, R., Gao, Y., Lu, Y., Bai, X., & Zhang, J. (2023). Fundamental roles of the Optineurin gene in the molecular pathology of Amyotrophic Lateral Sclerosis. *Frontiers in Neuroscience*, *17*, 1–8. DOI: 10.3389/fnins.2023.1319706 PMID: 38178841

Zhao, Y. G., Codogno, P., & Zhang, H. (2021). Machinery, regulation and pathophysiological implications of autophagosome maturation. *Nature Reviews. Molecular Cell Biology*, *22*(11), 733–750. DOI: 10.1038/s41580-021-00392-4 PMID: 34302147

Zhao, Z., Yan, J., Huang, L., & Yang, X. (2024). Phytochemicals targeting Alzheimer's disease via the AMP-activated protein kinase pathway, effects, and mechanisms of action. *Biomedicine and Pharmacotherapy*, *173*, 116373. DOI: 10.1016/j.biopha.2024.116373 PMID: 38442672

Zheng, Q., Chen, Y., Chen, D., Zhao, H., Feng, Y., Meng, Q., Zhao, Y., & Zhang, H. (2022). Calcium transients on the ER surface trigger liquid-liquid phase separation of FIP200 to specify autophagosome initiation sites. *Cell*, *185*(22), 4082–4098. DOI: 10.1016/j.cell.2022.09.001 PMID: 36198318

Zhou, F., Wu, Z., Zhao, M., Murtazina, R., Cai, J., Zhang, A., Li, R., Sun, D., Li, W., Zhao, L., Li, Q., Zhu, J., Cong, X., Zhou, Y., Xie, Z., Gyurkovska, V., Li, L., Huang, X., Xue, Y., & Segev, N. (2019). Rab5-dependent autophagosome closure by ESCRT. *The Journal of Cell Biology*, *218*(6), 1908–1927. DOI: 10.1083/jcb.201811173 PMID: 31010855

Chapter 3

# Modulating Neuroinflammation and Autophagy in Psychiatric Disorders Through Mechanisms and Therapies:

## Therapeutic Strategies and Pathways

**Reyed M. Reyed**
https://orcid.org/0000-0001-5027-3478

*Genetic Engineering and Biotechnology Research Institute (GEBRI), City of Scientific Research and Technological Applications (SRTA-City), Alexandria, Egypt*

**Pranav Kumar Prabhakar**
https://orcid.org/0000-0001-8130-1822

*Department of Biotechnology, School of Engineering and Technology, Nagaland University, Kohima, India*

## ABSTRACT

*The chapter presents the relationship between autophagy and neuroinflammation in major depression, bipolar disorder, and schizophrenia. It emphasizes that oxidative stress coupled with mitochondrial dysfunction can activate mTOR, AMPK, and FOXO1, thereby influencing synaptic function and behavior. The influence of*

DOI: 10.4018/979-8-3693-5908-2.ch003

*microglia, HMGB1, and Toll-like receptors is discussed. The chapter explores how altered gut microbiota and nutrition affect autophagy by regulating short-chain fatty acid levels and inflammatory signals. The effects of lifestyle modifications, such as time-restricted feeding and the use of key phytochemicals (curcumin, resveratrol, hydroxytyrosol, thymoquinone, saffron), on autophagy and inflammation levels are investigated. Summary tables and case studies support the results presented. Neuronal organoids and 3D-printed brain models are introduced as innovative tools to explore disease pathways for future clinical initiatives.*

## 1. INTRODUCTION

### 1.1. Overview of Inflammation and Autophagy in Neuropsychiatric Disorders

Major neurological, psychiatric, and neurodegenerative disorders are strongly associated with and regulated by neuroinflammation and autophagy (Absalyamova et al., 2025). In this chapter, therefore, three objectives are presented to assist in the main goal of explaining how these processes are related to the corresponding regulatory systems in view of potential treatment. Typically, stressors, including oxidative stress, mitochondrial dysfunction, and toxic protein aggregation, induce neuroinflammation in the CNS. A multitude of these pathogenic stimuli are related to aberrations in autophagy, a cellular process through which damaged proteins and organelles are cleared. Both of these mechanisms are thought to create feedback that worsens neuronal decay and the general pathology of the disease (Gadhave, et al., 2024).

### 1.2. Molecular Pathways Regulating Autophagy and Neuroinflammation

Several studies have shown that autophagy is modulated by different signaling networks involving the mechanistic target of rapamycin (mTOR) (Xu et al., 2025), AMP-activated protein kinase (AMPK) (Yang et al., 2025), and Forkhead box O1 (FOXO1) (Wang. et al., 2025). The fine-tuning of these pathways defines the ratio of cell survival to repair and is extensively regulated. For example, under ample energy conditions, mTOR suppresses autophagy, whereas under low energy conditions, AMPK promotes autophagy. These pathways sustain the progression of neurodegenerative disorders such as AD, PD, and ALS and prevent chronic neuroinflammation associated with these diseases (Ragupathi, et al. 2024; Sadeghi, et al. 2023). Autophagy is finely regulated through mTOR, AMPK, and FOXO1 signaling; however, its mechanisms and neuroinflammatory consequences are also shaped by epigenetic changes and interactions with the immune network of the CNS. Alterations in gene expression due to epigenetic changes regulate the synthesis of proteins important for autophagy and inflammation, while microglia are central to neuronal responses to inflammation and may be influenced by autophagy.

## 1.3. Epigenetic and Immune Modulation of Autophagy

In addition, neuroinflammation is also expected to affect CNS-associated immune cells, such as microglia. Toxic products such as proinflammatory cytokines and reactive oxygen species contribute to neuronal damage instigated by microglia (Adamu, et al., 2024). The NLRP3 inflammasome is a member of the innate immune system of the host defense response machinery (Alonaizan 2024). Datta et al. (2024) reported the subsequent activation of molecular forms of damage-associated molecular patterns (DAMPs), such as amyloid-β plaques, and another intracellular protein called high mobility group box 1 (HMGB1). A second level of control is produced, for example, by the interaction of noncoding RNAs and histones, which changes with inflammation and autophagy (Kumar, et al., 2024). For example, microRNAs (miRNAs) and long noncoding RNAs (lncRNAs) are known to play a role in the regulation of autophagy and inflammation within the scope of gene expression. Other posttranslational modifications, including histone acetylation and histone methylation, also affect autophagy, thereby illustrating the synergistic relationship between these regulatory pathways (Alpuche-Lazcano, et al., 2024). In the essay "Novel Pharmacological Regulators of Autophagy," published in Expert Opinion on Therapeutic Patents, Ha & Kim (2016) conducted a comprehensive review of patents from 2012--2015. These results highlight the growing focus on treatment strategies that address both autophagy and neuroinflammation. Importantly, various anti-inflammatory drugs, including pharmaceutical agents focused on mammalian target of rapamycin (mTOR), have demonstrated potential in enhancing autophagy, particularly via mechanisms such as autophosphorylation-mediated pathways. In the article "Cellular Reactive Oxygen Species and Antioxidants: Physiological and Pathological Roles," featured in a special issue on this subject, Kozlov et al. (2024) explore the dual functions of reactive oxygen species (ROS) and antioxidants in cellular health and disease. This article emphasizes the increasing number of studies on the capacity of antioxidants and anti-inflammatory drugs to regulate autophagy and combat oxidative stress, highlighting their potential for remedying the cellular damage caused by ROS. A sophisticated interplay exists between autophagy, neuroinflammation, and cellular regulatory mechanisms, which is further altered by systemic factors, particularly nutrition and the gut-brain connection. The interaction between gut microbiota and diet can significantly affect central nervous system inflammation and autophagy, indicating that these external factors must be considered when studying neuropsychiatric diseases.

## 1.4. Nutritional and Gut–Brain Axis Influences on Autophagy

In this context, the regulation of the gut–brain axis by probiotics and diet also supports the systemic basis of these pathways. Increasing evidence linking the gut microbiota and neuroinflammation advances the plausibility of unique nutritional and microbiota-directed treatments to restore the homeostasis of autophagy and decrease systemic inflammation) (Chandrasekaran, et al 2024). Among various types, intermittent fasting has been identified as one of the most effective nonpharmacological approaches for increasing autophagy (Sun, et al.2024). Studies suggest that fasting protocols, such as time-restricted feeding (e.g., 16:8), intermittent fasting (5:2 protocol) and alternating day fasting, can initiate AMPK and exhibit inhibitory effects on mTOR, thereby enhancing autophagy mechanisms (Shabkhizan, et al., 2024). These protocols have been associated with reduced oxidative stress, enhanced mitochondrial bioenergetics, and suppressed inflammatory cytokines, contributing to multifaceted techniques for neuroinflammation resolution (Gao, et al., 2024). Appropriate advanced models are crucial for discovering the intricate mechanisms linking nutritional factors, the gut-brain axis, autophagy, and neuroinflammation, as well as for converting this knowledge into effective clinical interventions. Advanced models provide scientists with the ability to thoroughly study underlying disease processes and examine new treatments in models that more accurately reflect the complexity of the human brain.

## 1.5. Advanced Models for Research and Therapeutic Perspectives

Advances in the development of neuronal organoids and even 3D-printed models present a marketplace of ideas that are ideal for deciphering the pathophysiology of diseases and identifying viable therapeutic targets (Giorgi, et al., 2024). For example, neuronal organoids allow for the assessment of autophagic flux and inflammation in three-dimensional cell cultures with the structural and functional characteristics of human postmortem brains (Cui et al 2024). This chapter underscores the necessity of adopting a bipolar comprehensive approach to address neuropsychiatric and neurodegenerative disorders. By incorporating lessons from cellular signaling, chemical modifications of DNA, fasting, and advanced research approaches, this study presents a solid platform for designing personalized treatments that alleviate neuroinflammation and enhance neuronal function. The subsequent sections offer a detailed discussion of the mechanisms regulating autophagy, the role of microglia, and the potential for novel interventions. This study not only enhances the understanding of diseases and their origins but also paves the way for new methods of molecular therapies and personalized treatments.

### 1.6 . Behavioral Relevance: Linking Molecular Pathways to Neuropsychiatric Symptoms

These cellular processes—autophagy and inflammation—not only affect neuronal integrity but also influence behaviorally relevant brain regions such as the prefrontal cortex, hippocampus, and amygdala. Disruptions in autophagic flux and chronic neuroinflammation have been associated with mood dysregulation, cognitive impairments, and behavioral rigidity seen in disorders such as depression, schizophrenia, and anxiety. From a mechanistic standpoint, these effects may stem from altered synaptic pruning, mitochondrial dysfunction, and epigenetic reprogramming—all of which are critical to emotional regulation and cognitive flexibility. Therefore, by examining the molecular crosstalk between autophagy and neuroinflammation, this chapter aims to provide new insights into the cellular foundations of neuropsychiatric symptoms and propose therapeutic frameworks rooted in brain-behavior integration.

## 2. REGULATION OF AUTOPHAGY: KEY PATHWAYS AND INFLUENCING FACTORS

An in-depth examination of the fundamental regulators that drive autophagy is essential for understanding the contributions of this disruption to behavioral pathologies in neuropsychiatric disorders. The process of autophagy is a regulated and continuous process that is crucial in the regulation of cellular turnover. It is crucial to understand where and what kind of autophagy occurs in a cell, and as such, the different types of autophagy processes are shown in Figure 1 (to illustrate that the processes are diverse and not glamorous). The major regulatory chemicals related to autophagy are mTOR, AMPK, and FOXO1. These pathways complement the external environment, bacterial cultures, and behavioral patterns, thereby synchronizing the rate of cell renewal and development of pathophysiological syndromes, including neuropsychiatric ones.

*Figure 1. Diverse Pathways of Cellular Recycling: A Commprehensive Overview of Autophagy Types*

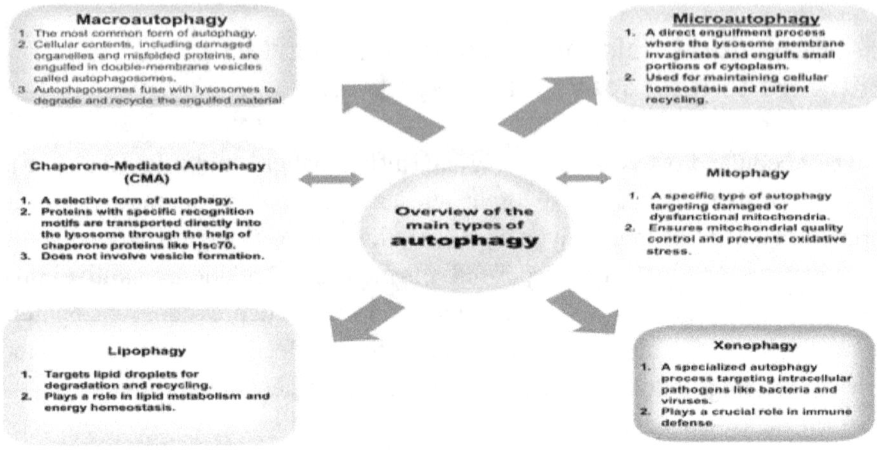


Fig. 1: Diverse Pathways of Cellular Recycling: A Comprehensive Overview of Autophagy Types

## 2.1. Key Pathways in Autophagy Regulation

### 2.1.1 mTOR Pathway:

The mammalian target of rapamycin (mTOR) pathway is often viewed as the gatekeeper of autophagy, a process it inhibits under conditions of nutrient abundance. mTOR inactivation upregulates the ULK complex (Unc-51-like autophagy-activating kinase complex); however, the activation of mTOR inhibits the formation of the autophagosome. With respect to neuropsychiatric disorders, dysregulation of mTOR can downregulate the process of autophagy, leading to the aberrant accumulation of damaged organelles and intracellular protein aggregates (Panwar, et al.,2023).

### 2.1.2 AMPK Pathway:

AMPK functions as an energy status reporter and drives autophagy during energy limitations by suppressing mTOR and phosphorylating proteins related to autophagy. AMPK activation has been reported to reverse cellular energy imbalances and prevent or repair oxidative damage; thus, AMPK is considered a target for the treatment of diseases involving mitochondrial dysfunction (Zong, et al.,2024).

## 2.2.3 FOXO1 Transcription Factor:

FOXO1 promotes autophagy by increasing the expression of genes associated with this process. Its activity is regulated by cellular stress, and it plays a critical role in increasing autophagic flux under stress. FOXO1 is particularly important in neuroinflammatory conditions because of its ability to decrease oxidative stress and inflammation (Tang, et al.,2024; Fig. 2).

*Figure 2. Key Pathways Regulating Autophagy and Their Role in Neuroinflammation.*

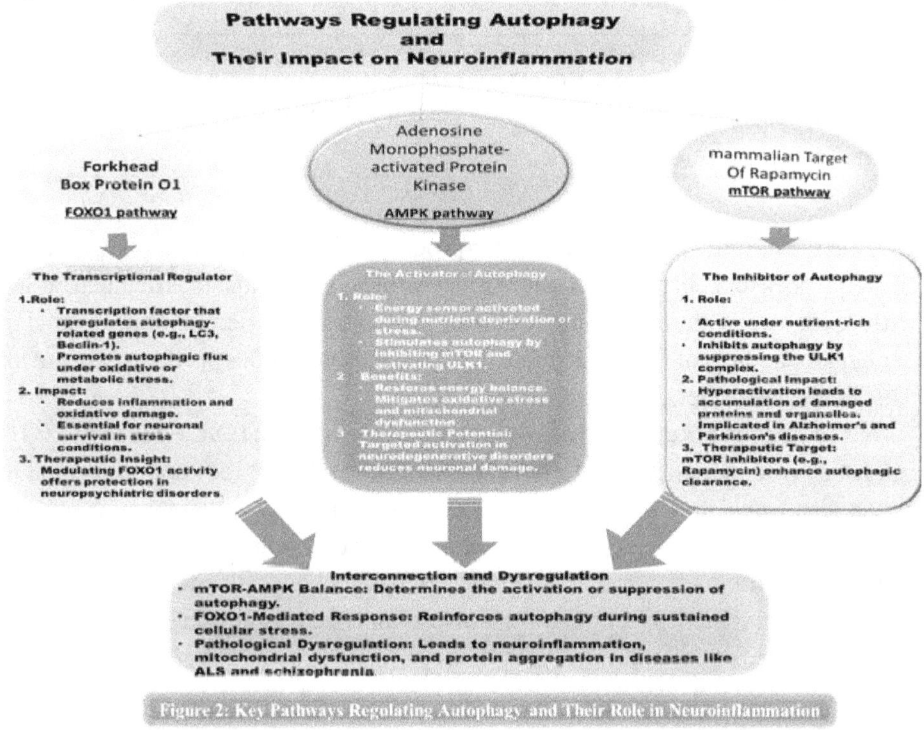

Figure 2: Key Pathways Regulating Autophagy and Their Role in Neuroinflammation

*Table 1. Pathways regulating autophagy and their impact on neuroinflammation.*

| Pathway | Mechanism | Effect on Autophagy | Impact on Neuroinflammation |
|---|---|---|---|
| **mTOR pathway (Xu, et al., 2025** | Dampens autophagy under nutrient-replete conditions. Although this pathway is inversely associated with the clearance of damaged proteins and organelles, it is relevant in selectively promoting autophagic clearance. | Inhibits autophagy under nutrient-rich conditions. | Reduces clearance of damaged proteins and organelles. |
| **AMPK pathway (Yang et al., 2025)** | Induces autophagy during energy starvation. | Activates autophagy during energy deprivation. | Helps replenish energy state and prevents free radical injuries. |
| **FOXO1 pathway (Wang,. et al., 2025)** | Regulates autophagy-related genes involved in antioxidant defense and anti-inflammation. | Upregulates autophagy-related genes. | Reduces oxidative damage and inflammation. |

This table outlines essential cellular processes that regulate autophagy and its consequences for neuroinflammation. Each pathway's mechanism is detailed, along with its impact on autophagic processes and neurological inflammation, enhancing our understanding of their potential therapeutic possibilities in the treatment of neurological disorders.

Dysregulation of key autophagy-modulating pathways (mTOR, AMPK, FOXO1) can have profoundly damaging effects on cellular health, as they are implicated in processes of oxidative stress and mitochondrial malfunction, which are central to the neuropath. The following section will discuss the complex relationship between neurological inflammation, oxidative stress, and mitochondrial function, elucidating the role of autophagy as a countermeasure against adverse mechanisms (Behura, et al., 2022).

## 3. NEUROLOGICAL INFLAMMATION, OXIDATIVE STRESS, AND THE MITOCHONDRIAL HYPERACTIVITY APPROACH

### 3.1 Oxidative Stress And Mitochondrial Function:

Since crucial molecular pathways regulating autophagy have been identified, it is important to consider the outcome of dysregulation. Oxidative stress and mitochondrial dysfunction are key adverse consequences due to damages in autophagy and are at the heart of neuroinflammation and neurodegeneration. This section will, therefore, discuss the interplay of these factors and the potential therapy implications associated with them. Proteins, lipids, and DNA, which include reactive oxygen species (ROS), are among the biological components whose activities are altered by oxidative stress. According to autophagy analysis, autophagy plays a significant role in mitophagy, which breaks down and recycles damaged mitochondria in the fight against oxidative stress (Yun, et al.2020). However, Li et al. (2024) reported that mitochondria further degrade and contribute to oxidative stress when autophagy is disrupted, advancing neurodegeneration.

## 3.2 . Neuroinflammation:

Autophagy is negatively impacted by microglia and proinflammatory cytokines associated with neuroinflammation. This imbalance exacerbates inflammation, causing the aggregation of inflammasome proteins such as NLRP3. It appears that autophagy is linked to neuroinflammation; when autophagy is regulated, it may lessen inflammatory damage in patients with neuropsychiatric disorders (Navabi, et al., 2024).

## 3.3. Gut Dysbiosis and Neuroinflammatory Disease Development

The gut microbiota is significantly involved in the regulation of body health through the gut–brain connection. Altered gut dysbiosis, disturbed microbiota composition, and diversity are related to neuroinflammatory disorders, including multiple sclerosis, Alzheimer's disease, and Parkinson's disease. These imbalances alter the communication between the gut and the brain, resulting in an increase in inflammation and neurodegeneration (Loh, et al., 2024). Microbiota metabolites synthesized by the host's gut microbiota, including SCFAs and neurotransmitters, extend to the brain level, affecting neurons and inflammation parameters. In separate studies, any decrease in the synthesis of SCFAs and other products of microbial metabolism was shown to play a role in the increased incidence of neuroinflammation (Sun, et al., 2024). The battle between immune cells and microbial stimuli is critical in the regulation of inflammation (Barturen, et al., 2022). Factors such as cytokines, T cells, and other inflammatory mediators are involved in this process (Fabisiak, et al. 2025). Abnormal activity of these components intensifies neuroinflammatory diseases (Liu, et al., 2024).

## 3.4. Inflammatory pathways

Disruptions in the gut microbial composition are associated with elevated rates of neuroinflammation through specific mechanisms. These pathways leading from gut microbiota imbalances to increased inflammation in the brain and spinal cord are well-documented in current literature (Hao, et al., 2024).

Several neuroinflammatory diseases have been extensively studied in relation to gut microbiota dysbiosis (Table 2).

*Table 2. Gut Dysbiosis and Autophagy Dysfunction in Neuroinflammatory Diseases.*

| Neuroinflammatory Disease | Role of Gut Dysbiosis | Autophagy Dysfunction | Research Study |
|---|---|---|---|
| **Multiple Sclerosis** | Disruption in microbial diversity linked to increased inflammation. Dysfunction in autophagy machinery impairs the clearance of misfolded proteins, driving neuroinflammation. | Defective autophagy fails to eliminate degraded proteins, contributing to neurodegeneration. | Cantoni, et al., 2025 |
| **Alzheimer's Disease** | Gut microbiota imbalance leads to higher levels of amyloid plaques and tau tangles, exacerbating neuroinflammation. Impaired autophagy fails to clear neurotoxic protein aggregates. | Compromised autophagic processes hinder the removal of toxic proteins, promoting neuronal death. | Delgadillo, et al., 2025 |
| **Parkinson's Disease** | Gut microbiota imbalance disrupts inflammatory signaling and autophagy processes. Dysfunction in autophagy impairs clearance of α-synuclein aggregates, exacerbating neuroinflammation. | Autophagy dysfunction hampers the degradation of α-synuclein, contributing to neurodegeneration. | Hamilton, et al., 2024 |

This table shows how gut dysbiosis produces neuroinflammatory disease severity by disrupting autophagy processes. Inflammatory processes and neuronal damage become more severe when multiple sclerosis, Alzheimer's disease, and Parkinson's disease cause imbalances in gut microbial composition. These diseases chronically affect brain functions, leading to memory loss alongside mood instability while causing motor problems due to synaptic damage, mitochondrial failure, and issues with clearing detrimental aggregates through autophagy. Such cellular disruptions produce pathological effects that explain the behavioral and neuropsychological symptoms observed in depression, anxiety, and schizophrenia, as well as in traditional neurodegenerative disorders.

- Multiple Sclerosis (MS): Several studies have noted that there is a significant relationship between MS and the gut microbiota, and research has demonstrated that low levels of microbial diversity are related to increased severity of MS (Cantoni, et al., 2025).

- Alzheimer's Disease (AD): Current literature indicates that dysbiosis has also been associated with the formation of amyloid plaques as well as tau pathology, which are characteristics of Alzheimer's disease (Delgadillo, et al., 2025).
- Parkinson's Disease (PD): Dysbiosis has been implicated in the worsening of both motor and cognitive symptoms in Parkinson's disease (Hamilton, et al., 2024).

## 3.5. Alcohol, Autophagy, and Neuroinflammation in Neurodegeneration

Besides imbalances in gut microbes, environmental and lifestyle factors, such as alcohol intake, prove to be important modulators of neuroinflammation and autophagy. As a result, these extrinsic factors may enhance the disintegration of the gut–brain process and fuel inflammation, thus promoting the progression of neurodegenerative diseases. Excessive alcohol consumption is sufficient to increase neuroinflammation, autophagy, and other age-related disorders in Alzheimer's disease and other neurodegenerative diseases. Alcohol impairs cell stability by causing oxidative damage, mitochondrial dysfunction, and dysregulation of autophagy, which is a crucial process required for neuronal integrity.

Alcohol stimulates microglia, which then release several toxins, including TNF-α, IL-1β, and CCL2, which are proinflammatory cytokines, evidence from clinical case studies presented in Table 3. This inflammatory context contributes to neuronal damage through the disruption of autophagy, the process by which cells dispose of damaged components. Abnormal autophagy leads to the accumulation of misfolded proteins and damaged mitochondria while exacerbating oxidative stress, which causatively contributes to neuroinflammation.

*Table 3. Impact of alcohol on autophagy and neurodegenerative diseases.*

| Condition | Model /System | Findings | Potential Intervention | Reference |
|---|---|---|---|---|
| **Alzheimer's Disease (AD)** | Murine model of AD | Chronic alcohol consumption increased amyloid-beta (Aβ) accumulation in the hippocampus. Reduced autophagic clearance of Aβ aggregates and elevated IL-6 and TNF-α levels were observed. | Rapamycin restored autophagy and reduced Aβ secretion. | Zhang et al., 2024 |
| **Aging and Cognitive Decline** | Longitudinal human study | Regular alcohol consumption accelerated cognitive decline, increased cerebrospinal fluid IL-1β, and reduced hippocampal volume. Postmortem analysis showed decreased LC3 expression. | Lifestyle changes and autophagy-enhancing agents. | Listabarth, et al., 2023 |
| **Parkinson's Disease (PD)** | Zebrafish PD model | Ethanol exposure inhibited mitophagy, causing dopaminergic neuron degeneration and mitochondrial ROS generation, leading to inflammation. | Spermidine supplementation reversed mitophagy deficits. | Razali et al., 2021 |

This table summarizes the effects of chronic alcohol consumption on autophagy, neuroinflammation, and neurodegenerative conditions on the basis of case studies and experimental models. This evaluation highlights the mechanistic pathways involved and potential therapeutic approaches.

Altogether, these pathways and environmental modulators converge to disrupt neuronal homeostasis, particularly in regions implicated in mood, memory, and cognitive control. Disrupted autophagy and chronic neuroinflammation can thus manifest in behavioral symptoms characteristic of major neuropsychiatric disorders.

## 4. ENVIRONMENTAL AND EPIGENETIC TRIGGERS

### 4.1 Oxidative Stress & Nanopollutants: Altered Autophagy and Neuroinflammation

Apart from lifestyle factors such as alcohol and the effects of gut microbiota, there is a wide variety of environmental influences and epigenetic changes that could also contribute significantly to the modulation of autophagy and neuroinflammation in the context of neuropsychiatric disorders. Importantly, diverse heavy metals and toxins, along with pollutants, are known inducers of cellular stress and neuroinflammation (Song, et al., 2025). Among these, nanoplastics and nanoparticles have become problematic owing to their high solubility in the environment and their ability to alter autophagic processes in cells. After illustrating how large plastics degrade to form tiny particles known as ultrasmall particles, many of which are byproducts of industry, these particles are now found in the air, water, and soil and can easily penetrate the human body through ingestion, inhalation, and skin contact (Fu et al.,2024).

NPs exhibit features that allow them to cross certain natural barriers, including the blood–brain barrier and the intestinal wall. Regardless of whether they cross the blood–brain barrier or gain access to the central nervous system through other mechanisms, they dysregulate antioxidant homeostasis, activate mitochondrial dysfunction, and trigger inflammatory molecular targets, all of which are considered to be involved in the pathophysiology of neuropsychiatric disorders. A process that promotes the removal of damaged organelles and misfolded proteins in cells, referred to as autophagy, is directly impacted by nanoparticle exposure. This disruption amplifies neuroinflammation and contributes to diseases such as Alzheimer's disease, Parkinson's disease, major depressive disorder, and multiple sclerosis (Wu et al., 2023).

## 4.2 Nanoparticles, Autophagy, and Neuroinflammation: Case Studies and Mechanisms

Owing to their unique physicochemical characteristics, nanoparticles have been established as potent modulators of cellular processes, including autophagy and neuroinflammation. General information about their effects is provided along with detailed descriptions of exemplary case studies referring to different models and systems, emphasizing mechanisms and pathways in Table 4

*Table 4. Impact of nanoparticles and microplastics on autophagy and neuroinflammation.*

| Study | Nanopollutant | Model/System | Findings | Reference |
|---|---|---|---|---|
| **The Effects of Polystyrene Nanoplastics on Neuronal Dysfunction** | Polystyrene nanoplastics | Cultured hippocampal neurons | Impaired autophagic flux, increased ROS, and synaptic changes. Mechanism involves oxidative stress disrupting lysosomes and inhibiting autophagy. | Paing, et al., 2024 |
| **Titanium Dioxide Nanoparticles in Alzheimer's Disease** | Titanium dioxide nanoparticles | APP/PS1 transgenic mice | Accelerated amyloid plaque formation, defective autophagic flux, disrupted mitophagy, and enhanced neuroinflammatory markers (IL-1β, TNF-α). | Zha, et al., 2024 |
| **Chitosan nanoparticles (ChNPs) and Parkinson's Disease** | Chitosan nanoparticles (ChNPs) | Zebrafish models | Suppression of LC3 lipidation disrupted autophagy, leading to dopaminergic neuron death characteristic of PD pathophysiology. | Abou-Saleh et al., 2019 |

continued on following page

*Table 4. Continued*

| Study | Nanopollutant | Model/System | Findings | Reference |
|---|---|---|---|---|
| **Airborne Nanoparticles Can Trigger Alzheimer's Disease** | Airborne nanoparticles | Alzheimer's Disease patient cohort | Increased serum levels of TNF-α and IL-1β, downregulation of autophagy-related proteins (Beclin-1, LC3), and progression of inflammation. | Maher 2019 |
| **Silver Nanoparticles in Animal Treatment** | Silver nanoparticles | C57BL/6 mice | Induced mitochondrial damage and oxidative stress; autophagy inducers like rapamycin preserved mitochondrial function and reduced IL-6 levels. | Alqahtani, et al., 2022 |
| **Microplastic Pollutants in Drinking Water and Human Neuroinflammation** | Microplastics | Human population analysis | Increased inflammatory biomolecules in neural tissues; lysosomal dysfunction hampered autophagy degradation and activated the NF-κB pathway. | Bruno et al., 2024 |
| **Nanoplastic Exposure in Marine Models of Contamination** | Nanoplastics | Sea urchin embryos | Dysregulated autophagy, defective autophagosome-lysosome fusion, cellular imbalance, and developmental defects due to stress signaling. | Paganos, et al., 2023 |

This table summarizes various studies on how nanoparticles and microplastics affect autophagy, neuroinflammation, and associated pathologies. The mechanisms, experimental models, and findings highlight the detrimental effects on cellular processes and potential disease progression.

## 4.2 Role of Microglia and NLRP3 Inflammasome

Activation of the NLRP3 inflammasome by microglial cells leads to neuroinflammatory effects that were initially discovered through peripheral diseases, including IBD (Liang et al., 2022). The CNS activation of TLR2/4 receptors by Aβ plaques and tau tangles drives microglia-mediated NLRP3 receptor complex formation, leading to caspase-1–dependent IL-1β/IL-18 protein secretion (Miao et al., 2023; Terzioglu & Young-Pearse, 2023). The inflammatory cascade leads to pyroptosis, along with inappropriate synaptic pruning in hippocampal and prefrontal circuits, which causes working memory deficits and executive dysfunction during the early stages of Alzheimer's disease (Kodi et al., 2024).

The same inflammasome axis functions similarly in IBD because intestinal macrophages release IL-1β/IL-18, which results in intestinal barrier breakdown (Zhen & Zhang, 2019; Weber et al., 2024). The connection between gut and brain signaling supports research into inflammasome-based treatments that address both physical infection and neurological symptoms (Dong et al., 2023).

This review establishes a connection between NLRP3 regulation in microglia and separate behavioral disability patterns while providing essential understanding of the neuropsychiatric effects of autophagy-inflammation interactions (see table 5).

*Table 5. Key case studies targeting the NLRP3 inflammasome in Alzheimer's disease (AD) and inflammatory bowel disease (IBD), summarizing model systems, interventions, and impacts on cognitive or tissue outcomes.*

| Case | Model/System | Key Finding | Ref. |
|---|---|---|---|
| **NLRP3 KO in AD** | APP/PS1 mouse | ↓ Aβ plaques, ↑ learning & memory, ↓ IL-1β | Valiukas et al., 2025 |
| **MCC950 treatment in AD mouse** | APP/PS1 mouse | ↓ microglial activation, ↑ synaptic density, ↑ spatial memory | Kodi et al., 2024 |
| **NLRP3 inhibition in IBD** | DSS-induced colitis in mouse | ↓ IL-1β/IL-18, ↑ mucosal healing | Weber et al., 2024 |
| **Probiotics + NLRP3 inhibitor** | IBD patients | ↓ gut inflammation, ↑ tight-junction proteins | Dong et al., 2023 |

## 4.3. Regulation of Inflammatory Signaling

Toll-like receptor (TLR) 2/4 recognizes danger-associated molecular patterns (DAMPs, including Aβ and HMGB1) to activate NF-κB and enhance the expression of NLRP3 and pro-inflammatory cytokines, according to Zhang & Yang (2023) and Li et al. (2025). Tissue damage releases the inflammatory factor HMGB1, which activates TLR4 and RAGE signaling to increase NLRP3 expression, leading to higher IL-1β/IL-18 production (Li et al., 2025). The signaling hub drives persistent synapse loss and dendrite atrophy while causing prefrontal network dysregulation and fear responses through the amygdala, which exist in the early stages of Alzheimer's disease and anxiety due to stress. This review establishes an improved model explaining how molecular initiators manifest into clinical outcomes by following receptor activity into inflammasome activation (Table 6).

## 4.4 Therapeutic Implications

A promising therapeutic approach exists in reducing TLR4/NF-κB–NLRP3 signaling through TLR4 antagonists and HMGB1 neutralizers that reduce microglial activation while preserving synaptic structure in rodent AD and IBD models (Zhang & Yang, 2023; Li et al., 2025). The combination of complementary lifestyle approaches using ketogenic diets and probiotic interventions has shown potential to control TLR4-associated inflammation when treating disorders affecting cognitive flexibility and emotional resilience.

## 4.5 Microglia-Driven Neuroinflammation in Viral Infections

Chronic microglial activation leads to executive dysfunction in prefrontal networks due to HIV-1 Tat/gp120 proteins, which both block mTOR-dependent autophagy and sustain NF-κB–mediated TNF-α/IL-6 production, according to Frericks et al. (2025) and Thompson et al. (2024). The SARS-CoV-2 ORF3a protein prevents autophagosome–lysosome fusion events, causing NLRP3 activation to persist while creating cognitive effects characteristic of long-COVID (Li et al., 2025). The identification of these maintained molecular pathways helps explain how viral disruption of autophagy mechanisms leads to sustained neuropsychiatric complications.

## 4.6. Chronic Inflammation and Microglia Activation and Their Dysregulation Processes

Neurotropic viruses, including HIV, HBV, HCV, and SARS-CoV-2, use cellular factors to evade recognition by the immune system and subvert microglial functions. Abnormal microglial activation is characterized by the continuous activation of signaling pathways, including but not limited to NF-κB, STAT3, and the NLRP3 inflammasome, thereby promoting inflammation. The following sections provide a more thorough analysis of these viruses and emphasize the specific viral processes that contribute to the development of complications.

*Table 6. Mechanisms of virus-induced neurotoxicity and associated clinical evidence.*

| Infection | Mechanism | Clinical Evidence |
|---|---|---|
| **HIV/AIDS: Chronic Cytokine-Induced Neurotoxicity** | 1. **Neurotoxic Proteins:** Tat and gp120 activate microglia and astrocytes to release TNF-α, IL-6, and IL-1β (Thompson et al., 2024).<br>2. **Downstream Events:** Trigger the MAPK signaling cascade, causing synaptic dysfunction and neuronal apoptosis.<br>3. **Oxidative Stress:** Increase ROS and nitric oxide ions (NO).<br>4. **BBB Breakdown:** Facilitates peripheral immune cell invasion into the CNS. | Examination of HIV-positive patients' brains after death shows increased microglial activation in the hippocampus and prefrontal cortex, correlating with synaptic loss and cognitive impairment (Frericks et al., 2025). |
| **HBV and HCV: NF-κB Signaling Pathway and Oxidative Stress** | 1. **NF-κB Pathway Activation:** Sustains microglial activation during chronic infection.<br>2. **Immunological Molecules:** Dysregulate neuronal homeostasis, producing IL-1β, TNF-α, and MCP-1.<br>3. **Oxidative Stress:** Increased ROS output leads to mitochondrial dysfunction.<br>4. **Apoptosis:** Caspase-3 and caspase-9 mediated cell death. | HBV patients show elevated oxidative stress markers (e.g., 8-OHdG) in cerebrospinal fluid compared to healthy individuals (Lee et al., 2024).<br>HCV patients exhibit increased pro-inflammatory cytokines and cortical atrophy, consistent with chronic neuroinflammation and cognitive impairment (Fasano et al., 2024). |
| **NLRP3 Inflammasome in SARS-CoV-2 and Cognitive Impairment** | 1. **Microglial Activation:** NLRP3 inflammasome stimulation by SARS-CoV-2 spike and ORF3a proteins leads to neuroinflammation and pyroptosis.<br>2. **Cytokine Storm:** Contributes to systemic and CNS-related cytokine storms, affecting neuronal circuits.<br>3. **Viral Persistence:** Chronic infection prolongs inflammasome activation and neuroinflammation.<br>4. **Synaptic Impairment:** IL-6 elevation impairs synaptic plasticity, explaining long COVID cognitive deficits. | Pathological analysis of COVID-19 patients shows microglial nodules and high inflammasome protein levels in the hippocampus and prefrontal cortex, areas responsible for memory and cognitive functions (Li et al., 2025). |

The primary ways in which numerous neurotropic viruses (HIV, HBV, HCV, and SARS-CoV-2) induce neurotoxicity are compiled in this table. This review describes the biological mechanisms underlying neurological disorders, inflammatory conditions, and the stimulation of microglia. Clinical data from studies on individuals affected by these viruses are also presented, emphasizing the effects of these processes on the areas of the brain involved in cognitive function and memory.

## 4.7 Microglia Dysregulation in Neurotropic Viral Infections

Neurotropic viruses disrupt autophagy functions in a sustained manner, activating microglial NF-κB/NLRP3 signaling pathways, which establishes direct correlations between molecular manipulation and discrete neurological impairment. The interaction of HIV-1 Tat/gp120 with mTOR-dependent autophagy pathways leads to the release of IL-6/TNF-α factors, resulting in HAND pathophysiology and executive cognitive decline (Frericks et al., 2025). The reactive oxygen species produced by HBV infection lead to synaptic destruction, causing prefrontal cortical atrophy and memory impairment (Lee et al., 2024). Caspase-3 becomes activated when HCV attacks, leading to mitochondrial failure while simultaneously causing a reduction in motor speed (Fasano et al., 2024). The SARS-CoV-2 ORF3a protein hinders autophagosome–lysosome fusion while amplifying NLRP3-driven IL-1β production, resulting in long-COVID cognitive difficulties (Li et al., 2025). These preserved channeling mechanisms serve as the foundation for developing therapies targeting microglia/autophagy to restore cognitive and motor function abilities (Table 7).

*Table 7. Comparison of Microglial Activation and Neuroinflammatory Outcomes in Neurotropic Viral Infections.*

| Virus | Primary Microglial Activation Pathway | Key Cytokines and Mediators | Impact on CNS | Clinical Outcome |
|---|---|---|---|---|
| **HIV/ AIDS** | NF-κB, MAPK | TNF-α, IL-6, IL-1β | Neuronal synaptic dysfunction, apoptosis | HAND, cognitive impairments, dementia |
| **HBV** | NF-κB, ROS | IL-1β, TNF-α, MCP-1 | Cortical thinning, increased neuroinflammation | Cognitive trauma, neuroinflammation |
| **HCV** | NF-κB, Caspase-3 | IL-6, IL-1β, TNF-α | Mitochondrial dysfunction, synaptic loss | MS-like symptoms, dementia |
| **SARS-CoV-2** | NLRP3 inflammasome | IL-6, IL-1β, IL-18, TNF-α | Cytokine storm, disrupted synaptic plasticity | Long COVID, encephalopathy |

HIV/AIDS activates NF-κB and MAPK, which release inflammatory cytokines such as TNF-α, IL-6, and IL-1β, leading to neuronal damage, cognitive impairments (HAND), and dementia. HBV activates NF-κB and ROS production, triggering the release of IL-1β, TNF-α, and MCP-1, causing cortical thinning and cognitive trauma, alongside heightened neuroinflammation. HCV also activates NF-κB and involves Caspase-3, leading to the release of proinflammatory cytokines (IL-6, IL-1β, and TNF-α), resulting in mitochondrial dysfunction and MS-like symptoms, including cognitive decline. SARS-CoV-2 induces the NLRP3 inflammasome, leading to a cytokine storm and disrupting synaptic plasticity, contributing to long-term neurological consequences such as long COVID-19 and encephalopathy

### 4.7.1 HIV-1 and HAND

Microglial cells containing HIV-1 proteins Tat and gp120 impede autophagy regulation via mTOR signaling to activate NF-κB, while releasing TNF-α and IL-6 into the surrounding environment. Dysfunction in the prefrontal circuit synapses leads to executive dysfunction, becoming a prominent symptom of HAND, or HIV-associated neurocognitive disorder.

### 4.7.2 HBV and Working-Memory Loss

During HBV infection, the cellular ROS levels rise, simultaneously activating NF-κB, which subsequently results in increased production of IL-1β, TNF-α, and MCP-1. Cortical thinning in the prefrontal lobe, resulting from long-term infections, causes working-memory difficulties in affected patients.

### 4.7.3 HCV and Motor Slowing

The activation of HCV-dependent caspase-3 pathways leads to mitochondrial breakdown while concurrently releasing immune signals IL-6 and IL-1β. Bioenergetic issues in neurons create conditions where patients experience motor deceleration alongside cognitive exhaustion.

### 4.7.4 SARS-CoV-2 and Long-COVID Cognitive Fog

By inhibiting autophagosome–lysosome fusion, the SARS-CoV-2 ORF3a protein facilitates ongoing NLRP3 inflammasome activity, resulting in the release of IL-1β/IL-18. Long-COVID patients experience enduring cognitive fog and poor concentration because SARS-CoV-2 damages the blood–brain barrier and disrupts synaptic plasticity.

Microglial responses can be balanced by using MCC950 alongside rapamycin, which protects synapses while restoring executive function and memory within these viral infection contexts.

## 4.8 Theoretical Structure: Viral Triggers → Molecular Pathways → Developmental Outcomes

*Table 8. Linking neurotropic viral triggers, microglial pathways, and specific neurobehavioral impairments.*

| Virus | Molecular Trigger | Pathway & Mediators | Neurobehavioral Outcome |
|---|---|---|---|
| HIV-1 | Tat/gp120 → mTOR-autophagy inhibition | NF-κB → TNF-α, IL-6 | Executive dysfunction (HAND) |
| HBV | ROS ↑ → NF-κB activation | IL-1β, TNF-α, MCP-1 | Working-memory deficits |
| HCV | Caspase-3 activation | IL-6, IL-1β | Motor slowing & cognitive fatigue |
| SARS-CoV-2 | ORF3a → autophagosome–lysosome fusion blockade | NLRP3 inflammasome → IL-1β, IL-18 | Long-COVID cognitive fog |

# 5 SEMLIKI FOREST VIRUS AS A NEUROINFLAMMATION MODEL

## 5.1 Neuroinvasion & Glial Activation

To demonstrate the effect of certain environmental triggers on neuroinflammation and autophagy, the Semliki Forest Virus (SFV) can be utilized as one model for studying these interactions. SFV entered scientific records in 1942 from Uganda's Semliki Forest, replicating as an enveloped alphavirus that gains access to neurons and glia across the blood–brain barrier through receptor-mediated transcytosis (Abdoullah et al., 2023). The cellular invasion of SFV into the CNS results in NLRP3 inflammasome activation of microglia, which triggers astrogliosis and subsequently enhances IL-1β, TNF-α, and IL-6 production (Badawi et al., 2023). SFV prevents autophagosomes from fusing with lysosomes, which expands neuroinflammatory processes and causes synaptic damage (Kim & Diamond, 2023).

## 5.2 Behavioral Correlates & Translational Prospects

When SFV infects newborn mice, it triggers acute brain inflammation and damages hippocampal neurons, along with reduced potentiation, which causes spatial memory impairment (Amor et al., 2014). The viral infection of SFV leads to tau hyperphosphorylation through oxidative stress mechanisms that produce synaptic dysfunction in cultured neurons, according to Fabiani and Antollini (2019). Pharmaceutical experts classify human occurrences of alphavirus encephalitis as rare since they exhibit symptoms matching those observed in animals, which include fever, confusion, seizures, and cortical inflammatory indications (de Oliveira Souza et al., 2024). Both rapamycin medications and MCC950 substances maintain synaptic markers and enhance motor and cognitive abilities in models that use SFV (Dedoni et al., 2023). SFV's CNS effects progress through three sequential phases that connect experimental observations to specific behavioral irregularities:

1. Blood-Brain Barrier Crossing: The entry of SFV into the brain occurs through receptor-mediated transcytosis, allowing the pathogen to access glial and neural cells.
2. Microglial Activation: The presence of viruses within brain tissue activates the NLRP3 inflammasome while simultaneously halting autophagosome-lysosome fusion, resulting in elevated secretion of IL-1β, TNF-α, and IL-6, which produce spatial memory deficits.

B3. Astrocyte Response: Fine motor coordination deficits emerge as NF-κB activation and the subsequent release of pro-inflammatory cytokines occur due to astrogliosis and glutamate dysregulation caused by SFV. The three-stage summary mapping illustrates the molecular effects of SFV on distinct neurobehavioral outcomes while identifying therapeutic intervention targets, including autophagy enhancement and NLRP3 interference for future drug development. Chronic inflammation affecting the nervous system, particularly due to unresolved viral infections, initiates classical immune responses and causes long-lasting changes in gene expression through epigenetic mechanisms. The stimulation of epigenetic changes by inflammatory cytokines, oxidative stress, and cellular damage is considered an important regulatory mechanism in modulating autophagy. By discovering the mechanisms of how these molecular adaptations relate to environmental stimuli and stable cellular reactions, researchers learn about the basic mechanisms of neurodegenerative and neuropsychiatric disorders.

## 6. EPIGENETIC REGULATION OF AUTOPHAGY

Multiple neuropsychiatric conditions, along with neurodegenerative disorders, show evidence of dysregulated epigenetic control of autophagy, leading to synaptic dysfunction, impaired neuroplasticity, and decreased cognitive performance (Yu et al., 2024). During the 1940s, Waddington first introduced epigenetics to denote gene expression alterations that are inheritable but do not modify DNA sequences (Meza-Menchaca et al., 2024). The three main epigenetic processes, which include DNA methylation, combined with histone modifications and noncoding ribonucleic acids (ncRNAs), serve as fundamental regulators of autophagic flux while influencing both cellular equilibrium and neural functioning.

## 6.1. Noncoding RNAs and Autophagy

The autophagy regulatory functions of two prominent ncRNAs are robustly demonstrated through microRNAs (miRNAs) (Mu, et al 2024) and long noncoding RNAs (lncRNAs) (Dong, et al., 2024). The molecular activities of miR-214 disrupt both LC3 lipidation processes and autophagosome formation mechanisms within mature osteoclast microorganisms, preventing damaged component clearance (Ghafouri-Fard et al., 2021). Scientists have utilized bone research to investigate these molecular pathways because they show similar effects during autophagy induction in neural cells under inflammatory and oxidative conditions. Autophagosome elongation suffers impairments in inflammatory cascades because miR-155 elevates expression in autoimmune and inflammatory diseases while suppressing the central autophagy gene ATG7 (Cichalewska-Studzinska et al., 2024). The behavioral deficits and synaptic loss in conditions such as depression and Alzheimer's disease may be caused by disruptions managed by miRNAs. The lncRNA HOTAIR and other similar elements direct histone-modifying complexes toward autophagy gene regulatory regions to modify transcriptional outcomes (Na Wu et al., 2024). The cancer bone metastasis model demonstrates that HOTAIR interacts with histone methyltransferases to alter autophagy levels, according to Dong et al. (2024). This epigenetic process might contribute to both CNS tumorigenesis and neurodegenerative diseases.

## 6.2. Histone Modifications and Autophagy:

The autophagy gene loci experience alterations in accessibility because histones undergo acetylation, methylation, phosphorylation, SUMOylation (SUMOs) and ubiquitination modifications. The gene expression process becomes more accessible due to histone acetylation by HATs; however, HDACs function to repress transcriptional processes. The epigenetic indicator HDAC6 serves as a vital regulatory point because blocking its activity enhances autophagy while promoting osteoblast survival under challenging conditions. Neuroprotective properties of HDAC6 inhibitors may arise since they boost both autophagy and reduce tau accumulation associated with Alzheimer's disease (Yalaev et al., 2024). Methylation of histones at H3K4 and H3K27 sites provides evidence that the process either activates or inactivates autophagy-related gene transcription. The precise equilibrium between these factors regulates autophagic activity during metabolic and oxidative stress conditions, offering potential therapeutic opportunities for neurodegenerative disease models (Liu et al., 2023).

## 6.3. Behavioral Implications and Clinical Relevance:

These epigenetic mechanisms combine to function as early regulatory elements that modify autophagy pathways, helping to maintain synaptic health alongside plasticity maintenance and neuron viability. Research shows that autophagy dysfunction connects to cognitive deterioration as well as mood and motor function problems in diseases including Alzheimer's disease, rheumatoid arthritis, and neuroinflammatory conditions. A malfunctioning autophagy pathway caused by either miR-155 or histone deacetylation leads to prefrontal cortex dysfunction, consistently resulting in working memory deficits in early Alzheimer's disease and executive dysfunction in autoimmune encephalopathies.

Novel therapeutic strategies that utilize emerging technologies for epigenetic checkpoint management enable the restoration of autophagy while controlling neuroinflammation through methods such as miRNA copy number adjustment and inhibition, along with the utilization of CRISPR/dCas9 and HDAC inhibitor substances. Rapamycin and HDAC6 inhibitors represent potential treatments that show promise for addressing both degenerative disorders and disorders affecting the immune system in the brain. Researchers are testing these agents against such neurological conditions. A comprehensive tabular overview presents case study based descriptions of these mechanisms, their significant molecular components, and therapeutic applications (Table 9).

*Table 9. Epigenetic regulation and autophagy in disease management*

| Disease | Mechanism | Key Molecular Players | Therapeutic Implications |
|---|---|---|---|
| **Osteoporosis Li, et al ., 2016** | Suppression of autophagy by miR-214 affects osteoclast bone-resorbing function, advancing disease progression. | miR-214 | miR-214-specific antagomiRs restore autophagy, improving bone mineralization and resilience. |
| **Rheumatoid Arthritis (RA) (Mueller, et al.,2021)** | Epigenetic alterations (DNA methylation, histone acetylation) affect RA pathogenesis. miR-155 inhibits autophagy genes, leading to enhanced inflammation. | miR-155, histone deacetylase (HDAC) | HDAC inhibitors reduce inflammation and restore autophagy in cellular and animal models of RA. |
| **Cancer Bone Metastasis (Mozdarani, et al.,2020)** | Histone-modifying enzymes, including HOTAIR, regulate autophagy-related genes, promoting metastasis to the bone. | HOTAIR, histone-modifying enzymes | Silencing HOTAIR and associated enzymes serves as a potential therapeutic approach to combat bone metastasis in cancers like breast and prostate cancer. |
| **Therapeutic Insights Dai, et al. (2024).** | Epigenetic modulation of autophagy presents new treatment options for various diseases. | miRNAs, HDAC inhibitors, histone modifiers | Emerging approaches include miRNA mimics/inhibitors, small molecules targeting histone modifications, and precise epigenome editing tools like CRISPR/dCas9 to fine-tune autophagy and gene expression. |
| **Conclusion** | Epigenetics and autophagy interplay is vital for maintaining cellular homeostasis. | Not applicable | Understanding these mechanisms allows for innovative therapies targeting unmet clinical needs in diseases like osteoporosis, RA, and metastatic cancers. |

This table summarizes key interactions between epigenetics and autophagy in specific diseases, such as osteoporosis, rheumatoid arthritis, and cancer bone metastasis. It highlights molecular mechanisms, therapeutic implications, and insights into innovative treatments and clinical applications. This design emphasizes clear segmentation of diseases, mechanisms, and therapeutic insights, providing a comprehensive yet concise framework for understanding epigenetic and autophagic contributions to pathology and treatment.

## 7. EPIGENETIC MOLECULAR MECHANISMS:

The epigenetic mechanisms that control autophagy consist of DNA methylation, histone modifications, and noncoding RNAs, which researchers have linked to synaptic deficits and cognitive problems in neuropsychiatric disorders (Yu et al., 2024; Meza-Menchaca et al., 2024).

## 7.1. DNA Methylation

DNA methylation at the promoters of autophagy-related (ATG) genes transforms these genes into silent states, producing diminished autophagic action. Research demonstrates that ATG5 and BECN1 gene promoter methylation occurs under chronic stress conditions, resulting in defective autophagosome production and greater sensitivities to depressive illness (Meza-Menchaca et al., 2024).

## 7.2. Histone Modifications and Autophagy

The chromatin accessibility-regulating mechanisms at ATG loci include multiple histone post-translational modifications, consisting of methylation, acetylation, phosphorylation, and ubiquitination, along with SUMOylation (Shin et al., 2016).

## 7.3. Arginine Methylation (CARM1)

The activity of CARM1 results in H3R17me2 demethylation, leading to enhanced ATG gene transcription and improved autophagic flux rates (Xie et al., 2024).

## 7.4. H3K4me3 vs. H3K27me3

The autophagy regulatory process is controlled by H3K4 trimethylation, which activates ATG loci, while H3K27me3 leads to their repression, thus enabling context-specific autophagy modulation (Menzies et al., 2017; He et al., 2024).

## 7.5. Histone Acetylation (HAT/HDAC)

The acetylation process activated by HATs uncovers ATG promoter sites; however, HDAC6 blockade both facilitates osteoblastic autophagy and exhibits neuroprotective advantages for Alzheimer's disease patients, thereby facilitating working memory enhancement (Boyle et al., 2017).

## 7.6. Noncoding RNAs and Autophagy

The autophagy regulation process controlled by miRNAs and lncRNAs involves transcript targeting of ATG proteins:

**miR-214** expression levels lead to the inhibition of LC3 lipidation, resulting in impaired autophagosome formation (Ghafouri-Fard et al., 2021).

**miR-155** reduces ATG7 expression, thus enhancing neuronal cell inflammation and detrimental autophagic dysfunction (Cichalewska-Studzinska et al., 2024).

Stress-related animal models of anxiety and depression show decreased autophagic capacity due to HOTAIR recruiting EZH2, which augments H3K27me3 at ATG promoters (Dong et al., 2024).

## 7.8. Behavioral and Clinical Implications

The following epigenetic pattern changes affect autophagy function:

1-Early stages of Alzheimer's disease cause prefrontal cortex dysregulation and impaired working memory function.
2-Biochemical processes in the amygdala activate fears while simultaneously creating elevated anxiety during stress exposure.
3-The epigenetic pathway of schizophrenia leads to cognitive rigidity, presumably through histone mark modifications targeting genes linked to synaptic plasticity (Jazieh et al., 2024).

### 7.8.1 Theoretical Model Connecting Epigenetic Marks, Autophagy Regulation and Behavioral Outcomes

The transcription of autophagy (ATG) genes is enhanced by CARM1-mediated H3R17me2 demethylation of histone H3 arginine 17, which leads to improved working memory performance (Xie et al., 2024; Shin et al., 2016). Histone deacetylase 6 (HDAC6) inhibition results in elevated autophagic flux, as demonstrated by increased LC3 lipidation, and produces corresponding reductions in amygdala fear behavior (Boyle et al., 2017). Upregulation of miR-155 causes a reduction in ATG7 protein through suppression, enhancing inflammation and reducing autophagic capability, which indicates increased anxiety-like tendencies (Cichalewska-Studzinska et al., 2024). Similarly, elevated levels of miR-214 interrupt LC3 lipidation and impair spatial memory processing (Ghafouri-Fard et al., 2021). When DNA hypermethylation targets the ATG5 promoter, it suppresses autophagosome development, leading to depressive-like behavioral expression (Meza-Menchaca et al., 2024). The expression of autophagy-related genes is repressed through H3K27 trimethylation by the long noncoding RNA HOTAIR, generating stress-induced anhedonia (Dong et al., 2024).

## 7.9. Therapeutic Perspectives

Autophagy-based neuroprotection methods have developed numerous approaches, which include the administration of the mTOR inhibitory drug rapamycin. This drug encourages the autophagic removal of toxic protein deposits while decreasing neuroinflammation, which helps restore cognitive abilities in Alzheimer's and

Parkinson's disease models (Wang et al., 2018). Short-chain fatty acids produced through gut–brain axis modulation via prebiotic and probiotic interventions boost autophagic flux to decrease microglial activation and treat multiple sclerosis symptoms, according to Sharon et al. (2020) and Campagnoli et al. (2024). The enhanced fusion rate of autophagosomes with lysosomes enables agents to clear damaged organelles while delivering additional neuroprotective outcomes, such as cathepsin modulators.

New therapeutic approaches for treating depression and schizophrenia along with post-COVID mental weariness might emerge from targeting the epigenetic regulatory mechanisms through CARM1 activators and HDAC6 inhibitors and DNA demethylating agents and ncRNA modulators to restore autophagic equilibrium. In addition to pharmacological and epigenetic strategies, there is increasing attention to nutraceutical interventions as viable complementary treatments for autophagy and neuroinflammation.

## 8. NUTRACEUTICAL APPROACHES: PHYTOCHEMICALS IN NEUROINFLAMMATION AND AUTOPHAGY

The pathogenesis of Alzheimer's disease along with Parkinson's disease and major depressive disorder originates from combined neuroinflammatory conditions and disrupted autophagic clearance mechanisms (Adamu et al., 2024; Ragupathi et al., 2024). The bidirectional communication between the gut and brain axis enables microbial metabolites to manage CNS immune activity and autophagic clearance patterns (Sharon et al., 2020; Sarb et al., 2024). Gut microbiota restoration through olive-oil polyphenols such as hydroxytyrosol and curcumin with essential oils leads to short-chain fatty acid production and AMPK activation and NF-κB/NLRP3 inflammasome pathway inhibition which enhances autophagy while decreasing proinflammatory cytokines (Papadopoulou et al., 2024; Zhu et al., 2024; Crocetta et al., 2024). Neuroprotection and cognitive preservation in neurodegenerative and psychiatric disorders seems attainable through the combination of gut integrity restoration with inflammatory signaling moderation and autophagy activation.

### 8.1. Gut–Brain Axis and Dysbiosis

The gut–brain axis is a bidirectional system that links the composition of intestinal microbiota to the stability of the central nervous system (CNS). Dysbiosis, a disturbance in the microbiota community, impairs the barrier function of the gut, allowing bacterial endotoxins, such as lipopolysaccharide (LPS) and other metabolites, to cross and activate microglial and astrocytic cells, causing sustained neuroinflammation (Sarb et al., 2024).

## 8.2. Nutritional Supplements and Their Effects on Gut–Brain Dysbiosis

Certain nutritive substances, such as polyphenols, dietary fibers, and essential oils, may restore microbial balance, increase SCFA production, particularly butyrate production, and support blood–brain barrier function, all of which influence neuronal autophagy and reduce inflammatory signaling.

## 8.3. Compound-Specific Nutraceutical Effects

### 8.3.1. Olive Oil Polyphenols

The presence of a large number of hydroxytyrosol and polyphenols in extra virgin olive oil inhibits NLRP3 inflammasome formation and stimulates autophagy by regulating AMPK–mTOR pathways, as shown in recent research (Papadopoulou et al., 2024). Evidence reported by Abrignani et al. (2024) indicates that the incorporation of EVOO into the diet of Mediterranean populations is associated with improved executive function and a reduced incidence of neurodegenerative diseases.

### 8.3.2. Curcumin (Turmeric)

Curcumin suppresses NF-κB activation and induces autophagy, positively modulating the gut microbiome by increasing the numbers of Lactobacillus and Bifidobacterium (Zhang et al., 2020). A randomized, placebo-controlled study revealed that an 18-month curcumin supplementation led to improved depressive symptoms and memory function, reinforcing its role in modulating mood via the gut–brain axis (Zhu et al., 2024).

### 8.3.3. Cinnamaldehyde (Cinnamon)

The compound cinnamaldehyde triggers lysosomal activity and autophagy, enhances the state of beneficial gut microorganisms, and reduces blood levels of IL-6 and TNF-α. Based on preclinical analyses, cinnamon can mitigate oxidative stress and protect neuronal function, which may help control neuroinflammation, as stated by Yan et al. (2024).

### 8.3.4.Thymoquinone (Black Seed, Nigella sativa)

The synthesis of SCFAs by thymoquinone strengthens the gut barrier and inhibits the release of peripheral inflammatory factors. This intervention is associated with lower CNS inflammation and enhanced neuroprotection in laboratory studies (Rashwan et al., 2023).

### 8.3.5. Myristicin (Nutmeg)

Myristicin has antioxidant and autophagy-promoting capacities and regulates the ratio of gut flora while limiting microglial overactivity and promoting neuronal restoration in inflammatory conditions (Fu et al., 2025).

### 8.3.6. Crocin and Safranal (Saffron)

Crocin and safranal protect neurons by inhibiting NLRP3 signaling and stimulating neuronal autophagy, which results in the clearance of toxic protein aggregates. Outcomes of meta-analyses support saffron's role in reducing depressive symptoms and inflammation in patients with major depressive disorder (Yang et al., 2023).

### 8.3.7.Cardamom Essential Oils

Cardamom oil encourages positive transformations in the gut microbiome, reduces oxidative stress levels, and induces autophagy in neurons. Increased cognitive performance in Alzheimer's rodent models and better maintenance of synaptic function were results of cardamom supplementation (Auti & Kulkarni, 2019).

## 8.4. Conclusion and Future Directions

These nutraceuticals demonstrate a coordinated strategy for controlling neuroinflammation (modulation of the gut–brain axis and autophagy) occurring simultaneously. Full-scale clinical trials are necessary to establish guidelines regarding the optimal dosing and integration of these phytocomplexes into more general neuroprotective approaches.

Neuroprotection in nutraceuticals containing these phytochemicals is achieved by restoring the diversity of gut bacteria, minimizing systemic inflammation, and promoting increased autophagy. Through their collective effort, nutraceuticals maintain gut–brain axis function (Adamu et al., 2024), protect the blood–brain barrier from dysregulation, and promote neuronal and cognitive robustness (Sarb et al., 2024). Greater knowledge of the potential of such nutraceuticals as therapeutics would entail exploring the molecular pathways through which they modulate autophagy and inflammation.

## 8.5. Mechanisms of Action:

Activation of autophagy and inhibition of inflammation through nutraceutical intervention. Nutraceuticals provide therapeutic benefits by affecting key cellular pathways:

### 8.5.1. Activation of AMPK:

Crucial bioactive elements, such as curcumin from turmeric and hydroxytyrosol from olive oil, activate AMPK—the main energy-sensitive enzyme, as observed by Papadopoulou et al. (2024)—resulting in increased autophagy and reduced neuroinflammation (Zhang et al., 2024).

### 8.5.2. NF-κB and NLRP3 Inflammasome Antagonism:

Zhu et al. (2024) demonstrated that curcumin and hydroxytyrosol can inhibit NF-κB and the NLRP3 inflammasome, leading to lower levels of cytokines and decreased inflammation, as previously indicated by Papadopoulou et al. (2024).

### 8.5.3. Promotion of SCFA Production:

Cinnamon and black seed contain prebiotics that promote the production of short-chain fatty acids like butyrate (Rashwan et al., 2023), benefiting gut barrier health and reducing systemic immune activation (Yan et al., 2024).

In conclusion, these mechanisms suggest that nutraceuticals provide a comprehensive approach to the neuroinflammatory response and autophagy stimulation (Crocetta et al., 2024). Through the maintenance of gut stability and reduction of inflammatory factors, they contribute to neuronal viability and optimal cognitive performance (Zhang et al., 2024). The clinical case studies depicted in Table 10 provide additional support for using nutraceuticals as a means of managing both neurodegenerative and neuropsychiatric disorders simultaneously.

*Table 10.*

| Study | Focus | Key Findings | Natural Substance | Reference |
|---|---|---|---|---|
| **Study 1** | Curcumin in Alzheimer's Disease | Curcumin supplementation (90 mg twice daily for 18 months) reduced amyloid-beta plaques, increased autophagic flux, decreased oxidative stress, and inhibited NF-κB. | Curcumin | Wei, et al. (2024) |
| **Study 2** | Resveratrol in Parkinson's Disease | Resveratrol activated AMPK-mTOR signaling, promoted autophagy-mediated alpha-synuclein degradation, reduced IL-6/ TNF-α, and improved motor function in animal models. | Resveratrol | Li, et al. (2024) |
| **Study 3** | Olive Oil Polyphenols in Cognitive Decline | Mediterranean diet with EVOO improved memory and executive function while modulating neuroinflammation via NLRP3 inhibition and enhancing autophagy through SIRT1 activation. | Olive Oil Polyphenols | Valls-Pedret et al. (2015) |
| **Study 4** | Black Seed in Multiple Sclerosis | Black seed extract improved fatigue and cognitive function while decreasing IL-1β and TNF-α levels. Thymoquinone activated SIRT1-AMPK pathways, reducing neuronal damage. | Nigella sativa (Black Seed) | Nikvarz et al. (2024) |
| **Study 5** | Saffron and Depression | Saffron's crocin and safranal reduced depression by modulating serotonergic pathways, decreasing inflammation, and enhancing autophagy and oxidative stress response. | Saffron | Marrone, et al.,2024 |

This structured representation underscores the therapeutic potential of nutraceuticals in managing neuroinflammation and promoting autophagy, with documented evidence supporting their role in neurodegenerative and psychiatric conditions. These findings indicate that nutraceuticals hold promising prospects for modulating neuroinflammation and influencing autophagy. They also highlight the importance of incorporating 'phytocomplexes' into dietary approaches for enhancing neuropro tection and cognitive function

## 9. COMPARATIVE THERAPEUTIC STRATEGIES TO ENHANCE AUTOPHAGY AND MITIGATE NEUROINFLAMMATION IN NEURODEGENERATIVE DISORDERS

Different treatment approaches have emerged to boost autophagy and reduce neuroinflammation in neurodegenerative conditions. The different therapeutic strategies function through separate pathways yet share the mutual purpose of boosting neuronal resistance and slowing disease advancement. Pharmacological agents focus on intracellular signaling pathways while gut–brain axis modulation together with nutraceuticals work on cellular homeostasis across the body (Table 11).

*Table 11. Approaches for enhancing autophagy in neurodegenerative diseases.*

| Therapeutic Approach | Description | Clinical Evidence/Study | Reference |
|---|---|---|---|
| **Pharmacological Treatments** | mTOR inhibitors (e.g., rapamycin) and lysosomal agents enhance autophagic clearance. | Improved autophagy and reduced neuroinflammation in Alzheimer's. | Xie, et al., 2024 |
| **Probiotics** | Live beneficial microorganisms that support gut health and boost autophagy. | Enhanced autophagic turnover in neurodegenerative disorders. | Kumar, et al., 2025 |
| **Prebiotics** | Nondigestible food ingredients that stimulate beneficial gut bacteria for improved autophagy. | Improved gut microbiota and autophagy in neurodegenerative diseases. | Sobral, et al., 2025 |
| **Postbiotics** | Byproducts from probiotic fermentation aiding in autophagy and reducing inflammation. | Improved autophagy in Alzheimer's patients. | Hijová (2024.) |
| **Synbiotics** | Combination of probiotics and prebiotics to enhance gut health and autophagy. | Enhanced gut-brain health and increased autophagic activity. | Wang, et al., 2025 |
| **Antibiotics and Gut-Brain Axis Modulation** | Modulation of gut-brain axis and reduction in dysbiosis to enhance autophagy. | Increased autophagic flux in multiple sclerosis patients. | Xin, et al., 2024 |
| **Nutraceutical Approaches** | Antioxidants and polyphenols (e.g., EGCG, fisetin, resveratrol) enhance autophagy. | Reduced neuroinflammation and cognitive decline in Alzheimer's. | Mamun, et al., 2024 |
| **Physical Activity/ Lifestyle** | Regular physical activity and a healthy lifestyle promote autophagy. | Improved autophagic processes and reduced inflammation. | Wu, et al., 2019 |

Several treatment strategies for improving autophagy in neurodegenerative diseases are listed in this table. Clinical data demonstrate the efficacy of each strategy in enhancing autophagic functions and reducing neuroinflammation. These methods show promise in managing and slowing disease progression in neurodegenerative conditions.

## 10. COMPARATIVE ANALYSIS AND FUTURE PERSPECTIVES

Studies have shown that autophagy intervention functions as an adaptable method for combating neurological inflammation while resolving cellular dysfunctions. New therapeutic strategies based on gut–brain axis management through probiotics and prebiotics, alongside synbiotics and nutraceuticals, provide comprehensive management benefits in addition to the precise acute targeting of mTOR inhibiting pharmaceuticals. Implementations at systemic levels enhance cellular durability throughout both CNS tissues and across the gut–brain interface, which reduces inflammation at multiple levels throughout the body.

Current research indicates that pharmacological agents can work in unison with microbiota-focused therapies and lifestyle adjustments to amplify autophagic flux, providing multilayered defense against progressive neurodegeneration. Tailored therapeutic approaches that match pathophysiological profiles and rely on autophagic activity biomarkers and inflammatory markers should lead to the best possible patient outcomes. The advancement of research must focus on developing standardized therapeutic protocols and optimizing strategies to protect the nervous system through autophagy modulation.

After exploring nutraceutical approaches through autophagy and neuroinflammation modulation of the gut–brain axis with phytocomplex interactions, researchers discovered intermittent fasting (IF) as a strong lifestyle-based intervention. Dietary compounds impact autophagy and neuroinflammation through specific molecular targets (curcumin and thymoquinone), whereas IF creates extensive metabolic changes that modify energy-sensing pathways along with autophagic flux and microbial balance. Homeostasis enhancement and inflammation reduction emerge as shared goals of both interventions; yet, IF provides unique timing control of autophagy, establishing a periodic resetting of physiological processes. The analysis presents mechanistic foundational elements and therapeutic applications of IF treatment as a solution for aging issues, along with neuroinflammation and neurodegeneration.

## 11. INTERMITTENT FASTING AND AUTOPHAGY: THERAPEUTIC POTENTIAL IN AGING, NEUROINFLAMMATION, AND NEURODEGENERATION

Intermittent fasting (IF) and nutrient starvation are gaining increasing attention as promising approaches for stimulating autophagy, which is an essential cellular mechanism for maintaining homeostasis. Autophagy is especially effective in controlling age-dependent cellular damage, toxic protein deposition, and inflammation, which are major factors in the development of neurodegenerative disorders. Fasting alters autophagy through molecular targets of the diet, which makes it an exciting approach to address aging and neuroinflammatory diseases (Diab, et al.,2024).

## 11.1 Autophagy and Fasting

Here, we describe various ways through which autophagy is induced during fasting. Fasting has been shown to activate autophagy, as it inhibits the mechanistic target of rapamycin (mTOR), a well-known negative regulator of autophagy. This inhibition activates Unc-51-like kinase 1 (ULK1) and begins the process of phagophore assembly for the destruction of damaged organelles and certain protein aggregates (Minciuna, et al.,2022). The present work highlights the preclinical use of fasting in increasing autophagic flux, improving cell health, and extending lifespan.

- **Aging**: Research in *C. elegans* and *D. melanogaster* has revealed a positive effect on lifespan and stress tolerance via the restoration of misfolded proteins through lysosome-dependent pathways. For example, DIF in Drosophila extends lifespan and reduces neurodegeneration by increasing neuronal autophagy (Berk, et al., 2024).
- **Neuroinflammation:** Defective autophagy exacerbates neuroinflammation, a characteristic of neurological disorders such as Alzheimer's disease. Dietary strategies such as time-restricted feeding (TRF) and alternate-day fasting (ADF) have been found to reduce inflammation and enable autophagy, in addition to eliminating neurotoxic aggregates (Dong, et al.,2024).

## 11.2. Neurodegenerative Diseases and Autophagy

### Alzheimer's Disease (AD)

When autophagy is altered in AD, it leads to the formation of Aβ plaques and tau tangles that increase neuroinflammation. Another study confirmed that fasting promotes lysosomal activity, thereby decreasing the formation of Aβ and improving learning ability. It has also been shown that TRF in mouse models of AD decreases neuroinflammation and cognitive decline (Zhang, et al., 2022).

### Parkinson's Disease (PD)

In other models of PD, fasting cycles promote a cleanup process that ameliorates aggregated alpha-synuclein proteins. Research has shown that periodic fasting benefits the preservation of dopaminergic neurons, enhances motor ability, and reduces neuroinflammation in mouse models (Tosefsky, et al., 2024).

## Multiple Sclerosis (MS)

MS is appropriately known for persistent inflammation and myelin destruction in the CNS. Fasting-mediated autophagy serves as a protective mechanism against inflammation and promotes the clearance of cellular waste products, contributing to myelin repair and optimal neuronal function (Langley et al.,2020).

## 11.3. Gut–Brain Connection and Neuroinflammation

The gut microbiota is involved in the pathogenesis of neuroinflammatory diseases through a network known as the gut–brain axis. An imbalance in the mucosal microbiota increases inflammation levels and impairs autophagy. Periodic fasting contributes to microbial homeostasis and increases the generation of SCFAs, which act as signaling molecules in autophagy pathways. Clinical trials, such as a 16:8 fasting regimen, demonstrated that an 8-day fasting approach in metabolic syndrome patients was associated with diminished inflammatory biomarkers and increased autophagy in immune cells (Mirrazavi, & Behrouz, 2024).

## 11.4 A View of the Main Differences in Fasting Protocols

To understand the varying impacts of fasting regimens on autophagy and neuroinflammation, the following strategies are highlighted. Several approaches to fasting have been discussed concerning their effects on autophagy and neuroinflammation. Moderate activation of autophagy is achieved with a 12-hour fast performed daily (Ma, 2024). This duration of fasting enhances the recycling process in cells and reduces inflammation, overall improving cellular health.

The 16-hour fast, commonly referred to as the 16:8 method, entails fasting for a total of 16 hours and then eating during an 8-hour window. This strategy increases autophagy and significantly decreases neuroinflammatory cytokines for neuropathology, including the cytokines IL-6 and TNF-α, which are linked to neurodegenerative diseases (Sukkriang, & Buranapin, 2024).

The 5:2 diet, which consists of five days of normal meals and two days of low-calorie meals, provides balanced activation of autophagy. This method is useful for inhibiting systemic and neuronal inflammation while enhancing cellular health (Brogi, et al., 2024).

ADF, which involves eating on one day followed by fasting on the next day, particularly triggers autophagy. This strategy has been reported to block neuroinflammation and promote cognitive resilience, thus serving as a potential key strategy for neuroprotective activity (Beveridge, et al., 2025).

Therefore, daily intermittent fasting, which involves eating a meal within an 8-hour window and fasting for 24 hours on the seventh day, strongly induces autophagy. This caloric restriction pattern minimizes neuroinflammation and protects neurons, suggesting that it can be adopted as a therapy for neurocognitive disorders and to improve brain function (Ren, et al., 2024).

## 12. MECHANISTIC INSIGHTS: "HUNGRY CELLS" AND CELLULAR TRASH BAGS – APPLICATION OF THE FAST THEORY PROTOCOL

Self-degradation, or autophagy, is an evolutionarily conserved process involved in regulating cellular quality control by eliminating damaged organelles, misfolded proteins, and other waste products. It is especially essential in neurodegenerative disorders, as impaired autophagy contributes to the accumulation of toxic proteins such as amyloid-beta and tau in Alzheimer's disease or alpha-synuclein in Parkinson's disease (Hussain et al., 2024) . These aggregations further augment neuroinflammation, oxidative stress, and neuronal loss.

The Fasting, Autophagy, and Stress Tolerance (FAST) theory outlines a rationale for applying structured fasting-induced autophagy to promote therapeutic treatments. The following is an expanded form of the FAST theory that aims to emphasize the implementation of the concept of promoting autophagy as a potential therapy for neurodegenerative disorders.

### 12.1. Objective of the FAST Theory Application Protocol

Slight increases in basal autophagy rates during fasting increase cell turnover, reduce inflammation in the brain, and halt neurodegenerative diseases.

#### 12.1.1 Preparation Phase

Possible investigations prior to the procedure include cognitive status tests, such as the Mini-Mental State Examination (MMSE) or Montreal Cognitive Assessment (MoCA); neuroinflammatory biomarkers, such as IL-6, TNF-α, and CRP; and radiological imaging, such as magnetic resonance imaging (MRI) or positron emission tomography (PET). The portion sizes should be reduced over a week to prepare individuals for fasting programs. Before screening, it is highly advisable to rule out patients with diabetes, eating disorders, or relatively unstable cardiovascular status (Wang, et al., 2022).

### 12.1.2. Implementation Phase

Extreme fasting regimens include intermittent fasting (with a 6-- to 8-hour eating period and a 16-- to 18-hour fasting period) alternating with normal eating, fasting for 2 days a week (with only 500 calories), and periodic fasting once a week with 24--48 hours of restricted food intake. The level of autophagy activity can also be determined through other biochemical analyses, such as those of LC3 II, Beclin 1, and p62. Differences in intervention efficacy have been analyzed on the basis of modifications in neuroinflammatory biomarkers such as IL-6 and TNF-α (Longo, et al., 2021).

## 12.4. Effectiveness of Different Types of Therapy

The consumption of foods and drinks high in omega-3 fatty acids, polyphenols, antioxidants, and ketogenic foods after fasting is recommended. Moreover, pharmacological promoters of autophagy, such as rapamycin (Kim et al., 2021), spermidine (Hofer, et al., 2024) and metformin (Zamanian), have been used quickly. Et al., 2024) to induce autophagy and enhance mitochondrial quality and engage in aerobic or resistance exercise training.

## 12.5. Evaluation Phase

Subsequent examinations consisted of cognitive status, neuroimaging features, and serum levels of inflammatory cytokines. Mental condition status, including cognitive status, should be screened via checklists such as the MoCA and MMSE before and after a treatment plan, with mood assessed before and after a treatment plan using the Hamilton Depression Rating Scale (HDRS). Assess biomarkers of the overarching processes of autophagy and neuroinflammation. Specific precautions include routine safety checks for side effects such as fatigue, electrolyte imbalance, and hypoglycemia (Andreatta Maduro, et al., 2024).

## 12.6. Long-Term Maintenance

The fasting times must be tailored to suit the individual's lifestyle to ensure lasting compliance. Inform participants of the need to adhere to certain dietary guidelines and recommend appropriate exercises. Meetings were scheduled after 3 to 6 months to adjust to fasting regimens on the basis of previous results related to fasting duration and frequency (O'Nea et al., 2023).

The FAST theory provides a framework for integrating fasting and autophagy activation into clinical interventions, offering potential benefits for neurodegenerative disease management. When combined with pharmacological agents, lifestyle modifications, and nutritional strategies, fasting can significantly increase cellular clearance and mitigate neuroinflammation. Future research should explore individualized approaches and long-term impacts to optimize therapeutic outcomes.

Thus, the implementation of structured fasting regimens allows for the restoration of cellular autophagy while simultaneously improving cognitive resilience and emotional regulation, alongside enhanced memory performance, to create a comprehensive therapeutic approach for neurodegenerative disorders.

New biotechnological exploration of neuronal organoids and 3D-printed systems enhance our understanding of autophagy and neuroinflammation mechanisms while lifestyle interventions like intermittent fasting provide therapeutic autophagy modulation. The forthcoming section examines how these innovations transform both research approaches and therapeutic methods in neurodegenerative disorders.

## 13. INNOVATIVE RESEARCH TECHNIQUES IN AUTOPHAGY: EXPLORING NEUROINFLAMMATION AND NEURODEGENERATION

The development of advanced research techniques has significantly improved our understanding of neuroinflammation and neurodegenerative disorders, particularly due to recent advances in autophagy mechanisms. By using neuronal organoids, 3D-printed models, and other in vitro or ex vivo methods, scientists have overcome substantial limitations in the study of autophagy mechanisms and the treatment of diseases.

### 13.1. Autophagic Flux and AD Studied Using Neuronal Organoids

Neuronal organoids are derived from hPSCs with self-assembly and morphogenic properties that produce three-dimensional tissues. Organoids recreate important aspects of the brain, such as cellular organization, thus providing a better environment for studying autophagy. Neuronal organoids more closely simulate the ongoing brain environment than conventional two-dimensional cultures, enabling scientists to test disease development and responses to treatment under conditions resembling human biology (Xu et al., 2023).

The assessment of autophagic flux, which includes the development, maturation, and circuit-breaking effects of autophagosomes, relies on the use of PCBs, as reported by Yan et al. (2023). Hou, Syn, Qian, et al. derived organoids from AD patients carrying PSEN1 mutations and associated the resultant outcomes with diminished autophagic flux and an accumulation of extracellular amyloid plaques (Maksour et al., 2024).

In contrast, exposure to autophagy-promoting drugs such as rapamycin, which has been shown to alleviate these issues by decreasing protein aggregates, further exemplifies the application of using organoid systems to test numerous compounds.

### 13.2. Fasting-Induced Modulation of Neuronal Autophagy

Investigators have now revealed fasting and dietary interventions as new means of manipulating autophagy in neuronal organoids. According to conclusions reached by Huynh et al. (2024), intermittent fasting stimulated autophagy, resulting in reduced tau levels in neuronal organoids. These findings suggest that dietary strategies could serve as an effective, drug-free approach to treating neuroinflammatory and neurodegenerative disorders.

### 13.4. Case Study: ALS and TDP-43 Pathology

TDP-43 protein aggregation was investigated in an experiment conducted by Tamaki et al. (2023), utilizing organoids derived from ALS patients. Previous studies have demonstrated that both the initiation of autophagy and lysosomal degradation, key processes responsible for neurocellular demise, are underperforming in organoids subjected to experimental treatments. Organoids-on-a-chip, when treated with rapamycin, exhibited a decreased number of protein aggregates and reduced neuronal death, indicating the efficacy of these organoids for identifying therapeutic approaches (Han et al., 2024).

## 14. 3D-PRINTED MODELS AND PERSONALIZED NEUROINFLAMMATION RESEARCH

A central method in this process, bioprinting, enables the creation of elaborate 3D models that resemble intricate human tissue structures. Utilizing these models, researchers gain the ability to personalize experiments, which serves as a powerful tool for studying disease mechanisms— including autophagic imbalances— in physiologically relevant three-dimensional frameworks (Orcheston-Findlay et al., 2021).

## 14.1. Neuroinflammatory Applications

The use of 3D-printed models promises new advancements in biomedical science, particularly concerning the adaptability and accuracy of the models. Such models are very effective in replicating disease-associated signaling pathways and provide investigators with new platforms to explore the intricate cellular interactions, particularly neuroinflammation and neurodegenerative disorders (Heinrich et al., 2024).

## 14.2. Personalized Disease Models

The use of artificial cellular structures built with brain-like substrates, such as 3D printed models, enables the observation of crucial processes that occur during development. Through the work of Logan et al. (2019), it was shown that 3D brain-like scaffolds provide a means to image autophagy, a primary process with applications for studying disease. Currently, hiPSC-derived organoids are regarded as the best approach for producing models of neurodegenerative diseases. When organoids are combined with 3D-printed matrices, they can reproduce significant neurophysiological niches, including communication between neurons and microglia, as well as between neurons and astrocytes (Mrza et al., 2024).

## 14.3. Clinical Case Studies Unfolding the Implementation of Established Ideas

### 14.3.1. Alzheimer's Disease (AD):

Liao et al. (2025) designed a three-dimensional printed amyloid-beta scaffold to analyze the formation of plaques and their effects on neighboring neuronal cells. This model offers an opportunity to observe how microglia interact with plaques and subsequently stimulate neuroinflammation, as reported by Balestri et al. (2024).

### 14.3.2. Parkinson's Disease (PD):

Amartumur et al. (2024) used 3D-printed dopaminergic neuronal scaffolds to determine how α-synuclein proteins aggregate. Enhanced autophagic breakdown of aggregates in the scaffold was observed due to environmental factors, thus demonstrating the applicability of autophagy as a therapy for PD.

### 14.3.3. Multiple Sclerosis (MS):

Hyung et al. (2021) applied 3D-printed myelin-like hydrogels to study demyelination and remyelination processes. This configuration simulated immune-oligodendrocyte interactions, critical for studying myelin repair and neuroinflammation.

### 14.3.4. Challenges and Limitations

Despite their significant potential, 3D-printed models face several limitations that hinder their further application. The high cost of equipment, the need for expert technical expertise, and the challenges in replicating the extracellular matrix (ECM) of the human brain remain major barriers (Liao et al., 2025; Yang et al., 2024). Complicating matters further is the difficulty in differentiating actual physiological responses from the simplified environment of in vitro models, indicating the need for additional animal studies (Kim et al., 2025).

## 14.4. Future Directions and Conclusions

New biomaterials, such as ECM-simulating smart materials, along with the incorporation of AI and automated design, are expected to alleviate existing challenges (Kantaros & Ganetsos, 2024). The 3D printing of neuroinflammation models holds great potential for a better understanding of disease mechanisms and testing new therapies (Logeshwaran et al., 2024). However, valuable data from these models is advancing our understanding of autophagy's role in neurodegenerative ailments.

# 15. APPLICATION OF IN VITRO AND EX VIVO SYSTEMS IN NEUROINFLAMMATION RESEARCH

The understanding of neuroinflammation and its mechanisms, including autophagy, has become clearer when in vitro and ex vivo models are employed. These models enable researchers to study cellular and molecular changes in simulated environments and assist in investigating neurodegenerative disorders, mood disorders, and POCD.

The next section explores the effects of these systems on autophagy proteins and neuroinflammatory pathways, as well as the use of withaferin A as a novel drug.

## 15.1. Deciphering Neuroinflammation: In Vitro Systems and Their Impacts on Autophagic Pathways

In vitro systems use cells in culture, especially in a two-dimensional environment, to study their biological actions separately. These systems are widely used because they are easy to implement, cost-effective, and allow for control over certain aspects. The key applications of in vitro systems include the following:

### 15.1.1 Autophagy Studies:

In vitro systems are critical for analyzing autophagic flux, which is the dynamic process of forming autophagosomes followed by degradation. For example, withaferin A, which is isolated from the root of Withania somnifera, has been found to elicit an autophagic response in cultured neuronal cells (Saha, et al 2024). Limanaqi et al. (2019) further demonstrated that withaferin A reduces neuroinflammation by activating the AMPK pathway for the autophagic degradation of misfolded proteins (Vilaboa, et al., 2023).

### 15.1.2. Lipid Droplets (LDs) and Neuroinflammation:

Both inflammatory cells and neurons display lipid bodies in response to stress, the significance of which remains unclear. It has also been shown in vitro that LDs may release inflammatory mediators and store them. In cultured microglia (Mallick, et al., 2024). Lee et al. (2021) reported that autophagy affects the turnover of LDs to decrease the secretion of proinflammatory cytokines, including IL-6 and TNF-α.

## 15.2. Mood and Neurodegenerative Disorders

Cell culture systems have been used to study the involvement of autophagy in depression and neurodegenerative disorders (Sánchez-Vidaña, et al.,2023). As explained by Gan et al. (2024), when autophagy enhancers protect neurons in models of depression by limiting oxidative stress and inflammation, such improvements indicate potential applications in depression treatment.

# 16. THE SIGNIFICANCE OF EX VIVO SYSTEMS IN EXAMINING NEUROINFLAMMATION AND AUTOPHAGY

Ex vivo systems encompass isolated organ or tissue samples that retain more intact structural functionality than traditional culture systems do (Monzani, et al,2022), making them more useful for the analysis of neuroinflammatory events. These systems aim to fill the gap between two-dimensional in vitro models and three-dimensional in vivo models, as well as to aid in understanding the behavior of cells in their appropriate microenvironment.

## 16.1. Autophagic Processes in Native Tissues

Studies in brain slices or using isolated hippocampi allow the assessment of autophagic flux in a less manipulated system than in isolated neurons or cell lines (Croft. For example, Hu et al. (2024) cultured organotypic hippocampal slices from rats to study how withaferin A lowers inflammation during postoperative cognitive dysfunction. This study revealed two main findings: the substance helps lower body inflammation and boosts self-cleaning processes.

## 16.2. Postoperative Cognitive Dysfunction (POCD)

POCD is best defined as a neuroinflammatory illness with a postoperative and anesthesia background (Zhang et al 2024). In vitro and ex vivo studies have shown that autophagy can help reduce neuroinflammation-associated damage (Zhang, et al., 2024). In their work on isolated rat brain slices, Peng et al. (2025) reported that improving autophagic function can minimize microglial and astrocyte activation and thus represents a new target for POCD intervention.

# 17. APPLICATIONS IN NEURODEGENERATIVE AND MOOD DISORDERS

With respect to neurodegenerative and mood disorder applications, through ex vivo and laboratory culture technology, scientists have made better discoveries about mood and neurodegenerative disorders.

## 17.1. Neurodegenerative Disorders

Parkinson's and Alzheimer's diseases grow from protein piles and ongoing brain inflammation. The requirement for in vitro models comes from studies that test patient-derived iPSCs and show disease-linked autophagic problems (Jusop, et al., 2023).

## 17.2. Depression and Mood Disorders

Data show that depression disrupts the ability of autophagy to repair cells. A research team led by Gukovskaya et al. (2022) reported that the pharmacological control of autophagy enhances mitochondrial performance while lowering the levels of inflammatory cytokines. For that reason, both in vitro and ex vivo systems are methods of significant importance in the study of neuroinflammation, providing valuable insights into the role of and potential therapeutic targets for autophagy. However, while it is possible to accurately model the activity of cells at the molecular level in vitro, ex vivo systems reflect the global interaction within a tissue. The development of these models, especially with the understanding of more reagents, such as withaferin A, has helped provide more information on neuroinflammation and its role in neurodegenerative diseases, mood disorders, and POCD.

# 18. CONCLUSION AND FUTURE PROSPECTS

In this chapter, researchers investigate the intricate relationship between autophagy and neuroinflammation, along with neuropsychiatric disorders, by explaining advanced research methodologies that facilitate deeper knowledge acquisition. The essential cellular mechanism of autophagy functions as a protection system and regulatory network to sustain neuronal health when cells experience stress and disease conditions. The pathway for autophagy maintenance enables the cellular removal of damaged elements and protein aggregates; however, disruptions caused by oxidative stress, protein misfolding, or systemic inflammation initiate neuroinflammatory responses that underlie disorders such as Alzheimer's disease (AD), Parkinson's disease (PD), and mood disorders.

The study of the effectiveness of SARS-CoV-2 (COVID-19) viral infection in activating neuroinflammation by damaging autophagic controls represents a new research emphasis highlighted throughout this chapter. Research findings confirm that viral interference with the autophagic system results in a reduced ability to eliminate damaged organelles and viral particles, leading to intensified inflammatory responses. Medical professionals must consider targeting autophagy mechanisms, as they show promise in serving as a new therapeutic approach for both persistent neurodegenerative conditions and acute viral pathogens that lead to enduring neuropsychiatric complications.

This section explores the vital relationships that exist between metabolic conditions, including diabetes, and neuroinflammatory processes. Prolonged high blood sugar levels, along with insulin resistance, create long-standing inflammation throughout the entire body that penetrates the CNS, triggering microglial activation and damaging brain cells. The dysregulation of autophagy in diabetic patients induces the accumulation of damaged cellular components, which intensifies neuroinflammatory processes. These relationships emphasize the need for healthcare solutions that target both metabolic aspects and neurological components to reduce disease progression.

This chapter highlights the vital therapeutic role that nutritional strategies play in controlling autophagy and inflammation. The activation of autophagic pathways through mTOR signaling inhibition and increased AMPK activity occurs when individuals experience intermittent fasting, cellular starvation, and dietary strategies involving caloric restriction. These mechanisms simultaneously clear cellular damage while controlling proinflammatory cytokine levels, producing combined advantages for tissue health and the whole-body inflammatory response. The practice of fasting leads to enhanced autophagic effects, which increase therapeutic outcomes in models of postoperative cognitive dysfunction (POCD) and neurodegenerative diseases by minimizing oxidative damage while protecting brain cells.

The gut–brain axis enhances this explanation. Research has described how dietary fibers combined with probiotics shape gut microbiota communities that produce short-chain fatty acids (SCFAs), which direct neural pathways for autophagy and inflammatory responses. Research shows that Lactobacillus and Bifidobacterium strains act as probiotics by controlling neuroinflammation through their ability to prevent microglia and astrocytes from flaring. Both cell types function as crucial elements during inflammatory responses. New treatment pathways emerge from uniting nutrition science with systemic health maintenance and neurologic well-being, thus becoming a bridge for therapeutic discoveries.

The development of organotypic brain slice cultures, along with 3D-printed tissue models and advanced imaging systems, has become central to our understanding of autophagy and neuroinflammation. Total control of disease mechanisms becomes possible through laboratory systems that replicate actual physiological conditions. The dynamic observation of autophagic flux and inflammatory responses has become feasible through the development of 3D-printed biological models. The potential of these novel technologies remains constrained by cost limitations, duplication concerns and scale-up obstacles until viable solutions can be developed for clinical use.

This chapter highlights the complex interplay between autophagy and neuroinflammation and presents a strong foundation for studying their responses to viral infections, metabolic problems, and nutrition-based interventions. This chapter aligns with the book's goal of building actionable strategies by integrating cellular biology with technological approaches and dietary changes to manage neuropsychological disorders. Autophagic therapeutic approaches show great promise by offering potential solutions for the simultaneous challenges of neuroinflammation and cellular dysfunction. This chapter serves as groundwork for future investigations and healthcare developments by emphasizing the value of both integrated and cross-disciplinary methods in confronting the causes of neuropsychiatric disorders. Future research should strive to establish individualized therapeutic approaches aimed at modulating specific autophagy-regulating pathways in major depressive disorder, bipolar disorder, and schizophrenia. Longitudinal studies are also required to properly understand the long-term effects of fasting and nutritional interventions on autophagy, neuroinflammation, and clinical outcomes in these conditions. Moreover, studying the specific mechanisms by which viruses undermine autophagy and continuously developing targeted antiviral treatments that promote autophagic removal are vital. Collaborative efforts among researchers, clinicians, and policymakers are necessary to translate the negative findings in autophagy and neuroinflammation research into appropriate clinical interventions for these detrimental neuropsychiatric disorders. Moreover, the significance of this research aligns with the overall scope of the UN's Sustainable Development Goals (SDGs). Specifically, improving our understanding of neuropsychiatric disorders, which is a focus of the chapter directly meets SDG 3. Good Health and Well-being. The focus on nutritional interventions and the use of the gut-brain axis correlate with SDG 2: Zero Hunger and SDG 12: Responsible Consumption and Production by discussing the importance of healthy diets for the brain. The SDG 9 benefits from the development of novel research techniques, including organoids and 3D-printed models. Industry, Innovation, and Infrastructure. Finally, integrated and interdisciplinary appeals emphasize the need for SDG 17: Partnerships for the Goals for effectively combating the multidimensional problems of neuropsychiatric conditions.

## REFERENCES


Abdoullah, B., Durand, G. A., Basco, L. K., El Bara, A., Bollahi, M. A., Bosio, L., Geulen, M., Briolant, S., & Boukhary, A. O. M. S. (2023). Seroprevalence of Alphaviruses (Togaviridae) among Urban Population in Nouakchott, Mauritania, West Africa. *Viruses, 15*(7), 1588. DOI: 10.3390/v15071588 PMID: 37515274

Abou-Saleh, H., Younes, N., Rasool, K., Younis, M. H., Prieto, R. M., Yassine, H. M., Mahmoud, K. A., Pintus, G., & Nasrallah, G. K. (2019). Impaired liver size and compromised neurobehavioral activity are elicited by chitosan nanoparticles in the zebrafish embryo model. *Nanomaterials (Basel, Switzerland), 9*(1), 122. DOI: 10.3390/nano9010122 PMID: 30669437

Abrignani, V., Salvo, A., Pacinella, G., & Tuttolomondo, A. (2024). The Mediterranean Diet, Its Microbiome Connections, and Cardiovascular Health: A Narrative Review. *International Journal of Molecular Sciences, 25*(9), 4942. DOI: 10.3390/ijms25094942 PMID: 38732161

Absalyamova, M., Traktirov, D., Burdinskaya, V., Artemova, V., Muruzheva, Z., & Karpenko, M. (2025). Molecular basis of the development of Parkinson's disease. *Neuroscience, 565*, 292–300. DOI: 10.1016/j.neuroscience.2024.12.009 PMID: 39653246

Adamu, A., Li, S., Gao, F., & Xue, G. (2024). The role of neuroinflammation in neurodegenerative diseases: Current understanding and future therapeutic targets. *Frontiers in Aging Neuroscience, 16*, 1347987. DOI: 10.3389/fnagi.2024.1347987 PMID: 38681666

Alonaizan, R. (2024). Molecular regulation of NLRP3 inflammasome activation during parasitic infection. *Bioscience Reports, 44*(5), BSR20231918. Advance online publication. DOI: 10.1042/BSR20231918 PMID: 38623843

Alpuche-Lazcano, S. P., Scarborough, R. J., & Gatignol, A. (2024). MicroRNAs and long noncoding RNAs during transcriptional regulation and latency of HIV and HTLV. *Retrovirology, 21*(1), 5. DOI: 10.1186/s12977-024-00637-y PMID: 38424561

Alqahtani, S., Xia, L., & Shannahan, J. H. (2022). Enhanced silver nanoparticle-induced pulmonary inflammation in a metabolic syndrome mouse model and resolvin D1 treatment. *Particle and Fibre Toxicology, 19*(1), 54. DOI: 10.1186/s12989-022-00495-6 PMID: 35933425

Amartumur, S., Nguyen, H., Huynh, T., Kim, T. S., Woo, R. S., Oh, E., Kim, K. K., Lee, L. P., & Heo, C. (2024). Neuropathogenesis-on-chips for neurodegenerative diseases. *Nature Communications*, *15*(1), 2219. DOI: 10.1038/s41467-024-46554-8 PMID: 38472255

Amor, S., Peferoen, L. A., Vogel, D. Y., Breur, M., van der Valk, P., Baker, D., & van Noort, J. M. (2014). Inflammation in neurodegenerative diseases—An update. *Immunology*, *142*(2), 151–166. DOI: 10.1111/imm.12233 PMID: 24329535

Andreatta Maduro, P., Guimarães, M. P., de Sousa Rodrigues, M., Pereira Rolim Coimbra Pinto, A. P., da Mota, A. A.Junior, Lima Rocha, A. S., Matoso, J. M. D., Bavaresco Gambassi, B., & Schwingel, P. A. (2024). Comparing the Efficacy of Two Cognitive Screening Tools in Identifying Gray and White Matter Brain Damage among Older Adults. *Journal of Aging Research*, *5527225*, 1–14. Advance online publication. DOI: 10.1155/2024/5527225 PMID: 38690079

Auti, S. T., & Kulkarni, Y. A. (2019). Neuroprotective Effect of Cardamom Oil Against Aluminum Induced Neurotoxicity in Rats. *Frontiers in Neurology*, *10*, 399. DOI: 10.3389/fneur.2019.00399 PMID: 31114535

Balestri, W., Sharma, R., da Silva, V. A., Bobotis, B. C., Curle, A. J., Kothakota, V., Kalantarnia, F., Hangad, M. V., Hoorfar, M., Jones, J. L., Tremblay, M.-È., El-Jawhari, J. J., Willerth, S. M., & Reinwald, Y. (2024). Modeling the neuroimmune system in Alzheimer's and Parkinson's diseases. *Journal of Neuroinflammation*, *21*(1), 32. DOI: 10.1186/s12974-024-03024-8 PMID: 38263227

Barturen, G., Carnero-Montoro, E., Martínez-Bueno, M., Rojo-Rello, S., Sobrino, B., Porras-Perales, Ó., Alcántara-Domínguez, C., Bernardo, D., & Alarcón-Riquelme, M. E. (2022). Whole blood DNA methylation analysis reveals respiratory environmental traits involved in COVID-19 severity following SARS-CoV-2 infection. *Nature Communications*, *13*(1), 4597. DOI: 10.1038/s41467-022-32357-2 PMID: 35933486

Brogi, S., Tabanelli, R., Puca, S., & Calderone, V. (2024). Intermittent Fasting: Myths, Fakes and Truth on This Dietary Regimen Approach. *Foods*, *13*(13), 1960. DOI: 10.3390/foods13131960 PMID: 38998465

Bruno, A., Dovizio, M., Milillo, C., Aruffo, E., Pesce, M., Gatta, M., Chiacchiaretta, P., Di Carlo, P., & Ballerini, P. (2024). Orally Ingested Micro- and Nano-Plastics: A Hidden Driver of Inflammatory Bowel Disease and Colorectal Cancer. *Cancers (Basel)*, *16*(17), 3079. DOI: 10.3390/cancers16173079 PMID: 39272937

Campagnoli, L. I. M., Marchesi, N., Varesi, A., Morozzi, M., Mascione, L., Ricevuti, G., Esposito, C., Galeotti, N., & Pascale, A. (2024). New therapeutic avenues in multiple sclerosis: Is there a place for gut microbiota-based treatments? *Pharmacological Research, 209*, 107456. DOI: 10.1016/j.phrs.2024.107456 PMID: 39389400

Cantoni, C., Smirnov, R. A., Firulyova, M., Andhey, P. S., Bradstreet, T. R., Esaulova, E., Terekhova, M., Schwarzkopf, E. A., Abdalla, N. M., Kleverov, M., Sabatino, J. J.Jr, Liu, K., Schwab, N., Meyer Zu Hörste, G., Cross, A. H., Artyomov, M. N., Edelson, B. T., & Wu, G. F. (2025). A single-cell compendium of human cerebrospinal fluid identifies disease-associated immune cell populations. *The Journal of Clinical Investigation, 135*(1), e177793. DOI: 10.1172/JCI177793 PMID: 39744938

Chandrasekaran, P., Weiskirchen, S., & Weiskirchen, R. (2024). Effects of Probiotics on Gut Microbiota: An Overview. *International Journal of Molecular Sciences, 25*(11), 6022. DOI: 10.3390/ijms25116022 PMID: 38892208

Cichalewska-Studzinska, M., Szymanski, J., Stec-Martyna, E., Perdas, E., Studzinska, M., Jerczynska, H., Kulczycka-Wojdala, D., Stawski, R., & Mycko, M. P. (2024). The Role of miR-155 in Modulating Gene Expression in CD4+ T Cells: Insights into Alternative Immune Pathways in Autoimmune Encephalomyelitis. *International Journal of Molecular Sciences, 25*(21), 11355. DOI: 10.3390/ijms252111355 PMID: 39518908

Crocetta, A., Liloia, D., Costa, T., Duca, S., Cauda, F., & Manuello, J. (2024). From gut to brain: Unveiling probiotic effects through a neuroimaging perspective-A systematic review of randomized controlled trials. *Frontiers in Nutrition, 11*, 1446854. DOI: 10.3389/fnut.2024.1446854 PMID: 39360283

Cui, X., Li, X., Zheng, H., Su, Y., Zhang, S., Li, M., Hao, X., Zhang, S., Hu, Z., Xia, Z., Shi, C., Xu, Y., & Mao, C. (2024). Human midbrain organoids: A powerful tool for advanced Parkinson's disease modeling and therapy exploration. *NPJ Parkinson's Disease, 10*(1), 189. DOI: 10.1038/s41531-024-00799-8 PMID: 39428415

Dai, W., Qiao, X., Fang, Y., Guo, R., Bai, P., Liu, S., Li, T., Jiang, Y., Wei, S., Na, Z., Xiao, X., & Li, D. (2024). Epigenetics-targeted drugs: Current paradigms and future challenges. *Signal Transduction and Targeted Therapy, 9*(332), 332. Advance online publication. DOI: 10.1038/s41392-024-02039-0 PMID: 39592582

Datta, S., Rahman, M. A., Koka, S., & Boini, K. M. (2024). High Mobility Group Box 1 (HMGB1): Molecular Signaling and Potential Therapeutic Strategies. *Cells, 13*(23), 1946. DOI: 10.3390/cells13231946 PMID: 39682695 de Oliveira Souza, R., Duarte Júnior, J. W. B., Della Casa, V. S., Santoro Rosa, D., Renia, L., & Claser, C. (2024). Unraveling the complex interplay: Immunopathology and immune evasion strategies of alphaviruses with emphasis on neurological implications. *Frontiers in Cellular and Infection Microbiology*, *14*, 1421571. DOI: 10.3389/fcimb.2024.1421571 PMID: 39211797

Dedoni, S., Scherma, M., Camoglio, C., Siddi, C., Dazzi, L., Puliga, R., Frau, J., Cocco, E., & Fadda, P. (2023). Overall view of the most common experimental models for multiple sclerosis. *Neurobiology of Disease*, *184*, 106230. DOI: 10.1016/j.nbd.2023.106230 PMID: 37453561

Delgadillo, D. R., Borelli, J. L., Mayer, E. A., Labus, J. S., Cross, M. P., & Pressman, S. D. (2025). Biological, environmental, and psychological stress and the human gut microbiome in healthy adults. *Scientific Reports*, *15*(1), 362. DOI: 10.1038/s41598-024-77473-9 PMID: 39747287

Diab, R., Dimachkie, L., Zein, O., Dakroub, A., & Eid, A. H. (2024). Intermittent fasting regulates metabolic homeostasis and improves cardiovascular health. *Cell Biochemistry and Biophysics*, *82*(3), 1583–1597. DOI: 10.1007/s12013-024-01314-9 PMID: 38847940

Dong, W., Peng, Q., Liu, Z., Xie, Z., Guo, X., Li, Y., & Chen, C. (2023). Estrogen plays an important role by influencing the NLRP3 inflammasome. *Biomedicine and Pharmacotherapy*, *167*, 115554. DOI: 10.1016/j.biopha.2023.115554 PMID: 37738797

Dong, Y., He, Y., Geng, Y., Wei, M., Zhou, X., Lian, J., & Hallajzadeh, J. (2024). Autophagy-related lncRNAs and exosomal lncRNAs in colorectal cancer: Focusing on lncRNA-targeted strategies. *Cancer Cell International*, *24*(1), 328. DOI: 10.1186/s12935-024-03503-1 PMID: 39342235

Fabiani, C., & Antollini, S. S. (2019). Alzheimer's Disease as a Membrane Disorder: Spatial Cross-Talk Among Beta-Amyloid Peptides, Nicotinic Acetylcholine Receptors and Lipid Rafts. *Frontiers in Cellular Neuroscience*, *13*, 309. DOI: 10.3389/fncel.2019.00309 PMID: 31379503

Fabisiak, A., Caban, M., Dudek, P., Strigáč, A., Małecka-Wojciesko, E., & Talar-Wojnarowska, R. (2025). Advancements in dual biologic therapy for inflammatory bowel diseases: Efficacy, safety, and future directions. *Therapeutic Advances in Gastroenterology*, *18*, 17562848241309871. DOI: 10.1177/17562848241309871 PMID: 39758970

Fu, Q., Qiu, R., Liang, J., Wu, S., Huang, D., Qin, Y., Li, Q., Shi, X., Xiong, X., Jiang, Z., Chen, Y., & Cheng, Y. (2025). Sugemule-7 alleviates oxidative stress, neuroinflammation, and cell death, promoting synaptic plasticity recovery in mice with postpartum depression. *Scientific Reports*, *15*(1), 1426. DOI: 10.1038/s41598-025-85276-9 PMID: 39789071

Fu, X., Han, H., Yang, H., Xu, B., Dai, W., Liu, L., He, T., Du, X., & Pei, X. (2024). Nrf2-mediated ferroptosis of spermatogenic cells involved in male reproductive toxicity induced by polystyrene nanoplastics in mice. *Journal of Zhejiang University. Science. B.*, *25*(4), 307–323. DOI: 10.1631/jzus.B2300138 PMID: 38584093

Gadhave, D. G., Sugandhi, V. V., Jha, S. K., Nangare, S. N., Gupta, G., Singh, S. K., Dua, K., Cho, H., Hansbro, P. M., & Paudel, K. R. (2024). Neurodegenerative disorders: Mechanisms of degeneration and therapeutic approaches with their clinical relevance. *Ageing Research Reviews*, *99*, 102357. DOI: 10.1016/j.arr.2024.102357 PMID: 38830548

Gan, H., Ma, Q., Hao, W., Yang, N., Chen, Z.-S., Deng, L., & Chen, J. (2024). Targeting autophagy to counteract neuroinflammation: A novel antidepressant strategy. *Pharmacological Research*, *202*, 107112. DOI: 10.1016/j.phrs.2024.107112 PMID: 38403256

Gao, L., Peng, L., Wang, J., Zhang, J. H., & Xia, Y. (2024). Mitochondrial stress: A key role of neuroinflammation in stroke. *Journal of Neuroinflammation*, *21*(1), 44. DOI: 10.1186/s12974-024-03033-7 PMID: 38321473

Giorgi, C., Lombardozzi, G., Ammannito, F., Scenna, M. S., Maceroni, E., Quintiliani, M., d'Angelo, M., Cimini, A., & Castelli, V. (2024). Brain Organoids: A Game-Changer for Drug Testing. *Pharmaceutics*, *16*(4), 443. DOI: 10.3390/pharmaceutics16040443 PMID: 38675104

Ha, J., & Kim, J. (2016). Novel pharmacological modulators of autophagy: An updated patent review (2012-2015). *Expert Opinion on Therapeutic Patents*, *26*(11), 1273–1289. DOI: 10.1080/13543776.2016.1217996 PMID: 27476990

Hamilton, A. M., Krout, I. N., White, A. C., & Sampson, T. R. (2024). Microbiome-based therapeutics for Parkinson's disease. *Neurotherapeutics; the Journal of the American Society for Experimental NeuroTherapeutics*, *21*(6), e00462. DOI: 10.1016/j.neurot.2024.e00462 PMID: 39393983

Han, X., Cai, C., Deng, W., Shi, Y., Li, L., Wang, C., Zhang, J., Rong, M., Liu, J., Fang, B., He, H., Liu, X., Deng, C., He, X., & Cao, X. (2024). Landscape of human organoids: Ideal model in clinics and research. *Innovation (Cambridge (Mass.))*, *5*(3), 100620. DOI: 10.1016/j.xinn.2024.100620 PMID: 38706954

Hao, W., Ma, Q., Wang, L., Yuan, N., Gan, H., He, L., Li, X., Huang, J., & Chen, J. (2024). Gut dysbiosis induces the development of depression-like behavior through abnormal synapse pruning in microglia, which is mediated by complement C3. *Microbiome*, *12*(34), 34. Advance online publication. DOI: 10.1186/s40168-024-01756-6 PMID: 38378622

He, R., Liu, Y., Fu, W., He, X., Liu, S., Xiao, D., & Tao, Y. (2024). Mechanisms and cross-talk of regulated cell death and their epigenetic modifications in tumor progression. *Molecular Cancer*, *23*(1), 267. DOI: 10.1186/s12943-024-02172-y PMID: 39614268

Heinrich, M. A., Huynh, N. T., Heinrich, L., & Prakash, J. (2024). Understanding glioblastoma stromal barriers against NK cell attack via a three-culture 3D spheroid model. *Heliyon*, *10*(3), e24808. DOI: 10.1016/j.heliyon.2024.e24808 PMID: 38317968

Hu, H., Cao, B., Huang, D., Lin, Y., Zhou, B., Ying, J., Huang, L., & Zhang, L. (2024). Withaferin a modulation of microglia autophagy mitigates neuroinflammation and enhances cognitive function in POCD. *Scientific Reports*, *14*(1), 26112. DOI: 10.1038/s41598-024-75284-6 PMID: 39478022

Hussain, M. S., Moglad, E., Afzal, M., Sharma, S., Gupta, G., Sivaprasad, G. V., Deorari, M., Almalki, W. H., Kazmi, I., Alzarea, S. I., Shahwan, M., Pant, K., Ali, H., Singh, S. K., Dua, K., & Subramaniyan, V. (2024). Autophagy-associated noncoding RNAs: Unraveling their impact on Parkinson's disease pathogenesis. *CNS Neuroscience & Therapeutics*, *30*(5), e14763. DOI: 10.1111/cns.14763 PMID: 38790149

Hyung, S., Lee, S. R., Kim, J., Kim, Y., Kim, S., Kim, H. N., & Jeon, N. L. (2021). A 3D disease and regeneration model of peripheral nervous system-on-a-chip. *Science Advances*, *7*(5), eabd9749. Advance online publication. DOI: 10.1126/sciadv.abd9749 PMID: 33514550

Jazieh, C., Arabi, T. Z., Asim, Z., Sabbah, B. N., Alsaud, A. W., Alkattan, K., & Yaqinuddin, A. (2024). Unraveling the epigenetic fabric of type 2 diabetes mellitus: Pathogenic mechanisms and therapeutic implications. *Frontiers in Endocrinology*, *15*, 1295967. DOI: 10.3389/fendo.2024.1295967 PMID: 38323108

Jusop, A. S., Thanaskody, K., Tye, G. J., Dass, S. A., Wan Kamarul Zaman, W. S., & Nordin, F. (2023). Development of brain organoid technology derived from iPSC for the neurodegenerative disease modeling: A glance through. *Frontiers in Molecular Neuroscience*, *16*, 1173433. DOI: 10.3389/fnmol.2023.1173433 PMID: 37602192

Kim, A. S., & Diamond, M. S. (2023). A molecular understanding of alphavirus entry and antibody protection. *Nature Reviews. Microbiology*, *21*(5), 396–407. DOI: 10.1038/s41579-022-00825-7 PMID: 36474012

Kim, H. R., Kim, J. C., Kang, S. Y., Kim, H. O., Park, C. W., & Chung, B. Y. (2021). Rapamycin Alleviates 2,3,7,8-Tetrachlorodibenzo-p-dioxin-Induced Aggravated Dermatitis in Mice with Imiquimod-Induced Psoriasis-Like Dermatitis by Inducing Autophagy. *International Journal of Molecular Sciences*, *22*(8), 3968. DOI: 10.3390/ijms22083968 PMID: 33921372

Kodi, T., Sankhe, R., Gopinathan, A., Nandakumar, K., & Kishore, A. (2024). New Insights on NLRP3 Inflammasome: Mechanisms of Activation, Inhibition, and Epigenetic Regulation. *Journal of Neuroimmune Pharmacology*, *19*(1), 7. DOI: 10.1007/s11481-024-10101-5 PMID: 38421496

Kozlov, A. V., Javadov, S., & Sommer, N. (2024). Cellular ROS and Antioxidants: Physiological and Pathological Role. *Antioxidants*, *13*(5), 602. DOI: 10.3390/antiox13050602 PMID: 38790707

Kumar, A., Yap, K. C., BharathwajChetty, B., Lyu, J., Hegde, M., Abbas, M., Alqahtani, M. S., Khadlikar, S., Zarrabi, A., Khosravi, A., Kumar, A. P., & Kunnumakkara, A. B. (2024). Regulating the regulators: Long noncoding RNAs as autophagic controllers in chronic disease management. *Journal of Biomedical Science*, *31*(1), 105. DOI: 10.1186/s12929-024-01092-9 PMID: 39716252

Kumar, D., Bishnoi, M., Kondepudi, K. K., & Sharma, S. S. (2025). Gut microbiota-based interventions for Parkinson's disease: Neuroprotective mechanisms and current perspective. *Probiotics and Antimicrobial Proteins*. Advance online publication. DOI: 10.1007/s12602-024-10433-x PMID: 39809955

Langley, M. R., Triplet, E. M., & Scarisbrick, I. A. (2020). Dietary influence on central nervous system myelin production, injury, and regeneration. Biochimica et Biophysica Acta (BBA) -. *Biochimica et Biophysica Acta. Molecular Basis of Disease*, *1866*(7), 165779. DOI: 10.1016/j.bbadis.2020.165779 PMID: 32224154

Lee, E.-J., Kim, H.-J., Choi, M. S., & Chang, J.-E. (2021). Crosstalk between Autophagy and Inflammatory Processes in Cancer. *Life (Chicago, Ill.)*, *11*(9), 903. DOI: 10.3390/life11090903 PMID: 34575052

Li, D., Liu, J., Guo, B., Liang, C., Dang, L., Lu, C., He, X., Cheung, H. Y., Xu, L., Lu, C., He, B., Liu, B., Shaikh, A. B., Li, F., Wang, L., Yang, Z., Au, D. W., Peng, S., Zhang, Z., & Zhang, G. (2016). Osteoclast-derived exosomal miR-214-3p inhibits osteoblastic bone formation. *Nature Communications*, *7*(1), 10872. DOI: 10.1038/ncomms10872 PMID: 26947250

Li, W., Chen, Q., Peng, C., Yang, D., Liu, S., Lv, Y., Jiang, L., Xu, S., & Huang, L. (2025). Roles of the Receptor for Advanced Glycation End Products and Its Ligands in the Pathogenesis of Alzheimer's Disease. *International Journal of Molecular Sciences*, *26*(1), 403. DOI: 10.3390/ijms26010403 PMID: 39796257

Li, Y., Zhang, W., Zhang, Q., Li, Y., Xin, C., Tu, R., & Yan, H. (2024). Oxidative stress of mitophagy in neurodegenerative diseases: Mechanism and potential therapeutic targets. *Archives of Biochemistry and Biophysics*, *110283*. Advance online publication. DOI: 10.1016/j.abb.2024.110283 PMID: 39743032

Li, Y. Y., Qin, Z. H., & Sheng, R. (2024). The multiple roles of autophagy in neural function and diseases. *Neuroscience Bulletin*, *40*(3), 363–382. DOI: 10.1007/s12264-023-01120-y PMID: 37856037

Liang, T., Zhang, Y., Wu, S., Chen, Q., & Wang, L. (2022). The Role of NLRP3 Inflammasome in Alzheimer's Disease and Potential Therapeutic Targets. *Frontiers in Pharmacology*, *13*, 845185. DOI: 10.3389/fphar.2022.845185 PMID: 35250595

Liao, W., Shi, Y., Li, Z., & Yin, X. (2025). Advances in 3D printing combined with tissue engineering for nerve regeneration and repair. *Journal of Nanobiotechnology*, *23*(1), 5. DOI: 10.1186/s12951-024-03052-9 PMID: 39754257

Limanaqi, F., Biagioni, F., Busceti, C. L., Ryskalin, L., Polzella, M., Frati, A., & Fornai, F. (2019). Phytochemicals Bridging Autophagy Induction and Alpha-Synuclein Degradation in Parkinsonism. *International Journal of Molecular Sciences*, *20*(13), 3274. DOI: 10.3390/ijms20133274 PMID: 31277285

Liu, R., Wu, J., Guo, H., Yao, W., Li, S., Lu, Y., Jia, Y., Liang, X., Tang, J., & Zhang, H. (2023). Posttranslational modifications of histones: Mechanisms, biological functions, and therapeutic targets. *MedComm*, *4*(3), e292. DOI: 10.1002/mco2.292 PMID: 37220590

Liu, Y., Pan, R., Ouyang, Y., Gu, W., Xiao, T., Yang, H., Tang, L., Wang, H., Xiang, B., & Chen, P. (2024). Pyroptosis in health and disease: Mechanisms, regulation, and clinical perspective. *Signal Transduction and Targeted Therapy*, *9*(1), 245. DOI: 10.1038/s41392-024-01958-2 PMID: 39300122

Logan, S., Arzua, T., Canfield, S. G., Seminary, E. R., Sison, S. L., Ebert, A. D., & Bai, X. (2019). Studying Human Neurological Disorders Using Induced Pluripotent Stem Cells: From 2D Monolayer to 3D Organoid and Blood Brain Barrier Models. *Comprehensive Physiology*, *9*(2), 565–611. DOI: 10.1002/j.2040-4603.2019.tb00076.x PMID: 30873582

Logeshwaran, A., Elsen, R., & Nayak, S. (2024). Artificial intelligence-based 3D printing strategies for bone scaffold fabrication and their application in preclinical and clinical investigations. *ACS Biomaterials Science & Engineering*, *10*(2), 677–696. DOI: 10.1021/acsbiomaterials.3c01368 PMID: 38252807

Loh, J. S., Mak, W. Q., Tan, L. K. S., Ng, C. X., Chan, H. H., Yeow, S. H., Foo, J. B., Ong, Y. S., How, C. W., & Khaw, K. Y. (2024). Microbiota–gut–brain axis and its therapeutic applications in neurodegenerative diseases. *Signal Transduction and Targeted Therapy*, *9*(37), 37. Advance online publication. DOI: 10.1038/s41392-024-01743-1 PMID: 38360862

Longo, V. D., Di Tano, M., Mattson, M. P., & Guidi, N. (2021). Intermittent and periodic fasting, longevity and disease. *Nature Aging*, *1*(1), 47–59. DOI: 10.1038/s43587-020-00013-3 PMID: 35310455

Ma, R. (2024). A detective story of intermittent fasting effect on immunity. *Immunology*, *173*(2), 227–247. DOI: 10.1111/imm.13829 PMID: 38922825

Maher, B. A. (2019). Airborne magnetite- and iron-rich pollution nanoparticles: Potential neurotoxicants and environmental risk factors for neurodegenerative disease, including Alzheimer's disease. *Journal of Alzheimer's Disease*, *71*(2), 361–375. DOI: 10.3233/JAD-190204 PMID: 31381513

Maksour, S., Finol-Urdaneta, R. K., Hulme, A. J., Cabral-da-Silva, M. E. C., Targa Dias Anastacio, H., Balez, R., Berg, T., Turner, C., Sanz Muñoz, S., Engel, M., Kalajdzic, P., Lisowski, L., Sidhu, K., Sachdev, P. S., Dottori, M., & Ooi, L. (2024). Alzheimer's disease induced neurons bearing PSEN1 mutations exhibit reduced excitability. *Frontiers in Cellular Neuroscience*, *18*, 1406970. DOI: 10.3389/fncel.2024.1406970 PMID: 39444394

Mallick, K., Paul, S., Banerjee, S., & Banerjee, S. (2024). Lipid Droplets and Neurodegeneration. *Neuroscience*, *549*, 13–23. DOI: 10.1016/j.neuroscience.2024.04.014 PMID: 38718916

Mamun, A. A., Shao, C., Geng, P., Wang, S., & Xiao, J. (2024). Polyphenols Targeting NF-κB Pathway in Neurological Disorders: What We Know So Far? *International Journal of Biological Sciences*, *20*(4), 1332–1355. DOI: 10.7150/ijbs.90982 PMID: 38385077

Menzies, F. M., Fleming, A., Caricasole, A., Bento, C. F., Andrews, S. P., Ashkenazi, A., Füllgrabe, J., Jackson, A., Jimenez Sanchez, M., Karabiyik, C., Licitra, F., Lopez Ramirez, A., Pavel, M., Puri, C., Renna, M., Ricketts, T., Schlotawa, L., Vicinanza, M., Won, H., & Rubinsztein, D. C. (2017). Autophagy and Neurodegeneration: Pathogenic Mechanisms and Therapeutic Opportunities. *Neuron, 93*(5), 1015–1034. DOI: 10.1016/j.neuron.2017.01.022 PMID: 28279350

Meza-Menchaca, T., Albores-Medina, A., Heredia-Mendez, A. J., Ruíz-May, E., Ricaño-Rodríguez, J., Gallegos-García, V., Esquivel, A., Vettoretti-Maldonado, G., & Campos-Parra, A. D. (2024). Revisiting Epigenetics Fundamentals and Its Biomedical Implications. *International Journal of Molecular Sciences*, *25*(14), 7927. DOI: 10.3390/ijms25147927 PMID: 39063168

Miao, J., Ma, H., Yang, Y., Liao, Y., Lin, C., Zheng, J., Yu, M., & Lan, J. (2023). Microglia in Alzheimer's disease: Pathogenesis, mechanisms, and therapeutic potentials. *Frontiers in Aging Neuroscience*, *15*, 1201982. DOI: 10.3389/fnagi.2023.1201982 PMID: 37396657

Minciuna, I., van Kleef, L. A., Stefanescu, H., & Procopet, B. (2022). Is Fasting Good When One Is at Risk of Liver Cancer? *Cancers (Basel)*, *14*(20), 5084. DOI: 10.3390/cancers14205084 PMID: 36291868

Mirrazavi, Z. S., & Behrouz, V. (2024). Various types of fasting diet and possible benefits in nonalcoholic fatty liver: Mechanism of actions and literature update. *Clinical Nutrition (Edinburgh, Lothian)*, *43*(2), 519–533. DOI: 10.1016/j.clnu.2024.01.005 PMID: 38219703

Monzani, R., Gagliardi, M., Clemente, N., Saverio, V., Pańczyszyn, E., Santoro, C., Yissachar, N., Visciglia, A., Pane, M., Amoruso, A., & Corazzari, M. (2022). The Gut-Ex-Vivo System (GEVS) Is a Dynamic and Versatile Tool for the Study of DNBS-Induced IBD in BALB/C and C57BL/6 Mice, Highlighting the Protective Role of Probiotics. *Biology (Basel)*, *11*(11), 1574. DOI: 10.3390/biology11111574 PMID: 36358275

Mozdarani, H., Ezzatizadeh, V., & Rahbar Parvaneh, R. (2020). The emerging role of the long noncoding RNA HOTAIR in breast cancer development and treatment. *Journal of Translational Medicine*, *18*(1), 152. DOI: 10.1186/s12967-020-02320-0 PMID: 32245498

Mrza, M. A., He, J., & Wang, Y. (2024). Integration of iPSC-Derived Microglia into Brain Organoids for Neurological Research. *International Journal of Molecular Sciences*, *25*(6), 3148. DOI: 10.3390/ijms25063148 PMID: 38542121

Mu, C., Gao, M., Xu, W., Sun, X., Chen, T., Xu, H., & Qiu, H. (2024). Mechanisms of microRNA-132 in central neurodegenerative diseases: A comprehensive review. *Biomedicine and Pharmacotherapy, 170*, 116029. DOI: 10.1016/j.biopha.2023.116029 PMID: 38128185

Mueller, A. L., Payandeh, Z., Mohammadkhani, N., Mubarak, S. M. H., Zakeri, A., Alagheband Bahrami, A., Brockmueller, A., & Shakibaei, M. (2021). Recent Advances in Understanding the Pathogenesis of Rheumatoid Arthritis: New Treatment Strategies. *Cells, 10*(11), 3017. DOI: 10.3390/cells10113017 PMID: 34831240

Navabi, S. P., Badreh, F., Shooshtari, M. K., Hajipour, S., Vastegani, S. M., & Khoshnam, S. E. (2024). Microglia-induced neuroinflammation in hippocampal neurogenesis following traumatic brain injury. *Heliyon, 10*(16), e35869. DOI: 10.1016/j.heliyon.2024.e35869 PMID: 39220913

Nikvarz, N., Sedighi, B., Ansari, M., Shahdizade, S., Shojaei, R., & Sharififar, F. (2024). Medicinal plants used in multiple sclerosis patients, prevalence and associated factors: A descriptive cross-sectional study. *BMC Complementary Medicine and Therapies, 24*(1), 278. DOI: 10.1186/s12906-024-04587-y PMID: 39039480

Orcheston-Findlay, L., Bax, S., Utama, R., Engel, M., Govender, D., & O'Neill, G. (2021). Advanced Spheroid, Tumouroid and 3D Bioprinted In-Vitro Models of Adult and Pediatric Glioblastoma. *International Journal of Molecular Sciences, 22*(6), 2962. DOI: 10.3390/ijms22062962 PMID: 33803967

Paganos, P., Ullmann, C. V., Gaglio, D., Bonanomi, M., Salmistraro, N., Arnone, M. I., & Jimenez-Guri, E. (2023). Plastic leachate-induced toxicity during sea urchin embryonic development: Insights into the molecular pathways affected by PVC. *The Science of the Total Environment, 864*, 160901. DOI: 10.1016/j.scitotenv.2022.160901 PMID: 36526210

Panwar, V., Singh, A., Bhatt, M., Tonk, R. K., Azizov, S., Raza, A. S., Sengupta, S., Kumar, D., & Garg, M. (2023). Multifaceted role of mTOR (mammalian target of rapamycin) signaling pathway in human health and disease. *Signal Transduction and Targeted Therapy, 8*(1), 375. DOI: 10.1038/s41392-023-01608-z PMID: 37779156

Papadopoulou, P., Polissidis, A., Kythreoti, G., Sagnou, M., Stefanatou, A., & Theoharides, T. C. (2024). Anti-Inflammatory and Neuroprotective Polyphenols Derived from the European Olive Tree, Olea europaea L., in Long COVID and Other Conditions Involving Cognitive Impairment. *International Journal of Molecular Sciences, 25*(20), 11040. DOI: 10.3390/ijms252011040 PMID: 39456822

Peng, X., Ju, J., Li, Z., Liu, J., Jia, X., Wang, J., Ren, J., & Gao, F. (2025). C3/C3aR Bridges Spinal Astrocyte-Microglia Crosstalk and Accelerates Neuroinflammation in Morphine-Tolerant Rats. *CNS Neuroscience & Therapeutics*, *31*(1), e70216. DOI: 10.1111/cns.70216 PMID: 39801259

Ragupathi, A., Kim, C., & Jacinto, E. (2024). The mTORC2 signaling network: Targets and cross-talks. *The Biochemical Journal*, *481*(2), 45–91. DOI: 10.1042/BCJ20220325 PMID: 38270460

Rashwan, H. K., Mahgoub, S., Abuelezz, N. Z., & Amin, H. K. (2023). Black Cumin Seed (Nigella sativa) in Inflammatory Disorders: Therapeutic Potential and Promising Molecular Mechanisms. *Drugs and Drug Candidates*, *2*(2), 516–537. DOI: 10.3390/ddc2020027

Razali, K., Othman, N., Mohd Nasir, M. H., Doolaanea, A. A., Kumar, J., Ibrahim, W. N., Mohamed Ibrahim, N., & Mohamed, W. M. Y. (2021). The Promise of the Zebrafish Model for Parkinson's Disease: Today's Science and Tomorrow's Treatment. *Frontiers in Genetics*, *12*, 655550. DOI: 10.3389/fgene.2021.655550 PMID: 33936174

Ren, L., Liang, H., Zhu, L., Yang, X., Zhang, H., Sun, N., Huang, D., Feng, J., Wu, Y., Xiong, L., Ke, X., Li, M., & Zhang, A. (2024). Dietary restriction improves perioperative neurocognitive disorders by inhibiting neuroinflammation and gut microbial dysbiosis. *Neuroscience*, *540*, 48–67. DOI: 10.1016/j.neuroscience.2024.01.012 PMID: 38272300

Sadeghi, A., Niknam, M., Momeni-Moghaddam, M. A., Shabani, M., Aria, H., Bastin, A., Teimouri, M., Meshkani, R., & Akbari, H. (2023). Crosstalk between autophagy and insulin resistance: Evidence from different tissues. *European Journal of Medical Research*, *28*(1), 456. DOI: 10.1186/s40001-023-01424-9 PMID: 37876013

Saha, P., Ajgaonkar, S., Maniar, D., Sahare, S., Mehta, D., & Nair, S. (2024). Current insights into transcriptional role(s) for the nutraceutical Withania somnifera in inflammation and aging. *Frontiers in Nutrition*, *11*, 1370951. DOI: 10.3389/fnut.2024.1370951 PMID: 38765810

Sánchez-Vidaña, D. I., Li, J., Abokyi, S., Chan, J. N.-M., Ngai, S. P.-C., & Lau, B. W.-M. (2023). In vitro methods in autophagy research: Applications in neurodegenerative diseases and mood disorders. *Frontiers in Molecular Neuroscience*, *16*, 1168948. DOI: 10.3389/fnmol.2023.1168948 PMID: 37122628

Sarb, O. F., Sarb, A. D., Iacobescu, M., Vlad, I. M., Milaciu, M. V., Ciurmarnean, L., Vacaras, V., & Tantau, A. I. (2024). From Gut to Brain: Uncovering Potential Serum Biomarkers Connecting Inflammatory Bowel Diseases to Neurodegenerative Diseases. *International Journal of Molecular Sciences, 25*(11), 5676. DOI: 10.3390/ijms25115676 PMID: 38891863

Shin, H. J., Kim, H., Oh, S., Lee, J. G., Kee, M., Ko, H. J., Kweon, M. N., Won, K. J., & Baek, S. H. (2016). AMPK-SKP2-CARM1 signaling cascade in transcriptional regulation of autophagy. *Nature, 534*(7608), 553–557. DOI: 10.1038/nature18014 PMID: 27309807

Sobral, J., Empadinhas, N., Esteves, A. R., & Cardoso, S. M. (2025). Impact of nutrition on the gut microbiota: Implications for Parkinson's disease. *Nutrition Reviews, 83*(4), 713–727. Advance online publication. DOI: 10.1093/nutrit/nuae208 PMID: 39812804

Song, M., Bai, Y., & Song, F. (2025). High-fat diet and neuroinflammation: The role of mitochondria. *Pharmacological Research, 107615*, 107615. Advance online publication. DOI: 10.1016/j.phrs.2025.107615 PMID: 39842474

Sukkriang, N., & Buranapin, S. (2024). Effect of intermittent fasting 16:8 and 14:10 compared with control-group on weight reduction and metabolic outcomes in obesity with type 2 diabetes patients: A randomized controlled trial. *Journal of Diabetes Investigation, 15*(9), 1297–1305. DOI: 10.1111/jdi.14186 PMID: 38932663

Sun, X., Shukla, M., & Wang, W. (2024). Unlocking gut-liver-brain axis communication metabolites: Energy metabolism, immunity and barriers. npj Biofilms and Microbiomes, 10(136). DOI: 10.1038/s41522-024-00610-9

Sun, Y., Zhang, C., Lei, T., Lin, F., Huang, J., Hu, Y., Wang, D., & Zhang, W. (2024). HIV-1 gp120 activates microglia via TLR2–NF-κB signaling to upregulate inflammatory cytokine expression and induce neuropathic pain. *Neuropharmacology, 260*, 110136. DOI: 10.1016/j.neuropharm.2024.110136 PMID: 39216684

Tamaki, Y., Ross, J. P., Alipour, P., Castonguay, C. É., Li, B., Catoire, H., Rochefort, D., Urushitani, M., Takahashi, R., Sonnen, J. A., Stifani, S., Dion, P. A., & Rouleau, G. A. (2023). Spinal cord extracts of amyotrophic lateral sclerosis spread TDP-43 pathology in cerebral organoids. *PLOS Genetics, 19*(2), e1010606. DOI: 10.1371/journal.pgen.1010606 PMID: 36745687

Tang, Y., Ju, W., Liu, Y., & Deng, Q. (2024). The role of SIRT1 in autophagy and drug resistance: Unveiling new targets and potential biomarkers in cancer therapy. *Frontiers in Pharmacology, 15*, 1469830. DOI: 10.3389/fphar.2024.1469830 PMID: 39403142

Terzioglu, G., & Young-Pearse, T. L. (2023). Microglial function, INPP5D/SHIP1 signaling, and NLRP3 inflammasome activation: Implications for Alzheimer's disease. *Molecular Neurodegeneration*, *18*(89), 89. Advance online publication. DOI: 10.1186/s13024-023-00674-9 PMID: 38017562

Thompson, L. J.-P., Genovese, J., Hong, Z., Singh, M. V., & Singh, V. B. (2024). HIV-Associated Neurocognitive Disorder: A Look into Cellular and Molecular Pathology. *International Journal of Molecular Sciences*, *25*(9), 4697. DOI: 10.3390/ijms25094697 PMID: 38731913

Tosefsky, K. N., Zhu, J., Wang, Y. N., Lam, J. S. T., Cammalleri, A., & Appel-Cresswell, S. (2024). The Role of Diet in Parkinson's Disease. *Journal of Parkinson's Disease*, *14*(s1), S21–S34. DOI: 10.3233/JPD-230264 PMID: 38251061

Valiukas, Z., Tangalakis, K., Apostolopoulos, V., & Feehan, J. (2025). Microglial activation states and their implications for Alzheimer's disease. *The Journal of Prevention of Alzheimer's Disease*, *12*(1), 100013. DOI: 10.1016/j.tjpad.2024.100013 PMID: 39800461

Valls-Pedret, C., Sala-Vila, A., Serra-Mir, M., Corella, D., de la Torre, R., Martínez-González, M. Á., Martínez-Lapiscina, E. H., Fitó, M., Pérez-Heras, A., Salas-Salvadó, J., Estruch, R., & Ros, E. (2015). Mediterranean Diet and Age-Related Cognitive Decline: A Randomized Clinical Trial. *JAMA Internal Medicine*, *175*(7), 1094–1103. DOI: 10.1001/jamainternmed.2015.1668 PMID: 25961184

Wang, G., Estrella, A., Hakim, O., Milazzo, P., Patel, S., Pintagro, C., Li, D., Zhao, R., Vance, D. E., & Li, W. (2022). Mini-Mental State Examination and Montreal Cognitive Assessment as Tools for Following Cognitive Changes in Alzheimer's Disease Neuroimaging Initiative Participants. *Journal of Alzheimer's Disease*, *90*(1), 263–270. DOI: 10.3233/JAD-220397 PMID: 36093696

Wang, X., Hu, J., Zhang, H., & Zhou, P. (2025). Probiotics encapsulated via biological macromolecule for neurological therapy and functional food: A review. *Probiotics and Antimicrobial Proteins*, *17*(3), 1754–1768. Advance online publication. DOI: 10.1007/s12602-025-10453-1 PMID: 39821001

Wāng, Y. (2025). Ambient fine particulate matter provokes multiple modalities of cell death via perturbation of subcellular structures. *Environment International*, *195*, 109193. DOI: 10.1016/j.envint.2024.109193 PMID: 39721566

Weber, S., Sitte, S., Voegele, A.-L., Sologub, L., Wilfer, A., Rath, T., Nägel, A., Zundler, S., Franchi, L., Opipari, A. W., Sonnewald, S., Reid, S., Hartmann, A., Eichhorn, P., Handtrack, C., Weber, K., Grützmann, R., Neufert, C., Schellerer, V. S., & Atreya, R. (2024). NLRP3 inhibition leads to impaired mucosal fibroblast function in patients with inflammatory bowel diseases. *Journal of Crohn's and Colitis*, *18*(3), 446–461. DOI: 10.1093/ecco-jcc/jjad164 PMID: 37748021

Wei, Y., Li, H., Li, Y., Zeng, Y., Quan, T., Leng, Y., Chang, E., Bai, Y., Bian, Y., & Hou, Y. (2024). Advances of curcumin in nervous system diseases: The effect of regulating oxidative stress and clinical studies. *Frontiers in Pharmacology*, *15*, 1496661. DOI: 10.3389/fphar.2024.1496661 PMID: 39555102

Wu, D., Chen, Q., Chen, X., Han, F., Chen, Z., & Wang, Y. (2023). The blood–brain barrier: Structure, regulation and drug delivery. *Signal Transduction and Targeted Therapy*, *8*(217), 217. Advance online publication. DOI: 10.1038/s41392-023-01481-w PMID: 37231000

Wu, N., Zheng, W., Zhou, Y., Tian, Y., Tang, M., Feng, X., Ashrafizadeh, M., Wang, Y., Niu, X., Tambuwala, M., Wang, L., Tergaonkar, V., Sethi, G., Klionsky, D., Huang, L., & Gu, M. (2024). Autophagy in aging-related diseases and cancer: Principles, regulatory mechanisms and therapeutic potential. *Ageing Research Reviews*, *100*, 102428. DOI: 10.1016/j.arr.2024.102428 PMID: 39038742

Wu, N. N., Tian, H., Chen, P., Wang, D., Ren, J., & Zhang, Y. (2019). Physical Exercise and Selective Autophagy: Benefit and Risk on Cardiovascular Health. *Cells*, *8*(11), 1436. DOI: 10.3390/cells8111436 PMID: 31739509

Xie, Z., Tian, Y., Guo, X., & Xie, N. (2024). The emerging role of CARM1 in cancer. Cellular oncology (Dordrecht, Netherlands), 47(5), 1503–1522. /DOI: 10.1007/s13402-024-00943-9

Xin, Z., Xin, C., Huo, J., Liu, Q., Dong, H., Li, R., & Liu, Y. (2024). Neuroprotective effect of a multistrain probiotic mixture in SOD1G93A mice by reducing SOD1 aggregation and targeting the microbiota-gut-brain axis. *Molecular Neurobiology*, *61*(11), 10051–10071. DOI: 10.1007/s12035-024-03988-x PMID: 38349516

Xu, J., Wang, G., Hou, Y., Sun, K., Zheng, Z., Guo, Z., Hou, L., Zhang, X., Ruan, Z., Ye, Y., & Guo, F. (2025). RICTOR-mediated activation of AKT/mTOR signaling and autophagy inhibition promote osteoarthritis. *International Immunopharmacology*, *144*, 113681. DOI: 10.1016/j.intimp.2024.113681 PMID: 39591826

Xu, Z., Yang, J., Xin, X., Liu, C., Li, L., Mei, X., & Li, M. (2023). Merits and challenges of iPSC-derived organoids for clinical applications. *Frontiers in Cell and Developmental Biology*, *11*, 1188905. DOI: 10.3389/fcell.2023.1188905 PMID: 37305682

Yalaev, B., Tyurin, A., Akhiiarova, K., & Khusainova, R. (2024). Hypomethylation of the RUNX2 Gene Is a New Potential Biomarker of Primary Osteoporosis in Men and Women. *International Journal of Molecular Sciences*, *25*(13), 7312. DOI: 10.3390/ijms25137312 PMID: 39000419

Yan, B., Li, Z., Su, H., Xue, H., Qiu, D., Xu, Z., & Tan, G. (2023). Regulatory mechanisms of autophagy-related ncRNAs in bone metabolic diseases. *Frontiers in Pharmacology*, *14*, 1178310. DOI: 10.3389/fphar.2023.1178310 PMID: 38146458

Yan, H., Xing, Q., Xiao, X., Yu, B., He, J., Mao, X., Yu, J., Zheng, P., Luo, Y., Wu, A., Pu, J., Lu, P., Wei, M., Khafipour, E., & Chen, D. (2024). Effect of *Saccharomyces cerevisiae* Postbiotics and Essential Oil on Growth Performance and Intestinal Health of Weanling Pigs During K88 ETEC Infection. *Journal of Animal Science*, *102*, skae007. Advance online publication. DOI: 10.1093/jas/skae007 PMID: 38198728

Yang, C., Xiong, W., Dong, J., Zhao, X., Liang, G., & Zheng, W. (2025). Artemisinin protected human bronchial epithelial cells from amiodarone-induced oxidative damage via 5′-AMP-activated protein kinase (AMPK) activation. *Redox Report : Communications in Free Radical Research*, *30*(1), 2447721. DOI: 10.1080/13510002.2024.2447721 PMID: 39803706

Yang, W., Qiu, X., Wu, Q., Chang, F., Zhou, T., Zhou, M., & Pei, J. (2023). Active constituents of saffron (Crocus sativus L.) and their prospects in treating neurodegenerative diseases [Review]. *Experimental and Therapeutic Medicine*, *25*(5), 235. DOI: 10.3892/etm.2023.11934 PMID: 37114174

Yang, Y., He, H., Miao, F., Yu, M., Wu, X., Liu, Y., Fu, J., Chen, J., Ma, L., Chen, X., Peng, X., You, Z., & Zhou, C. (2024). 3D-printed PCL framework assembling an ECM-inspired multilayer mineralized GO-Col-HAp microscaffold for in situ mandibular bone regeneration. *Journal of Translational Medicine*, *22*(1), 224. DOI: 10.1186/s12967-024-05020-1 PMID: 38429799

Yu, X., Zhao, H., Wang, R., Chen, Y., Ouyang, X., Li, W., Sun, Y., & Peng, A. (2024). Cancer epigenetics: From laboratory studies and clinical trials to precision medicine. *Cell Death Discovery*, *10*(28), 28. Advance online publication. DOI: 10.1038/s41420-024-01803-z PMID: 38225241

Yun, H. R., Jo, Y. H., Kim, J., Shin, Y., Kim, S. S., & Choi, T. G. (2020). Roles of Autophagy in Oxidative Stress. *International Journal of Molecular Sciences*, *21*(9), 3289. DOI: 10.3390/ijms21093289 PMID: 32384691

Zha, S., Liu, H., Li, H., Li, H., Wong, K.-L., & Homayoun, A. (2024). Functionalized nanomaterials capable of crossing the blood–brain barrier. *ACS Nano*, *18*(3), 1820–1845. DOI: 10.1021/acsnano.3c10674 PMID: 38193927

Zhang, C., & Yang, Y. (2023). Targeting toll-like receptor 4 (TLR4) and the NLRP3 inflammasome: Novel and emerging therapeutic targets for hyperuricemia nephropathy. *Biomolecules and Biomedicine*, *24*(4), 688–697. DOI: 10.17305/bb.2023.9838 PMID: 38041694

Zhang, W., Xu, C., Sun, J., Shen, H. M., Wang, J., & Yang, C. (2022). Impairment of the autophagy-lysosomal pathway in Alzheimer's disease: Pathogenic mechanisms and therapeutic potential. *Acta Pharmaceutica Sinica. B*, *12*(3), 1019–1040. DOI: 10.1016/j.apsb.2022.01.008 PMID: 35530153

Zhang, Y., Yang, H., Hou, S., Xia, Y., & Wang, Y. (2024). Influence of the brain-gut axis on neuroinflammation in cerebral ischemia-reperfusion injury [Review]. *International Journal of Molecular Medicine*, *53*(3), 30. DOI: 10.3892/ijmm.2024.5354 PMID: 38299236

Zhang, Z., Yang, W., Wang, L., Zhu, C., Cui, S., Wang, T., Gu, X., Liu, Y., & Qiu, P. (2024). Unraveling the role and mechanism of mitochondria in postoperative cognitive dysfunction: A narrative review. *Journal of Neuroinflammation*, *21*(1), 293. DOI: 10.1186/s12974-024-03285-3 PMID: 39533332

Zhen, Y., & Zhang, H. (2019). NLRP3 Inflammasome and Inflammatory Bowel Disease. *Frontiers in Immunology*, *10*, 276. DOI: 10.3389/fimmu.2019.00276 PMID: 30873162

Zong, Y., Li, H., Liao, P., Chen, L., Pan, Y., Zheng, Y., Zhang, C., Liu, D., Zheng, M., & Gao, J. (2024). Mitochondrial dysfunction: Mechanisms and advances in therapy. *Signal Transduction and Targeted Therapy*, *9*(1), 124. DOI: 10.1038/s41392-024-01839-8 PMID: 38744846

# Chapter 4
# Autophagy Pathways in Major Depressive Disorder

**Golnaz Shafiei**
*Anatomical Sciences Research Center, Institute for Basic Sciences, Kashan University of Medical Sciences, Kashan, Iran*

**Maryam Saadat**
https://orcid.org/0000-0003-0000-1524
*Department of Anatomical Sciences, School of Medicine, Semnan University of Medical Sciences, Semnan, Iran*

**Shima Farajpour**
*Department of Anatomy, School of Medicine, Tehran University of Medical Sciences, Tehran, Iran*

**Shima Mohammadi**
https://orcid.org/0000-0001-5199-5958
*Department of Neuroscience, School of Medicine, Shahroud University of Medical Sciences, Shahroud, Iran*

**Nasrin Dashti**
*Medical Laboratory Sciences, Allied Medical Sciences, Tehran University of Medical Sciences, Tehran, Iran*

**Tahmineh Mokhtari**
https://orcid.org/0000-0002-4126-7102
*Department of Entomology and Nematology, University of California, Davis, USA*

## ABSTRACT

*Major depressive disorder (MDD) is a chronic condition with high prevalence, especially in individuals with physical illnesses. This chapter explores autophagy's involvement in MDD, connecting it to oxidative stress, increased pro-inflammatory cytokine levels, and impaired mitochondrial function. Dysregulated autophagy contributes to MDD pathophysiology, offering insights into novel therapeutic strategies. The authors review preclinical and clinical evidence, the effects of conventional antidepressants on autophagy, and the potential of herbal products in modulating these pathways. Understanding autophagy's role in MDD underscores the need for*

DOI: 10.4018/979-8-3693-5908-2.ch004

*further research to develop targeted treatments.*

## 1. INTRODUCTION

Major depressive disorder (MDD) is a widespread and persistent psychiatric illness, impacting nearly 4.7% of people worldwide and representing a significant burden on global mental health (Ferrari et al., 2013). Its prevalence is notably higher in individuals suffering from long-term physical health conditions, among whom the incidence of MDD ranges from 20% to 40%—a rate that is two to three times greater than that observed in the general population (Read et al., 2017). Chronic diseases such as diabetes, cancer, arthritis, cardiovascular conditions, and neurodegenerative disorders frequently co-occur with MDD, reflecting the intricate connection between mental and physical well-being (Read et al., 2017; Almeida et al., 2020). Individuals with MDD commonly display symptoms like reduced enjoyment in previously rewarding activities, persistent low mood, and cognitive difficulties (Otte et al., 2016) . These manifestations extend beyond emotional distress, profoundly disrupting quality of life and overall functioning by impairing mood regulation and cognitive processes (Voros et al., 2020). Neurologically, MDD is associated with dysfunction in the critical brain areas like the hippocampus (HIP) and medial prefrontal cortex (mPFC), critical for emotional and cognitive functions (Belleau et al., 2019; Mokhtari et al., 2019). Unravelling the broader mechanisms driving MDD could open doors to innovative treatment strategies.

Research highlights MDD as a pathological state involving nitrosative and oxidative stress, mitochondrial dysfunction, and impaired autophagy, with studies noting increased pro-inflammatory mediators and enhanced cell-based immune mechanisms (Maes et al., 2011; Gassen & Rein, 2019). Understanding autophagy's role in MDD not only enhances insight into its biological foundations but also suggests new therapeutic possibilities. This chapter reviews preclinical and clinical evidence supporting autophagy's involvement in MDD's pathophysiology, emphasizing its medical implications for advancing treatment approaches

## 2. AUTOPHAGY PATHWAYS

Autophagy is essential for maintaining cellular homeostasis and is crucial to tissue and organismal function in both physiological and pathological conditions (Rubinsztein et al., 2012; Kroemer et al., 2010). This evolutionarily conserved mechanism facilitates the breakdown of damaged cytosolic components, including pathogens, proteins, organelles, and aggregates, through a meticulously orchestrated sequence (Gómez-Virgilio et al., 2022). The process begins with the assembly of a phagophore, a membrane structure that engulfs cytoplasmic contents, eventually maturing into a double-membrane autophagosome (Lamark & Johansen, 2021). This vesicle then merges with lysosomes, where lysosomal hydrolases degrade the encapsulated materials into basic molecular components for reuse (Wang et al., 2019; Wang, Nolan, et al., 2020; Lee et al., 2014). Autophagy's regulation involves a complex interplay of proteins critical to phagosomal and endosomal pathways, alongside autophagy-related genes (ATG) (Li et al., 2021). Notably, ATG8, often termed light-chain 3 (LC3), serves as a key marker for detecting and measuring autophagosomes, highlighting the sophisticated control underpinning cellular maintenance (Wang, Nolan, et al., 2020).

In normal conditions, autophagy operates at a baseline level but can be markedly activated by triggers such as nutrient scarcity, infections, organelle impairment, or protein misfolding (Lee et al., 2014; Sorbara et al., 2018; Tannous et al., 2008). Various pharmacological agents and environmental stressors also potently induce autophagy, positioning it as a cornerstone of the cellular stress response (Kroemer et al., 2010; Galluzzi et al., 2017). While typically protective, excessive autophagy can, in certain contexts, yield harmful effects (Maiuri et al., 2007; Fabrizi et al., 2014). Notably, autophagy is distinguished from apoptosis, or Type I programmed cell death (Kroemer et al., 2010). Yet, some studies contest this separation, suggesting that overactive autophagy may lead to Type II cell death, marked by the accumulation of large autophagic vacuoles (Booth et al., 2014; Lockshin & Zakeri, 2004). This nuanced role underscores autophagy's dual nature in cellular survival and pathology.

## 3. PATHOPHYSIOLOGY OF MDD

MDD is a complex, heterogeneous condition with diverse risk factors, symptom profiles, and treatment responses across individuals (Khawagi et al., 2024; More et al., 2024). Its Its development stems from a intricate blend of genetic vulnerabilities, family dynamics, and societal stressors, driven by environmental and biological factors that disrupt molecular and synaptic functions (Liu et al., 2023; Tartt et al., 2022). Key to MDD's pathology are brain changes, including atrophy in the HIP and frontal cortex, alongside compromised neuroplasticity (Su et al., 2022; Andrus et al., 2012; Bhatt et al., 2020). Central to the disorder is impaired synthesis of monoamines like dopamine and serotonin, coupled with altered neurotransmitter metabolism, such as norepinephrine, partly due to reduced catechol-O-methyl transferase activity (Bhatt et al., 2020; Pandya et al., 2013; Aggarwal & Mortensen, 2017). Dysregulation of the monoamine system, modulated by neuropeptides, cytokines, and neurotrophic factors, fuels neuropsychiatric impairments (Khawagi et al., 2024;

Ogłodek et al., 2014). Overactivity in the locus coeruleus-norepinephrine system and hypothalamic-pituitary-adrenal (HPA) axis (Dudek et al., 2021; Wood et al., 2017), combined with excessive activation of serotonin autoreceptors (5-HT1A and 5-HT2A) (Khawagi et al., 2024; Bhatt et al., 2020), suppresses serotonin release (Bhatt et al., 2020; Schieber & Chandel, 2014).

Dysfunctional serotonin transporters and tryptophan metabolism further intensify MDD symptoms (Cruz-Pereira et al., 2020; Georgin-Lavialle et al., 2016). This cascade depletes serotonin, disrupts glutamate signaling, alters receptor functionality, and lowers brain-derived neurotrophic factor (BDNF) levels, contributing to mood instability and cognitive deficits (25, 40). An imbalance between BDNF and inflammatory mediators like nuclear factor kappa B (NF-κB) aggravates MDD by impairing neuroprotection and promoting neurodegeneration (Bhatt et al., 2020). Impaired hippocampal neurogenesis also disrupts HPA axis regulation, leading to excessive glucocorticoid release, dendritic shrinkage, and reduced neuronal adaptability (Cruz-Pereira et al., 2020; Sapolsky, 2015). Elevated inflammatory markers are common in MDD patients, and anti-inflammatory therapies have shown promise in reducing symptoms (Kohler et al., 2016). Human studies indicate higher ATG expression in blood mononuclear cells of individuals with MDD (Alcocer-Gómez et al., 2017). Murine models of chronic stress further demonstrate increased autophagic activity, highlighting the interplay of stress, inflammation, and depression's pathology (Nestler & Hyman, 2010; Slattery & Cryan, 2017; Woo et al., 2018; Xiao et al., 2018). Elucidating these mechanisms, particularly autophagy's role, is vital for crafting precise therapies to address MDD's varied presentations and enhance treatment efficacy.

## 4. ROLE OF AUTOPHAGY IN PATHOPHYSIOLOGY OF MDD

MDD is intricately linked to neuronal apoptosis in the HIP, suggesting a pivotal role for autophagy in its pathophysiology (Shelton et al., 2011). Elevated autophagic markers often correlate with the amelioration of behavioral deficits, reinforcing the idea that autophagy functions as a protective element in the stress response (Kroemer et al., 2010; Xiao et al., 2018). However, emerging evidence complicates this view, indicating that autophagy induction may also precipitate depressive-like behaviors and cognitive deficits, particularly in prenatal stress models (Zhang, Shang, et al., 2017). Notably, inhibiting autophagy has been found to alleviate depressive symptoms triggered by ecstasy in rodents, highlighting its context-dependent effects (Shih et al., 2019). Autophagy operates as a distinct mechanism in MDD, independent of other biological pathways, with disruptions in neuronal autophagic processes directly tied to depressive symptoms (Scaini et al., 2022; Abelaira et al., 2014). Its regulation is orchestrated by extracellular and intracellular signals through three primary cascades: (1) the mTOR-mediated pathway, involving the Atg1/unc-51-like kinase complex downstream of mTOR complex 1 (mTORC1); (2) the phosphatidylinositol 3-kinase (PI3K)/beclin1 cascade; and (3) the calcium signaling cascade (Polajnar & Žerovnik, 2014).

The mTORC1 complex integrates signals from energy, growth factors, nutrients, and amino acids, suppressing autophagy by inhibiting the Atg1 kinase complex, while the beclin complex promotes autophagic activity (Zeng & Zhou, 2008; Pattingre et al., 2008). Calcium ($Ca^{2+}$)'s role in autophagy regulation remains ambiguous, acting as both a positive and negative modulator (Decuypere et al., 2011). The PI3K-Akt-mTOR axis is involved in MDD, with decreased AKT1/mTOR mRNA expression noted in short-term bipolar disorder (Jernigan et al., 2011; Machado-Vieira et al., 2015). Neuronal stimulation via N-methyl-D-aspartate receptors (NMDARs) can trigger autophagy through this pathway in long-term depression (LTD) models, while postmortem analyses reveal impaired mTOR signaling in the PFC of MDD patients (Machado-Vieira et al., 2015; Lee et al., 2013). Additionally, a post-mortem analysis indicated impaired mTOR signaling in the PFC of individuals with MDD (58). Activation of metabotropic glutamate receptors enhances PI3K-mTOR signaling, driving synaptic protein synthesis in LTD models, and inhibiting either protein synthesis or mTOR disrupts mGluR-dependent LTD (Lee et al., 2013; Thomas & Huganir, 2004). This underscores mTOR's critical role at the nexus of protein synthesis and autophagy impairment in neurodegenerative processes (Hoeffer & Klann, 2010; Hands et al., 2009). mTOR activation supports localized protein synthesis at synapses, essential for synapse formation and function, but simultaneously inhibits autophagy, a key process for protein degradation (Hoeffer & Klann, 2010; Wong & Holzbaur, 2015). ysregulation of the PI3K-Akt-mTOR pathway is also linked to autophagy deficits in conditions like Alzheimer's disease and mild cognitive impairment (Tramutola et al., 2015), highlighting mTOR's dual role in neuroprotection via autophagy and neuroregeneration through protein synthesis across central nervous system (CNS) disorders (Abelaira et al., 2014).

BDNF is crucial for neuronal plasticity, and its downregulation is a hallmark of MDD animal models (Numakawa et al., 2014). BDNF exerts neuroprotective effects by inhibiting apoptosis through caspase suppression and promoting neuronal survival by modulating autophagy (Smith et al., 2014). These effects are partly mediated via the PI3K-Akt-mTOR pathway, reinforcing autophagy's significance in MDD (Chen et al., 2013). The induction of LTD requires only a transient postsynaptic $Ca^{2+}$ influx, mediated by NMDARs or voltage-dependent $Ca^{2+}$ channels (68). Cellular stress from reactive oxygen species (ROS) or other factors can lead to protein misfolding and aggregation, overwhelming autophagy or the ubiquitin-proteasome system (Le Guerroué & Youle, 2021). Impaired autophagy hinders the clearance of these aggregates and damaged organelles, exacerbating mitochondrial dysfunction, excitotoxicity, and pore formation, which elevate intracellular calcium and trigger necrosis or apoptosis (Rikiishi, 2012). Conversely, excessive autophagy can also initiate apoptosis, illustrating its delicate balance in neuronal health (Marino et al., 2014).

Animal models, though limited in replicating MDD's complexity, provide critical insights into autophagy's role (Nestler & Hyman, 2010). Maternal separation increases autophagic markers in the PFC but not the HIP, a pattern echoed by corticosterone's differential effects on astrocytes from these regions (MacQueen et al., 2003; Liu et al., 2018; Xia et al., 2018). Prenatal stress markedly elevates hippocampal autophagy markers in male offspring, while chronic unpredictable stress (CUMS) reduces these markers (Zhang, Shang, et al., 2017; Yang et al., 2017; Huang et al., 2018). Lipopolysaccharide (LPS) treatment and CUMS induce depressive-like behaviors in rodents alongside decreased autophagic marker expression (Zhao et al., 2017; Dhillon et al., 2023). Inhibiting Beclin1, an autophagy initiator, also triggers depression-like behaviors in mice (Levine et al., 2015; Gulbins et al., 2018). These conflicting findings—some showing enhanced autophagy, others reduced—complicate the narrative, as functional autophagy assessments, such as flux assays, are challenging in vivo (Zhang et al., 2020; Zhou et al., 2021). Additional studies report diminished autophagic responses in CUMS and chronic restraint stress models, with LPS and certain genetic modifications further reducing autophagic marker expression, impacting cognitive and neural outcomes (Ali et al., 2020; Kallergi et al., 2022). However, preclinical studies have not yet provided clear histopapathological evidence regarding whether autophagy is increased or decreased in MDD.

Recent research highlights a synergistic interplay between autophagy and inflammation in MDD, offering insights into its molecular underpinnings. Neurodevelopmental and mood disorders often feature abnormal protein aggregation, including autophagy-related proteins. For instance, elevated levels of LC3-II have been observed in cellular models of chronic MDD. Post-mortem analyses of MDD patients' brains revealed that enhanced autophagy, apoptosid, and stress-associated biomarker. Furthermore, diminished mRNA expression in the AKT/mTOR pathway, linked to autophagy regulation, has been noted in short-term bipolar disorder, implicating the PI3K-AKT-mTOR cascade in emotional dysregulation (Martinelli et al., 2021). A bidirectional model of autophagic homeostasis posits that ATG upregulate in response to stressors, enhancing autophagy to degrade harmful and protective proteins and organelles (Zhang et al., 2025).

Autophagy, akin to the broader stress response, is protective up to a threshold but may become insufficient in pathological states, necessitating therapeutic enhancement. Short-term calorie restriction, a potent autophagy inducer (Bagherniya et al., 2018), exhibits antidepressant effects in animal models and humans, though the benefits of long-term restriction are debated (Zhang et al., 2015). Physical exercise, which enhances autophagy, also helps alleviate depressive symptoms in humans (He et al., 2012; Brosse et al., 2002). However, due to the broad effects of both calorie restriction and exercise, these studies offer only indirect evidence for a potential link between autophagy and MDD. Analysis of the human autophagy database has uncovered potential diagnostic markers for MDD, pointing to the influence of the immune microenvironment (Levine et al., 2015). This emphasizes the complex interaction between MDD, autophagy, and neuroinflammation, highlighting their roles in the persistence and development of depressive symptoms. Ongoing research is vital to unravel these mechanisms and develop targeted therapies for MDD.

## 5. ANTIDEPRESSANTS FOR REGULATION OF AUTOPHAGY

Autophagy may function as a double-edged sword in MDD, which could explain why some patients do not respond to certain antidepressant therapies. Numerous studies have indicated that markers of cellular autophagy are upregulated following treatment with conventional antidepressants (Zschocke & Rein, 2011). One of the initial signs of this relationship was the detection of autophagy-related structures within the cytoplasm after treating the clomipramine, a tricyclic antidepressant (Daley et al., 2005). This observation could result from either the activation of autophagy or a blockage in autophagic flux, which would impede functional autophagy. Notably, active autophagy is frequently determined only by the presence of autophagic biomarkers, which can be misleading without experiments that assess the turnover of long-lived proteins or autophagic flux (Klionsky et al., 2021). In contrast, another tricyclic antidepressant, amitriptyline (AMI), has been shown to enhance autophagy in primary astrocytes and neurons, similar to the effects observed with citalopram (CIT), a selective serotonin reuptake inhibitor (SSRI). Nevertheless, the SSRIs and serotonin-norepinephrine reuptake inhibitors (SNRIs) venlafaxine did not influence autophagy (Zschocke & Rein, 2011; Sabino et al., 2011). These findings suggest that antidepressants can have varied effects on functional autophagy, potentially influenced by the specific cell types involved.

### 5.1 mTOR-Dependent Pathway

The PI3K-Akt-mTOR signaling pathway, a key regulator of cellular functions, is modulated by various antidepressants, influencing autophagy and neuronal function in MDD (Jeon et al., 2011). This pathway is targeted by a range of antidepressants, including tricyclic antidepressants (e.g., desipramine (DMI), nortriptyline, imipramine), noradrenergic and serotonergic agents (mirtazapine), tetracyclic antidepressants (maprotiline, mianserin), atypical antidepressants such as agomelatine, SSRIs like sertraline, fluoxetine (FLX), and paroxetine, SNRIs like desvenlafaxine, and mood stabilizers including valproic acid and lithium (Motoi et al., 2014; Pagnin et al., 2008; Otabe et al., 2014; Ma et al., 2013). For instance, FLX enhances serotonin availability, promoting neurogenesis and neuronal survival by upregulating phosphorylated AKT, a central component of the PI3K-Akt-mTOR cascade (Kitagishi et al., 2012). Conversely, DMI suppresses this pathway, a negative regulator of autophagy, while activating endoplasmic reticulum stress pathways (PERK-eIF2α and ATF6) to trigger autophagy in C6 glioma cells(Ma et al., 2013). Similarly, imipramine triggers autophagy and exerts anticancer impacts in glioma cells by suppressing PI3K-Akt-mTOR cascade, promoting autophagic cell death (Jeon et al., 2011).

In vitro studies further reveal that antidepressants stimulate dendritic growth and synaptic protein expression through mTOR signaling (Park et al., 2014). For instance, combining FLX with methylphenidate enhances mTOR activity in animal models (Warren et al., 2011), while ketamine, a NMDAR antagonist rapid- and acting antidepressant, stimulates mTOR, increasing synaptic protein levels in the rat prefrontal cortex (Li et al., 2010). Rapamycin, a potent autophagy inducer and mTOR inhibitor, exhibits antidepressant-like effects in animal models (Kara et al., 2018; Cleary et al., 2008), underscoring the mTOR pathway's significance (Ryskalin et al., 2018). Other antidepressants and compounds with antidepressant properties also modulate autophagy across various models, with research emphasizing mTOR's role in antidepressant-induced synaptogenesis (Duman & Li, 2012). Given mTOR's pivotal function in balancing protein synthesis and autophagy disruption in neurodegenerative disorders (Hoeffer & Klann, 2010; Hands et al., 2009), its pathways may illuminate antidepressant mechanisms. Notably, while antidepressants like sertraline activate mTOR, rapamycin's antidepressant effects highlight the pathway's complexity (Cleary et al., 2008). Understanding the interaction between autophagy and mTOR signaling is essential for creating targeted and effective treatments for MDD.

*Table 1. Potential antidepressants via modulation of the mTOR-dependent pathway*

| Authors (year) | Agents | Study Models | Autophagy Outcomes |
|---|---|---|---|
| Ma et al., 2013 | Desipramine | **In vivo**<br>C6 glioma cells | Inhibition of the mTOR pathway<br>Increase in the levels of Beclin1, LC3, and autophagosomes |
| Sundaramurthy et al., 2013 | Nortriptyline | **In vitro**<br>High-content chemical screening in HeLa cells | Induction of autophagy via increased LC3II/I and flux |
| Jeon et al., 2011 | Maprotiline | **In vitro**<br>Burkitt's lymphoma cell line | Upregulation of Beclin1, more cytoplasmic vacuoles |

continued on following page

*Table 1. Continued*

| Authors (year) | Agents | Study Models | Autophagy Outcomes |
|---|---|---|---|
| Alcocer-Gómez et al., 2017 | Mianserin | **In vitro**<br>THP-1 cells<br><br>**Clinical**<br>Depressed patients | Increased levels of LC3 and Beclin1 mRNA |
| Alcocer-Gómez et al., 2017 | Mirtazapine | **In vitro**<br>THP-1 cells<br>ATG5-/- MEFs<br><br>**Clinical**<br>Depressed patients | Increased levels of LC3 and Beclin1 mRNA<br>Elevated LC3II/I |
| Bo Kyung Sun et al., 2015 | Fluoxetine | **In vitro**<br>Human breast cancer cell lines<br>human adipose-derived stem cells<br>mature adipocytes<br>brain injury in rats<br>stress model in rats | Increased levels of LC3II/I, Beclin1, ATG5, ATG7, and ATG12; more LC3 punctae; decreased SQSTM1/p62 |
| Di Xia et al., 2017 | Sertraline | **In vitro**<br>Burkitt's lymphoma cell line<br>Non-small cell lung cancer cells<br>AML cell lines | Upregulation of Beclin1, more cytoplasmic vacuoles, increased LC3II, enhanced flux, formation of autolysosomes, increased LC3II/I |
| Alcocer-Gómez et al., 2017 | Desvenlafaxine | **In vitro**<br>THP-1 cells,<br><br>**Clinical**<br>depressed patients | Increased levels of LC3 and Beclin1 mRNA |
| Alcocer-Gómez et al., 2017 | Agomelatine | **In vitro**<br>THP-1 cells<br><br>**Clinical**<br>Depressed patients | Increased levels of LC3 and Beclin1 mRNA |

AML: Acute myeloid leukemia, ATG: Autophagy-related gene, HeLa: Human cervical cancer cell line, LC3: Microtubule-associated protein 1A/1B light chain 3, MEFs: Mouse embryonic fibroblasts, mTOR: Mechanistic target of rapamycin, SQSTM1: Sequestosome 1, THP-1: Human monocytic leukemia cell line

## 5.2 mTOR-Independent Pathways

Several medications influence autophagy through mTOR-independent pathways, offering novel insights into MDD treatment (Gassen & Rein, 2019). Compounds such as trehalose, rosiglitazone, geldanamycin, dapsone, and α-tocopherol modulate autophagy independently of mTOR signaling. Elevated autophagy markers, including beclin1, are observed in the brains of mice treated with antidepressants, indicating enhanced autophagic activity (Gassen et al., 2014). For instance, amitriptyline (AMI) and citalopram (CIT) induce functional autophagy, mediated partly by class III PI3K and ROS-dependent pathways, with beclin1 pathways also contributing, as 3-methyladenine partially mitigates AMI's effects (91). Both AMI and CIT up-regulate autophagy markers LC3-II and beclin1, unlike venlafaxine, which shows no such effect (Zschocke et al., 2011). Dapsone enhances learning and memory in elderly rats post-propofol anesthesia by boosting hippocampal autophagy, increasing LC3B-II/LC3B-I and beclin1 levels (Yang et al., 2017). Trehalose produces antidepressant-like effects by enhancing autophagy and reducing the p62/beclin-1 ratio in the frontal cortex (Kara et al., 2013). FK506 binding protein 51 (FKBP51) enhances antidepressant effects by binding to beclin1, altering its phosphorylation, and boosting autophagy biomarkers and flux, thus activating autophagic pathways (Gassen et al., 2015). Long-term paroxetine treatment in stress-related depression models demonstrates FKBP51's role in modulating behavioral and autophagic responses (Gassen et al., 2014). Antidepressants used to treat MDD in cancer patients also promote autophagy. In C6 glioma cells, DMI increases beclin-1 and redistributes LC3, triggering autophagic cell death via autophagosome formation (Ma et al., 2013). Intracellular $Ca^{2+}$ critically regulates autophagy (Decuypere et al., 2011), and $Ca^{2+}$ channel blockade alters antidepressant-induced interactions between α1- and β-adrenergic receptors (119). Maprotiline inhibits $Ca^{2+}$ increases triggered by γ-aminobutyric acid and NMDA in human prostate cancer cells and rat cortical neurons (Takebayashi et al., 2000; Hsu et al., 2004). Lithium, a long-standing bipolar disorder treatment, induces autophagy, clearing mutant huntingtin and alpha-synuclein in experimental models(Heiseke et al., 2009; Cloonan & Williams, 2011). Rosiglitazone reduces inflammation-related cytokines (IL-6, TNFα) and corrects cytokine secretion abnormalities (IGF-1, CNTF, NF-α1) in chronic unpredictable mild stress (CUMS) models, promoting autophagy via AMPKα activation in neurons and astrocytes, contributing to its antidepressant effects (Zhao et al., 2017). These findings underscore that mTOR-independent autophagic pathways are integral to the therapeutic effects of antidepressants in MDD. Targeting autophagy initiation mechanisms holds promise as a pharmacological strategy to enhance MDD treatment. However, further research is needed to clarify the specific autophagic pathways involved and their precise contributions to antidepressant mechanisms.

*Table 2. Potential antidepressants via modulation of the mTOR-independent pathway*

| Authors (year) | Agents | Study Models | Main results |
|---|---|---|---|
| Sarkar et al., 2007 | Trehalose | **In vivo**<br>Mice<br><br>**In vitro**<br>various mammalian cells, ATG5-/- MEFs | Enhanced flux<br>Down regulation of p62/beclin1 ratio in the frontal cortex<br>Up regulation of LC3II/I levels |
| Zhao et al., 2017 | Rosiglitazone | **In vivo**<br>Mice<br><br>**In vitro**<br>N2a cells,<br>Primary neurons | Regulation of Beclin1, ULK1, LC3II/I, pAMPK, and pAKT1<br>Reduction of p62 in stressed mice |
| Song et al., 2017 | Silibinin | **In vivo**<br>Mice | Regulation of LC3II/I |
| Ning Yang et al., 2017 | Dapsone | **In vivo**<br>Rats | Regulation of LC3II/I and Beclin1 levels,<br>Reduction of p62 |
| Zare et al., 2015 | Geldanamycin | **In vivo**<br>Rats | Increased levels of ATG12, ATG7, and LC3II/I |
| Zschocke et al., 2011 | Citalopram | **In vitro**<br>Primary rat astrocytes and neurons | Increased levels of LC3II/I and Beclin1 |

AKT1: Protein kinase B, AMPK: Adenosine monophosphate–activated protein kinase, ATG: Autophagy-related gene, LC3: Microtubule-associated protein 1A/1B-light chain 3, pAKT1: Phosphorylated protein kinase B, pAMPK: Phosphorylated adenosine monophosphate–activated protein kinase, ULK1: Unc-51 like autophagy activating kinase

## 6. HERBAL PRODUCTS FOR TREATING DEPRESSION

The literature affirms the established role of herbal medicines in managing MDD(Liu et al., 2015). An expanding body of preclinical studies supports the therapeutic potential of various herbs and their bioactive constituents in alleviating depressive-like behaviors in animal models (Dai et al., 2022; Talei et al., 2021). These natural compounds demonstrate antidepressant properties primarily through their antioxidant, anti-inflammatory, and neuroprotective actions (Anjomshoa et al., 2020). Multiple studies have confirmed that plant-derived compounds alleviate depressive-like behaviors by enhancing antioxidant defenses—this includes lowering malondialdehyde (MDA) levels to reduce lipid peroxidation, suppressing the generation of ROS and nitric oxide (NO), inhibiting mitochondrial-driven apoptotic pathways activating the antioxidant regulator Nrf2, and modulating NMDAR signaling (Amiri et al., 2016; Cui et al., 2020; Morvaridzadeh et al., 2021).

## 6.1 Apigenin

A wide range of plant-derived natural compounds exhibit antidepressant potential, notably through their ability to modulate autophagy—one such compound is Apigenin (APG), a flavonoid predominantly found in plants from the *Compositae* and *Umbelliferae* families (Mokhtari, 2022). APG is known for its broad spectrum of biological and pharmacological effects, encompassing antioxidant (Sánchez-Marzo et al., 2019), anti-inflammatory (Chen et al., 2020), neuroprotective (Ling et al., 2020), neurotrophic (Zhao et al., 2013), and anticancer properties (Imran et al., 2020). In the context of depression, APG has been shown to suppress proinflammatory cytokine release—such as IL-1β and TNF-α—in the PFC, while downregulating the expression of inflammation-related genes, including NF-κB, inducible nitric oxide synthase (iNOS), and cyclooxygenase-2 (COX-2), supporting its anti-inflammatory role in producing antidepressant-like effects (Li et al., 2015). Additionally, evidence from chronic restraint stress models in mice indicates that oral administration of APG confers antidepressant benefits by enhancing autophagic activity through modulation of the TOR/AMPK/ULK1 signaling pathway (Zhang et al., 2019).

## 6.2 Resveratrol

Resveratrol (Res), a nonflavonoid polyphenol abundant in red wine, grape skins, and peanuts, exhibits diverse biological effects, including antitumor activity and mood disorder protection (Wang et al., 2018; Hurley et al., 2014; Shayganfard, 2020). Studies demonstrate that Res mitigates stress-induced depression-like behaviors in rodents and alleviates LPS-induced depressive symptoms and chronic glucocorticoid-related depression (Ardianto et al., 2021; Liu et al., 2016; Liu et al., 2019). A key mechanism involves inducing autophagy by suppressing the AKT/mTOR pathway, essential for cell growth and survival (Wang et al., 2018). Recent findings emphasize Res's role in ameliorating postpartum depressive symptomes by regulating autophagy. Res activates SIRT1, a vital mediator in reducing postpartum depressive symptoms, enhancing autophagy while inhibiting AKT/mTOR signaling. In the HIP of postpartum depressed mice, Res upregulates autophagy-asscoaited proteins like ATG5, LC3B, and Beclin-1, while reducing p62 levels (Ye et al., 2023). These actions highlight Res's potential as a therapeutic agent for mood disorders linked to neuroinflammation and autophagic dysregulation, suggesting its value in clinical interventions for MDD.

## 6.3 Salvianolic acid B

Salvianolic acid B (SalB), a water-soluble phenolic acid derived from the dried roots and rhizomes of *Salvia miltiorrhiza*, is esteemed for treating cardiovascular and cerebrovascular conditions (Li et al., 2020). Beyond these applications, SalB exerts significant effects in the CNS, particularly in addressing depressive disorders. It enhances autophagy and facilitates NLRP3 inflammasome clearance, delivering neuroprotective and antidepressant benefits (Jiang et al., 2017). SalB mitigates depressive-like behaviors by balancing pro-inflammatory mediators (IL-1β, TNF-α) and anti-inflammatory mediators (TGF-β, IL-10) in the HIP and cortex of chronic mild stress mice. In a study, SalB doses of 20 or 40 mg/kg, administered from weeks four to six of a six-week CMS protocol, alleviated depression-like symptoms by modulating hippocampal NLRP3 inflammasome activity and reducing oxidative stress (Huang et al., 2019). Additionally, a 28-day regimen of 30 mg/kg SalB daily lowered brain inflammatory mediators (NF-κB p65, TNF-α, IκB, IL-1β, IL-6) in CMS models, preventing depression onset. These effects, linked to enhanced autophagy (evidenced by elevated BECN1, LC3II, LC3I levels) and NLRP3 regulation, underscore SalB's potential as a therapeutic agent for depression (Jiang et al., 2017).

## 6.4 Isoliquiritin

Isoliquiritin (ILQ), a flavonoid glycoside from Glycyrrhiza uralensis, is renowned for its diverse pharmacological effects, including antiangiogenic, anti-neurotoxic, and antitumor properties (Liu et al., 2019). ILQ protects against corticosterone-induced neurotoxicity, a factor in stress-related disorders. It boosts antioxidant enzyme activity (superoxide dismutase and catalase), reduces ROS and MDA levels, and mitigates oxidative stress. ILQ also alleviates intracellular calcium overload and regulates apoptotic proteins, decreasing caspase-3, Bax, and cytochrome C while increasing Bcl proteins in PC12 cells (Zhou et al., 2017). Recent studies highlight ILQ's antidepressant-like effects, evidenced by reduced immobility in behavioral tests, key indicators of behavioral despair in depression models (Wang et al., 2008). Mechanistically, ILQ enhances NeuN expression, a marker of neuronal survival, and curbs pyroptosis-related neuronal death. By inhibiting NLRP3-mediated pyroptosis via the microRNA 27a/SYK/NF-κB pathway, ILQ mitigates depressive symptoms, bolstering neuronal structure and function to promote survival and prevent cell death (Li et al., 2021). ILQ's neuroprotective and antidepressant properties position it as a promising therapeutic for depression.

## 6.5 Nobiletin

Nobiletin (NOB), a non-toxic flavonoid glycoside extracted from citrus peels, is celebrated for its diverse biological properties, including anti-atherogenic, anti-inflammatory, antidiabetic, antioxidant, hepatoprotective, and antitumor effects (Wang et al., 2023). NOB exhibits antidepressant-like effects, likely mediated through interactions with serotonergic, noradrenergic, and dopaminergic systems (Yi et al., 2011). It effectively mitigates depressive behaviors in CUMS models, with mechanisms potentially involving the BDNF-TrkB signaling pathway, critical for neuronal survival and plasticity (Li et al., 2013). In vivo and in vitro studies demonstrate NOB's ability to curb neuroinflammation by modulating NLRP3 inflammasome activity, a central inflammatory regulator. Furthermore, NOB promotes autophagy by regulating the AMPK signaling pathway in the HIP of LPS-treated rats and LPS-stimulated BV2 microglial cells (Wang et al., 2020). By simultaneously reducing neuroinflammation and enhancing autophagy, NOB emerges as a potent neuroprotective candidate for depression treatment.

## 6.6 Andrographolide

Andrographolide (ANRDO), a key extract from *Andrographis paniculata*, a traditional Chinese herb, is widely used in Asia to relieve cold symptoms (Peng et al., 2018; Burgos et al., 2020). ANRDO exerts diverse CNS benefits, including proneurogenic, antitumor, antiviral, antifibrotic, anti-neuroinflammatory, and neuroprotective effects (Xu et al., 2016; Zhang et al., 2021). Emerging evidence underscores ANRDO's regulatory role in the hippocampal BDNF signaling pathway in mice exposed to CUMS (Zhang et al., 2019). It attenuates NLRP3 inflammasome activation by downregulating NLRP3, ASC, and caspase-1, thereby diminishing proinflammatory mediators such as COX-2, TNF-α, NF-κB, nitric oxide (NO), IL-1β, IL-6, and iNOS in the prefrontal cortex. In parallel, ANRDO enhances autophagic activity by increasing LC3 II levels, reducing p62 accumulation, and upregulating Beclin1 in a dose-dependent manner. This dual action—autophagy induction and inflammation suppression—contributes to its antidepressant effects in a depression model (Geng et al., 2019), highlighting ANRDO as a promising neuroprotective candidate for therapeutic intervention in mood disorders.

## 6.7 Oridonin

Oridonin (Ori), a bioactive diterpenoid from the Chinese herb *Rabdosia rubescens*, is celebrated for its broad therapeutic potential (He et al., 2018). This natural diterpenoid is recognized for its multifaceted pharmacological effects, including antibacterial, antimutagenic, antitumor, anti-inflammatory, and neuroprotective properties, as well as its ability to scavenge RSO (Wu & Zhou, 2023). The efficacy of oridonin has been validated in various animal models, demonstrating its protective effects in conditions such as traumatic brain injury (Yan et al., 2019), ischemic stroke (Xu et al., 2018), Parkinson's disease, seizure (Lin et al., 2019), AD (Wang et al., 2016), and depression (Liu & Du, 2020). A key mechanism underlying Ori's anti-inflammatory effects is its regulation of the NLRP3 inflammasome, which is crucial for the inflammatory response (He et al., 2018). Ori suppresses MAPK and NF-κB pathways, reducing pro-inflammatory cytokines such as TNF-α and IL-6 (Huang et al., 2018). Additionally, it modulates autophagy-related signaling, vital for cellular homeostasis (Zhao et al., 2021). In the CUMS model, Ori lowers thioredoxin-interacting protein, NLRP3, and caspase-1 levels in the HIP (Wang et al., 2021). It alleviates depressive-like behaviors by curbing neuroinflammatory cytokines (IL-1β, IL-18, caspase-1) and normalizing autophagy-related proteins (LC3B, Beclin-1, Atg5, p62) in CUMS rats, partly by disrupting NLRP3-NEK7 interactions in LPS-activated BV2 microglia (Liang et al., 2022). Ori's modulation of neuroinflammation and autophagy markers (BECN1, p62, Atg5, LC3B) underscores its promise as a therapeutic agent for mood disorders, enhancing autophagic processes to address neurodegenerative and stress-related conditions (Liang et al., 2022). Therefore, these findings suggest that Ori's ability to enhance autophagic processes makes it a promising candidate for developing therapies for neurodegenerative and stress-related disorders.

## 6.8 Luteolin

Luteolin (LUT), a versatile flavonoid, is renowned for its antibacterial, antioxidant, antitumor, and anti-inflammatory properties (Luo et al., 2017). Its neuroprotective effects are particularly compelling, stemming from its capacity to mitigate oxidative stress, regulate microglia-driven inflammation, and promote neuronal survival (Kempuraj et al., 2021; Zhu et al., 2011). LUT modulates the NLRP3 inflammasome, a pivotal innate immune component. TLR-mediated NF-κB signaling activates NLRP3 transcription, driving inflammasome assembly and IL-1β production. LUT disrupts this cascade by inhibiting NF-κB activation, thus blocking upstream inflammasome recruitment (Caporali et al., 2022). Additionally, LUT regulates autophagy through the LC3-II/I ratio and the p62/Keap1/Nrf2 pathway, essential for cellular homeostasis under oxidative stress (Tan et al., 2020). In rat models of late-onset depression (LOD), LUT ameliorates depressive-like behaviors by modulating glycerophospholipid metabolism in the HIP and PFC. In LOD rats, the HIP shows a reduced LC3-II/I ratio and elevated p62 expression, signaling diminished autophagy, while the PFC exhibits heightened autophagic activity. LUT treatment curbs this PFC autophagy (Wu et al., 2024). Thus, LUT's multifaceted role in neuroprotection and inflammation modulation underscores its therapeutic promise for mood disorders and neurodegenerative conditions.

## 6.9 Curcumin

Curcumin (CuC), the principal active compound derived from Curcuma longa, is recognized for its significant antioxidant, anti-inflammatory, antitumor, and antimicrobial properties (Matias et al., 2021). Its neuroprotective effects are particularly noteworthy, as CuC modulates autophagic processes by influencing the PI3K/Akt/mTOR signaling pathway (Huang et al., 2018). This modulation is crucial for reducing oxidative stress, which is achieved through the regulation of autophagy-related proteins, including Atg-5, Atg-7, LC3II, and Becn1, as well as the Keap1-Nrf2-antioxidant response element (ARE) signaling pathway in the HIP. These mechanisms are especially relevant in the context of doxorubicin (DOX)-induced neurotoxicity and the manifestation of depression-like behaviors in rat models (Liao et al., 2023). CuC mitigates behavioral deficits caused by corticosterone (CORT) by facilitating mitochondrial biogenesis. This process may counteract the adverse effects of CORT on key regulatory proteins such as NRF1, TFAM, and PGC-1α, as well as the relative MT-CO1/SDH-A ratio. Furthermore, CuC enhances mitophagy by increasing the expression levels of proteins including LC3II/LC3I, NIX, Parkin, Beclin-1, and PTEN-induced putative kinase 1 (PINK1), thereby improving the relative mt-Keima signal (Zhang et al., 2022). Overall, CUR appears to regulate mitochondrial biogenesis, oxidative stress, and mitophagy through PGC-1α, contributing to its anti-depressive-like effects and enhancing our understanding of how CUR alleviates depressive symptoms.

## 6.10 Quercetin

Quercetin (QuR, 3,5,7,3′,4′-pentahydroxyflavone) is a flavonoid compound renowned for its diverse biological activities, including antioxidant, antiviral, antibacterial, and anti-inflammatory properties, and is commonly found in a variety of fruits and vegetables (Di Petrillo et al., 2022). Recent studies have demonstrated that QuR significantly alleviates anxio-depressive-like behavior and mitigates brain oxidative stress, while also suppressing the excess levels of corticosterone that arise following adriamycin treatment (Merzoug et al., 2014). The antidepressant effects of QuR are attributed to its neuroprotective capabilities, which promote the expression of mitophagy-related proteins such as PTEN-induced PINK1, Parkin, and Bcl-2/adenovirus E1B 19 kDa interacting protein 3.These proteins play a crucial role in inhibiting the activation of the NLRP3 inflammasome, which is mediated by mitochondrial ROS in microglial cells (Han et al., 2021). Furthermore, QuR has been shown to attenuate depressive symptoms by modulating the expression of key proteins, including Akt, PI3K, Nrf2, and Heme Oxygenase-1 (HO-1) within the HIP. In addition to these effects, QuR effectively restores the balance between oxidative damage and antioxidant defences disrupted by CUMS, while also reducing the elevated levels of NO and iNOS in the mouse HIP, thereby reinforcing its potential as a therapeutic agent for depression (Guan et al., 2021). QuR combined with improved depressive symptoms in an animal model via modulation of various biochemical markers. QuR regulated the oxidative stress indicators such as hydrogen peroxide, MDA, and glutathione, alongside inflammatory cytokines like IL-1β and TNF-α. Additionally, the study assessed caspase-3 activity, which is crucial for understanding neural apoptosis, and the expression of mitophagy-related proteins, including Parkin and PINK1, which are vital for mitochondrial health. Furthermore, the impact on brain neurotransmitters such as BDNF and serotonin was evaluated, providing insights into the neuroprotective pathways underlying the antidepressant effects of QuR and exercise (Abdallah et al., 2024).

## 7. CONCLUSION

MDD represents a complex interplay between biological, psychological, and environmental factors, significantly impacting individuals' quality of life. The evidence presented in this chapter underscores the critical roles of autophagy in the pathophysiology of MDD, suggesting that disruptions in these processes may contribute to the onset and persistence of depressive symptoms. Furthermore, the modulation of autophagy through conventional antidepressants and herbal remedies presents a promising avenue for therapeutic intervention. Given the variability in individual responses to treatment, a deeper understanding of the underlying mechanisms of autophagy and inflammation in MDD is essential for developing personalized treatment strategies. Future research should focus on elucidating the intricate molecular pathways involved and exploring innovative therapeutic approaches that target these mechanisms, ultimately aiming to improve outcomes for individuals suffering from MDD.

## REFERENCES


Abdallah, H. M., El-Gohary, R. M., Khattab, H., Farghal, E. E., Ghabrial, M. M., & Hashad, A. F. (2024). Potential Synergestic effect of quercetin and exercise on depressive like behavior in male albino rats: Targeting oxidative stress, inflammation, neural apoptosis and Mitophagy. *Bulletin of Egyptian Society for Physiological Sciences.*, *44*(2), 54–68. DOI: 10.21608/besps.2023.243730.1157

Abelaira, H. M., Réus, G. Z., Neotti, M. V., & Quevedo, J. (2014). The role of mTOR in depression and antidepressant responses. *Life Sciences*, *101*(1-2), 10–14. DOI: 10.1016/j.lfs.2014.02.014 PMID: 24582593

Alcocer-Gómez, E., Casas-Barquero, N., Núñez-Vasco, J., Navarro-Pando, J. M., & Bullón, P. (2017). Psychological status in depressive patients correlates with metabolic gene expression. *CNS Neuroscience & Therapeutics*, *23*(10), 843–845. DOI: 10.1111/cns.12755 PMID: 28879683

Ali, T., Rahman, S. U., Hao, Q., Li, W., Liu, Z., Ali Shah, F., Murtaza, I., Zhang, Z., Yang, X., Liu, G., & Li, S. (2020). Melatonin prevents neuroinflammation and relieves depression by attenuating autophagy impairment through FOXO3a regulation. *Journal of Pineal Research*, *69*(2), e12667. DOI: 10.1111/jpi.12667 PMID: 32375205

Almeida, S. S., Zizzi, F. B., Cattaneo, A., Comandini, A., Di Dato, G., Lubrano, E., Pellicano, C., Spallone, V., Tongiani, S., & Torta, R. (2020). Management and treatment of patients with major depressive disorder and chronic diseases: A multidisciplinary approach. *Frontiers in Psychology*, *11*, 542444. DOI: 10.3389/fpsyg.2020.542444 PMID: 33101117

Amiri, S., Alijanpour, S., Tirgar, F., Haj-Mirzaian, A., Amini-Khoei, H., Rahimi-Balaei, M., Rastegar, M., Ghaderi, M., Ghazi-Khansari, M., & Zarrindast, M.-R. (2016). NMDA receptors are involved in the antidepressant-like effects of capsaicin following amphetamine withdrawal in male mice. *Neuroscience*, *329*, 122–133. DOI: 10.1016/j.neuroscience.2016.05.003 PMID: 27167081

Andrus, B., Blizinsky, K., Vedell, P., Dennis, K., Shukla, P., Schaffer, D., Radulovic, J., Churchill, G. A., & Redei, E. E. (2012). Gene expression patterns in the hippocampus and amygdala of endogenous depression and chronic stress models. *Molecular Psychiatry*, *17*(1), 49–61. DOI: 10.1038/mp.2010.119 PMID: 21079605

Anjomshoa, M., Boroujeni, S. N., Ghasemi, S., Lorigooini, Z., Amiri, A., Balali-Dehkordi, S., & Amini-khoei, H. (2020). Rutin via increase in the CA3 diameter of the hippocampus exerted antidepressant-like effect in mouse model of maternal separation stress: Possible involvement of NMDA receptors. *Behavioural Neurology, 2020*(1), 4813616. DOI: 10.1155/2020/4813616 PMID: 32587637

Ardianto, C., Budiatin, A. S., Sumartha, I. N. B., Nurrahmi, N., Rahmadi, M., & Khotib, J. (2021). Resveratrol ameliorates physical and psychological stress-induced depressive-like behavior. *Journal of Basic and Clinical Physiology and Pharmacology, 32*(4), 335–340. DOI: 10.1515/jbcpp-2020-0437 PMID: 34214312

Bagherniya, M., Butler, A. E., Barreto, G. E., & Sahebkar, A. (2018). The effect of fasting or calorie restriction on autophagy induction: A review of the literature. *Ageing Research Reviews, 47*, 183–197. DOI: 10.1016/j.arr.2018.08.004 PMID: 30172870

Belleau, E. L., Treadway, M. T., & Pizzagalli, D. A. (2019). The impact of stress and major depressive disorder on hippocampal and medial prefrontal cortex morphology. *Biological Psychiatry, 85*(6), 443–453. DOI: 10.1016/j.biopsych.2018.09.031 PMID: 30470559

Bhatt, S., Nagappa, A. N., & Patil, C. R. (2020). Role of oxidative stress in depression. *Drug Discovery Today, 25*(7), 1270–1276. DOI: 10.1016/j.drudis.2020.05.001 PMID: 32404275

Booth, L. A., Tavallai, S., Hamed, H. A., Cruickshanks, N., & Dent, P. (2014). The role of cell signalling in the crosstalk between autophagy and apoptosis. *Cellular Signalling, 26*(3), 549–555. DOI: 10.1016/j.cellsig.2013.11.028 PMID: 24308968

Brosse, A. L., Sheets, E. S., Lett, H. S., & Blumenthal, J. A. (2002). Exercise and the treatment of clinical depression in adults: Recent findings and future directions. *Sports Medicine (Auckland, N.Z.), 32*(12), 741–760. DOI: 10.2165/00007256-200232120-00001 PMID: 12238939

Burgos, R. A., Alarcón, P., Quiroga, J., Manosalva, C., & Hancke, J. (2020). Andrographolide, an anti-inflammatory multitarget drug: All roads lead to cellular metabolism. *Molecules (Basel, Switzerland), 26*(1), 5. DOI: 10.3390/molecules26010005 PMID: 33374961

Caporali, S., De Stefano, A., Calabrese, C., Giovannelli, A., Pieri, M., Savini, I., Tesauro, M., Bernardini, S., Minieri, M., & Terrinoni, A. (2022). Anti-inflammatory and active biological properties of the plant-derived bioactive compounds luteolin and luteolin 7-glucoside. *Nutrients, 14*(6), 1155. DOI: 10.3390/nu14061155 PMID: 35334812

Chen, A., Xiong, L.-J., Tong, Y., & Mao, M. (2013). Neuroprotective effect of brain-derived neurotrophic factor mediated by autophagy through the PI3K/Akt/mTOR pathway. *Molecular Medicine Reports*, *8*(4), 1011–1016. DOI: 10.3892/mmr.2013.1628 PMID: 23942837

Chen, P., Huo, X., Liu, W., Li, K., Sun, Z., & Tian, J. (2020). Apigenin exhibits anti-inflammatory effects in LPS-stimulated BV2 microglia through activating GSK3β/Nrf2 signaling pathway. *Immunopharmacology and Immunotoxicology*, *42*(1), 9–16. DOI: 10.1080/08923973.2019.1688345 PMID: 31760890

Cleary, C., Linde, J., Hiscock, K., Hadas, I., Belmaker, R., Agam, G., Flaisher-Grinberg, S., & Einat, H. (2008). Antidepressive-like effects of rapamycin in animal models: Implications for mTOR inhibition as a new target for treatment of affective disorders. *Brain Research Bulletin*, *76*(5), 469–473. DOI: 10.1016/j.brainresbull.2008.03.005 PMID: 18534253

Cloonan, S. M., & Williams, D. C. (2011). The antidepressants maprotiline and fluoxetine induce Type II autophagic cell death in drug-resistant Burkitt's lymphoma. *International Journal of Cancer*, *128*(7), 1712–1723. DOI: 10.1002/ijc.25477 PMID: 20503272

Cruz-Pereira, J. S., Rea, K., Nolan, Y. M., O'Leary, O. F., Dinan, T. G., & Cryan, J. F. (2020). Depression's unholy trinity: Dysregulated stress, immunity, and the microbiome. *Annual Review of Psychology*, *71*(1), 49–78. DOI: 10.1146/annurev-psych-122216-011613 PMID: 31567042

Cui, X., Lin, Q., & Liang, Y. (2020). Plant-derived antioxidants protect the nervous system from aging by inhibiting oxidative stress. *Frontiers in Aging Neuroscience*, *12*, 209. DOI: 10.3389/fnagi.2020.00209 PMID: 32760268

Dai, W., Feng, K., Sun, X., Xu, L., Wu, S., Rahmand, K., Jia, D., & Han, T. (2022). Natural products for the treatment of stress-induced depression: Pharmacology, mechanism and traditional use. *Journal of Ethnopharmacology*, *285*, 114692. DOI: 10.1016/j.jep.2021.114692 PMID: 34742864

Daley, E., Wilkie, D., Loesch, A., Hargreaves, I., Kendall, D., Pilkington, G., & Bates, T. E. (2005). Chlorimipramine: A novel anticancer agent with a mitochondrial target. *Biochemical and Biophysical Research Communications*, *328*(2), 623–632. DOI: 10.1016/j.bbrc.2005.01.028 PMID: 15694394

Decuypere, J.-P., Bultynck, G., & Parys, J. B. (2011). A dual role for Ca2+ in autophagy regulation. *Cell Calcium*, *50*(3), 242–250. DOI: 10.1016/j.ceca.2011.04.001 PMID: 21571367

Dhillon, S. K., Simoens, C., Cuypers, L., Bode, J., Bonde, J., Corbisier, P., Cocuzza, C. E., Van Ranst, M., & Arbyn, M. (2023). Assessment of the clinical and analytical performance of the Aptima SARS-CoV-2 assay using the VALCOR protocol. *Virology Journal*, *20*(1), 35. DOI: 10.1186/s12985-023-01986-4 PMID: 36829164

Di Petrillo, A., Orrù, G., Fais, A., & Fantini, M. C. (2022). Quercetin and its derivates as antiviral potentials: A comprehensive review. *Phytotherapy Research*, *36*(1), 266–278. DOI: 10.1002/ptr.7309 PMID: 34709675

Dudek, K. A., Dion-Albert, L., Kaufmann, F. N., Tuck, E., Lebel, M., & Menard, C. (2021). Neurobiology of resilience in depression: Immune and vascular insights from human and animal studies. *The European Journal of Neuroscience*, *53*(1), 183–221. DOI: 10.1111/ejn.14547 PMID: 31421056

Duman, R. S., & Li, N. (2012). A neurotrophic hypothesis of depression: Role of synaptogenesis in the actions of NMDA receptor antagonists. *Philosophical Transactions of the Royal Society of London. Series B, Biological Sciences*, *367*(1601), 2475–2484. DOI: 10.1098/rstb.2011.0357 PMID: 22826346

Fabrizi, C., De Vito, S., Somma, F., Pompili, E., Catizone, A., Leone, S., Lenzi, P., Fornai, F., & Fumagalli, L. (2014). Lithium improves survival of PC12 pheochromocytoma cells in high-density cultures and after exposure to toxic compounds. *International Journal of Cell Biology*, *2014*(1), 135908. DOI: 10.1155/2014/135908 PMID: 24563652

Ferrari, A. J., Somerville, A. J., Baxter, A. J., Norman, R., Patten, S. B., Vos, T., & Whiteford, H. A. (2013). Global variation in the prevalence and incidence of major depressive disorder: A systematic review of the epidemiological literature. *Psychological Medicine*, *43*(3), 471–481. DOI: 10.1017/S0033291712001511 PMID: 22831756

Galluzzi, L., Bravo-San Pedro, J. M., Levine, B., Green, D. R., & Kroemer, G. (2017). Pharmacological modulation of autophagy: Therapeutic potential and persisting obstacles. *Nature Reviews. Drug Discovery*, *16*(7), 487–511. DOI: 10.1038/nrd.2017.22 PMID: 28529316

Gassen, N. C., Hartmann, J., Schmidt, M. V., & Rein, T. (2015). FKBP5/FKBP51 enhances autophagy to synergize with antidepressant action. *Autophagy*, *11*(3), 578–580. DOI: 10.1080/15548627.2015.1017224 PMID: 25714272

Gassen, N. C., Hartmann, J., Zschocke, J., Stepan, J., Hafner, K., Zellner, A., Kirmeier, T., Kollmannsberger, L., Wagner, K. V., Dedic, N., Balsevich, G., Deussing, J. M., Kloiber, S., Lucae, S., Holsboer, F., Eder, M., Uhr, M., Ising, M., Schmidt, M. V., & Rein, T. (2014). Association of FKBP51 with priming of autophagy pathways and mediation of antidepressant treatment response: Evidence in cells, mice, and humans. *PLoS Medicine*, *11*(11), e1001755. DOI: 10.1371/journal.pmed.1001755 PMID: 25386878

Gassen, N. C., & Rein, T. (2019). Is There a Role of Autophagy in Depression and Antidepressant Action? *Frontiers in Psychiatry*, *10*, 337. DOI: 10.3389/fpsyt.2019.00337 PMID: 31156481

Geng, J., Liu, J., Yuan, X., Liu, W., & Guo, W. (2019). Andrographolide triggers autophagy-mediated inflammation inhibition and attenuates chronic unpredictable mild stress (CUMS)-induced depressive-like behavior in mice. *Toxicology and Applied Pharmacology*, *379*, 114688. DOI: 10.1016/j.taap.2019.114688 PMID: 31340160

Georgin-Lavialle, S., Moura, D. S., Salvador, A., Chauvet-Gélinier, J.-C., Launay, J.-M., Damaj, G., Côté, F., Soucié, E., Chandesris, M.-O., Barète, S., Grandpeix-Guyodo, C., Bachmeyer, C., Alyanakian, M.-A., Aouba, A., Lortholary, O., Dubreuil, P., Teyssier, J.-R., Trojak, B., Haffen, E., & Gaillard, R. (2016). Mast cells' involvement in inflammation pathways linked to depression: Evidence in mastocytosis. *Molecular Psychiatry*, *21*(11), 1511–1516. DOI: 10.1038/mp.2015.216 PMID: 26809839

Gómez-Virgilio, L., Silva-Lucero, M.-C., Flores-Morelos, D.-S., Gallardo-Nieto, J., Lopez-Toledo, G., Abarca-Fernandez, A.-M., Zacapala-Gómez, A.-E., Luna-Muñoz, J., Montiel-Sosa, F., Soto-Rojas, L. O., Pacheco-Herrero, M., & Cardenas-Aguayo, M.-C. (2022). Autophagy: a key regulator of homeostasis and disease: an overview of molecular mechanisms and modulators. *Cells*, *11*(15), 2262. DOI: 10.3390/cells11152262 PMID: 35892559

Guan, Y., Wang, J., Wu, X., Song, L., Wang, Y., Gong, M., & Li, B. (2021). Quercetin reverses chronic unpredictable mild stress-induced depression-like behavior in vivo by involving nuclear factor-E2-related factor 2. *Brain Research*, *1772*, 147661. DOI: 10.1016/j.brainres.2021.147661 PMID: 34529966

Gulbins, A., Schumacher, F., Becker, K. A., Wilker, B., Soddemann, M., Boldrin, F., Müller, C. P., Edwards, M. J., Goodman, M., Caldwell, C. C., Kleuser, B., Kornhuber, J., Szabo, I., & Gulbins, E. (2018). Antidepressants act by inducing autophagy controlled by sphingomyelin–ceramide. *Molecular Psychiatry*, *23*(12), 2324–2346. DOI: 10.1038/s41380-018-0090-9 PMID: 30038230

Han, X., Xu, T., Fang, Q., Zhang, H., Yue, L., Hu, G., & Sun, L. (2021). Quercetin hinders microglial activation to alleviate neurotoxicity via the interplay between NLRP3 inflammasome and mitophagy. *Redox Biology*, *44*, 102010. DOI: 10.1016/j.redox.2021.102010 PMID: 34082381

Hands, S. L., Proud, C. G., & Wyttenbach, A. (2009). mTOR's role in ageing: Protein synthesis or autophagy? *Aging (Albany NY)*, *1*(7), 586–597. DOI: 10.18632/aging.100070 PMID: 20157541

He, C., Bassik, M. C., Moresi, V., Sun, K., Wei, Y., Zou, Z., An, Z., Loh, J., Fisher, J., Sun, Q., Korsmeyer, S., Packer, M., May, H. I., Hill, J. A., Virgin, H. W., Gilpin, C., Xiao, G., Bassel-Duby, R., Scherer, P. E., & Levine, B. (2012). Exercise-induced BCL2-regulated autophagy is required for muscle glucose homeostasis. *Nature*, *481*(7382), 511–515. DOI: 10.1038/nature10758 PMID: 22258505

He, Y., She, H., Zhang, T., Xu, H., Cheng, L., Yepes, M., Zhao, Y., & Mao, Z. (2018). p38 MAPK inhibits autophagy and promotes microglial inflammatory responses by phosphorylating ULK1. *The Journal of Cell Biology*, *217*(1), 315–328. DOI: 10.1083/jcb.201701049 PMID: 29196462

Heiseke, A., Aguib, Y., Riemer, C., Baier, M., & Schätzl, H. M. (2009). Lithium induces clearance of protease resistant prion protein in prion-infected cells by induction of autophagy. *Journal of Neurochemistry*, *109*(1), 25–34. DOI: 10.1111/j.1471-4159.2009.05906.x PMID: 19183256

Hoeffer, C. A., & Klann, E. (2010). mTOR signaling: At the crossroads of plasticity, memory and disease. *Trends in Neurosciences*, *33*(2), 67–75. DOI: 10.1016/j.tins.2009.11.003 PMID: 19963289

Hsu, S. S., Chen, W. C., Lo, Y. K., Cheng, J. S., Yeh, J. H., Cheng, H. H., Chen, J.-S., Chang, H.-T., Jiann, B.-P., Huang, J.-K., & Jan, C.-R. (2004). Effect of the antidepressant maprotiline on Ca2+ movement and proliferation in human prostate cancer cells. *Clinical and Experimental Pharmacology & Physiology*, *31*(7), 444–449. DOI: 10.1111/j.1440-1681.2004.04024.x PMID: 15236632

Huang, Q., Ye, X., Wang, L., & Pan, J. (2019). Salvianolic acid B abolished chronic mild stress-induced depression through suppressing oxidative stress and neuro-inflammation via regulating NLRP3 inflammasome activation. *Journal of Food Biochemistry*, *43*(3), e12742. PMID: 31353549

Huang, W., Huang, M., Ouyang, H., Peng, J., & Liang, J. (2018). Oridonin inhibits vascular inflammation by blocking NF-κB and MAPK activation. *European Journal of Pharmacology*, *826*, 133–139. DOI: 10.1016/j.ejphar.2018.02.044 PMID: 29518395

Huang, X., Wu, H., Jiang, R., Sun, G., Shen, J., Ma, M., Ma, C., Zhang, S., Huang, Z., Wu, Q., Chen, G., & Tao, W. (2018). The antidepressant effects of α-tocopherol are related to activation of autophagy via the AMPK/mTOR pathway. *European Journal of Pharmacology*, *833*, 1–7. DOI: 10.1016/j.ejphar.2018.05.020 PMID: 29782858

Hurley, L. L., Akinfiresoye, L., Kalejaiye, O., & Tizabi, Y. (2014). Antidepressant effects of resveratrol in an animal model of depression. *Behavioural Brain Research*, *268*, 1–7. DOI: 10.1016/j.bbr.2014.03.052 PMID: 24717328

Imran, M., Aslam Gondal, T., Atif, M., Shahbaz, M., Batool Qaisarani, T., Hanif Mughal, M., Salehi, B., Martorell, M., & Sharifi-Rad, J. (2020). Apigenin as an anticancer agent. *Phytotherapy Research*, *34*(8), 1812–1828. DOI: 10.1002/ptr.6647 PMID: 32059077

Jeon, S.-H., Kim, S. H., Kim, Y., Kim, Y. S., Lim, Y., Lee, Y. H., & Shin, S. Y. (2011). The tricyclic antidepressant imipramine induces autophagic cell death in U-87MG glioma cells. *Biochemical and Biophysical Research Communications*, *413*(2), 311–317. DOI: 10.1016/j.bbrc.2011.08.093 PMID: 21889492

Jernigan, C. S., Goswami, D. B., Austin, M. C., Iyo, A. H., Chandran, A., Stockmeier, C. A., & Karolewicz, B. (2011). The mTOR signaling pathway in the prefrontal cortex is compromised in major depressive disorder. *Progress in Neuro-Psychopharmacology & Biological Psychiatry*, *35*(7), 1774–1779. DOI: 10.1016/j.pnpbp.2011.05.010 PMID: 21635931

Jiang, P., Guo, Y., Dang, R., Yang, M., Liao, D., Li, H., Sun, Z., Feng, Q., & Xu, P. (2017). Salvianolic acid B protects against lipopolysaccharide-induced behavioral deficits and neuroinflammatory response: Involvement of autophagy and NLRP3 inflammasome. *Journal of Neuroinflammation*, *14*(1), 1–10. DOI: 10.1186/s12974-017-1013-4 PMID: 29212498

Kallergi, E., Daskalaki, A.-D., Kolaxi, A., Camus, C., Ioannou, E., Mercaldo, V., Haberkant, P., Stein, F., Sidiropoulou, K., Dalezios, Y., Savitski, M. M., Bagni, C., Choquet, D., Hosy, E., & Nikoletopoulou, V. (2022). Dendritic autophagy degrades postsynaptic proteins and is required for long-term synaptic depression in mice. *Nature Communications*, *13*(1), 680. DOI: 10.1038/s41467-022-28301-z PMID: 35115539

Kara, N., Toker, L., Agam, G., Anderson, G. W., Belmaker, R., & Einat, H. (2013). Trehalose induced antidepressant-like effects and autophagy enhancement in mice. *Psychopharmacology*, *229*(2), 367–375. DOI: 10.1007/s00213-013-3119-4 PMID: 23644913

Kara, N. Z., Flaisher-Grinberg, S., Anderson, G. W., Agam, G., & Einat, H. (2018). Mood-stabilizing effects of rapamycin and its analog temsirolimus: Relevance to autophagy. *Behavioural Pharmacology*, *29*(4), 379–384. DOI: 10.1097/FBP.0000000000000334 PMID: 28777104

Kempuraj, D., Thangavel, R., Kempuraj, D. D., Ahmed, M. E., Selvakumar, G. P., Raikwar, S. P., Zaheer, S. A., Iyer, S. S., Govindarajan, R., Chandrasekaran, P. N., & Zaheer, A. (2021). Neuroprotective effects of flavone luteolin in neuroinflammation and neurotrauma. *BioFactors (Oxford, England)*, *47*(2), 190–197. DOI: 10.1002/biof.1687 PMID: 33098588

Khawagi, W. Y., Al-kuraishy, H. M., Hussein, N. R., Al-Gareeb, A. I., Atef, E., Elhussieny, O., Alexiou, A., Papadakis, M., Jabir, M. S., Alshehri, A. A., Saad, H. M., & Batiha, G. E.-S. (2024). Depression and type 2 diabetes: A causal relationship and mechanistic pathway. *Diabetes, Obesity & Metabolism*, *26*(8), 3031–3044. DOI: 10.1111/dom.15630 PMID: 38802993

Kitagishi, Y., Kobayashi, M., Kikuta, K., & Matsuda, S. (2012). Roles of PI3K/AKT/GSK3/mTOR pathway in cell signaling of mental illnesses. *Depression Research and Treatment*, *2012*(1), 752563. DOI: 10.1155/2012/752563 PMID: 23320155

Kohler, O., Krogh, J., Mors, O., & Benros, M. E. (2016). Inflammation in Depression and the Potential for Anti-Inflammatory Treatment. *Current Neuropharmacology*, *14*(7), 732–742. DOI: 10.2174/1570159X14666151208113700 PMID: 27640518

Kroemer, G., Mariño, G., & Levine, B. (2010). Autophagy and the integrated stress response. *Molecular Cell*, *40*(2), 280–293. DOI: 10.1016/j.molcel.2010.09.023 PMID: 20965422

Lamark, T., & Johansen, T. (2021). Mechanisms of selective autophagy. *Annual Review of Cell and Developmental Biology*, *37*(1), 143–169. DOI: 10.1146/annurev-cellbio-120219-035530 PMID: 34152791

Le Guerroué, F., & Youle, R. J. (2021). Ubiquitin signaling in neurodegenerative diseases: An autophagy and proteasome perspective. *Cell Death and Differentiation*, *28*(2), 439–454. DOI: 10.1038/s41418-020-00667-x PMID: 33208890

Lee, J. M., Wagner, M., Xiao, R., Kim, K. H., Feng, D., Lazar, M. A., & Moore, D. D. (2014). Nutrient-sensing nuclear receptors coordinate autophagy. *Nature*, *516*(7529), 112–115. DOI: 10.1038/nature13961 PMID: 25383539

Lee, K.-M., Hwang, S.-K., & Lee, J.-A. (2013). Neuronal autophagy and neurodevelopmental disorders. *Experimental Neurobiology*, *22*(3), 133–142. DOI: 10.5607/en.2013.22.3.133 PMID: 24167408

Levine, B., Liu, R., Dong, X., & Zhong, Q. (2015). Beclin orthologs: Integrative hubs of cell signaling, membrane trafficking, and physiology. *Trends in Cell Biology*, *25*(9), 533–544. DOI: 10.1016/j.tcb.2015.05.004 PMID: 26071895

Li, C., Liu, B., Wang, Z.-y., Xie, F., Qiao, W., Cheng, J., Kuang, J., Wang, Y., Zhang, M., & Liu, D. (2020). Salvianolic acid B improves myocardial function in diabetic cardiomyopathy by suppressing IGFBP3. *Journal of Molecular and Cellular Cardiology*, *139*, 98–112. DOI: 10.1016/j.yjmcc.2020.01.009 PMID: 31982427

Li, J., Zhou, Y., Liu, B. B., Liu, Q., Geng, D., Weng, L. J., & Yi, L.-T. (2013). Nobiletin Ameliorates the Deficits in Hippocampal BDNF, TrkB, and Synapsin I Induced by Chronic Unpredictable Mild Stress. *Evidence-Based Complementary and Alternative Medicine : eCAM*, *2013*, 359682. DOI: 10.1155/2013/359682 PMID: 23573124

Li, N., Lee, B., Liu, R.-J., Banasr, M., Dwyer, J. M., Iwata, M., Li, X.-Y., Aghajanian, G., & Duman, R. S. (2010). mTOR-dependent synapse formation underlies the rapid antidepressant effects of NMDA antagonists. *Science*, *329*(5994), 959–964. DOI: 10.1126/science.1190287 PMID: 20724638

Li, R., Zhao, D., Qu, R., Fu, Q., & Ma, S. (2015). The effects of apigenin on lipopolysaccharide-induced depressive-like behavior in mice. *Neuroscience Letters*, *594*, 17–22. DOI: 10.1016/j.neulet.2015.03.040 PMID: 25800110

Li, W., He, P., Huang, Y., Li, Y.-F., Lu, J., Li, M., Kurihara, H., Luo, Z., Meng, T., Onishi, M., Ma, C., Jiang, L., Hu, Y., Gong, Q., Zhu, D., Xu, Y., Liu, R., Liu, L., Yi, C., & Feng, D. (2021). Selective autophagy of intracellular organelles: Recent research advances. *Theranostics*, *11*(1), 222–256. DOI: 10.7150/thno.49860 PMID: 33391472

Li, Y., Song, W., Tong, Y., Zhang, X., Zhao, J., Gao, X., Yong, J., & Wang, H. (2021). Isoliquiritin ameliorates depression by suppressing NLRP3-mediated pyroptosis via miRNA-27a/SYK/NF-κB axis. *Journal of Neuroinflammation*, *18*(1), 1. DOI: 10.1186/s12974-020-02040-8 PMID: 33402173

Liang, L., Wang, H., Hu, Y., Bian, H., Xiao, L., & Wang, G. (2022). Oridonin relieves depressive-like behaviors by inhibiting neuroinflammation and autophagy impairment in rats subjected to chronic unpredictable mild stress. *Phytotherapy Research*, *36*(8), 3335–3351. DOI: 10.1002/ptr.7518 PMID: 35686337

Liang, L., Wang, H., Hu, Y., Bian, H., Xiao, L., & Wang, G. (2022). Oridonin relieves depressive-like behaviors by inhibiting neuroinflammation and autophagy impairment in rats subjected to chronic unpredictable mild stress. *Phytotherapy Research*, *36*(8), 3335–3351. DOI: 10.1002/ptr.7518 PMID: 35686337

Liao, D., Shangguan, D., Wu, Y., Chen, Y., Liu, N., Tang, J., Yao, D., & Shi, Y. (2023). Curcumin protects against doxorubicin induced oxidative stress by regulating the Keap1-Nrf2-ARE and autophagy signaling pathways. *Psychopharmacology*, *240*(5), 1179–1190. DOI: 10.1007/s00213-023-06357-z PMID: 36949340

Lin, K.-H., Li, C.-Y., Hsu, Y.-M., Tsai, C.-H., Tsai, F.-J., Tang, C.-H., Yang, J.-S., Wang, Z.-H., & Yin, M.-C. (2019). Oridonin, A natural diterpenoid, protected NGF-differentiated PC12 cells against MPP+-and kainic acid-induced injury. *Food and Chemical Toxicology*, *133*, 110765. DOI: 10.1016/j.fct.2019.110765 PMID: 31430510

Ling, C., Lei, C., Zou, M., Cai, X., Xiang, Y., Xie, Y., Li, X., Huang, D., & Wang, Y. (2020). Neuroprotective effect of apigenin against cerebral ischemia/reperfusion injury. *The Journal of International Medical Research*, *48*(9). DOI: 10.1177/0300060520945859 PMID: 32993408

Liu, C., Hao, S., Zhu, M., Wang, Y., Zhang, T., & Yang, Z. (2018). Maternal separation induces different autophagic responses in the hippocampus and prefrontal cortex of adult rats. *Neuroscience*, *374*, 287–294. DOI: 10.1016/j.neuroscience.2018.01.043 PMID: 29391188

Liu, L., Liu, C., Wang, Y., Wang, P., Li, Y., & Li, B. (2015). Herbal medicine for anxiety, depression and insomnia. *Current Neuropharmacology*, *13*(4), 481–493. DOI: 10.2174/1570159X1304150831122734 PMID: 26412068

Liu, L., Wang, H., Chen, X., Zhang, Y., Zhang, H., & Xie, P. (2023). Gut microbiota and its metabolites in depression: From pathogenesis to treatment. *EBioMedicine*, *90*, 90. DOI: 10.1016/j.ebiom.2023.104527 PMID: 36963238

Liu, L., Zhang, Q., Cai, Y., Sun, D., He, X., Wang, L., Yu, D., Li, X., Xiong, X., Xu, H., Yang, Q., & Fan, X. (2016). Resveratrol counteracts lipopolysaccharide-induced depressive-like behaviors via enhanced hippocampal neurogenesis. *Oncotarget*, *7*(35), 56045–56059. DOI: 10.18632/oncotarget.11178 PMID: 27517628

Liu, P., & Du, J. (2020). Oridonin is an antidepressant molecule working through the PPAR-γ/AMPA receptor signaling pathway. *Biochemical Pharmacology*, *180*, 114136. DOI: 10.1016/j.bcp.2020.114136 PMID: 32628930

Liu, T., Ma, Y., Zhang, R., Zhong, H., Wang, L., Zhao, J., Yang, L., & Fan, X. (2019). Resveratrol ameliorates estrogen deficiency-induced depression-and anxiety-like behaviors and hippocampal inflammation in mice. *Psychopharmacology*, *236*(4), 1385–1399. DOI: 10.1007/s00213-018-5148-5 PMID: 30607478

Liu, Y., Xu, X., Xu, R., & Zhang, S. (2019). Renoprotective effects of isoliquiritin against cationic bovine serum albumin-induced membranous glomerulonephritis in experimental rat model through its anti-oxidative and anti-inflammatory properties. *Drug Design, Development and Therapy, 13*, 3735–3751. DOI: 10.2147/DDDT.S213088 PMID: 31802848

Lockshin, R. A., & Zakeri, Z. (2004). Apoptosis, autophagy, and more. *The International Journal of Biochemistry & Cell Biology, 36*(12), 2405–2419. DOI: 10.1016/j.biocel.2004.04.011 PMID: 15325581

Luo, Y., Shang, P., & Li, D. (2017). Luteolin: A flavonoid that has multiple cardioprotective effects and its molecular mechanisms. *Frontiers in Pharmacology, 8*, 692. DOI: 10.3389/fphar.2017.00692 PMID: 29056912

Machado-Vieira, R., Zanetti, M. V., Teixeira, A. L., Uno, M., Valiengo, L. L., Soeiro-de-Souza, M. G., Oba-Shinjo, S. M., de Sousa, R. T., Zarate, C. A.Jr, Gattaz, W. F., & Marie, S. K. N. (2015). Decreased AKT1/mTOR pathway mRNA expression in short-term bipolar disorder. *European Neuropsychopharmacology, 25*(4), 468–473. DOI: 10.1016/j.euroneuro.2015.02.002 PMID: 25726893

MacQueen, G. M., Ramakrishnan, K., Ratnasingan, R., Chen, B., & Young, L. T. (2003). Desipramine treatment reduces the long-term behavioural and neurochemical sequelae of early-life maternal separation. *The International Journal of Neuropsychopharmacology, 6*(4), 391–396. DOI: 10.1017/S1461145703003729 PMID: 14641986

Maes, M., Kubera, M., Obuchowiczwa, E., Goehler, L., & Brzeszcz, J. (2011). Depression's multiple comorbidities explained by (neuro) inflammatory and oxidative & nitrosative stress pathways. *Neuroendocrinology Letters, 32*(1), 7–24. PMID: 21407167

Maiuri, M. C., Zalckvar, E., Kimchi, A., & Kroemer, G. (2007). Self-eating and self-killing: Crosstalk between autophagy and apoptosis. *Nature Reviews. Molecular Cell Biology, 8*(9), 741–752. DOI: 10.1038/nrm2239 PMID: 17717517

Marino, G., Niso-Santano, M., Baehrecke, E. H., & Kroemer, G. (2014). Self-consumption: The interplay of autophagy and apoptosis. *Nature Reviews. Molecular Cell Biology, 15*(2), 81–94. DOI: 10.1038/nrm3735 PMID: 24401948

Martinelli, S., Anderzhanova, E. A., Bajaj, T., Wiechmann, S., Dethloff, F., Weckmann, K., Heinz, D. E., Ebert, T., Hartmann, J., Geiger, T. M., Döngi, M., Hafner, K., Pöhlmann, M. L., Jollans, L., Philipsen, A., Schmidt, S. V., Schmidt, U., Maccarrone, G., Stein, V., & Gassen, N. C. (2021). Stress-primed secretory autophagy promotes extracellular BDNF maturation by enhancing MMP9 secretion. *Nature Communications*, *12*(1), 4643. DOI: 10.1038/s41467-021-24810-5 PMID: 34330919

Matias, J. N., Achete, G., Campanari, G. S. S., Guiguer, É. L., Araújo, A. C., Buglio, D. S., & Barbalho, S. M. (2021). A systematic review of the antidepressant effects of curcumin: Beyond monoamines theory. *The Australian and New Zealand Journal of Psychiatry*, *55*(5), 451–462. DOI: 10.1177/0004867421998795 PMID: 33673739

Merzoug, S., Toumi, M. L., & Tahraoui, A. (2014). Quercetin mitigates Adriamycin-induced anxiety- and depression-like behaviors, immune dysfunction, and brain oxidative stress in rats. *Naunyn-Schmiedeberg's Archives of Pharmacology*, *387*(10), 921–933. DOI: 10.1007/s00210-014-1008-y PMID: 24947870

Mokhtari, T. (2022). Targeting autophagy and neuroinflammation pathways with plant-derived natural compounds as potential antidepressant agents. *Phytotherapy Research*, *36*(9), 3470–3489. DOI: 10.1002/ptr.7551 PMID: 35794794

Mokhtari, T., Tu, Y., & Hu, L. (2019). Involvement of the hippocampus in chronic pain and depression. *Brain Science Advances.*, *5*(4), 288–298. DOI: 10.26599/BSA.2019.9050025

More, S., Kaleem, M., Kharwade, R., Almutairy, A. F., Shahzad, N., & Mujtaba, M. A.. (2024). Depression unveiled: Insights into etiology and animal models for behavioral assessment, exploring the multifactorial nature and treatment of depression. *Brain Research*, 149313. PMID: 39515744

Morvaridzadeh, M., Agah, S., Dulce Estêvão, M., Hosseini, A. S., Heydari, H., Toupchian, O., Abdollahi, S., Persad, E., Abu-Zaid, A., Rezamand, G., & Heshmati, J. (2021). Effect of saffron supplementation on oxidative stress parameters: A systematic review and meta-analysis of randomized placebo-controlled trials. *Food Science & Nutrition*, *9*(10), 5809–5819. DOI: 10.1002/fsn3.2463 PMID: 34646548

Motoi, Y., Shimada, K., Ishiguro, K., & Hattori, N. (2014). Lithium and autophagy. *ACS Chemical Neuroscience*, *5*(6), 434–442. DOI: 10.1021/cn500056q PMID: 24738557

Nestler, E. J., & Hyman, S. E. (2010). Animal models of neuropsychiatric disorders. *Nature Neuroscience*, *13*(10), 1161–1169. DOI: 10.1038/nn.2647 PMID: 20877280

Numakawa, T., Richards, M., Nakajima, S., Adachi, N., Furuta, M., Odaka, H., & Kunugi, H. (2014). The role of brain-derived neurotrophic factor in comorbid depression: Possible linkage with steroid hormones, cytokines, and nutrition. *Frontiers in Psychiatry*, *5*, 136. DOI: 10.3389/fpsyt.2014.00136 PMID: 25309465

Ogłodek, E., Szota, A., Just, M., Moś, D., & Araszkiewicz, A. (2014). The role of the neuroendocrine and immune systems in the pathogenesis of depression. *Pharmacological Reports*, *66*(5), 776–781. DOI: 10.1016/j.pharep.2014.04.009 PMID: 25149980

Otabe, H., Nibuya, M., Shimazaki, K., Toda, H., Suzuki, G., Nomura, S., & Shimizu, K. (2014). Electroconvulsive seizures enhance autophagy signaling in rat hippocampus. *Progress in Neuro-Psychopharmacology & Biological Psychiatry*, *50*, 37–43. DOI: 10.1016/j.pnpbp.2013.11.012 PMID: 24316174

Otte, C., Gold, S. M., Penninx, B. W., Pariante, C. M., Etkin, A., Fava, M., Mohr, D. C., & Schatzberg, A. F. (2016). Major depressive disorder. *Nature Reviews. Disease Primers*, *2*(1), 1–20. DOI: 10.1038/nrdp.2016.65 PMID: 27629598

Pagnin, D., de Queiroz, V., Pini, S., & Cassano, G. B. (2008). Efficacy of ECT in depression: A meta-analytic review. *Focus - American Psychiatric Publishing*, *6*(1), 155–162. DOI: 10.1176/foc.6.1.foc155

Pandya, C. D., Howell, K. R., & Pillai, A. (2013). Antioxidants as potential therapeutics for neuropsychiatric disorders. *Progress in Neuro-Psychopharmacology & Biological Psychiatry*, *46*, 214–223. DOI: 10.1016/j.pnpbp.2012.10.017 PMID: 23123357

Park, S. W., Lee, J. G., Seo, M. K., Lee, C. H., Cho, H. Y., Lee, B. J., Seol, W., & Kim, Y. H. (2014). Differential effects of antidepressant drugs on mTOR signalling in rat hippocampal neurons. *The International Journal of Neuropsychopharmacology*, *17*(11), 1831–1846. DOI: 10.1017/S1461145714000534 PMID: 24901414

Pattingre, S., Espert, L., Biard-Piechaczyk, M., & Codogno, P. (2008). Regulation of macroautophagy by mTOR and Beclin 1 complexes. *Biochimie*, *90*(2), 313–323. DOI: 10.1016/j.biochi.2007.08.014 PMID: 17928127

Peng, Y., Wang, Y., Tang, N., Sun, D., Lan, Y., Yu, Z., Zhao, X., Feng, L., Zhang, B., Jin, L., Yu, F., Ma, X., & Lv, C. (2018). Andrographolide inhibits breast cancer through suppressing COX-2 expression and angiogenesis via inactivation of p300 signaling and VEGF pathway. *Journal of Experimental & Clinical Cancer Research : CR*, *37*(1), 1–14. DOI: 10.1186/s13046-018-0926-9 PMID: 30314513

Polajnar, M., & Žerovnik, E. (2014). Impaired autophagy: A link between neurodegenerative and neuropsychiatric diseases. *Journal of Cellular and Molecular Medicine, 18*(9), 1705–1711. DOI: 10.1111/jcmm.12349 PMID: 25139375

Read, J. R., Sharpe, L., Modini, M., & Dear, B. F. (2017). Multimorbidity and depression: A systematic review and meta-analysis. *Journal of Affective Disorders, 221*, 36–46. DOI: 10.1016/j.jad.2017.06.009 PMID: 28628766

Rikiishi, H. (2012). Novel insights into the interplay between apoptosis and autophagy. *International Journal of Cell Biology, 2012*(1), 317645. DOI: 10.1155/2012/317645 PMID: 22496691

Rubinsztein, D. C., Shpilka, T., & Elazar, Z. (2012). Mechanisms of autophagosome biogenesis. *Current Biology, 22*(1), R29–R34. DOI: 10.1016/j.cub.2011.11.034 PMID: 22240478

Ryskalin, L., Limanaqi, F., Frati, A., Busceti, C. L., & Fornai, F. (2018). mTOR-related brain dysfunctions in neuropsychiatric disorders. *International Journal of Molecular Sciences, 19*(8), 2226. DOI: 10.3390/ijms19082226 PMID: 30061532

Sabino, V., Cottone, P., Blasio, A., Iyer, M. R., Steardo, L., Rice, K. C., Conti, B., Koob, G. F., & Zorrilla, E. P. (2011). Activation of σ-receptors induces binge-like drinking in Sardinian alcohol-preferring rats. *Neuropsychopharmacology : Official Publication of the American College of Neuropsychopharmacology, 36*(6), 1207–1218. DOI: 10.1038/npp.2011.5 PMID: 21346735

Sánchez-Marzo, N., Pérez-Sánchez, A., Ruiz-Torres, V., Martínez-Tébar, A., Castillo, J., Herranz-López, M., & Barrajón-Catalán, E. (2019). Antioxidant and photoprotective activity of apigenin and its potassium salt derivative in human keratinocytes and absorption in Caco-2 cell monolayers. *International Journal of Molecular Sciences, 20*(9), 2148. DOI: 10.3390/ijms20092148 PMID: 31052292

Sapolsky, R. M. (2015). Stress and the brain: Individual variability and the inverted-U. *Nature Neuroscience, 18*(10), 1344–1346. DOI: 10.1038/nn.4109 PMID: 26404708

Scaini, G., Mason, B. L., Diaz, A. P., Jha, M. K., Soares, J. C., Trivedi, M. H., & Quevedo, J. (2022). Dysregulation of mitochondrial dynamics, mitophagy and apoptosis in major depressive disorder: Does inflammation play a role? *Molecular Psychiatry, 27*(2), 1095–1102. DOI: 10.1038/s41380-021-01312-w PMID: 34650203

Schieber, M., & Chandel, N. S. (2014). ROS function in redox signaling and oxidative stress. *Current Biology, 24*(10), R453–R62. DOI: 10.1016/j.cub.2014.03.034 PMID: 24845678

Shayganfard, M. (2020). Molecular and biological functions of resveratrol in psychiatric disorders: A review of recent evidence. *Cell & Bioscience*, *10*(1), 128. DOI: 10.1186/s13578-020-00491-3 PMID: 33292508

Shelton, R., Claiborne, J., Sidoryk-Wegrzynowicz, M., Reddy, R., Aschner, M., Lewis, D., & Mirnics, K. (2011). Altered expression of genes involved in inflammation and apoptosis in frontal cortex in major depression. *Molecular Psychiatry*, *16*(7), 751–762. DOI: 10.1038/mp.2010.52 PMID: 20479761

Shih, J.-H., Chiu, C.-H., Ma, K.-H., Huang, Y.-S., Shiue, C.-Y., Yeh, T.-Y., Kao, L.-T., Lin, Y.-Y., & Li, I.-H. (2019). Autophagy inhibition plays a protective role against 3, 4-methylenedioxymethamphetamine (MDMA)-induced loss of serotonin transporters and depressive-like behaviors in rats. *Pharmacological Research*, *142*, 283–293. DOI: 10.1016/j.phrs.2019.02.026 PMID: 30826457

Slattery, D. A., & Cryan, J. F. (2017). Modelling depression in animals: At the interface of reward and stress pathways. *Psychopharmacology*, *234*(9-10), 1451–1465. DOI: 10.1007/s00213-017-4552-6 PMID: 28224183

Smith, E. D., Prieto, G. A., Tong, L., Sears-Kraxberger, I., Rice, J. D., Steward, O., & Cotman, C. W. (2014). Rapamycin and interleukin-1β impair brain-derived neurotrophic factor-dependent neuron survival by modulating autophagy. *The Journal of Biological Chemistry*, *289*(30), 20615–20629. DOI: 10.1074/jbc.M114.568659 PMID: 24917666

Su, Y., D'Arcy, C., Li, M., O'Donnell, K., Caron, J., Meaney, M., & Meng, X. (2022). Specific and cumulative lifetime stressors in the aetiology of major depression: A longitudinal community-based population study. *Epidemiology and Psychiatric Sciences*, *31*, e3. DOI: 10.1017/S2045796021000779 PMID: 35078547

Takebayashi, M., Kagaya, A., Inagaki, M., Kozuru, T., Jitsuiki, H., Kurata, K., Okamoto, Y., & Yamawaki, S. (2000). Effects of antidepressants on γ-aminobutyric acid-and N-methyl-D-aspartate-induced intracellular Ca2+ concentration increases in primary cultured rat cortical neurons. *Neuropsychobiology*, *42*(3), 120–126. DOI: 10.1159/000026681 PMID: 11015029

Talei, S., Mokhtari, T., Asadi, I., Arab, L., Hassanzadeh, N., Hosseinjani, E., & Hassanzadeh, G. (2021). Flaxseed oil (Linum usitatissimum) attenuates restraint stress-induced depressive-like behavior: Upregulation of neurotrophic factors in CA1 region of hippocampus. *Journal of Contemporary Medical Sciences.*, *7*(1), 61–65. DOI: 10.22317/jcms.v7i1.945

Tan, X., Yang, Y., Xu, J., Zhang, P., Deng, R., Mao, Y., He, J., Chen, Y., Zhang, Y., Ding, J., Li, H., Shen, H., Li, X., Dong, W., & Chen, G. (2020). Luteolin exerts neuroprotection via modulation of the p62/Keap1/Nrf2 pathway in intracerebral hemorrhage. *Frontiers in Pharmacology*, *10*, 1551. DOI: 10.3389/fphar.2019.01551 PMID: 32038239

Tannous, P., Zhu, H., Nemchenko, A., Berry, J. M., Johnstone, J. L., Shelton, J. M., Miller, F. J.Jr, Rothermel, B. A., & Hill, J. A. (2008). Intracellular protein aggregation is a proximal trigger of cardiomyocyte autophagy. *Circulation*, *117*(24), 3070–3078. DOI: 10.1161/CIRCULATIONAHA.107.763870 PMID: 18541737

Tartt, A. N., Mariani, M. B., Hen, R., Mann, J. J., & Boldrini, M. (2022). Dysregulation of adult hippocampal neuroplasticity in major depression: Pathogenesis and therapeutic implications. *Molecular Psychiatry*, *27*(6), 2689–2699. DOI: 10.1038/s41380-022-01520-y PMID: 35354926

Thomas, G. M., & Huganir, R. L. (2004). MAPK cascade signalling and synaptic plasticity. *Nature Reviews. Neuroscience*, *5*(3), 173–183. DOI: 10.1038/nrn1346 PMID: 14976517

Tramutola, A., Triplett, J. C., Di Domenico, F., Niedowicz, D. M., Murphy, M. P., Coccia, R., Perluigi, M., & Butterfield, D. A. (2015). Alteration of mTOR signaling occurs early in the progression of Alzheimer disease (AD): Analysis of brain from subjects with pre-clinical AD, amnestic mild cognitive impairment and late-stage AD. *Journal of Neurochemistry*, *133*(5), 739–749. DOI: 10.1111/jnc.13037 PMID: 25645581

Voros, V., Martin Gutierrez, D., Alvarez, F., Boda-Jorg, A., Kovacs, A., Tenyi, T., Fekete, S., & Osvath, P. (2020). The impact of depressive mood and cognitive impairment on quality of life of the elderly. *Psychogeriatrics*, *20*(3), 271–277. DOI: 10.1111/psyg.12495 PMID: 31811691

Wang, H., Guo, Y., Qiao, Y., Zhang, J., & Jiang, P. (2020). Nobiletin ameliorates NLRP3 inflammasome-mediated inflammation through promoting autophagy via the AMPK pathway. *Molecular Neurobiology*, *57*(12), 5056–5068. DOI: 10.1007/s12035-020-02071-5 PMID: 32833187

Wang, H.-H., Zhang, Y., Qu, T.-Q., Sang, X.-Q., Li, Y.-X., Ren, F.-Z., Wen, P.-C., & Sun, Y.-N. (2023). Nobiletin improves D-galactose-induced aging mice skeletal muscle atrophy by regulating protein homeostasis. *Nutrients*, *15*(8), 1801. DOI: 10.3390/nu15081801 PMID: 37111020

Wang, J., Li, J., Cao, N., Li, Z., Han, J., & Li, L. (2018). Resveratrol, an activator of SIRT1, induces protective autophagy in non-small-cell lung cancer via inhibiting Akt/mTOR and activating p38-MAPK. *OncoTargets and Therapy*, *11*, 7777–7786. DOI: 10.2147/OTT.S159095 PMID: 30464525

Wang, L., Huang, Z., Ping, J., Liu, X., Cai, Z., & Dai, C. (2021). Oridonin ameliorates depressive-like behaviors induced by chronic unpredictable mild stress in mice via TXNIP/NLRP3 signaling pathway. *Tropical Journal of Pharmaceutical Research*, *20*(3), 531–536. DOI: 10.4314/tjpr.v20i3.13

Wang, P., Nolan, T. M., Yin, Y., & Bassham, D. C. (2020). Identification of transcription factors that regulate ATG8 expression and autophagy in Arabidopsis. *Autophagy*, *16*(1), 123–139. DOI: 10.1080/15548627.2019.1598753 PMID: 30909785

Wang, R., Tan, J., Chen, T., Han, H., Tian, R., Tan, Y., Wu, Y., Cui, J., Chen, F., Li, J., Lv, L., Guan, X., Shang, S., Lu, J., & Zhang, Z. (2019). ATP13A2 facilitates HDAC6 recruitment to lysosome to promote autophagosome–lysosome fusion. *The Journal of Cell Biology*, *218*(1), 267–284. DOI: 10.1083/jcb.201804165 PMID: 30538141

Wang, S., Yu, L., Yang, H., Li, C., Hui, Z., Xu, Y., & Zhu, X. (2016). Oridonin attenuates synaptic loss and cognitive deficits in an Aβ1–42-induced mouse model of Alzheimer's disease. *PLoS One*, *11*(3), e0151397. DOI: 10.1371/journal.pone.0151397 PMID: 26974541

Wang, W., Hu, X., Zhao, Z., Liu, P., Hu, Y., Zhou, J., Zhou, D., Wang, Z., Guo, D., & Guo, H. (2008). Antidepressant-like effects of liquiritin and isoliquiritin from Glycyrrhiza uralensis in the forced swimming test and tail suspension test in mice. *Progress in Neuro-Psychopharmacology & Biological Psychiatry*, *32*(5), 1179–1184. DOI: 10.1016/j.pnpbp.2007.12.021 PMID: 18289754

Warren, B. L., Iniguez, S. D., Alcantara, L. F., Wright, K. N., Parise, E. M., Weakley, S. K., & Bolanos-Guzman, C. A. (2011). Juvenile administration of concomitant methylphenidate and fluoxetine alters behavioral reactivity to reward-and mood-related stimuli and disrupts ventral tegmental area gene expression in adulthood. *The Journal of Neuroscience : The Official Journal of the Society for Neuroscience*, *31*(28), 10347–10358. DOI: 10.1523/JNEUROSCI.1470-11.2011 PMID: 21753012

Wong, Y. C., & Holzbaur, E. L. (2015). Autophagosome dynamics in neurodegeneration at a glance. *Journal of Cell Science*, *128*(7), 1259–1267. DOI: 10.1242/jcs.161216 PMID: 25829512

Woo, H., Hong, C. J., Jung, S., Choe, S., & Yu, S.-W. (2018). Chronic restraint stress induces hippocampal memory deficits by impairing insulin signaling. *Molecular Brain, 11*(1), 1–13. DOI: 10.1186/s13041-018-0381-8 PMID: 29970188

Wood, C. S., Valentino, R. J., & Wood, S. K. (2017). Individual differences in the locus coeruleus-norepinephrine system: Relevance to stress-induced cardiovascular vulnerability. *Physiology & Behavior, 172*, 40–48. DOI: 10.1016/j.physbeh.2016.07.008 PMID: 27423323

Wu, S., & Zhou, W. (2023). Antimicrobial activity of oridonin. *Food Science and Technology (Campinas), 43*, e110222. DOI: 10.1590/fst.110222

Wu, X., Xu, H., Zeng, N., Li, H., Yao, G., Liu, K., Yan, C., & Wu, L. (2024). Luteolin alleviates depression-like behavior by modulating glycerophospholipid metabolism in the hippocampus and prefrontal cortex of LOD rats. *CNS Neuroscience & Therapeutics, 30*(3), e14455. DOI: 10.1111/cns.14455 PMID: 37715585

Xia, C.-Y., Wang, Z.-Z., Zhang, Z., Chen, J., Wang, Y.-Y., Lou, Y.-X., Gao, Y., Luo, P., Ren, Q., Du, G.-H., & Chen, N.-H. (2018). Corticosterone impairs gap junctions in the prefrontal cortical and hippocampal astrocytes via different mechanisms. *Neuropharmacology, 131*, 20–30. DOI: 10.1016/j.neuropharm.2017.12.003 PMID: 29223529

Xiao, X., Shang, X., Zhai, B., Zhang, H., & Zhang, T. (2018). Nicotine alleviates chronic stress-induced anxiety and depressive-like behavior and hippocampal neuropathology via regulating autophagy signaling. *Neurochemistry International, 114*, 58–70. DOI: 10.1016/j.neuint.2018.01.004 PMID: 29339018

Xu, F., Wu, H., Zhang, K., Lv, P., Zheng, L., & Zhao, J. (2016). Pro-neurogenic effects of andrographolide on RSC96 Schwann cells in vitro. *Molecular Medicine Reports, 14*(4), 3573–3580. DOI: 10.3892/mmr.2016.5717 PMID: 27599453

Xu, J., Wold, E. A., Ding, Y., Shen, Q., & Zhou, J. (2018). Therapeutic potential of oridonin and its analogs: From anticancer and antiinflammation to neuroprotection. *Molecules (Basel, Switzerland), 23*(2), 474. DOI: 10.3390/molecules23020474 PMID: 29470395

Yan, J., Luo, A., Gao, J., Tang, X., Zhao, Y., & Zhou, B. (2019). The role of SIRT1 in neuroinflammation and cognitive dysfunction in aged rats after anesthesia and surgery. *American Journal of Translational Research, 11*(3), 1555. PMID: 30972182

Yang, N., Li, L., Li, Z., Ni, C., Cao, Y., Liu, T., Tian, M., Chui, D., & Guo, X. (2017). Protective effect of dapsone on cognitive impairment induced by propofol involves hippocampal autophagy. *Neuroscience Letters*, *649*, 85–92. DOI: 10.1016/j.neulet.2017.04.019 PMID: 28411068

Yang, Y., Hu, Z., Du, X., Davies, H., Huo, X., & Fang, M. (2017). miR-16 and fluoxetine both reverse autophagic and apoptotic change in chronic unpredictable mild stress model rats. *Frontiers in Neuroscience*, *11*, 428. DOI: 10.3389/fnins.2017.00428 PMID: 28790887

Ye, S., Fang, L., Xie, S., Hu, Y., Chen, S., Amin, N., Fang, M., & Hu, Z. (2023). Resveratrol alleviates postpartum depression-like behavior by activating autophagy via SIRT1 and inhibiting AKT/mTOR pathway. *Behavioural Brain Research*, *438*, 114208. DOI: 10.1016/j.bbr.2022.114208 PMID: 36356720

Yi, L. T., Xu, H. L., Feng, J., Zhan, X., Zhou, L. P., & Cui, C. C. (2011). Involvement of monoaminergic systems in the antidepressant-like effect of nobiletin. *Physiology & Behavior*, *102*(1), 1–6. DOI: 10.1016/j.physbeh.2010.10.008 PMID: 20951716

Zeng, M., & Zhou, J.-N. (2008). Roles of autophagy and mTOR signaling in neuronal differentiation of mouse neuroblastoma cells. *Cellular Signalling*, *20*(4), 659–665. DOI: 10.1016/j.cellsig.2007.11.015 PMID: 18207367

Zhang, H., Li, S., Si, Y., & Xu, H. (2021). Andrographolide and its derivatives: Current achievements and future perspectives. *European Journal of Medicinal Chemistry*, *224*, 113710. DOI: 10.1016/j.ejmech.2021.113710 PMID: 34315039

Zhang, H., Shang, Y., Xiao, X., Yu, M., & Zhang, T. (2017). Prenatal stress-induced impairments of cognitive flexibility and bidirectional synaptic plasticity are possibly associated with autophagy in adolescent male-offspring. *Experimental Neurology*, *298*, 68–78. DOI: 10.1016/j.expneurol.2017.09.001 PMID: 28882411

Zhang, J. J., Gao, T. T., Wang, Y., Wang, J. L., Guan, W., Wang, Y. J., Wang, C.-N., Liu, J.-F., & Jiang, B. (2019). Andrographolide Exerts Significant Antidepressant-Like Effects Involving the Hippocampal BDNF System in Mice. *The International Journal of Neuropsychopharmacology*, *22*(9), 585–600. DOI: 10.1093/ijnp/pyz032 PMID: 31181145

Zhang, W., Wu, Y., Guo, Y., Wang, C., Long, X., & Jiang, P. (2022). Curcumin relieves corticosterone-induced behaviour deficits via PGC-1α-mediated mitochondrial biogenesis and mitophagy. *Clinical and Translational Discovery*, *2*(4), e161. DOI: 10.1002/ctd2.161 PMID: 36712172

Zhang, W., Zhou, R., Lei, X., Wang, M., Duan, Q., Miao, Y., Zhang, T., Li, X., Zutong, Z., Wang, L., Jones, O. D., Xu, M., Bryant, J., Ma, J., Liu, Y., & Xu, X. (2025). Molecular mechanism on autophagy associated cardiovascular dysfunction in Drosophila melanogaster. *Frontiers in Cell and Developmental Biology*, *13*, 1512341. DOI: 10.3389/fcell.2025.1512341 PMID: 40099194

Zhang, X., Bu, H., Jiang, Y., Sun, G., Jiang, R., Huang, X., Duan, H., Huang, Z., & Wu, Q. (2019). The antidepressant effects of apigenin are associated with the promotion of autophagy via the mTOR/AMPK/ULK1 pathway. *Molecular Medicine Reports*, *20*(3), 2867–2874. DOI: 10.3892/mmr.2019.10491 PMID: 31322238

Zhang, Y., Liu, C., Zhao, Y., Zhang, X., Li, B., & Cui, R. (2015). The effects of calorie restriction in depression and potential mechanisms. *Current Neuropharmacology*, *13*(4), 536–542. DOI: 10.2174/1570159X13666150326003852 PMID: 26412073

Zhang, Z., Cai, X., Yao, Z., Wen, F., Fu, Z., Zhang, J., Zhong, Z., Huang, Y., & Qu, S. (2020). EA ameliorated depressive behaviors in CUMS rats and was related to its suppressing autophagy in the hippocampus. *Neural Plasticity*, *2020*(1), 8860968. DOI: 10.1155/2020/8860968 PMID: 33029121

Zhao, L., Wang, J.-L., Liu, R., Li, X.-X., Li, J.-F., & Zhang, L. (2013). Neuroprotective, anti-amyloidogenic and neurotrophic effects of apigenin in an Alzheimer's disease mouse model. *Molecules (Basel, Switzerland)*, *18*(8), 9949–9965. DOI: 10.3390/molecules18089949 PMID: 23966081

Zhao, X., Zhang, Q., Wang, Y., Li, S., Yu, X., Wang, B., & Wang, X. (2021). Oridonin induces autophagy-mediated cell death in pancreatic cancer by activating the c-Jun N-terminal kinase pathway and inhibiting phosphoinositide 3-kinase signaling. *Annals of Translational Medicine*, *9*(13), 1084. DOI: 10.21037/atm-21-2630 PMID: 34422996

Zhao, Z., Zhang, L., Guo, X.-D., Cao, L.-L., Xue, T.-F., Zhao, X.-J., Yang, D.-D., Yang, J., Ji, J., Huang, J.-Y., & Sun, X.-L. (2017). Rosiglitazone exerts an anti-depressive effect in unpredictable chronic mild-stress-induced depressive mice by maintaining essential neuron autophagy and inhibiting excessive astrocytic apoptosis. *Frontiers in Molecular Neuroscience*, *10*, 293. DOI: 10.3389/fnmol.2017.00293 PMID: 28959186

Zhou, Y., Li, X., Gong, W., Tian, J., Gao, X., Gao, L., Zhang, X., Du, G., & Qin, X. (2017). Protective effect of isoliquiritin against corticosterone-induced neurotoxicity in PC12 cells. *Food & Function*, *8*(3), 1235–1244. DOI: 10.1039/C6FO01503D PMID: 28229156

Zhou, Y., Yan, M., Pan, R., Wang, Z., Tao, X., Li, C., Xia, T., Liu, X., & Chang, Q. (2021). Radix Polygalae extract exerts antidepressant effects in behavioral despair mice and chronic restraint stress-induced rats probably by promoting autophagy and inhibiting neuroinflammation. *Journal of Ethnopharmacology*, *265*, 113317. DOI: 10.1016/j.jep.2020.113317 PMID: 32861821

Zhu, L.-H., Bi, W., Qi, R., Wang, H., & Lu, D. (2011). Luteolin inhibits microglial inflammation and improves neuron survival against inflammation. *The International Journal of Neuroscience*, *121*(6), 329–336. DOI: 10.3109/00207454.2011.569040 PMID: 21631167

Zschocke, J., & Rein, T. (2011). Antidepressants encounter autophagy in neural cells. *Autophagy*, *7*(10), 1247–1248. DOI: 10.4161/auto.7.10.16520 PMID: 21642768

Zschocke, J., Zimmermann, N., Berning, B., Ganal, V., Holsboer, F., & Rein, T. (2011). Antidepressant drugs diversely affect autophagy pathways in astrocytes and neurons—Dissociation from cholesterol homeostasis. *Neuropsychopharmacology : Official Publication of the American College of Neuropsychopharmacology*, *36*(8), 1754–1768. DOI: 10.1038/npp.2011.57 PMID: 21508931

# Chapter 5
# Autophagy in Spinal Cord Injury:
## A Double-Edged Sword in Preclinical Studies of Synthetic and Natural Compounds

**Tahmineh Mokhtari**
https://orcid.org/0000-0002-4126-7102
*Department of Entomology and Nematology, University of California, Davis, USA*

**Farzaneh Majidi**
https://orcid.org/0009-0005-1908-7606
*Department of Anatomy, Tehran University of Medical Sciences, Tehran, Iran*

**Fatemeh Sheikhbahaei**
*Department of Anatomical Sciences, Kerman University of Medical Sciences, Kerman, Iran*

**Parvaneh Behi Shahreza**
https://orcid.org/0000-0002-1196-1224
*Department of Anatomy, Tehran University of Medical Sciences, Tehran, Iran*

**Mahnaz Poorhassan**
https://orcid.org/0000-0002-7684-1637
*Department of Anatomical Sciences and Cognitive Neuroscience, Tehran Medical Sciences, Islamic Azad University, Tehran, Iran*

## ABSTRACT

*Autophagy is a crucial cellular process involved in spinal cord injury (SCI) patho-*

DOI: 10.4018/979-8-3693-5908-2.ch005

*physiology. It can protect neural tissues by removing damaged cellular components, thereby promoting recovery. However, excessive autophagy may lead to neuronal apoptosis, inflammation, and disrupted cellular balance. This chapter explores the balance between autophagic activity and cellular health, emphasizing the need for precise regulation to harness autophagy's therapeutic potential. Recent studies highlight how autophagy interacts with mitochondrial dysfunction, reactive oxygen species, and inflammatory responses in SCI. Moreover, it discusses how natural compounds and nutraceuticals can modulate autophagy, offering promising strategies for managing consequences of SCI. Understanding autophagy is vital for advancing therapeutic interventions in SCI.*

## 1. INTRODUCTION

Spinal cord injury (SCI) is a complex condition that results in sensory, motor, and autonomic dysfunction, often leading to long-term disabilities and significant healthcare burdens (Cunha et al., 2023). Traumatic SCI (tSCI) results from external physical forces, such as falls, sports injuries, and vehicular accidents (Grassner et al., 2016). Current treatment strategies include both surgical and conservative approaches. Recent studies suggest that targeting molecular pathways involved in SCI may offer therapeutic benefits by modulating cellular responses to injury (Ray, 2020; Luo & Tao, 2020). SCI progresses through two phases: the primary injury, caused by mechanical trauma leading to neural tissue compression and axonal disruption, and the secondary injury, characterized by oxidative stress, inflammation, and excitotoxicity, which exacerbate neuronal damage (Ghaffari et al., 2024; Ahuja et al., 2017). A key pathological feature of secondary injury is the activation of the NLRP3 inflammasome, contributing to immune responses and programmed cell death (Mokhtari & Uludag, 2023). Autophagy, a crucial cellular process, plays a dual role in SCI, with evidence suggesting both neuroprotective and detrimental effects (Vargova et al., 2021), serving as an evolutionarily conserved process where lysosomes degrade cellular components to maintain metabolic balance. It progresses through four stages: initiation, autophagosome formation, maturation, and degradation (Chen et al., 2024), ultimately preserving cellular homeostasis and preventing disease (Yu et al., 2018). While autophagy supports neuroprotection and functional recovery in SCI by reducing apoptosis and inflammation (Tang et al., 2014; Liao et al., 2021), excessive activation may contribute to cellular damage, highlighting the need for precise regulation regulation (Wang et al., 2024; Wu et al., 2024).

Neurons, due to their postmitotic nature and specialized morphology, are particularly vulnerable to autophagy–lysosomal dysfunction. These dysfunctions intensify with age and are associated with neurodegenerative diseases, including

Parkinson's disease, Alzheimer's disease, and frontotemporal dementia (Nixon, 2024). Several key regulators modulate autophagy, including the Beclin1 complex and the inhibitory effect of mammalian target of rapamycin (mTOR) on adenosine monophosphate-activated protein kinase (AMPK) (Wang et al., 2022). Notably, studies report Beclin1 overexpression and autophagosome accumulation in SCI lesions, further emphasizing autophagy's complex role in injury response (Yu et al., 2018; Liao et al., 2021). This chapter explores the multifaceted role of autophagy in SCI, examining its protective and detrimental effects.

## 2. AUTOPHAGY AND ITS ROLE IN SPINAL CORD

Autophagy, meaning "self-eating," is a fundamental cellular process that degrades and recycles damaged organelles and proteins to maintain homeostasis (Pattison & Korolchuk, 2019). It occurs through microautophagy, chaperone-mediated autophagy (CMA), and macroautophagy, the predominant form (Assaye & Gizaw, 2022). Macroautophagy progresses through four key stages: initiation, autophagosome formation, maturation, and degradation, regulated by cellular stressors such as oxidative stress, hypoxia, and energy deprivation (Gao, 2019). The initiation phase begins with phagophore formation at the phagophore assembly site (PAS), involving the ULK complex and regulated by TOR inhibition and AMPK activation (Alers et al., 2012). Once activated, the ULK complex stimulates class III phosphatidylinositol 3-kinase (PtdIns3K), which plays a central role in autophagy by regulating phagophore nucleation through Beclin1 (BECN1), NRBF2, ATG14, and UVRAG (Itakura et al., 2008; Burman & Ktistakis, 2010). T The maturation phase involves ubiquitin-like conjugation systems, including ATG5-ATG12 and LC3 lipidation, facilitating autophagosome formation (Mannack & Lane, 2015). Selective autophagy relies on cargo adapter proteins such as P62 and NBR1, which direct specific cellular components into the autophagosome (Johansen & Lamark, 2011). The final step, fusion of the autophagosome with the lysosome to form an autolysosome, is mediated by critical proteins such as PLEKHM1, EPG5, Rab7, and the HOPS complex, leading to enzymatic degradation of sequestered cellular components (Ke, 2024). Autophagy plays a dual role in SCI, where its activation can provide neuroprotection by reducing apoptosis and inflammation, while excessive autophagy may contribute to cellular damage, necessitating precise regulation (Liao et al., 2021). The involvement of mediators in the regulation of autophagy is summarized in Figure 1.

*Figure 1. Overview of various cellular stressors that modulate the activity of autophagy*

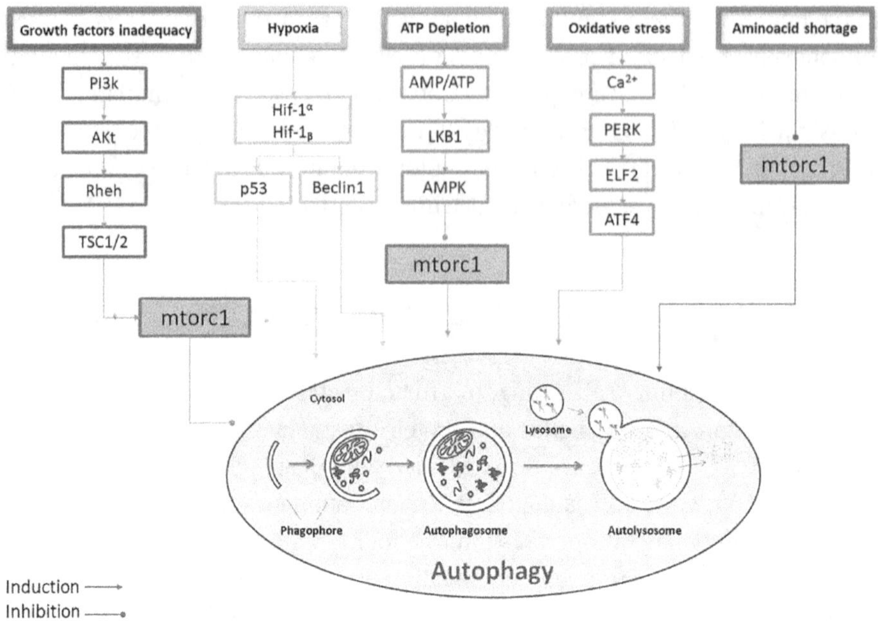


Despite extensive research on autophagy as a therapeutic target for experimental SCI treatments, its baseline activity in the uninjured spinal cord remains incompletely understood (Lipinski et al., 2015). The presence of detectable LC3-II levels provides strong evidence for the existence of autophagic structures, indicating that autophagy functions constitutively within the intact spinal cord (Chua et al., 2022). Immunofluorescence assays reveal significant variations in autophagic activity among primary neural cell types in the spinal cord. In neurons, LC3 staining is most pronounced in axonal tracts, while its presence in neuronal somas varies (Muñoz-Galdeano et al., 2018). The increased LC3 staining in axonal tracts suggests heightened autophagosome formation in distal axonal segments. This observation is further supported by Beclin-1 staining, which highlights the retrograde transport of autophagosomes toward the soma (Maday, 2016). In contrast, the weaker LC3 signal in neuronal somas likely reflects efficient degradation of transported autophagosomes (Lee, 2012).

Oligodendrocytes exhibit basal autophagic activity similar to neurons. However, astrocytes show a distinct pattern, with minimal autophagic structures. LC3 staining is largely absent in astrocytes throughout the intact spinal cord, except in those near the pia mater (Hou et al., 2014). This lack of autophagic structures suggests a disruption in early autophagy stages, as evidenced by the failure to activate Beclin-

1-dependent signaling pathways (Muñoz-Galdeano et al., 2018; Hou et al., 2014) . Thus, autophagic flux in spinal cord astrocytes appears to be significantly low or inactive under normal physiological conditions. The regulation of autophagy in SCI is summarized in Figure 2.

*Figure 2. Summary of the regulatory mechanisms of autophagy in SCI*

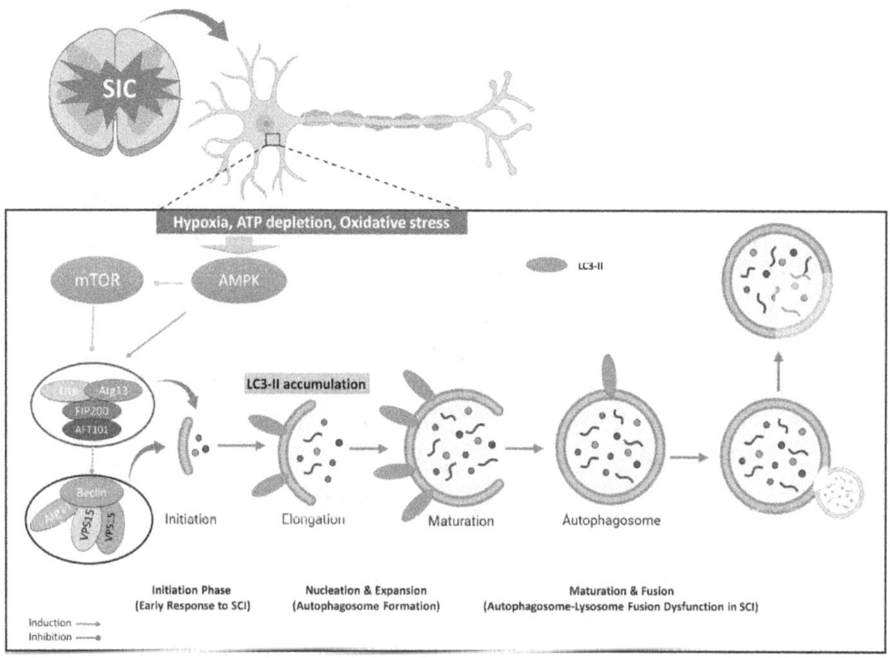


## 2.1 Post-SCI Autophagy Activation

The activation and extent of autophagy play complex roles in modulating the pathogenesis of SCI. However, its exact function in SCI pathophysiology remains a topic of debate, with many mechanisms yet to be fully defined (Lipinski et al., 2015). Following SCI, autophagy activity increases rapidly, rising as early as 0.5 hours post-injury, peaking at 24 hours, and remaining elevated for up to 72 hours after the initial trauma (Wang et al., 2014; Sakai et al., 2014). Moreover, studies in both animal models and human subjects indicate that autophagy markers can persist for weeks to months after the injury (Chen et al., 2012). In the early stages of secondary injury following CNS trauma, experimental studies in murine models have shown a significant increase in autophagy-related markers, including elevated expression of

Beclin-1 and LC3B-II, as well as an accumulation of autophagosomes (Kanno et al., 2011). Similar findings have been observed in human autopsy samples (Sakai et al., 2014). A more recent quantitative cellular analysis of complete spinal cord sections during the subacute phase (7 days post-injury) in a mouse contusion SCI model revealed significantly increased LC3 staining in neurons compared to uninjured controls. This suggests that autophagy remains highly active during the subacute phase following SCI (Muñoz-Galdeano et al., 2018). Overall, the findings show significant autophagy activation in the early phases of SCI across various models.

## 2.2 Post-SCI Autophagy Flux

Autophagy is a vital intracellular degradation pathway that maintains cellular equilibrium and functional stability within the CNS. Animal studies indicate that autophagic flux is disrupted following SCI, suggesting an incomplete autophagy process (Liu et al., 2015; Liu et al., 2018). Recent studies suggest that impaired autophagy flux post-SCI is closely associated with endoplasmic reticulum (ER) stress and motor neuron apoptosis (Liu et al., 2015; Zhou et al., 2016). Impaired motor neurons co-express ER stress markers, such as Caspase-12 (Casp-12) and cleaved Caspase-3 (Casp-3), indicating that SCI induces lysosomal dysfunction, disrupts autophagy, triggers ER stress, and ultimately leads to motor neuron apoptosis (Liu et al., 2015). Although the precise mechanisms behind autophagy flux disruption are unclear, intracellular calcium ($Ca^{2+}$) overload—a hallmark of acute SCI—is believed to play a significant role. Excessive $Ca^{2+}$ may cause lysosomal rupture and autophagy flux failure, potentially mediated by calpain, which inhibits the chaperone protein heat shock protein 70 (HSP70) by cleaving it (Yamashima & Oikawa, 2009). The extent of autophagic flux disruption after SCI varies based on injury severity and model type. (Lipinski et al., 2015; Lipinski et al., 2015). Studies have demonstrated increased levels of autophagic biomarkers, such as Sequestosome 1 (SQSTM1) and LC3-II, in multiple SCI animal models (Lipinski et al., 2015). However, the degree of autophagy flux impairment differs depending on the injury model, including: 1) Acute contusion SCI models (Liu et al., 2015); 2) Chronic spinal cord compression models (Tanabe et al., 2011; Chen et al., 2014); and 3) Spinal nerve ligation models (neuropathic pain in rats) (Berliocchi et al., 2011). A proposed classification of preclinical SCI models categorizes injuries as mild, moderate, or severe, suggesting that injury severity differentially regulates autophagy flux and inflammation. Mild SCI may enhance autophagy initiation and preserve autophagy flux, potentially exerting neuroprotective effects (Verma et al., 2019).

In moderate SCI, impaired autophagic flux may lead to neurodegeneration unless therapeutic interventions restore and enhance autophagy pathways (Zhang et al., 2017; Zhou et al., 2016). In contrast, severe SCI can result in autophagy hyperactivation, causing autophagic neuronal death during the secondary injury phase (Lipinski et al., 2015). In moderate SCI, autophagy flux may be inhibited during the acute phase but hyperactivated in the chronic phase, exacerbating progressive neurodegeneration (Ray, 2020; Anjum et al., 2020). This dual-phase dysregulation likely contributes to long-term neurological deterioration (Liao et al., 2021). Mechanisms of autophagy flux inhibition during the acute phase include: 1) Mitochondrial dysfunction due to excessive ROS production; 2) Calpain-mediated cleavage of HSP70, compromising lysosomal integrity; 3) Bax-mediated lysosomal membrane damage, disrupting autolysosome formation; and 4) ER stress, destabilizing autophagosomal membranes (Ray, 2020. These disruptions result in enhanced apoptosis in oligodendrocytes and neurons, ultimately contributing to neurological impairments (Ray, 2020; Fleming et al., 2022). In severe SCI, autophagy flux overactivation is more likely to drive autophagic neuronal death during the secondary injury phase (Ray, 2020; Lipinski et al., 2015). The progression of autophagy flux varies depending on the type and severity of injury. Severe contusion injuries tend to block autophagy flux, whereas hemisection injuries enhance it. Compression injuries exhibit differential effects, with severe compression (15 g for 1 min) inhibiting autophagy flux, while moderate compression (30 g for 1 min) facilitates autophagy activation (Zhao et al., 2017). These findings underscore the complex regulatory mechanisms governing autophagy in SCI. Studies suggest that autophagy flux disruption post-SCI exacerbates ER stress and neuronal cell death (Liu et al., 2015). Therefore, a deeper understanding of injury-specific autophagy flux regulation is crucial for developing targeted therapeutic strategies to improve SCI outcomes.

## 2.3 Post-SCI Autophagy Cell Type Specificity

In SCI, neurons initiate autophagy, or mitophagy, as a protective response to clear damaged mitochondria and harmful proteins, which enhances neuroprotection within the first 24 hours post-injury (Tang et al., 2014). Although many studies primarily focus on ventral horn motor neurons, recent research indicates a greater accumulation of autophagic markers, such as SQSTM1 and LC3, in these motor neurons compared to dorsal horn sensory neurons, despite sensory neurons being affected over a broader area (Liu et al., 2015). This suggests that motor neurons may be particularly vulnerable to disruptions in autophagic flux. Dorsal sensory neurons are also impacted, as seen in neuropathic pain models where spinal nerve ligation impairs autophagy flux within the dorsal horn (Berliocchi et al., 2011; Wu, Pajoohesh-Ganji, et al., 2012).

Following neuronal injury, significant responses occur, including chromatolysis, axonal membrane sealing, and Wallerian degeneration (Bradke et al., 2012). Axonal regeneration and growth cone formation require extensive cytoplasmic and axonal restructuring, which involves local protein synthesis and degradation (Willis & Twiss, 2006). Cytoskeletal reorganization is fundamental to axonal development and regeneration (Bradke et al., 2012; Hur & Zhou, 2012). Notably, autophagosome accumulation has been observed in axons following both traumatic brain injury (TBI) and SCI (Wu & Lipinski, 2019). Induced autophagy stabilizes microtubules by degrading destabilizing proteins, thereby promoting post-injury axonal regeneration (He et al., 2016). Moreover, the administration of microtubule stabilizers enhances axonal regeneration and improves locomotor activity (Ruschel et al., 2015; Hellal et al., 2011).

Other CNS cells, including astrocytes and oligodendrocytes, also exhibit increased autophagic activity after SCI (Kong et al., 2024). Similarly, oligodendrocytes display elevated autophagy levels following SCI (Muñoz-Galdeano et al., 2018). Microglia, the resident immune cells of the CNS, upregulate LC3 expression following SCI and TBI (Liu et al., 2024). In TBI, microglia with the highest CD68 upregulation and amoeboid morphology are the most affected. This suggests that autophagy may regulate microglial activation and function after injury (Wu, Pajoohesh-Ganji, et al., 2012). The accumulation of autophagosomes in these glial cells correlates with their distance from the injury site. Specifically, activated microglia and macrophages post-SCI accumulate autophagosomes and experience autophagy flux inhibition. In Becn1$^{+/-}$ mice, which exhibit reduced autophagic activity, neuroinflammation increases, and microglia produce higher levels of pro-inflammatory mediators. This results in worse functional recovery and greater tissue damage. Conversely, treatments that restore autophagy flux help mitigate neuroinflammation and improve recovery outcomes after SCI (Li, Lei, et al., 2022). Notably, autophagosome accumulation in glial cells typically begins around day 3 post-injury (Lipinski et al., 2015). This temporal difference suggests that glial cells may have varying sensitivities to autophagy disruptions or that different mechanisms are involved compared to neurons.

## 2.4 Autophagy Dual Role in SCI

Autophagy plays a dual role in SCI. On one hand, it supports cellular survival by eliminating damaged organelles and proteins, thereby protecting neurons. On the other hand, under certain conditions, autophagy can induce cell death, potentially worsening CNS damage. As a result, autophagy is often described as a "double-edged sword" in SCI.

### 2.4.1 Autophagy Neuroprotective Role in SCI

Maintaining effective autophagic flux during the acute phase of SCI is crucial for clearing damaged cells, organelles, and debris. This process promotes neuroprotection and aids functional recovery (Li, Li, & Wu, 2022). Autophagy operates through two primary intracellular protein degradation pathways: the ubiquitin-proteasome system (UPS) and the autophagy-lysosome system. These systems work together to enhance neuroprotection (Nedelsky et al., 2008). Following SCI, autophagy activation helps counteract homeostatic disturbances, such as ER stress. In contrast, disruptions in autophagic flux can intensify ER stress, exacerbating apoptosis (Xu et al., 2013). An inhibitor of the Ser/Thr kinase mTOR signaling pathway has been shown to enhance autophagy by upregulating Beclin-1 and LC3b-II. This leads to reduced neuronal apoptosis and significantly improves motor function in rats with acute SCI (Wang et al., 2014). Basal autophagic activity is essential for cellular homeostasis, particularly in terminally differentiated cells like oligodendrocytes and neurons (Aber et al., 2022). The importance of this process is underscored by studies showing that knocking out key autophagy genes, such as Atg5 or Atg7, in neural tissue results in severe neurodegeneration, motor dysfunction, and reflex abnormalities. The inhibition of autophagic flux primarily leads to cellular damage and death due to the accumulation of toxic protein aggregates and dysfunctional organelles (Bové et al., 2011).

Enhancing autophagy after SCI in rats has shown neuroprotective effects by preserving spinal cord structure, reducing neuronal apoptosis, and suppressing immune responses. These effects ultimately improve functional recovery, primarily through the inhibition of the Akt/mTOR signaling pathway (Li, Yao, et al., 2021). Additionally, regulated autophagy activation helps maintain blood-brain barrier integrity post-SCI, reducing inflammation, limiting blood cell infiltration, and minimizing neuronal death, thereby supporting neuronal function restoration (Zhou et al., 2016). Autophagy also modulates inflammation after SCI by suppressing the activation of the NLRP3 inflammasome and promoting the clearance of damaged mitochondria. This reduces ROS accumulation and mitigates inflammation-induced tissue damage (Liu et al., 2024; Cao et al., 2019). Metformin protects against SCI and preserves motor neurons to enhance recovery and pyroptosis by activating AMPK, suppressing NLRP3 inflammasome activation, and reducing inflammatory mediators (IL-6, IL-1β, and TNF-α) (Yuan et al., 2022).

Furthermore, direct administration of vascular endothelial growth factor (VEGF) in rat models of trauma-induced SCI has been shown to lower inflammatory cytokine levels, including IL-1β, TNF-α, and IL-10 (Wu et al., 2018). These findings emphasize autophagy's vital role in promoting neuroprotection and facilitating functional recovery following SCI.

## 2.4.2 Autophagy Pathogenic Role in SCI

Properly regulated autophagy is a crucial cellular process that supplies essential components for protein synthesis, cellular restructuring, and ATP production. It alleviates ER stress and promotes cell survival. However, excessive autophagy activation can lead to cellular damage and contribute to pathological conditions. Membrane trafficking systems, including autophagy, must maintain a balance between activation and inhibition to ensure precise cargo delivery in both space and time. This underscores the necessity of strict autophagy regulation (Liang, 2010). When dysregulated, autophagy disrupts cellular homeostasis, impairs metabolic processes, and can induce cell death via autophagic lysis (Hagberg et al., 2014). Numerous studies have established a correlation between autophagy and inflammation, particularly regarding upstream inflammatory factors across various diseases (Netea-Maier et al., 2016; Heneka et al., 2015). Both autophagy and apoptosis increase shortly after an injury, highlighting their interconnected roles in cellular stress responses (Hiebert et al., 2015; Tobin et al., 2014). An excessive rise in autophagy flux during severe SCI progression can lead to neuronal cell death, either through uncontrolled autophagic activity (autophagic death) or by triggering apoptosis and other cell death pathways (Huang et al., 2023). This overactivation of autophagy also hampers axonal regeneration, impairing neural function recovery.

Autophagic cell death presents distinct morphological features compared to apoptosis. A key characteristic is the significant accumulation of autophagic vacuoles, which are absent in apoptosis (Dupont et al., 2011). Additionally, excessive autophagy disrupts axonal regeneration and neural function recovery (Nagayach & Wang, 2024). Unlike apoptosis, which is associated with nuclear shrinkage and fragmentation, autophagic cell death does not alter nuclear structure, leaving it intact (Clarke, 1990; Diskin et al., 2005). A recent study demonstrated increased LC3-II protein expression and the presence of LC3-positive cells in astrocytes, oligodendrocytes, and neurons after SCI, accompanied by a rise in autophagic vacuoles within damaged cells. Notably, terminal deoxynucleotidyl transferase dUTP Nick-End Labeling (TUNEL)-positive cells with intact, rounded nuclei in LC3-positive cells indicated a strong link to autophagic cell death, distinct from apoptotic nuclear alterations (Kanno et al., 2011). These findings emphasize the critical role of autophagy in neuronal preservation and functional restoration following SCI. Therefore, developing therapeutics that modulate autophagic pathways is essential for enhancing neuroprotection and improving functional recovery after SCI.

## 2.5 Therapeutic Approaches Targeting Autophagy

With no FDA-approved treatments for SCI, research is exploring synthetic and herbal agents to regulate autophagy and mitigate secondary injury. These compounds offer promising therapeutic potential by modulating key autophagic pathways, aiming to prevent neuronal loss and improve SCI outcomes.

### 2.5.1 Synthetic components

Zinc is known to be one of the most pervasive metal ions in the CNS, where it plays an essential role in both physiological and pathophysiological processes (Wang, Fang, & Chen, 2023). In SCI, zinc has been shown to induce autophagy while inhibiting apoptosis by modulating the AMPK/mTOR pathway through key factors such as p-AMPK, LC3B, Beclin-1, phosphorylated p70 ribosomal protein S6 kinase (p-p70S6K), and p-mTOR (Lin et al., 2020). Additionally, zinc has been suggested to aid in functional recovery in SCI by improving neuronal survival, minimizing lesion volume, and inhibiting focal apoptosis in vivo. Neuroprotection is achieved through zinc's ability to inhibit Casp3 and NLRP3 inflammasome-related cytokines, including TNF-α, IL-1β, and IL-6. Moreover, zinc has been found to stimulate autophagy through SIRT3, regulating LC3B and Atg5 expression while also inhibiting NLRP3 inflammasome activation and reducing ROS generation in both injured spinal cord tissues and cultured neurons (Xu et al., 2023).

Melatonin (MT) is a neurohormone primarily secreted by the pineal gland, where it plays a key role in regulating circadian rhythms. MLT supports neuroprotection by reducing oxidative stress, preserving mitochondria, and restoring autophagic balance (Boga et al., 2019). In SCI, MT has been shown to enhance autophagy and reduce apoptosis, thereby promoting locomotor recovery through the regulation of the PI3K/AKT/mTOR signaling pathway. Specifically, MT upregulates PI3K expression while downregulating AKT and mTOR levels following SCI. Furthermore, MT promotes the expression of autophagy-related proteins such as Beclin-1, LC3-II/LC3-I, and p62, while simultaneously decreasing the expression of apoptosis-related proteins like Casp-3, Casp-9, and Bax in neurons after SCI (Li, Guo, et al., 2019). The same study found that MT reduces apoptosis through the SIRT1/AMPK signaling pathway, leading to increased levels of Beclin-1, LC-3B, SIRT1, and p-AMPK in spinal cord tissues. However, this effect is blocked by EX527, a SIRT1 inhibitor, indicating that MT-induced autophagy is mediated through the SIRT1/AMPK pathway in SCI (Gao et al., 2020).

The Wnt family of glycoproteins, which includes Wnt-1, Wnt-3, Wnt-3a, and Wnt-5a, plays a crucial role in regulating neural cell proliferation, composition, and survival (González-Fernández et al., 2014). Wnt-3a, in particular, serves as a key activator of the classical Wnt/β-catenin signaling pathway, which significantly influences the proliferation and differentiation of spinal cord neurons (David et al., 2010). It also supports the continued differentiation of spinal cord precursor cells while inhibiting the expression of protein kinase inhibitors (Park et al., 2013). Studies on Wnt-3a's neuroprotective mechanisms in SCI demonstrated that it significantly increases LC3-II and Beclin-1 levels in spinal cord neurons, promoting neuronal autophagy by downregulating mTOR activity. This effect is inhibited by MHY-1485, and co-treatment with MHY-1485 reduces Wnt-3a's neuroprotective effects, underscoring the critical role of autophagy in these mechanisms (Gao et al., 2020).

Recent advancements in compound development for SCI therapy focus on targeting various pathophysiological pathways, especially through autophagy modulation. These efforts involve specific activators and inhibitors that enhance cellular survival and repair in the injured spinal cord, ultimately improving neuronal recovery and functional outcomes (Haeck, 2023; Abbaszadeh et al., 2020). The endocannabinoid system (ECS) primarily comprises two receptor types: cannabinoid receptor type 1 (CB1R) and CB2R (Zou & Kumar, 2018). It also includes enzymes involved in the production, transport, and breakdown of endocannabinoids, such as 2-arachidonoyl glycerol and anandamide, which are crucial for various physiological processes and pathological conditions (Lu & Mackie, 2021). Notably, activation of CB2R within the CNS has been shown to exert neuroprotective effects in neurodegenerative diseases, primarily due to its role in immune regulation (Vuic et al., 2022). Research suggests that CB2R activation can influence the transition of macrophages from M1 to M2 in response to inflammation and regulate autophagy-related pathways (Liu et al., 2022). CB2R helps reduce neuroinflammation by facilitating the autophagic degradation of the NLRP3 inflammasome following SCI (Jiang et al., 2022a). It promotes the differentiation of M2 microglia and regulates the expression of cytokines such as IL-10, IL-1β, and IL-6. Furthermore, CB2R activation enhances NLRP3 inflammasome ubiquitination and interacts with autophagy-related proteins like p62 and LC3 (Jiang et al., 2022b). Administration of the CB2R agonist JWH-133 has been shown to prevent myelin degradation, reduce neuronal apoptosis, and limit glial scar formation, ultimately aiding in the recovery of hindlimb function. In contrast, treatment with the CB2R antagonist AM630 led to adverse effects, highlighting the critical role of CB2R activation in neuroprotection (Jiang et al., 2022a).

Lysine-specific demethylase 1 (LSD1), also known as KDM1A, is an essential enzyme involved in several physiological processes, including epithelial-mesenchymal transition (EMT), cellular proliferation, cell cycle regulation, chromosome segregation, and cellular motility (Bennani-Baiti et al., 2012). Dysregulation of LSD1 has been strongly linked to tumor development (Noce et al., 2023). Additionally, LSD1 deficiency can lead to severe consequences, including paralysis and extensive neurodegeneration in the hippocampus and cortex, ultimately resulting in learning and memory impairments (Christopher et al., 2017). Treatment with the LSD1 inhibitor SP2509 has been shown to significantly reduce SCI severity in rat models by enhancing autophagy and upregulating apoptotic markers, including Beclin1, LC3II/I, ATG5, ATG7, p62, Bax, Bcl-2, and Casp-3 (Gu et al., 2021).

Cyclin-dependent kinase 1 (CDK1) is a serine/threonine kinase that plays a vital role in regulating the cell cycle, especially in actively dividing cells (Massacci et al., 2023). CDK1 activity is closely associated with apoptotic cell death, with elevated levels typically observed during apoptosis. Inhibition of CDK1, through mutations or chemical inhibitors, is crucial for inducing apoptosis (Castedo et al., 2002). For instance, CDK1 downregulation in glioblastoma cells suppresses cell cycle-related genes while activating growth-related genes (Xie et al., 2005). In SCI, CDK1 inhibition reduces neuronal apoptosis and promotes autophagy, leading to neuroprotective effects. Downregulation of CDK1 using CDK1 shRNA lentivirus significantly decreases microglial and astrocyte activation, enhances motor neuron regeneration, and reduces neuronal apoptosis, with these outcomes linked to changes in pro-apoptotic and autophagy-related proteins, such as Casp-3, Bax, Bcl-2, Beclin1, and LC3 (Nie et al., 2022). These results are summarized in Table 1.

*Table 1. Regulation of autophagy pathways via synthetic components the management of SCI*

| Author (year) | Model | Chemical/Dose | Main results |
|---|---|---|---|
| Lin et al. (2019) [87] | - C57 mice<br>Weight-drop trauma method | - Zink<br>30 mg/kg | - Inhibition of apoptosis<br>- Promotion of autophagy<br>Regulation of key proteins involved in these processes, including p-AMPK, LC3B, Beclin-1, p-p70S6K, and p-mTOR. |
| Xu et al. (2023) [88] | **In vitro**<br>- Murine neurons<br>PC12 cells<br>**In vivo**<br>- C57 mice<br>Weight-drop trauma method | - Zink<br>- 100 μM<br><br><br>30 mg/kg | - Promotion of functional recovery<br>- Increasing neuronal survival and reducing lesion size and focal apoptosis levels.<br>- Reduced mitochondrial ROS production<br>- Inhibition of NLRP3 inflammasome activation, leading to decreased levels of pro-inflammatory cytokines IL-1β, TNF-α, and IL-6. .<br>Promotion of SIRT3-mediated autophagy evidenced by significantly increased levels of LC3B and modulation of autophagy-related proteins such as Atg5. |
| Li et al. (2019) [90] | **In vitro**<br>- Primary neurons from the spinal cord of fetal rats<br><br>**In vivo**<br>- Rat<br>Weight-drop trauma method | - Melatonin<br>- 0, 5, 10, 20, 40, 80, and 100 μM<br><br><br>- 12.5 mg/kg/day | - Improved motor function recovery.<br>- Upregulation of PI3K expression and downregulation of AKT and mTOR<br>- Enhanced autophagy via regulation of LC3-II/LC3-I, Beclin-1, and p62<br>Reduced apoptosis via regulation of Casp-3, Casp-9, and Bax |
| Gao et al. (2020) [91] | - Rat<br>Weight-drop trauma method | - Melatonin<br>10 mg/kg | - Improved hind limb motor function indicated by BBB scores.<br>- Increased survival of motoneurons in the anterior horn of the spinal cord.<br>- Regulation of Casp-3 and Casp-9, leading to decreased apoptosis in neurons and<br>Upregulation of autophagy-associated proteins, including LC3B, Beclin-1, SIRT1, and p-AMPK |
| Gao et al. (2020) [95] | **In vitro**<br>- BV2 microglia and HEK-293T<br><br>**In vivo**<br>- Rat<br>Weight-drop trauma method | - Wnt-3a<br>- 0, 10, 20, 30, 40, 50, 60, 70, 80, 90, and 100 ng/mL<br><br>-50 ng/mL | - Enhanced autophagy via improvement of expression of LC3-II/I and Beclin-1in neurons<br>- Reduction of p70S6K phosphorylation levels while increasing p70S6K protein levels<br>Autophagy initiation via suppressing the mTOR signaling cascade |

continued on following page

*Table 1. Continued*

| Author (year) | Model | Chemical/Dose | Main results |
|---|---|---|---|
| Jiang et al. (2022) [102] | **In vitro**<br>- BV2 microglia and HEK-293T<br><br>**In vivo**<br>- C57BL/6J mice Weight-drop trauma method | - JWH-133<br>A CB2R activator | - Reduction in myelin loss, neuronal apoptosis, and glial scarring, resulting in enhanced recovery of hind limb function.<br>- Promotion of M2 differentiation and modulation of IL-10, IL-1β, and IL-6 expression.<br>Enhanced ubiquitination of the NLRP3 inflammasome through interactions with autophagy-assciated molecules such as p62 and LC3. |
| Gu et al. (2021) [107]. | **In vitro**<br>- PC12 cells<br><br>**In vivo**<br>- Rat<br>Weight-drop trauma method | - SP2509<br>A LSD1 inhibitor | - Contribution to neuronal survival and functional recovery<br>- Inhibition of apoptosis in the injured region via regulation of Bax, Bcl-2, and Casp-3<br>Promotion of autophagy via regulation of Beclin1, LC3II/I, ATG5, ATG7, and p62 |
| Nie et al. (2022) [111] | - Rat<br>Drilling and Screw Placement | CDK1 shRNA lentivirus | - Improved NSS and MEPs<br>- Promotion of motor neuron repair and inhibition of neuronal apoptosis.<br>- Improvement of neuronal apoptosis and autophagy.<br>- Suppression of the astrocytes and microglial activity in the anterior horn.<br>Modulation of pro-apoptotic and autophagy-related proteins, including Casp-3, Bax, Bcl-2, Beclin1, and LC3 |

Akt: Protein Kinase B, ATG: Autophagy-related gene, Bax: Bcl-2-associated X protein, BBB: Basso, Beattie and Bresnahan, Casp: Caspase, HEK: Human embryonic kidney, IL: Interleukin, LC3: Light Chain 3, NLRP3: NOD-like receptor protein 3,-NSS: Neurological severity scores, MEPs: Motor evoked potentials, mTOR: mammalian Target of Rapamycin, PI3K: Ohosphatidylinositol 3-kinase, SIRT1: Sirtuin 1,

## 2.5.2 Herbal Ingredients

Polyphenols, along with other phytochemicals, terpenes, and alkaloids, have shown significant potential in modulating autophagy after SCI due to their anti-inflammatory, anti-apoptotic, and antioxidant properties, making them promising candidates for therapeutic strategies in SCI and other CNS disorders. Resveratrol (RES), a polyphenolic compound found in grape skins, peanuts, and bilberries, has demonstrated the ability to reduce oxidative stress and neuroinflammation, emphasizing its importance in health research (Komorowska et al., 2023). In SCI, RES alleviates complications by suppressing the NF-κB signaling pathway (Xu et al., 2018). When administered intraperitoneally at 100 mg/kg post-SCI, RES enhances the expression of SIRT1, p-AMPK, Bcl-2, LC3, and Beclin-1 while decreasing p62, Casp-3, Casp-9, and Bax. These findings indicate that RES provides neuroprotection by modulating apoptosis and autophagy through the SIRT1/AMPK signaling pathway (Zhao et al., 2017; Yan et al., 2017). Additionally, RES supports recovery by improving locomotor function, reducing neuroinflammation, and activating autophagy through increased LC3-II/LC3-I ratios and Beclin-1 expression. It plays a crucial role in regulating the AMPK/mTOR pathway, vital for neuroprotection and cellular repair (Meng et al., 2018). Another study revealed that RES enhances functional recovery after SCI by promoting neuronal autophagic flux and reducing apoptosis, effectively protecting motor neurons and leading to improved recovery outcomes in Sprague-Dawley rats. The neuroprotective effects of RES are associated with its activation of the LKB1/AMPK/mTOR/p70S6K signaling pathway, essential for regulating autophagic flux (Wang et al., 2018).

Quercetin (QRT), a naturally occurring flavonol found in plants such as nuts, seeds, flowers, and bark, exhibits neuroprotective properties through several pathways, including p38 MAPK, PI3K, the Nrf2 pathway, and modulation of the NLRP3 inflammasome (Carrillo-Martinez et al., 2024; Moratilla-Rivera et al., 2023). In SCI treatment, nanoparticles encapsulating QRT within an acellular scaffold have shown potential for promoting functional recovery and neuroregeneration, associated with an increase in nestin- and GFAP-positive cells and modulation of the NLRP3 inflammasome at the injury site (Ebrahimi et al., 2024). In addition to its anti-inflammatory effects, QRT regulates autophagy, with in vivo studies demonstrating improved neurological outcomes by reducing lesion size and cavity formation while enhancing axonal regeneration. These benefits are linked to increased serotonin-positive nerve fibers and neurofilament 200-positive neurons, alongside a reduction in reactive astrocytes. QRT also supports spinal cord energy metabolism and activates autophagy via the Akt/mTOR/p70S6K pathway, highlighting its neuroprotective potential and role in functional recovery after SCI (Wang et al., 2021).

Curcumin (CC), a polyphenolic compound derived from turmeric (*Curcuma longa*), exhibits antioxidant effects by activating the Nrf2 pathway and reducing ROS through the NF-κB pathway, contributing to its anti-inflammatory properties (Zhong et al., 2016). In the context of SCI, the mTOR pathway is a crucial target for CC (Lin et al., 2017). CC upregulates CDGSH iron sulfur domain 2 (CISD2), a survival gene vital for calcium metabolism and anti-apoptotic functions in SCI. By promoting the interaction between Beclin-1 and Bcl-2, CISD2 helps maintain the balance between autophagy and apoptosis, preventing mitochondrial dysfunction and cell death. Under stress, CISD2 reduces excitotoxic calcium levels in the ER and inhibits nitric oxide (NO) signaling, protecting neurons from apoptosis. The neuroprotective effects of CC in SCI are closely linked to its ability to modulate NO levels via CISD2 (Lin & Lin, 2016). Additionally, CC enhances functional recovery, prevents neuronal apoptosis, supports remyelination, and reduces inflammation by regulating pro-inflammatory cytokines such as TNF-α, IL-6, and IL-1β. It also influences autophagy through the LC3-II/LC3-I ratio and the Akt/mTOR pathway in SCI models (Li, Yao, et al., 2021).

Oxymatrine (OMT), an alkaloid derived from the traditional Chinese medicinal plant *Sophora flavescens* (Zhang et al., 2014), demonstrated to alleviate brain injury in neonatal rats subjected to hypoxia-ischemia (Huang et al., 2018). The neuroprotective effects of OMT are primarily attributed to its ability to regulate the PI3K/Akt/mTOR signaling pathway (Wei et al., 2021). In experimental models of SCI, OMT has been shown to support motor neuron survival, significantly reduce lesion size, and improve motor function. Additionally, OMT enhances autophagic activity by modulating key proteins such as p62, Beclin-1, LC3B, and the LC3B-II/I ratio. It also inhibits neuronal apoptosis by regulating proteins like Casp-3, Bcl-2, and Bcl-x while increasing the expression of p-AMPK and SIRT1(Li et al., 2023).

Acteoside, a bioactive phenylpropanoid glycoside found in various plant species, has been shown to promote autophagy and inhibit Casp-3, along with the modulation of key signaling pathways such as CCL2, CXCL10, ICAM1, VEGF, HIF-1α, NF-κB, and IL-10/STAT3 (Wu et al., 2022; Xiao et al., 2022). In an SCI rat model, acteoside treatment led to significant improvements in locomotor function, with effects observed as early as the second week post-treatment and sustained through the fourth week. Acteoside also increased the expression of MAP1LC3A and MAP-2 while significantly reducing GFAP levels. Additionally, it decreased apoptosis by regulating Bax, Casp-3, and Bcl-2 expression. The treatment further reduced inflammatory markers such as iNOS, TNF-α, and IL-1β, along with a decline in ROS and malondialdehyde (MDA) levels. Conversely, Nrf2 and Keap-1 expression increased, indicating an enhanced antioxidant response. Molecular docking studies revealed strong interactions between acteoside and targets such as MAP1LC3A, TNF-α, and Casp-3 (Ning et al., 2024).

Diosgenin, a steroidal triterpenoid sapogenin, is primarily extracted from plant sources such as *Dioscorea alata, Smilax china, and Trigonella foenum-graecum* (Arya et al., 2023). This bioactive compound has gained attention for its anti-inflammatory properties, which contribute to its neuroprotective potential. Studies have shown that diosgenin mitigates inflammation by suppressing pro-inflammatory cytokines, including TNF-α, IL-6, and IL-1β. Additionally, it exerts neuroprotective effects by reducing apoptosis, as evidenced by the downregulation of Casp-3 and the Bax/Bcl-2 ratio following SCI (Zhang et al., 2017). Recent research by Chen et al. has further explored the neuroprotective mechanisms of diosgenin. It was found to inhibit the

Fas signaling pathway while simultaneously enhancing Bcl-2-mediated PI3K/Akt survival pathways, reinforcing its role in neuronal protection (Luo & Tao, 2020). Another study by Chen et al. demonstrated that diosgenin suppresses p62 expression while activating the Ras/mTOR signaling pathway, leading to reduced apoptosis and enhanced autophagy following SCI (Chen et al., 2018).

Ginsenoside Rh2 (GRh2), an active phytochemical associated with ginseng, has been reported to enhance specific immune responses and influence the anti-inflammatory NLRP3 inflammasome signaling pathway (Jin et al., 2022). Studies have also investigated its role in autophagy regulation (Wang et al., 2020). Both ginsenosides and ginseng extract have demonstrated the ability to improve neurological function and provide antioxidant benefits in animal models of SCI (Sng et al., 2022). In a mouse model of SCI, GRh2 promoted functional recovery by inhibiting pyroptosis through enhanced autophagy—an effect that was reversed by the autophagy inhibitor 3-MA. Moreover, GRh2 facilitated the nuclear translocation and dephosphorylation of transcription factor EB (TFEB), modulating its activity through the EGFR-MAPK pathway. Notably, the presence of NSC228155, an EGFR activator, partially blocked GRh2's effects on this pathway (Liu et al., 2024). Collectively, these findings underscore the potential of herbal ingredients as neuroprotective agents that can facilitate recovery following SCI. A summary of these findings is presented in Table 2.

*Table 2. Regulation of autophagy pathways via herbal ingredients for the management of SCI*

| Author (year) | Model | Agent/Dose | Main results |
|---|---|---|---|
| Zhao et al. (2017) [52] | - Rat<br>Weight-drop trauma method | - Resveratrol 100 mg/kg | - Promotion of functional recovery through improved BBB scoring<br>- Reduction in motor neuron loss and lesion size in the spinal cord.<br>- Enhancement of neuroprotection by modulating autophagy and apoptosis pathways.<br>- Regulation of SIRT1, p-AMPK, Beclin-1, LC3-B, and Bcl-2 expression<br>Inhibition of p62, Bax, Cleaved Casp-3, and Casp-9, expression. |
| Yan et al. (2017) [114] | - Rat<br>Weight-drop trauma method | - Resveratrol 100 mg/kg | - Improvement in cellular response post-trauma<br>- Activation of the AMPK/SIRT1 pathway<br>- Alleviation of Cell Damage.<br>- Promotion of autophagy flux via regulation of of LC3II/LC3I ratio and p62 level |
| Meng et al. (2018) [115] | - Rat<br>Weight-drop trauma method | - Resveratrol 100 mg/kg | - Functional Improvement in Locomotor Activity<br>- Reduction of Neuroinflammation through Decreased Levels of Pro-inflammatory Factors<br>- Activation of Autophagy through Increased Ratios of LC3-II/LC3-I and Elevated Expression of Beclin-1<br>Regulation of the AMPK/mTOR Pathway |
| Wang et al. (2018) [138] | **In vitro**<br>- PC12 cells culture<br>**In vivo**<br>- Rat<br>- Weight-drop trauma method | - Resveratrol 200 mg/kg | - Reduction of Neuronal Apoptosis<br>- Activation of Autophagic flux via regulation of of LKB1/AMPK/mTOR/p70s6k signalling pathway, |
| Li et al. (2023) [127] | - Rat<br>a sterile metal impounder | - Oxymatrine (N1835)<br>40 mg/mL | - Reduction of lesion size and motor dysfunction<br>- Enhancement autophagy activity by regulating key proteins such as p62, Beclin-1, LC3B, and the ratio of LC3B-II/I.<br>- Inhibition of apoptosis: by modulating the expression of proteins such as Casp-3, Bcl-2, and Bcl-x, contributing to neuronal survival.<br>Increased expression of SIRT1 and p-AMPK, which are important for cellular energy regulation and autophagy. |

continued on following page

*Table 2. Continued*

| Author (year) | Model | Agent/Dose | Main results |
|---|---|---|---|
| Ning et al. (2024) [130] | - Rat Weight-drop trauma method | - Acteoside 20, 40, 60, and 10 mg/kg/day | - Improvement in locomotor activity<br>- Increased expression of autophagy-related proteins MAP-1, LC3A and MAP-2<br>- Reduction in the levels of GFAP, indicating decreased astrocytic activation and potentially reduced glial scarring.<br>- Decreased the expression of inflammatory markers, including iNOS, TNF-α, and IL-1β<br>- Modulation of apoptotic proteins such as Bax, Casp-3, and Bcl-2, promoting neuronal survival.<br>- Reduction in levels of ROS and MDA<br>- Increase in the expression of Nrf2 and Keap-1<br>Strong affinity with targets MAP1LC3A, TNF-α, and Casp-3 |
| Chen et al. (2018) [133] | - Rat Weight-drop trauma method | - Diosgenin 100 mg/kg | - Significant functional recovery<br>- Reduced tissue damage and edema<br>- Modulation of p62 expression and the Rheb/mTOR signalling pathway<br>Decreased levels of miR-155-3p |
| Wang et al. (2018) [120] | - Rat Weight-drop trauma method | - Quercetin 20 mg/kg | - Improved neurological outcomes through reduced latency in response times and increased amplitudes of somatosensory and motor evoked potentials.<br>- Decreased lesion size and cavity formation, leading to improved functional recovery.<br>- Enhanced axonal regeneration.<br>- Increased levels of serotonin-positive nerve fibers.<br>- Elevated levels of neurofilament 200-positive neurons, indicating successful neuronal survival and regeneration.<br>- Reduction in reactive astrocytes, resulting in decreased inflammation and a more favorable environment for neuronal repair and regeneration.<br>Facilitated spinal cord energy metabolism and activated autophagy via the Akt/mTOR/p70S6K pathway. |

continued on following page

*Table 2. Continued*

| Author (year) | Model | Agent/Dose | Main results |
|---|---|---|---|
| Li et al. (2021) [70] | - Rat Weight-drop trauma method | - Curcumin 60 mg/ml/kg | - Promotion of functional recovery through improved BBB scoring<br>- Enhanced spinal cord integrity<br>- Increased re-myelination<br>- Suppressed inflammatory response via NF-α, IL-6, and IL-1β<br>Enhanced autophagy through regulation of LC3-II/LC-I ratio and Akt/mTOR pathway |
| Liu et al. (2024) [137] | - Mice Weight-drop trauma method | - Ginsenoside-Rh2 10, 20, and 40 mg/kg | - Promoted functional recovery.<br>- Inhibition of pyroptosis by enhancing autophagy.<br>- Promotion of TFEB nuclear translocation and dephosphorylation.<br>Modulation of the EGFR-MAPK signalling pathway. |

Akt: Protein Kinase B, AMPK: AMP-Activated Protein Kinase, Bax: Bcl-2-associated X protein, BBB: Basso, Beattie and Bresnahan, Bcl-2: B-cell lymphoma 2, Casp: Caspase, LC3-B: Microtubule-associated protein 1A/1B-light chain 3 beta, GFAP: Glial Fibrillary Acidic Protein, IL: Interleukin, LC3: Light Chain 3, iNOS: Inducible nitric oxide synthase, KB1: Liver kinase B1, Keap-1: Kelch-like ECH-associated protein 1, mTOR: Mammalian Target of Rapamycin, MDA: Malondialdehyde, MAP: Microtubule-associated protein, NF-α: Nuclear Factor Alpha, Nrf2: Nuclear factor (erythroid-derived 2)-like 2, SCI: Spinal Cord Injury, SIRT1: Sirtuin 1, TFEB: Transcription factor EB, TNF-α: Tumor Necrosis Factor-alpha, ROS: Reactive Oxygen Species,

## 6. CONCLUSION

Understanding the interaction between autophagy, mitochondrial dysfunction, and inflammation has provided valuable insights into its role in SCI progression. Pharmacological and natural compounds that regulate autophagy show promise in reducing secondary injury and promoting neuroprotection. However, challenges remain in translating these findings to clinical applications due to inconsistencies in injury models, treatment timing, and drug bioavailability. Additionally, prolonged autophagy modulation may pose risks, such as immune suppression or disruptions in cellular homeostasis. Given its dual role as both protective and detrimental, refining therapeutic strategies is crucial. Advancing preclinical and clinical research will be essential for optimizing autophagy-targeted treatments, ultimately improving recovery and quality of life for SCI patients.

## REFERENCES


Abbaszadeh, F., Fakhri, S., & Khan, H. (2020). Targeting apoptosis and autophagy following spinal cord injury: Therapeutic approaches to polyphenols and candidate phytochemicals. *Pharmacological Research, 160*, 105069. DOI: 10.1016/j.phrs.2020.105069 PMID: 32652198

Aber, E. R., Griffey, C. J., Davies, T., Li, A. M., Yang, Y. J., Croce, K. R., Goldman, J. E., Grutzendler, J., Canman, J. C., & Yamamoto, A. (2022). Oligodendroglial macroautophagy is essential for myelin sheath turnover to prevent neurodegeneration and death. *Cell Reports*, *41*(3), 111480. DOI: 10.1016/j.celrep.2022.111480 PMID: 36261002

Ahuja, C. S., Wilson, J. R., Nori, S., Kotter, M., Druschel, C., Curt, A., & Fehlings, M. G. (2017). Traumatic spinal cord injury. *Nature Reviews. Disease Primers*, *3*(1), 1–21. DOI: 10.1038/nrdp.2017.18 PMID: 28447605

Alers, S., Löffler, A. S., Wesselborg, S., & Stork, B. (2012). Role of AMPK-mTOR-Ulk1/2 in the regulation of autophagy: Cross talk, shortcuts, and feedbacks. *Molecular and Cellular Biology*, *32*(1), 2–11. DOI: 10.1128/MCB.06159-11 PMID: 22025673

Anjum, A., Yazid, M. D., Fauzi Daud, M., Idris, J., Ng, A. M. H., Selvi Naicker, A., Ismail, O. H. R., Athi Kumar, R. K., & Lokanathan, Y. (2020). Spinal cord injury: Pathophysiology, multimolecular interactions, and underlying recovery mechanisms. *International Journal of Molecular Sciences*, *21*(20), 7533. DOI: 10.3390/ijms21207533 PMID: 33066029

Arya, P., Munshi, M., & Kumar, P. (2023). Diosgenin: Chemistry, extraction, quantification and health benefits. *Food Chemistry Advances*, *2*, 100170. DOI: 10.1016/j.focha.2022.100170

Assaye, M. A., & Gizaw, S. T. (2022). Chaperone-mediated autophagy and its implications for neurodegeneration and cancer. *International Journal of General Medicine*, *15*, 5635–5649. DOI: 10.2147/IJGM.S368364 PMID: 35734200

Bennani-Baiti, I. M., Machado, I., Llombart-Bosch, A., & Kovar, H. (2012). Lysine-specific demethylase 1 (LSD1/KDM1A/AOF2/BHC110) is expressed and is an epigenetic drug target in chondrosarcoma, Ewing's sarcoma, osteosarcoma, and rhabdomyosarcoma. *Human Pathology*, *43*(8), 1300–1307. DOI: 10.1016/j.humpath.2011.10.010 PMID: 22245111

. Berliocchi L, Russo R, Maiarù M, Levato A, Bagetta G, Corasaniti MT. Autophagy impairment in a mouse model of neuropathic pain. Molecular pain. 2011;7:1744-8069-7-83.

Boga, J. A., Caballero, B., Potes, Y., Perez-Martinez, Z., Reiter, R. J., Vega-Naredo, I., & Coto-Montes, A. (2019). Therapeutic potential of melatonin related to its role as an autophagy regulator: A review. *Journal of Pineal Research, 66*(1), e12534. DOI: 10.1111/jpi.12534 PMID: 30329173

Bové, J., Martínez-Vicente, M., & Vila, M. (2011). Fighting neurodegeneration with rapamycin: Mechanistic insights. *Nature Reviews. Neuroscience, 12*(8), 437–452. DOI: 10.1038/nrn3068 PMID: 21772323

Bradke, F., Fawcett, J. W., & Spira, M. E. (2012). Assembly of a new growth cone after axotomy: The precursor to axon regeneration. *Nature Reviews. Neuroscience, 13*(3), 183–193. DOI: 10.1038/nrn3176 PMID: 22334213

Burman, C., & Ktistakis, N. T. (2010). Regulation of autophagy by phosphatidylinositol 3-phosphate. *FEBS Letters, 584*(7), 1302–1312. DOI: 10.1016/j.febslet.2010.01.011 PMID: 20074568

Cao, Z., Wang, Y., Long, Z., & He, G. (2019). Interaction between autophagy and the NLRP3 inflammasome. *Acta Biochimica et Biophysica Sinica, 51*(11), 1087–1095. DOI: 10.1093/abbs/gmz098 PMID: 31609412

Carrillo-Martinez, E. J., Flores-Hernández, F. Y., Salazar-Montes, A. M., Nario-Chaidez, H. F., & Hernández-Ortega, L. D. (2024). Quercetin, a flavonoid with great pharmacological capacity. *Molecules (Basel, Switzerland), 29*(5), 1000. DOI: 10.3390/molecules29051000 PMID: 38474512

Castedo, M., Perfettini, J. L., Roumier, T., & Kroemer, G. (2002). Cyclin-dependent kinase 1: Linking apoptosis to cell cycle and mitotic catastrophe. *Cell Death and Differentiation, 9*(12), 1287–1293. DOI: 10.1038/sj.cdd.4401130 PMID: 12478465

Chen, H.-C., Fong, T.-H., Lee, A.-W., & Chiu, W.-T. (2012). Autophagy is activated in injured neurons and inhibited by methylprednisolone after experimental spinal cord injury. *Spine, 37*(6), 470–475. DOI: 10.1097/BRS.0b013e318221e859 PMID: 21587101

Chen, L., Gao, T., Zhou, P., Xia, W., Yao, H., Xu, S., & Xu, J. (2024). Recent advances of vacuolar protein-sorting 34 inhibitors targeting autophagy. *Bioorganic Chemistry, 143*, 107039. DOI: 10.1016/j.bioorg.2023.107039 PMID: 38134519

Chen, X.-B., Wang, Z.-L., Yang, Q.-Y., Zhao, F.-Y., Qin, X.-L., Tang, X.-E., Du, J.-L., Chen, Z.-H., Zhang, K., & Huang, F.-J. (2018). Diosgenin glucoside protects against spinal cord injury by regulating autophagy and alleviating apoptosis. *International Journal of Molecular Sciences, 19*(8), 2274. DOI: 10.3390/ijms19082274 PMID: 30072674

Chen, Z., Fu, Q., Shen, B., Huang, X., Wang, K., He, P., Li, F., Zhang, F., & Shen, H. (2014). Enhanced p62 expression triggers concomitant autophagy and apoptosis in a rat chronic spinal cord compression model. *Molecular Medicine Reports*, *9*(6), 2091–2096. DOI: 10.3892/mmr.2014.2124 PMID: 24715058

Christopher, M. A., Myrick, D. A., Barwick, B. G., Engstrom, A. K., Porter-Stransky, K. A., Boss, J. M., Weinshenker, D., Levey, A. I., & Katz, D. J. (2017). LSD1 protects against hippocampal and cortical neurodegeneration. *Nature Communications*, *8*(1), 805. DOI: 10.1038/s41467-017-00922-9 PMID: 28993646

Chua, J. P., De Calbiac, H., Kabashi, E., & Barmada, S. J. (2022). Autophagy and ALS: Mechanistic insights and therapeutic implications. *Autophagy*, *18*(2), 254–282. DOI: 10.1080/15548627.2021.1926656 PMID: 34057020

Clarke, P. G. (1990). Developmental cell death: Morphological diversity and multiple mechanisms. *Anatomy and Embryology*, *181*(3), 195–213. DOI: 10.1007/BF00174615 PMID: 2186664

Cunha, N. S. C., Malvea, A., Sadat, S., Ibrahim, G. M., & Fehlings, M. G. (2023). Pediatric Spinal Cord Injury: A Review. *Children (Basel, Switzerland)*, *10*(9), 1456. DOI: 10.3390/children10091456 PMID: 37761417

David, M. D., Cantí, C., & Herreros, J. (2010). Wnt-3a and Wnt-3 differently stimulate proliferation and neurogenesis of spinal neural precursors and promote neurite outgrowth by canonical signaling. *Journal of Neuroscience Research*, *88*(14), 3011–3023. DOI: 10.1002/jnr.22464 PMID: 20722074

Diskin, T., Tal-Or, P., Erlich, S., Mizrachy, L., Alexandrovich, A., Shohami, E., & Pinkas-Kramarski, R. (2005). Closed head injury induces upregulation of Beclin 1 at the cortical site of injury. *Journal of Neurotrauma*, *22*(7), 750–762. DOI: 10.1089/neu.2005.22.750 PMID: 16004578

Dupont, N., Jiang, S., Pilli, M., Ornatowski, W., Bhattacharya, D., & Deretic, V. (2011). Autophagy-based unconventional secretory pathway for extracellular delivery of IL-1β. *The EMBO Journal*, *30*(23), 4701–4711. DOI: 10.1038/emboj.2011.398 PMID: 22068051

Ebrahimi, B., Mokhtari, T., Ghaffari, N., Adabi, M., & Hassanzadeh, G. (2024). Acellular spinal cord scaffold containing quercetin-encapsulated nanoparticles plays an anti-inflammatory role in functional recovery from spinal cord injury in rats. *Inflammopharmacology*, *32*(4), 1–20. DOI: 10.1007/s10787-024-01478-z PMID: 38702577

Fleming, A., Bourdenx, M., Fujimaki, M., Karabiyik, C., Krause, G. J., Lopez, A., Martín-Segura, A., Puri, C., Scrivo, A., Skidmore, J., Son, S. M., Stamatakou, E., Wrobel, L., Zhu, Y., Cuervo, A. M., & Rubinsztein, D. C. (2022). The different autophagy degradation pathways and neurodegeneration. *Neuron, 110*(6), 935–966. DOI: 10.1016/j.neuron.2022.01.017 PMID: 35134347

Gao, K., Niu, J., & Dang, X. (2020). Neuroprotection of melatonin on spinal cord injury by activating autophagy and inhibiting apoptosis via SIRT1/AMPK signaling pathway. *Biotechnology Letters, 42*(10), 2059–2069. DOI: 10.1007/s10529-020-02939-5 PMID: 32514788

Gao, K., Niu, J., & Dang, X. (2020). Wnt-3a improves functional recovery through autophagy activation via inhibiting the mTOR signaling pathway after spinal cord injury. *Neuroscience Letters, 737*, 135305. DOI: 10.1016/j.neulet.2020.135305 PMID: 32818590

. Gao Q. Oxidative stress and autophagy. Autophagy: biology and diseases: basic science. 2019:179-98.

Ghaffari, N., Mokhtari, T., Adabi, M., Ebrahimi, B., Kamali, M., Gholaminejhad, M., & Hassanzadeh, G. (2024). Neurological recovery and neurogenesis by curcumin sustained-release system cross-linked with an acellular spinal cord scaffold in rat spinal cord injury: Targeting NLRP3 inflammasome pathway. *Phytotherapy Research, 38*(6), 2669–2686. DOI: 10.1002/ptr.8179 PMID: 38500263

González-Fernández, C., Fernández-Martos, C. M., Shields, S. D., Arenas, E., & Javier Rodriguez, F. (2014). Wnts are expressed in the spinal cord of adult mice and are differentially induced after injury. *Journal of Neurotrauma, 31*(6), 565–581. DOI: 10.1089/neu.2013.3067 PMID: 24367909

Grassner, L., Wutte, C., Klein, B., Mach, O., Riesner, S., Panzer, S., Vogel, M., Bühren, V., Strowitzki, M., Vastmans, J., & Maier, D. (2016). Early decompression (< 8 h) after traumatic cervical spinal cord injury improves functional outcome as assessed by spinal cord independence measure after one year. *Journal of Neurotrauma, 33*(18), 1658–1666. DOI: 10.1089/neu.2015.4325 PMID: 27050499

Gu, Y., Chen, D., Zhou, L., Zhao, X., Lin, J., Lin, B., Lin, T., Chen, Z., Chen, Z., Wang, Z., & Liu, W. (2021). Lysine-specific demethylase 1 inhibition enhances autophagy and attenuates early-stage post-spinal cord injury apoptosis. *Cell Death Discovery, 7*(1), 69. DOI: 10.1038/s41420-021-00455-7 PMID: 33824301

Hagberg, H., Mallard, C., Rousset, C. I., & Thornton, C. (2014). Mitochondria: Hub of injury responses in the developing brain. *Lancet Neurology, 13*(2), 217–232. DOI: 10.1016/S1474-4422(13)70261-8 PMID: 24457191

He, M., Ding, Y., Chu, C., Tang, J., Xiao, Q., & Luo, Z.-G. (2016). Autophagy induction stabilizes microtubules and promotes axon regeneration after spinal cord injury. *Proceedings of the National Academy of Sciences of the United States of America, 113*(40), 11324–11329. DOI: 10.1073/pnas.1611282113 PMID: 27638205

Hellal, F., Hurtado, A., Ruschel, J., Flynn, K. C., Laskowski, C. J., Umlauf, M., Kapitein, L. C., Strikis, D., Lemmon, V., Bixby, J., Hoogenraad, C. C., & Bradke, F. (2011). Microtubule stabilization reduces scarring and causes axon regeneration after spinal cord injury. *Science, 331*(6019), 928–931. DOI: 10.1126/science.1201148 PMID: 21273450

Heneka, M. T., Carson, M. J., El Khoury, J., Landreth, G. E., Brosseron, F., & Feinstein, D. L.. (2015). Neuroinflammation in Alzheimer's disease. *Lancet Neurology, 14*(4), 388–405. DOI: 10.1016/S1474-4422(15)70016-5 PMID: 25792098

Hiebert, J. B., Shen, Q., Thimmesch, A. R., & Pierce, J. D. (2015). Traumatic brain injury and mitochondrial dysfunction. *The American Journal of the Medical Sciences, 350*(2), 132–138. DOI: 10.1097/MAJ.0000000000000506 PMID: 26083647

Hou, H., Zhang, L., Zhang, L., & Tang, P. (2014). Acute spinal cord injury in rats should target activated autophagy. *Journal of Neurosurgery. Spine, 20*(5), 568–577. DOI: 10.3171/2014.1.SPINE13237 PMID: 24559459

Huang, Y., Meng, S., Wu, B., Shi, H., Wang, Y., Xiang, J., Li, J., Shi, Z., Wu, G., Lyu, Y., Jia, X., Hu, J., Xu, Z.-X., & Gao, Y. (2023). HSPB2 facilitates neural regeneration through autophagy for sensorimotor recovery after traumatic brain injury. *JCI Insight, 8*(16), e168919. DOI: 10.1172/jci.insight.168919 PMID: 37606039

Hur, E.-M., & Zhou, F.-Q. (2012). Growing the growth cone: Remodeling the cytoskeleton to promote axon regeneration. *Trends in Neurosciences, 35*(3), 164–174. DOI: 10.1016/j.tins.2011.11.002 PMID: 22154154

Itakura, E., Kishi, C., Inoue, K., & Mizushima, N. (2008). Beclin 1 forms two distinct phosphatidylinositol 3-kinase complexes with mammalian Atg14 and UVRAG. *Molecular Biology of the Cell, 19*(12), 5360–5372. DOI: 10.1091/mbc.e08-01-0080 PMID: 18843052

Jiang, F., Xia, M., Zhang, Y., Chang, J., Cao, J., Zhang, Z., Qian, Z., & Yang, L. (2022). Cannabinoid receptor-2 attenuates neuroinflammation by promoting autophagy-mediated degradation of the NLRP3 inflammasome post spinal cord injury. *Frontiers in Immunology, 13*, 993168. DOI: 10.3389/fimmu.2022.993168 PMID: 36238284

Jiang, F., Xia, M., Zhang, Y., Chang, J., Cao, J., Zhang, Z., Qian, Z., & Yang, L. (2022). Cannabinoid receptor-2 attenuates neuroinflammation by promoting autophagy-mediated degradation of the NLRP3 inflammasome post spinal cord injury. *Frontiers in Immunology*, *13*, 993168. DOI: 10.3389/fimmu.2022.993168 PMID: 36238284

Jin, G.-N., Lu, J.-M., Lan, H.-W., Lu, Y.-N., Shen, X.-Y., Xu, X., & Piao, L.-X. (2022). Protective effect of ginsenoside Rh2 against Toxoplasma gondii infection-induced neuronal injury through binding TgCDPK1 and NLRP3 to inhibit microglial NLRP3 inflammasome signaling pathway. *International Immunopharmacology*, *112*, 109176. DOI: 10.1016/j.intimp.2022.109176 PMID: 36067653

Kanno, H., Ozawa, H., Sekiguchi, A., Yamaya, S., & Itoi, E. (2011). Induction of autophagy and autophagic cell death in damaged neural tissue after acute spinal cord injury in mice. *Spine*, *36*(22), E1427–E34. DOI: 10.1097/BRS.0b013e3182028c3a PMID: 21304420

Ke, P.-Y. (2024). Molecular Mechanism of Autophagosome–Lysosome Fusion in Mammalian Cells. *Cells*, *13*(6), 500. DOI: 10.3390/cells13060500 PMID: 38534345

Komorowska, J., Wątroba, M., Bednarzak, M., Grabowska, A. D., & Szukiewicz, D. (2023). The Role of Glucose Concentration and Resveratrol in Modulating Neuroinflammatory Cytokines: Insights from an In Vitro Blood-Brain Barrier Model. *Medical Science Monitor*, *29*, e941044. DOI: 10.12659/MSM.941044 PMID: 37817396

Kong, J., Zhang, Q., Zheng, H., Tang, D., Fang, L., An, S., Li, J., & Fan, Z. (2024). TGN-020 ameliorates motor dysfunction post-spinal cord injury via enhancing astrocyte autophagy and mitigating inflammation by activating AQP4/PPAR-γ/mTOR pathway. *Experimental Neurology*, *382*, 114975. DOI: 10.1016/j.expneurol.2024.114975 PMID: 39326822

Lee, J. A. (2012). Neuronal autophagy: A housekeeper or a fighter in neuronal cell survival? *Experimental Neurobiology*, *21*(1), 1–8. DOI: 10.5607/en.2012.21.1.1 PMID: 22438673

Li, J., Cao, Y., Li, L.-N., Chu, X., Wang, Y.-S., Cai, J.-J., Zhao, J., Ma, S., Li, G., & Fan, Z.-K. (2023). Neuroprotective effects of oxymatrine via triggering autophagy and inhibiting apoptosis following spinal cord injury in rats. *Molecular Neurobiology*, *60*(8), 4450–4471. DOI: 10.1007/s12035-023-03364-1 PMID: 37115405

Li, W., Yao, S., Li, H., Meng, Z., & Sun, X. (2021). Curcumin promotes functional recovery and inhibits neuronal apoptosis after spinal cord injury through the modulation of autophagy. *The Journal of Spinal Cord Medicine*, *44*(1), 37–45. DOI: 10.1080/10790268.2019.1616147 PMID: 31162984

Li, Y., Guo, Y., Fan, Y., Tian, H., Li, K., & Mei, X. (2019). Melatonin enhances autophagy and reduces apoptosis to promote locomotor recovery in spinal cord injury via the PI3K/AKT/mTOR signaling pathway. *Neurochemical Research, 44*(8), 2007–2019. DOI: 10.1007/s11064-019-02838-w PMID: 31325156

Li, Y., Lei, Z., Ritzel, R. M., He, J., Li, H., Choi, H. M. C., Lipinski, M. M., & Wu, J. (2022). Impairment of autophagy after spinal cord injury potentiates neuroinflammation and motor function deficit in mice. *Theranostics, 12*(12), 5364–5388. DOI: 10.7150/thno.72713 PMID: 35910787

Li, Y., Li, S., & Wu, H. (2022). Ubiquitination-Proteasome System (UPS) and Autophagy Two Main Protein Degradation Machineries in Response to Cell Stress. *Cells, 11*(5), 851. DOI: 10.3390/cells11050851 PMID: 35269473

Liang, C. (2010). Negative regulation of autophagy. *Cell Death and Differentiation, 17*(12), 1807–1815. DOI: 10.1038/cdd.2010.115 PMID: 20865012

Liao, H.-Y., Wang, Z.-Q., Ran, R., Zhou, K.-S., Ma, C.-W., & Zhang, H.-H. (2021). Biological functions and therapeutic potential of autophagy in spinal cord injury. *Frontiers in Cell and Developmental Biology, 9*, 761273. DOI: 10.3389/fcell.2021.761273 PMID: 34988074

Liao, H. Y., Wang, Z. Q., Ran, R., Zhou, K. S., Ma, C. W., & Zhang, H. H. (2021). Biological Functions and Therapeutic Potential of Autophagy in Spinal Cord Injury. *Frontiers in Cell and Developmental Biology, 9*, 761273. DOI: 10.3389/fcell.2021.761273 PMID: 34988074

Lin, C.-C., & Lin, M.-S. (2016). New insight into curcumin-based therapy in spinal cord injuries: CISD2 regulation. *Neural Regeneration Research, 11*(2), 222–223. DOI: 10.4103/1673-5374.177718 PMID: 27073364

Lin, J., Huo, X., & Liu, X. (2017). "mTOR signaling pathway": A potential target of curcumin in the treatment of spinal cord injury. *BioMed Research International, 2017*(1), 1634801. DOI: 10.1155/2017/1634801 PMID: 28691015

Lin, S., Tian, H., Lin, J., Xu, C., Yuan, Y., Gao, S., Song, C., Lv, P., & Mei, X. (2020). Zinc promotes autophagy and inhibits apoptosis through AMPK/mTOR signaling pathway after spinal cord injury. *Neuroscience Letters, 736*, 135263. DOI: 10.1016/j.neulet.2020.135263 PMID: 32682846

Lipinski, M. M., Wu, J., Faden, A. I., & Sarkar, C. (2015). Function and mechanisms of autophagy in brain and spinal cord trauma. *Antioxidants & Redox Signaling, 23*(6), 565–577. DOI: 10.1089/ars.2015.6306 PMID: 25808205

Liu, Q.-R., Aseer, K. R., Yao, Q., Zhong, X., Ghosh, P., O'Connell, J. F., & Egan, J. M. (2022). Anti-inflammatory and pro-autophagy effects of the cannabinoid receptor CB2R: Possibility of modulation in type 1 diabetes. *Frontiers in Pharmacology*, *12*, 809965. DOI: 10.3389/fphar.2021.809965 PMID: 35115945

Liu, R., Jiang, L., Chen, Y., Shao, J., Chen, K., Li, X., Lv, J., Cai, W., Cai, H., Zhu, Z., Wang, C., Zhou, K., Huang, J., Xiao, J., Ni, W., & Wu, C. (2024). Ginsenoside-Rh2 promotes functional recovery after spinal cord Injury by enhancing TFEB-Mediated autophagy. *Journal of Agricultural and Food Chemistry*, *72*(26), 14727–14746. DOI: 10.1021/acs.jafc.4c02379 PMID: 38907713

Liu, S., Li, Y., Choi, H. M., Sarkar, C., Koh, E. Y., Wu, J., & Lipinski, M. M. (2018). Lysosomal damage after spinal cord injury causes accumulation of RIPK1 and RIPK3 proteins and potentiation of necroptosis. *Cell Death & Disease*, *9*(5), 476. DOI: 10.1038/s41419-018-0469-1 PMID: 29686269

Liu, Y., Chu, W., Ma, H., Peng, W., Li, Q., Han, L., Wang, H., Wang, L., Zhang, B., Yang, J., & Lu, X. (2024). Fisetin orchestrates neuroinflammation resolution and facilitates spinal cord injury recovery through enhanced autophagy in pro-inflammatory glial cells. *International Immunopharmacology*, *130*, 111738. DOI: 10.1016/j.intimp.2024.111738 PMID: 38428149

Lu, H. C., & Mackie, K. (2021). Review of the Endocannabinoid System. *Biolog ical Psychiatry: Cognitive Neuroscience and Neuroimaging*, *6*(6), 607–615. DOI: 10.1016/j.bpsc.2020.07.016 PMID: 32980261

Luo, C., & Tao, L. (2020). The function and mechanisms of autophagy in spinal cord injury. Autophagy: Biology and Diseases. *Clinical Science*, 649–654.

Mannack, L. V. J. C., & Lane, J. D. (2015). The autophagosome: Current understanding of formation and maturation. *Research and Reports in Biochemistry*, 39–58.

Massacci, G., Perfetto, L., & Sacco, F. (2023). The Cyclin-dependent kinase 1: More than a cell cycle regulator. *British Journal of Cancer*, *129*(11), 1707–1716. DOI: 10.1038/s41416-023-02468-8 PMID: 37898722

Meng, H. Y., Shao, D. C., Li, H., Huang, X. D., Yang, G., Xu, B., & Niu, H.-Y. (2018). Resveratrol improves neurological outcome and neuroinflammation following spinal cord injury through enhancing autophagy involving the AMPK/mTOR pathway. *Molecular Medicine Reports*, *18*(2), 2237–2244. DOI: 10.3892/mmr.2018.9194 PMID: 29956767

Mokhtari, T., & Uludag, K. (2023). Role of NLRP3 inflammasome in post-spinal-cord-injury anxiety and depression: Molecular mechanisms and therapeutic implications. *ACS Chemical Neuroscience*, *15*(1), 56–70. DOI: 10.1021/acschemneuro.3c00596 PMID: 38109051

Moratilla-Rivera, I., Sánchez, M., Valdés-González, J. A., & Gómez-Serranillos, M. P. (2023). Natural products as modulators of Nrf2 signaling pathway in neuroprotection. *International Journal of Molecular Sciences*, *24*(4), 3748. DOI: 10.3390/ijms24043748 PMID: 36835155

Muñoz-Galdeano, T., Reigada, D., Del Águila, Á., Velez, I., Caballero-López, M. J., Maza, R. M., & Nieto-Díaz, M. (2018). Cell Specific Changes of Autophagy in a Mouse Model of Contusive Spinal Cord Injury. *Frontiers in Cellular Neuroscience*, *12*, 164. DOI: 10.3389/fncel.2018.00164 PMID: 29946241

Nagayach, A., & Wang, C. (2024). Autophagy in neural stem cells and glia for brain health and diseases. *Neural Regeneration Research*, *19*(4), 729–736. DOI: 10.4103/1673-5374.382227 PMID: 37843206

Nedelsky, N. B., Todd, P. K., & Taylor, J. P. (2008). Autophagy and the ubiquitin-proteasome system: Collaborators in neuroprotection. Biochimica et Biophysica Acta (BBA)-. *Biochimica et Biophysica Acta. Molecular Basis of Disease*, *1782*(12), 691–699. DOI: 10.1016/j.bbadis.2008.10.002 PMID: 18930136

Netea-Maier, R. T., Plantinga, T. S., van de Veerdonk, F. L., Smit, J. W., & Netea, M. G. (2016). Modulation of inflammation by autophagy: Consequences for human disease. *Autophagy*, *12*(2), 245–260. DOI: 10.1080/15548627.2015.1071759 PMID: 26222012

Nie, B.-X., Zhao, G., Yuan, X.-F., Yu, L.-X., Zhang, J., Yuan, Y., Liu, Y., Hu, J., Song, E., Zhou, Y.-C., & Shu, J. (2022). Inhibition of CDK1 attenuates neuronal apoptosis and autophagy and confers neuroprotection after chronic spinal cord injury in vivo. *Journal of Chemical Neuroanatomy*, *119*, 102053. DOI: 10.1016/j.jchemneu.2021.102053 PMID: 34839004

Ning, S., Chen, Y., Shao, J., Zhu, H., Zhang, Z., & Miao, J. (2024). The effects of acteoside on locomotor recovery after spinal cord injury – The role of autophagy and apoptosis signaling pathway. *Biomedicine and Pharmacotherapy*, *175*, 116607. DOI: 10.1016/j.biopha.2024.116607 PMID: 38692056

Nixon, R. A. (2024). Autophagy–lysosomal-associated neuronal death in neurodegenerative disease. *Acta Neuropathologica*, *148*(1), 42. DOI: 10.1007/s00401-024-02799-7 PMID: 39259382

Noce, B., Di Bello, E., Fioravanti, R., & Mai, A. (2023). LSD1 inhibitors for cancer treatment: Focus on multi-target agents and compounds in clinical trials. *Frontiers in Pharmacology*, *14*, 1120911. DOI: 10.3389/fphar.2023.1120911 PMID: 36817147

Park, J. H., Min, J., Baek, S. R., Kim, S. W., Kwon, I. K., & Jeon, S. R. (2013). Enhanced neuroregenerative effects by scaffold for the treatment of a rat spinal cord injury with Wnt3a-secreting fibroblasts. *Acta Neurochirurgica*, *155*(5), 809–816. DOI: 10.1007/s00701-013-1663-7 PMID: 23456240

Ray, S. K. (2020). Modulation of autophagy for neuroprotection and functional recovery in traumatic spinal cord injury. *Neural Regeneration Research*, *15*(9), 1601–1612. DOI: 10.4103/1673-5374.276322 PMID: 32209759

Ruschel, J., Hellal, F., Flynn, K. C., Dupraz, S., Elliott, D. A., Tedeschi, A., Bates, M., Sliwinski, C., Brook, G., Dobrindt, K., Peitz, M., Brüstle, O., Norenberg, M. D., Blesch, A., Weidner, N., Bunge, M. B., Bixby, J. L., & Bradke, F. (2015). Systemic administration of epothilone B promotes axon regeneration after spinal cord injury. *Science*, *348*(6232), 347–352. DOI: 10.1126/science.aaa2958 PMID: 25765066

Sakai, K., Fukuda, T., & Iwadate, K. (2014). Immunohistochemical analysis of the ubiquitin proteasome system and autophagy lysosome system induced after traumatic intracranial injury: Association with time between the injury and death. *The American Journal of Forensic Medicine and Pathology*, *35*(1), 38–44. DOI: 10.1097/PAF.0000000000000067 PMID: 24317096

Sng, K. S., Li, G., Zhou, L.-y., Song, Y., Chen, X., Wang, Y., Yao, M., & Cui, X. (2022). Ginseng extract and ginsenosides improve neurological function and promote antioxidant effects in rats with spinal cord injury: A meta-analysis and systematic review. *Journal of Ginseng Research*, *46*(1), 11–22. DOI: 10.1016/j.jgr.2021.05.009 PMID: 35058723

Tanabe, F., Yone, K., Kawabata, N., Sakakima, H., Matsuda, F., Ishidou, Y., Maeda, S., Abematsu, M., Komiya, S., & Setoguchi, T. (2011). Accumulation of p62 in degenerated spinal cord under chronic mechanical compression: Functional analysis of p62 and autophagy in hypoxic neuronal cells. *Autophagy*, *7*(12), 1462–1471. DOI: 10.4161/auto.7.12.17892 PMID: 22082874

Tang, P., Hou, H., Zhang, L., Lan, X., Mao, Z., Liu, D., He, C., Du, H., & Zhang, L. (2014). Autophagy reduces neuronal damage and promotes locomotor recovery via inhibition of apoptosis after spinal cord injury in rats. *Molecular Neurobiology*, *49*(1), 276–287. DOI: 10.1007/s12035-013-8518-3 PMID: 23954967

Tobin, M. K., Bonds, J. A., Minshall, R. D., Pelligrino, D. A., Testai, F. D., & Lazarov, O. (2014). Neurogenesis and inflammation after ischemic stroke: What is known and where we go from here. *Journal of Cerebral Blood Flow and Metabolism*, *34*(10), 1573–1584. DOI: 10.1038/jcbfm.2014.130 PMID: 25074747

Vargova, I., Machova Urdzikova, L., Karova, K., Smejkalova, B., Sursal, T., Cimermanova, V., Turnovcova, K., Gandhi, C. D., Jhanwar-Uniyal, M., & Jendelova, P. (2021). Involvement of mTOR pathways in recovery from spinal cord injury by modulation of autophagy and immune response. *Biomedicines*, *9*(6), 593. DOI: 10.3390/biomedicines9060593 PMID: 34073791

Verma, R., Virdi, J. K., Singh, N., & Jaggi, A. S. (2019). Animals models of spinal cord contusion injury. *The Korean Journal of Pain*, *32*(1), 12–21. DOI: 10.3344/kjp.2019.32.1.12 PMID: 30671199

Vuic, B., Milos, T., Tudor, L., Konjevod, M., Nikolac Perkovic, M., Jazvinscak Jembrek, M., Nedic Erjavec, G., & Svob Strac, D. (2022). Cannabinoid cb2 receptors in neurodegenerative proteinopathies: New insights and therapeutic potential. *Biomedicines*, *10*(12), 3000. DOI: 10.3390/biomedicines10123000 PMID: 36551756

Wang, B., Fang, T., & Chen, H. (2023). Zinc and central nervous system disorders. *Nutrients*, *15*(9), 2140. DOI: 10.3390/nu15092140 PMID: 37432243

Wang, H.-D., Lv, C.-L., Feng, L., Guo, J.-X., Zhao, S.-Y., & Jiang, P. (2024). The role of autophagy in brain health and disease: Insights into exosome and autophagy interactions. *Heliyon*, *10*(21), e38959. DOI: 10.1016/j.heliyon.2024.e38959 PMID: 39524893

Wang, J., Bian, S., Wang, S., Yang, S., Zhang, W., Zhao, D., Liu, M., & Bai, X. (2020). Ginsenoside Rh2 represses autophagy to promote cervical cancer cell apoptosis during starvation. *Chinese Medicine*, *15*(1), 1–14. DOI: 10.1186/s13020-020-00396-w PMID: 33292331

Wang, P., Jiang, L., Zhou, N., Zhou, H., Liu, H., Zhao, W., Zhang, H., Zhang, X., & Hu, Z. (2018). Resveratrol ameliorates autophagic flux to promote functional recovery in rats after spinal cord injury. *Oncotarget*, *9*(9), 8427–8440. DOI: 10.18632/oncotarget.23877 PMID: 29492205

Wang, S., Li, H., Yuan, M., Fan, H., & Cai, Z. (2022). Role of AMPK in autophagy. *Frontiers in Physiology*, *13*, 1015500. DOI: 10.3389/fphys.2022.1015500 PMID: 36505072

Wang, X., Li, X., Zuo, X., Liang, Z., Ding, T., Li, K., Ma, Y., Li, P., Zhu, Z., Ju, C., Zhang, Z., Song, Z., Quan, H., Zhang, J., Hu, X., & Wang, Z. (2021). Photobiomodulation inhibits the activation of neurotoxic microglia and astrocytes by inhibiting Lcn2/JAK2-STAT3 crosstalk after spinal cord injury in male rats. *Journal of Neuroinflammation*, *18*(1), 1–20. DOI: 10.1186/s12974-021-02312-x PMID: 34740378

Wang, Z. Y., Liu, W. G., Muharram, A., Wu, Z. Y., & Lin, J. H. (2014). Neuroprotective effects of autophagy induced by rapamycin in rat acute spinal cord injury model. *Neuroimmunomodulation*, *21*(5), 257–267. DOI: 10.1159/000357382 PMID: 24603048

Wei, W., Lu, M., Lan, X., Liu, N., Su, W., Dushkin, A. V., & Yu, J. (2021). Neuroprotective effects of oxymatrine on PI3K/Akt/mTOR pathway after hypoxic-ischemic brain damage in neonatal rats. *Frontiers in Pharmacology*, *12*, 642415. DOI: 10.3389/fphar.2021.642415 PMID: 33927621

Willis, D. E., & Twiss, J. L. (2006). The evolving roles of axonally synthesized proteins in regeneration. *Current Opinion in Neurobiology*, *16*(1), 111–118. DOI: 10.1016/j.conb.2006.01.002 PMID: 16418002

Wu, F., Wei, X., Wu, Y., Kong, X., Hu, A., Tong, S., Liu, Y., Gong, F., Xie, L., Zhang, J., Xiao, J., & Zhang, H. (2018). Chloroquine promotes the recovery of acute spinal cord injury by inhibiting autophagy-associated inflammation and endoplasmic reticulum stress. *Journal of Neurotrauma*, *35*(12), 1329–1344. DOI: 10.1089/neu.2017.5414 PMID: 29316847

Wu, J., & Lipinski, M. M. (2019). Autophagy in neurotrauma: Good, bad, or dysregulated. *Cells*, *8*(7), 693. DOI: 10.3390/cells8070693 PMID: 31295858

Wu, J., Pajoohesh-Ganji, A., Stoica, B. A., Dinizo, M., Guanciale, K., & Faden, A. I. (2012). Delayed expression of cell cycle proteins contributes to astroglial scar formation and chronic inflammation after rat spinal cord contusion. *Journal of Neuroinflammation*, *9*(1), 1–12. DOI: 10.1186/1742-2094-9-169 PMID: 22784881

Wu, M., Gao, Y., & Chen, B. (2022). Mechanism of acteoside-activated let-7g-5P attenuating Aβ-induced increased permeability and apoptosis of brain microvascular endothelial cells based on experimental and network pharmacology. *Neuroreport*, *33*(16), 714–722. DOI: 10.1097/WNR.0000000000001837 PMID: 36165002

Wu, Y., Li, L., Ning, Z., Li, C., Yin, Y., Chen, K., Li, L., Xu, F., & Gao, J. (2024). Autophagy-modulating biomaterials: Multifunctional weapons to promote tissue regeneration. *Cell Communication and Signaling : CCS*, *22*(1), 124. DOI: 10.1186/s12964-023-01346-3 PMID: 38360732

Xiao, Y., Ren, Q., & Wu, L. (2022). The pharmacokinetic property and pharmacological activity of acteoside: A review. *Biomedicine and Pharmacotherapy, 153*, 113296. DOI: 10.1016/j.biopha.2022.113296 PMID: 35724511

Xie, X. H., An, H. J., Kang, S., Hong, S., Choi, Y. P., Kim, Y. T., Choi, Y. D., & Cho, N. H. (2005). Loss of Cyclin B1 followed by downregulation of Cyclin A/Cdk2, apoptosis and antiproliferation in Hela cell line. *International Journal of Cancer, 116*(4), 520–525. DOI: 10.1002/ijc.21056 PMID: 15818617

Xu, C., Zhou, Z., Zhao, H., Lin, S., Zhang, P., Tian, H., & Mei, X. (2023). Zinc promotes spinal cord injury recovery by blocking the activation of NLRP3 inflammasome through SIRT3-mediated autophagy. *Neurochemical Research, 48*(2), 435–446. DOI: 10.1007/s11064-022-03762-2 PMID: 36152137

Xu, L., Botchway, B. O. A., Zhang, S., Zhou, J., & Liu, X. (2018). Inhibition of NF-κB signaling pathway by resveratrol improves spinal cord injury. *Frontiers in Neuroscience, 12*, 690. DOI: 10.3389/fnins.2018.00690 PMID: 30337851

Xu, Y., Yuan, J., & Lipinski, M. M. (2013). Live imaging and single-cell analysis reveal differential dynamics of autophagy and apoptosis. *Autophagy, 9*(9), 1418–1430. DOI: 10.4161/auto.25080 PMID: 23748697

Yamashima, T., & Oikawa, S. (2009). The role of lysosomal rupture in neuronal death. *Progress in Neurobiology, 89*(4), 343–358. DOI: 10.1016/j.pneurobio.2009.09.003 PMID: 19772886

Yan, P., Bai, L., Lu, W., Gao, Y., Bi, Y., & Lv, G. (2017). Regulation of autophagy by AMP-activated protein kinase/sirtuin 1 pathway reduces spinal cord neurons damage. *Iranian Journal of Basic Medical Sciences., 20*(9), 1029–1036. PMID: 29085598

Yu, L., Chen, Y., & Tooze, S. A. (2018). Autophagy pathway: Cellular and molecular mechanisms. *Autophagy, 14*(2), 207–215. DOI: 10.1080/15548627.2017.1378838 PMID: 28933638

Yuan, Y., Fan, X., Guo, Z., Zhou, Z., & Gao, W. (2022). Metformin Protects against Spinal Cord Injury and Cell Pyroptosis via AMPK/NLRP3 Inflammasome Pathway. *Analytical Cellular Pathology (Amsterdam), 2022*, 3634908. DOI: 10.1155/2022/3634908 PMID: 35387358

Zhang, B., Chen, Y., Yang, B., Jiang, F., Zhao, J., & Ouyang, Z. (2017). Neuroprotective effects of diosgenin in rats with experimental spinal cord injury via promotion of autophagy. *International Journal of Clinical and Experimental Medicine, 10*(8), 11655–11663.

Zhang, D., Xuan, J., Zheng, B., Zhou, Y., Lin, Y., Wu, Y., Zhou, Y., Huang, Y., Wang, Q., Shen, L., Mao, C., Wu, Y., Wang, X., Tian, N., Xu, H.-Z., & Zhang, X. (2017). Metformin improves functional recovery after spinal cord injury via autophagy flux stimulation. *Molecular Neurobiology*, *54*(5), 3327–3341. DOI: 10.1007/s12035-016-9895-1 PMID: 27167128

Zhang, Y., Sun, S., Chen, J., Ren, P., Hu, Y., Cao, Z., Sun, H., & Ding, Y. (2014). Oxymatrine induces mitochondria dependent apoptosis in human osteosarcoma MNNG/HOS cells through inhibition of PI3K/Akt pathway. *Tumour Biology : The Journal of the International Society for Oncodevelopmental Biology and Medicine*, *35*(2), 1619–1625. DOI: 10.1007/s13277-013-1223-z PMID: 24078465

Zhao, H., Chen, S., Gao, K., Zhou, Z., Wang, C., Shen, Z., Guo, Y., Li, Z., Wan, Z., Liu, C., & Mei, X. (2017). Resveratrol protects against spinal cord injury by activating autophagy and inhibiting apoptosis mediated by the SIRT1/AMPK signaling pathway. *Neuroscience*, *348*, 241–251. DOI: 10.1016/j.neuroscience.2017.02.027 PMID: 28238848

Zhong, W., Qian, K., Xiong, J., Ma, K., Wang, A., & Zou, Y. (2016). Curcumin alleviates lipopolysaccharide induced sepsis and liver failure by suppression of oxidative stress-related inflammation via PI3K/AKT and NF-κB related signaling. *Biomedicine and Pharmacotherapy*, *83*, 302–313. DOI: 10.1016/j.biopha.2016.06.036 PMID: 27393927

Zhou, Y., Zhang, H., Zheng, B., Ye, L., Zhu, S., Johnson, N. R., Wang, Z., Wei, X., Chen, D., Cao, G., Fu, X., Li, X., Xu, H.-Z., & Xiao, J. (2016). Retinoic acid induced-autophagic flux inhibits ER-stress dependent apoptosis and prevents disruption of blood-spinal cord barrier after spinal cord injury. *International Journal of Biological Sciences*, *12*(1), 87–99. DOI: 10.7150/ijbs.13229 PMID: 26722220

Zhou, Y., Zheng, B., Ye, L., Zhang, H., Zhu, S., Zheng, X., Xia, Q., He, Z., Wang, Q., Xiao, J., & Xu, H. (2016). RETRACTED ARTICLE: Retinoic Acid Prevents Disruption of Blood-Spinal Cord Barrier by Inducing Autophagic Flux after Spinal Cord Injury. *Neurochemical Research*, *41*(4), 813–825. DOI: 10.1007/s11064-015-1756-1 PMID: 26582233

Zou, S., & Kumar, U. (2018). Cannabinoid Receptors and the Endocannabinoid System: Signaling and Function in the Central Nervous System. *International Journal of Molecular Sciences*, *19*(3), 833. DOI: 10.3390/ijms19030833 PMID: 29533978

# Chapter 6
# Autophagy and Inflammation in Neuropsychological Disorders:
## Autophagy in the Neurological-Visual Disease – Mechanisms and Therapeutic Potential

**Fatemeh Azimi**
https://orcid.org/0000-0001-9995-0142
*Iran University of Medical Sciences, Tehran, Iran*

## ABSTRACT

*Vision is a highly complex neural process that begins in the retina and is transmitted to the brain through the optic nerve, relying on the precise function of highly specialized neural and glial cells. It is of great importance for the integrity of vision to maintain the health of these cells. Autophagy, a lysosome-dependent degradation pathway, plays a crucial role in maintaining cellular homeostasis by eliminating misfolded proteins and damaged organelles. Dysregulation of autophagy has been recognized as a pivotal factor in the pathogenesis of multiple neurological-visual diseases, such as glaucoma, age-related macular degeneration (AMD), and inherited retinal dystrophies (IRDs). This chapter studies how autophagy is essentially governed at the molecular level during physiological/pathophysiological processes, particularly in the retina and optic nerve. Finally, emerging therapeutic approaches that aim to regulate autophagy are discussed to offer potential avenues to slow disease progression and preserve vision in neurodegenerative eye disorders.*

DOI: 10.4018/979-8-3693-5908-2.ch006

## INTRODUCTION

### Neurological-Visual Diseases: Definition and Scope

Neurological-visual diseases including both the central nervous system (CNS) and visual system are a category of eye disorders, particularly affecting the optic nerve, retina, and visual cortex. These include glaucoma, age-related macular degeneration (AMD), and inherited retinal dystrophies (IRDs), which lead to causes of irreversible vision loss worldwide (Ahmad et al., 2020; Jindal, 2015), commonly defined by the gradual decline in retinal ganglion cells (RGCs), retinal pigment epithelium (RPE), and optic nerve axons, via mechanisms involving apoptosis and autophagy dysfunction (Golestaneh et al., 2018; Rodríguez-Muela et al., 2012).

The retina and the CNS are composed mainly of post-mitotic neurons, which have a limited regenerative capacity and are long-lived. These neurons are particularly dependent on autophagy, a lysosome-dependent mechanism of cellular digestion, to clear damaged organelles and accumulated proteins (Gao et al., 2018; Scrivo et al., 2018). Dysregulation of autophagy has been increasingly recognized as a contributing factor in the pathogenesis of neurological-visual disorders (Jiménez-Loygorri et al., 2023).

### Autophagy: A Critical Cellular Mechanism

Autophagy is a highly conserved intracellular process essential for maintaining neural and retinal homeostasis. This process of decomposition and recycling of damaged organelles, misfolded protein complexes, and proteins is particularly applied to differentiated cells such as RGCs and photoreceptors that cannot dilute toxic substances through cell division (Jiménez-Loygorri et al., 2023).

In eye tissues, degradation and quality control mechanisms operate through autophagy-dependent pathways, such as macroautophagy (De Duve & Wattiaux, 1966; Yang & Klionsky, 2010). Key molecular players like LC3, Beclin-1 (BECN1), autophagy-related 5/7 (ATG5/7) factors, and PINK1 (PTEN-induced kinase 1) / Parkin regulate mitophagy in RPE and photoreceptors, protecting against AMD and glaucoma (Chen et al., 2013; Greene et al., 2012; Nishida et al., 2009; Park et al., 2016; Sinha et al., 2017) Autophagy-independent degradation pathways are also vital for ocular physiology (Morishita, 2022). In contrast, autophagy-independent pathways, especially the UPS (ubiquitin-proteasome system) (Ross & Pickart, 2004) and PLAAT (phospholipase A and acyltransferase)-mediated organelle degradation contribute to lens transparency and protein quality control without involving canonical autophagy machinery (Morishita et al., 2021) (Table 1).

This complex mechanism comprises several fundamental steps that are strictly regulated by access to nutrients, oxidative stress, and intracellular damage.

## A. Initiation: Triggering the Autophagic Cascade

This process is activated in response to various stimuli, including nutrient scarcity, oxidative stress, or the accumulation of misfolded proteins. The activation begins with the unc-51-like autophagy activating kinase-1 (ULK1), which is homologous to yeast Atg1. It forms a crucial complex with ATG1,ATG13,ATG17, ATG101, and FIP200 (also referred to as RB1-inducible coiled-coil 1 - RB1CC1) (Levine & Kroemer, 2019).

Under nutrient-rich conditions, mTOR, a serine/threonine kinase, inhibits ULK1, whereas under stress or energy depletion, AMPK signaling is activated and mTOR is suppressed, releasing ULK1 to initiate autophagy Dysregulated mTOR-ULK1 signaling in the context of retinal degeneration impairs autophagosome formation, promoting toxic protein aggregates, mitochondrial dysfunction and neurodegeneration (Kim et al., 2011; Mizushima, 2020).

## B. Nucleation: Formation of the Phagophore

After the initial step, ULK1 kinase activity initiates phosphatidylinositol 3-kinase (PI3K) complexes at phagophore initiation locations, resulting in the production of phosphatidylinositol 3-phosphate (PI3P). Beclin 1 (Atg6 ortholog) and vacuolar protein sorting 34 (VPS34), a membrane precursor to the autophagosome, facilitate the formation of the phagophore (Mizushima & Komatsu, 2011; Mizushima & Levine, 2010), along with LC3, ATG2, Double FYVE-containing protein 1 (DFCP1). Mutations in Beclin 1 complexes are associated with impaired autophagy regulation in various neurodegenerative diseases (Gómez-Virgilio et al., 2022; Mellén et al., 2008).

## C. Elongation (expansion) and Maturation: Building the Autophagosome

Phagophore elongation involves two ubiquitin-like conjugation systems that engulf targeted cellular debris:

- ATG12–ATG5–ATG16L1 complex: Facilitates membrane expansion (Parzych & Klionsky, 2014).
- LC3-I (processed by ATG4, ATG7, and ATG3): Binding to phosphatidylethanolamine (PE) forms Lc3-II, which integrates into the autophagosomal

membrane and acts as a marker for the progression of autophagy (Gómez-Virgilio et al., 2022).

By autophagy cargo receptors (ACRs), the cargo recognition and selectivity are mediated, including:

- p62/SQSTM1 (p62, sequestosome, autophagy cargo adaptor protein): is a known ACR that binds ubiquitinated proteins and LC3 through its LIR motif (Lamark et al., 2009; Manley et al., 2013; Pankiv et al., 2007).
- Calcium, coiled-coil domain 2 (CALCOCO2 or NDP52), and optineurin (OPTN): Especially important in mitophagy as well as xenophagy, they recruit ULK1 and TANK-binding kinase 1 (TBK1) to enhance the degradation of damaged mitochondria, particularly in RGC survival. These receptors are particularly relevant in the retina, where mutation or downregulation of ACRs can lead to total accumulation and cell death (Ravenhill et al., 2019; Vargas et al., 2019).

## D. Fusion: Autophagosomes meet the Lysosomes

At the end of autophagy, the autophagosomes merge with lysosomes to form an autolysosome, the activity there being carried out by other lysosomal hydrolases and lipases; thereafter follows the degradation of autophagic cargoes. The most functionally important regulators in terms of autophagosome-lysosome fusion are mediated by:

- Soluble NSF attachment protein receptor (SNARE) proteins (e.g., SNARE protein syntaxin-17 (STX17), Vesicle-associated membrane protein 8 (VAMP8), synaptosomal-associated protein 29 (SNAP29) (Gómez-Virgilio et al., 2022).
- Rab7 GTPase and the homotypic fusion and protein sorting complex (HOPS) complex (VPS16, VPS33A, and VPS39), with the crucial STX17 situated at the autophagosome (Yu et al., 2018).

Defective lysosomal degradation leads to the generation of lipofuscin in animal models deficient in bA3/A1-crystallin, which is implicated in AMD (Sinha et al., 2008). The latter protein plays a key role in modulating V-ATPase activity through the AKT/mTORC1 pathway and therefore regulates lysosomal degradation. Defective lysosomal function disturbs the autophagic degradation of toxic components and misfolded proteins (Valapala, Edwards, et al., 2014; Valapala, Wilson, et al., 2014).

## E. Degradation and Recycling: Cellular Cleanup

Within the autolysosome, lysosomal enzymes cause the breakdown of autophagic cargoes into amino acids, fatty acids, and recycled sugars for cellular metabolism. Efficient lysosomal function is especially crucial in photoreceptors and RPE cells that are under constant metabolic demand and oxidative stress. With lysosomal dysfunction and autophagic blockage, RPE degradation occurs in AMD and TM dysfunction in glaucoma (Kaarniranta et al., 2023; Schneider & Cuervo, 2014).

Autophagy was previously viewed as a non-selective degradation pathway, but it is currently known to be highly selective, targeting selected cargoes in the cell such as damaged mitochondria, misfolded proteins, and invalid intracellular components (Johansen & Lamark, 2020). This selectivity is mediated by LC3 (microtubule-associated protein light chain 3) - interacting regains (LIRs) domain on cargo receptors, which recognize substrates marked for degradation. Selective autophagy includes lipophagy, aggrephagy, xenophagy, and organelle-specific macroautophagic processes such as ER-phagy, peroxyphagy, mitophagy, and lysophagy (Johansen & Lamark, 2020; Lamark & Johansen, 2021; Tekirdag & Cuervo, 2018). In response to cellular stress, transcription factor EB (TFEB) is activated and translocated into the nucleus, promoting the transcription of lysosomal and autophagy genes (Puertollano et al., 2018).

Under normal conditions, TFEB is retained in the cytoplasm; however, upon nutrient starvation, oxidative stress, or mTOR inhibition, it becomes activated (Colacurcio & Nixon, 2016; Puertollano et al., 2018). To date, over 40 genes have been shown to be vital for autophagic regulation (Mizushima, 2020).

There are three major autophagy pathways that differ in how substrates are delivered to lysosomes for degradation (Figure 1):

1. Macroautophagy engulfs cytoplasmic materials with membrane-bound dual structures known as autophagosomes, which then integrate with lysosomes for degradation (Figure 2).
2. Chaperone-mediated autophagy (CMA) specifically degrades proteins containing a defined amino acid sequence that is recognized by the Hsc70 chaperone protein and is only found in mammalian cells.
3. Micro-autophagy is for material taken for degradation involving the membrane penetration by lysosomal or endosomal membranes and is not well characterized (Kaushik & Cuervo, 2018).

*Figure 1. Main autophagy pathways*

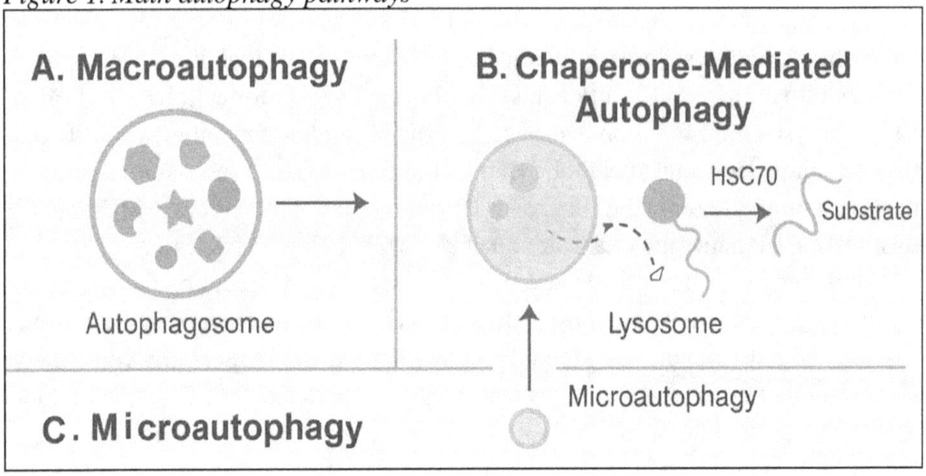


Figure 2. Role of autophagy process and related eye diseases: A schematic representation of the different pathways associated with macroautophagy, LC3-related phagocytosis, and PLAAT-mediated organelle degradation-act in different eye tissues, with their dysfunction associated with diseases such as glaucoma, cataracts, AMD, retinitis pigmentosa, and diabetic retinopathy.

Hallmarks of many neurodegenerative and retinal diseases in the visual system are due to impaired autophagy, which causes protein accumulation, defective mitochondrial turnover, and oxidative stress. Autophagy is particularly essential for maintaining retinal homeostasis, including the clearance of damaged mitochondria and the outer part of photoreceptors (Liton et al., 2023).

This chapter provides an in-depth review of the molecular mechanisms regulating autophagy, its influence on nerve and retinal cells, and the contribution to the pathogenesis of neurological-visual diseases. Hence, we discuss the updated therapies to modify autophagy, possibly halting disease progression and saving vision.

## PATHOPHYSIOLOGY OF NEUROLOGICAL-VISUAL DISEASES RELATED TO AUTOPHAGY DYSFUNCTION

Neurological-visual diseases, such as glaucoma and AMD, are defined by progressive degeneration of RGCs, photoreceptors, and optic nerve axons (Golestaneh et al., 2018; Rodríguez-Muela et al., 2012). These disorders are unified by the common pathological features, including oxidative stress, chronic inflammation, mitochondrial dysfunction, and the aggregation of misfolded proteins (Levine & Kroemer, 2019).

In glaucoma, elevated intraocular pressure (IOP) induces mechanical stress and apoptosis in RGCs. Defective autophagic flux in RGCs accelerates axonal degeneration and optic nerve damage. Ultimately, impaired autophagy as a cellular quality-control system by failing to remove toxic aggregates and dysfunctional mitochondria drives chronic neurodegeneration, leading to irreversible visual impairment (Ahmad et al., 2020; Jindal, 2015).

In AMD, chronic inflammation and lipofuscin accumulation damage the RPE, and impaired mitophagy leads to elevated reactive oxygen species (ROS), which further damage the RPE (Golestaneh et al., 2018; Rodríguez-Muela et al., 2012).

In IRDs such as retinitis pigmentosa (RP), genetic mutations cause the accumulation of misfolded proteins that overload the ubiquitin-proteasome system (UPS) (Illing et al., 2002). In this case, autophagy becomes a vital compensatory mechanism, and if disturbed, it cannot clear these misfolded proteins, thereby leading to the acceleration of photoreceptor degradation (Zhou et al., 2015).

## PHYSIOLOGICAL FUNCTIONS OF AUTOPHAGY IN THE EYE SYSTEM

Autophagy, which is activated in response to various stressors such as nutrient deprivation, hypoxia, oxidative stress, and microbial infections, is a fully regulated and lysosome-dependent process (Mizushima & Klionsky, 2007; Pankiv et al., 2007). It plays an important role in maintaining cellular homeostasis by removing damaged organelles, accumulated proteins, intracellular pathogens, and other harmful components of the cytoplasm. In the absence of efficient autophagy, ROS, and damaged intracellular structures accumulate in the cell, which can lead to cellular dysfunction and death (Jiménez-Loygorri et al., 2023; Pankiv et al., 2007).

Autophagy is also essential during the perinatal period to maintain energy balance in the tissues of adults and neonates, especially during the stage of neonatal starvation when foreign nutrients are suddenly restricted (Kuma et al., 2004). In addition to metabolic role, autophagy is also involved in the regulation of immunity: Facilitating the processing and delivery of antigens, regulation of homeostasis of T cells and B cells, and help control inflammation by destroying inflammatory components (Hara et al., 2006; Takamura et al., 2011).

In the eye system, autophagy is critical to maintaining the function of retinal cells, especially in the oxidative environment of the eye. Selective autophagy processes such as mitophagy are important for mitochondrial quality control (Kuma et al., 2004; Jiménez-Loygorri et al., 2023). In eye tissues daily autophagy regulation has been observed: the rhythm of autophagy-related protein expression (e.g., LC3, BECN1) is in harmony with the light cycle of the receptor and energy demand and decreases in the dark. This function highlights the physiological role of autophagy in maintaining retinal integrity and visual function (McWilliams et al., 2019; Reme et al., 1986).

## AUTOPHAGY IN THE DEVELOPING NEUROLOGICAL-VISUAL DISEASES

During ocular development, the need for autophagy is demonstrated by embryonic lethality in autophagy knockout animals, including *Becn1*−/− and *Ambra1*−/− (autophagy and Beclin 1 regulator 1) (Mellén et al., 2008), as well as ocular dysfunction in *Ambra1* knockout zebrafish, which results in a reduction of ocular size and embryonic lethality (Skobo et al., 2014). High expression of autophagy proteins was observed during key embryonic stages (e.g., at E16.5 in mice), where autophagy provides the preservation of ATP and cleansing of apoptotic cells (McWilliams et al., 2019). Deficient Atg5 in animals exhibits increased cellular death in the retinal neurons (Esteban-Martínez et al., 2017; Qu et al., 2007). Also, Autophagy activity plays a key role in cell death during vessel-hyaloid regression dependent on autophagy (Kim et al., 2010).

Table 2 shows the role of autophagy in corneal resiliency, lens, conventional outflow pathway, neural retina and glaucoma, RPE and AMD, photoreceptors and retinal diseases, and inherited retinal diseases.

### 1. Autophagy in Corneal Homeostasis

Autophagy plays an important role in the regeneration of the corneal epithelium and in the maintenance of the transparency of corneal layers (Dias-Teixeira et al., 2022). Limbal stem cells at the cornea-conjunctiva junction are important for corneal epithelial turnover, and MIR103-107 micro-RNA stimulates autophagy, which is vital for their regenerative activity (Peng et al., 2017). Autophagy-defective mice lead to the loss of stem cells and their reduced expression of proliferation markers (Kaplan et al., 2019).

## 2. Autophagy in the Lens: Proteostasis and Cataract Prevention

Autophagy is a vital mechanism that regulates the quality of lens fiber cells and keeps them transparent. Autophagy has been linked to the organelles of cells, with the use of autophagy-associated genes raising concern for cataract formation through a buildup of damaged proteins (Matsui et al., 2006; Morishita et al., 2013).

Disruption of autophagy-related genes such as Atg5 and Fyco1 leads to cataract formation due to the accumulation of misfolded complex crystallins (Morishita et al., 2013). While the precise mechanisms remain obscure, autophagy modulation is posited as a strategy for therapy, with some autophagy-inducing drugs, such as rapamycin, having somewhat of a paradoxical influence (Wilkinson et al., 2012).

## 3. Autophagy in the Conventional Outflow Pathway and Glaucoma

The trabecular meshwork (TM) and Schlemm's canal (SC) are the conventional outflow paths controlling IOP and experiencing cyclic mechanical and oxidative stress. The presence of autophagy deficiency in TM cells has impacted TGFβ-induced pathways (Nettesheim et al., 2019), resulting in increased TGFβ2 expression and altered fibrotic marker levels. Fibrosis may be regulated by autophagy by degrading BMP and Activin Membrane Bound Inhibitor (BAMBI), which is an antagonist of TGFβ signaling (Xavier et al., 2010).

Glaucoma essentially means progressive RGC loss and degeneration of the optic nerve (Li et al., 2022). In the glaucomatous model, surviving RGCs fail to sufficiently degrade damaged mitochondria or protein aggregates (Lin & Kuang, 2014). The apoptotic RGCs are thus damaged and stressed. Autophagy protects RGCs from stress in their environment by decreasing oxidative stress and balancing the function of mitochondria (Dixon et al., 2023; Lin & Kuang, 2014).

## 4. Autophagy in RGCs and Glaucoma

RGCs rely on mitochondrial quality control. Under chronic IOP elevation, dysfunctional autophagy leads to the accumulation of inefficient mitochondria and gradual axonal degeneration. The DBA/2J mouse model of glaucoma simulates human glaucoma through elevated IOP, causing optic neuropathy characterized by the degeneration of RGCs (Schlamp et al., 2006). Modulation of autophagy is a functional response of RGCs to insults related to glaucoma (Li et al., 2022). The process of autophagy in RGCs can either support neuroprotection or trigger neuronal death, depending on the type of experiment carried out and several other relevant factors (Liton et al., 2023).

## 5. Autophagy in the RPE and AMD

The RPE is a monolayer of polarized cells that is extremely vital for the maintenance of the photoreceptors (Lakkaraju et al., 2020). Autophagy is required in the RPE for quality control due to the primarily terminally differentiated nature of these cells. Autophagy regulates the homeostasis of visual pigment balance, phagocytosis, metabolic homeostasis, and organelle function. Aberrant autophagy exacerbates AMD (Liton et al., 2023).

Autophagic dysfunction is related to the RPE via lipofuscin and drusen accumulation pertinent to AMD. Restoring autophagy could halt the degeneration of the RPE and slow the progression of AMD (Dhingra et al., 2018; Golestaneh et al., 2018; Mitter et al., 2014). Increasing lysosomal biogenesis and restoring TFEB function are promising approaches to reverse this decline (Puertollano et al., 2018)

## 6. Autophagy in Photoreceptors and Circadian Regulation

Photoreceptors show a high basal autophagy rate due to metabolic demand. Circadian control affects autophagy rhythms in the neural retina and RPE, with increased autophagosome formation occurring during the light phase (Liton et al., 2023; Remé & Sulser, 1977). In studies of mice under normal light/dark conditions, it was revealed that photoreceptors undergo a biphasic autophagic flux and there is a correlation with outer segment ingestion in the RPE (Yao et al., 2014). DR and AMD are two diseases that are closely linked to dysregulated autophagy, while circadian rhythm disturbance is connected to diabetes and metabolism. Diurnal rhythmicity and overall levels of autophagic proteins, particularly ATG7, ATG9, LC3, and BECN1, were reduced in mice with type 1 diabetes and rats with type 2 diabetes. Autophagic proteins were suppressed in T2D rats and rhythms were disrupted. With aging and diabetes, the rhythm and level of vital autophagy proteins such as ATG9, LC3, and BECN1 decrease, disrupting metabolic homeostasis and increasing sensitivity to retinal degeneration (Qi et al., 2020).

## 7. Autophagy in IRDs

Autophagy plays a double-aged sword: on one hand, it can effectively clear toxic misfolded proteins or exacerbate degeneration if overactivated, as seen with rapamycin-induced autophagy restoring photoreceptor morphology in RP models

(Zhao et al., 2022). On the other hand, excessive autophagic activity may trigger apoptosis, requiring finely tuned modulation strategies (Liu et al., 2022).

Genetic mutations that affect photoreceptors' essential roles lead to IRDs such as RP (W. Liu et al., 2022; Newton & Megaw, 2020). The two-way function of autophagy in IRD-associated retinal degeneration has been elaborated: autophagy can promote cell survival by allowing the degradation of certain toxic proteins, while in some cases, increased autophagy leads to cell death. Preclinical models of IRDs have shown promise in autophagy levels (Lin & Xu, 2019; W. Liu et al., 2022).

## AUTOPHAGY RELATED TO LC3-ASSOCIATED PHAGOCYTOSIS (LAP) AND THE VISUAL CYCLE

While canonical autophagy requires the ULK1 complex, Beclin-1-VPS34 nucleation, and LC3 conjugation pathways to form autophagosomes sequentially, LAP represents a distinct but overlapping process. In LAP, LC3-II associates with the phagosome membrane to assist lysosomal degradation. LAP is a vital component of normal physiology and homeostasis in various systems. Unlike macroautophagy, LAP is independent of the autophagy preinitiation complex, but it still uses original autophagy machinery (e.g., Beclin-1, ATG5, ATG7, and LC3) (Green & Levine, 2014). This distinction allows LAP to selectively process swallowed extracellular material, especially in epithelial and immune cells, including those of the RPE. In the eye, LAP plays a critical role in RPE cells, which eliminate the daily phagocytosis of the outer parts of the photoreceptor. Precise degradation of these segments via LC3-II-coated phagosomes is vital for oxidative balance, lipid homeostasis, and avoidance of lipofuscin aggregation, a hallmark of AMD (Dhingra et al., 2018).

Furthermore, the RPE is an integral part of the visual cycle, particularly the recycling of the retinal chromophore 11-cis. LAP also contributes to this process by recovering vitamin A, which is critical for vision. Reduced autophagy is claimed to be an indicator of aging, and hence the extreme difficulties for older individuals in seeing at night. A complementary degradation and recycling network is formed by canonical autophagy and LAP for RPE homeostasis and visual function. Disruption in the regulation of both pathways endangers photoreceptor support and contributes to the pathogenesis of retinal diseases (Perusek et al., 2015; Tang et al., 2013).

## THERAPEUTIC TARGETING OF AUTOPHAGY IN NEUROLOGICAL-VISUAL DISEASES

Modulating autophagy could be a promising therapeutic approach for conditions like glaucoma, AMD, and IRDs, given its central role in cellular homeostasis, mitochondrial quality control, and protein aggregate clearance in retinal and neuronal cells. There are several new strategies that target key regulators in the autophagic pathway to restore balance and prevent neurodegeneration. Autophagy may be modulated in several ways being considered for therapeutic gain (Table 3):

### 1. Pharmacological Activators of Autophagy

The mTOR signaling pathway, which inhibits autophagy in nutrient-rich conditions, can be targeted to stimulate autophagy pharmacologically. Inhibition of mTOR can be achieved via:

- Rapamycin: These compounds work by creating autophagosomes through the removal of mTOR-mediated suppression from the ULK1 complex. Rapamycin in RP and glaucoma models restores photoreceptor integrity and slows degeneration (Eisenberg et al., 2009; Rubinsztein et al., 2011; Song et al., 2025; Zhao et al., 2022).
- Spermidine: A natural polyamine promotes autophagy independently of mTOR and affects the regulation of transcription of autophagy-related genes by increasing histone acetylation. These protective effects have been observed in models of neurodegenerative diseases (Eisenberg et al., 2009; Ghosh et al., 2020; Han et al., 2022; Madeo et al., 2010; Zhao et al., 2022).

### 2. Gene Therapy

Gene delivery systems employing adeno-associated virus (AAV) vectors are used to deliver genes that promote endogenous autophagy via overexpressing key autophagy regulators:

- BECN1: The overexpression of BECN1 induces autophagy and protects photoreceptor cells from death in models of RP by promoting the overexpression of autophagosome nucleation, it increases cellular cleansing capacity (Mellén et al., 2008).
- In addition, the delivery of cathepsin D (CTSD) in AAV vectors has been proposed as a method to restore CTSD enzymatic activity close to wild-type

levels and markedly attenuate retinal degeneration by a pronounced rescue of photoreceptor cells and rod bipolar cells (J. Liu et al., 2022).

In preclinical models of hereditary retinal diseases, these strategies have been tested, where advanced autophagy minimizes photoreceptor apoptosis, improves retinal morphology, and maintains visual function (Esteban-Martínez et al., 2017; Mellen et al., 2008; Zhao et al., 2022). Autophagy is sensitive to metabolic signals, making lifestyle modifications a non-invasive and adjunctive therapeutic option.

### 3. Lifestyle Interventions

Enhancing autophagy can complement pharmacological approaches using lifestyle modifications:

- Caloric restriction (CR) decreases nutrient signaling through mTOR and insulin pathways, which is akin to starvation-induced autophagy. In multiple species, CR has been suggested to enhance cellular resilience and lifespan, and it may boost retinal mitochondrial quality control (Colman et al., 2014).
- Exercise causes AMPK signaling to be activated, which leads to mTOR inhibition and autophagy initiation. In both systemic and ocular tissues, this aids in resisting oxidative stress and maintaining metabolic homeostasis (He et al., 2012).

## CHALLENGES AND FUTURE PROSPECTS

Although autophagy targeting holds promise, there are still challenges, such as achieving cell-specific modulation and preventing excessive autophagy induction, which can be pernicious. The focus of future research should be on understanding the context-dependent roles of autophagy and designing targeted therapies for neurological and visual diseases.

## CONCLUSIONS

Autophagy is an evolutionarily conserved lysosome-dependent degradation pathway that maintains cellular homeostasis in the retina and optic nerve, so its dysfunction significantly contributes to the pathogenesis of neurodegeneration such as glaucoma, AMD, and IRDs. There is promising therapeutic potential in autophagy modulation; however, challenges such as achieving cell-specific regulation and preventing over-activation, which may have detrimental effects, remain unresolved.

Future research should focus on clarifying the roles of autophagy in different eye tissues and developing targeted interventions that restore its balance. Addressing these challenges, new treatments may emerge to stop the progression of the disease and maintain vision in patients with these debilitating conditions.

*Table 1. Autophagy Mechanism in Neurological-Visual Diseases*

| Mechanism | Functions in the Eye | Key Molecular Players | Example Tissues | Associated Disorders |
|---|---|---|---|---|
| Autophagy-Dependent | -Clearance of damaged organelles (e.g., mitophagy)<br>- Degradation of protein aggregates<br>- Stress response and survival | LC3, Beclin-1, ATG5/7, PINK1, Parkin | RPE, retina, trabecular meshwork | AMD, glaucoma, IRDs |
| Autophagy-Independent | Protein degradation (UPS)<br>- Organelle clearance without autophagosomes (e.g., lens denucleation) | Proteasome (20S/26S), PLAAT family, LAMP2A | Lens, cornea, conjunctiva | Cataract, corneal dystrophies |

*Table 2. Role of autophagy in ocular disease*

| Disorders | Sub- Disorders | Key Points |
|---|---|---|
| Corneal Resiliency | Autophagy in Corneal Epithelial Diseases | Dry Eye: Lacritin promotes autophagy and mitochondrial function via FOXO3 acetylation (Vijmasi et al., 2014).<br>Keratoconus: Decreased autophagy markers linked to oxidative stress and (nuclear factor erythroid 2–related factor 2) NRF2 signaling(Shinde et al., 2020)<br>Microbial Infections: Pathogens like Herpes Simplex Virus -1 (HSV-1) (Shukla et al., 1999) and Pseudomonas (Kroken et al., 2022) manipulate autophagy |
| | Autophagy in Corneal Stromal Diseases | TGFBI gene mutations impair autophagic clearance in epithelial-stromal dystrophies (Choi et al., 2012).<br>Macular corneal dystrophy shows reduced autophagy and pyroptosis (Zheng et al., 2020) |
| | Autophagy in Corneal Endothelial Dystrophies | Dysregulated autophagy in endothelial dystrophies:<br>- Lithium carbonate restores endothelial cell density in some models (Kim et al., 2013).<br>- Fuchs corneal dystrophy involves decreased mitochondrial mass and activated mitophagy (Meng et al., 2013) . |

continued on following page

*Table 2. Continued*

| Disorders | Sub- Disorders | Key Points |
|---|---|---|
| Autophagy in the Lens | cataracts | Cataractogenesis involves the aggregation of misfolded crystalline, with mutations in crystallin genes linked to reduced autophagy (Moreau & King, 2012). Other non-crystallin gene mutations also associated with cataracts, indicating a complex relationship between autophagy suppression and cataract development (Chen et al., 2011). |
| Autophagy in the Conventional Outflow Pathway | Autophagy in Aging | Aging increases autophagy but diminishes lysosomal function (Kaufman, 2005). |
| | Autophagy in Glaucoma | Glaucomatous TM cells show impaired autophagy and mTOR activation (Schneider & Cuervo, 2014). |
| Autophagy in the Neural Retina and Glaucoma | Autophagy in RGC Death | Autophagy can be neuroprotective or contribute to cell death (Liton et al., 2023).<br>DBA/2J mice show increased autophagosomes but decreased autophagic flux (Schlamp et al., 2006) |
| | Autophagy in Axonal Degeneration | Autophagy inhibition attenuates axonal degeneration (Knöferle et al., 2010).<br>Chronic IOP elevation increases autophagic figures (Park et al., 2018; Park et al., 2012). |
| Autophagy in the RPE and AMD | Autophagy in RPE Metabolic Homeostasis | Autophagy is essential for energy supply and photoreceptor outer segment ingestion (Reyes-Reveles et al., 2017).<br>Defective autophagy leads to glucose reliance (Shi et al., 2015). |
| | Autophagy in Inflammation | Autophagy inhibition activates inflammasomes and oxidative stress (Kaarniranta et al., 2023).<br>Autophagy degrades damaged mitochondria (Nakahira et al., 2011). |
| | Autophagy and AMD | Reduced autophagy is linked to AMD progression (Zhang et al., 2020).<br>Mitophagy dysfunction contributes to AMD (Hyttinen et al., 2018). |
| Autophagy in Photoreceptors and Retinal Diseases | Autophagy in Photoreceptor Cells | HIF-1α activates autophagy during early response signals to activate autophagy and decrease PR apoptosis (Chinskey et al., 2014)<br>During RD, the RPE are separated from their source of nutrition, and autophagy is activated (Besirli et al., 2011; Xiao et al., 2021) |
| Inherited Retinal Diseases | Retinitis pigmentosa (RP) | Activation of autophagy via rapamycin treatment improved retinal function and morphology in a mouse model of RP.<br>- RP is a group of inherited vision disorders causing progressive retinal degeneration (Moreno et al., 2018; Zhao et al., 2022). |

*Table 3. Emerging Therapeutic Strategies Targeting Autophagy in Neurological-Visual Diseases*

| Strategy | Target | Mechanism | Disease Link | Status |
|---|---|---|---|---|
| Rapamycin | mTOR | Inhibition → ULK1 activation | RP, Glaucoma | Preclinical |
| Spermidine | Histone acetylation | mTOR-independent autophagy | Glaucoma, AMD, CNS disorders | Preclinical |
| AAV- CTSD / Beclin-1 | Autophagy genes | Restores autophagic flux | RP, AMD, DR | Preclinical |
| Caloric Restriction | mTOR, AMPK | Nutrient sensing → ↑ autophagy | AMD, DR | Observational/ animal |
| Exercise | AMPK activation | mTOR inhibition → ↑ autophagy | Systemic and retinal aging | Observational/ animal |

## REFERENCES


Ahmad, I., Teotia, P., Erickson, H., & Xia, X. (2020). Recapitulating developmental mechanisms for retinal regeneration. *Progress in Retinal and Eye Research*, *76*, 100824. DOI: 10.1016/j.preteyeres.2019.100824 PMID: 31843569

Besirli, C. G., Chinskey, N. D., Zheng, Q.-D., & Zacks, D. N. (2011). Autophagy activation in the injured photoreceptor inhibits fas-mediated apoptosis. *Investigative Ophthalmology & Visual Science*, *52*(7), 4193–4199. DOI: 10.1167/iovs.10-7090 PMID: 21421874

Chen, J., Ma, Z., Jiao, X., Fariss, R., Kantorow, W. L., Kantorow, M., Pras, E., Frydman, M., Pras, E., Riazuddin, S., Riazuddin, S. A., & Hejtmancik, J. F. (2011). Mutations in FYCO1 cause autosomal-recessive congenital cataracts. *American Journal of Human Genetics*, *88*(6), 827–838. DOI: 10.1016/j.ajhg.2011.05.008 PMID: 21636066

Chen, Y., Sawada, O., Kohno, H., Le, Y.-Z., Subauste, C., Maeda, T., & Maeda, A. (2013). Autophagy protects the retina from light-induced degeneration. *The Journal of Biological Chemistry*, *288*(11), 7506–7518. DOI: 10.1074/jbc.M112.439935 PMID: 23341467

Chinskey, N. D., Zheng, Q.-D., & Zacks, D. N. (2014). Control of photoreceptor autophagy after retinal detachment: The switch from survival to death. *Investigative Ophthalmology & Visual Science*, *55*(2), 688–695. DOI: 10.1167/iovs.13-12951 PMID: 24408986

Choi, S.-I., Kim, B.-Y., Dadakhujaev, S., Oh, J.-Y., Kim, T.-I., Kim, J. Y., & Kim, E. K. (2012). Impaired autophagy and delayed autophagic clearance of transforming growth factor β-induced protein (TGFBI) in granular corneal dystrophy type 2. *Autophagy*, *8*(12), 1782–1797. DOI: 10.4161/auto.22067 PMID: 22995918

Colacurcio, D. J., & Nixon, R. A. (2016). Disorders of lysosomal acidification—The emerging role of v-ATPase in aging and neurodegenerative disease. *Ageing Research Reviews*, *32*, 75–88. DOI: 10.1016/j.arr.2016.05.004 PMID: 27197071

Colman, R. J., Beasley, T. M., Kemnitz, J. W., Johnson, S. C., Weindruch, R., & Anderson, R. M. (2014). Caloric restriction reduces age-related and all-cause mortality in rhesus monkeys. *Nature Communications*, *5*(1), 3557. DOI: 10.1038/ncomms4557 PMID: 24691430

De Duve, C., & Wattiaux, R. (1966). Functions of lysosomes. *Annual Review of Physiology*, *28*(1), 435–492. DOI: 10.1146/annurev.ph.28.030166.002251 PMID: 5322983

Dhingra, A., Bell, B. A., Peachey, N. S., Daniele, L. L., Reyes-Reveles, J., Sharp, R. C., Jun, B., Bazan, N. G., Sparrow, J. R., Kim, H. J., Philp, N. J., & Boesze-Battaglia, K. (2018). Microtubule-associated protein 1 light chain 3B,(LC3B) is necessary to maintain lipid-mediated homeostasis in the retinal pigment epithelium. *Frontiers in Cellular Neuroscience*, *12*, 351. DOI: 10.3389/fncel.2018.00351 PMID: 30349463

Dias-Teixeira, K. L., Gh, M. S., Romano, J., Norouzi, F., & Laurie, G. W. (2022). Autophagy in the normal and diseased cornea. *Experimental Eye Research*, *225*, 109274. DOI: 10.1016/j.exer.2022.109274 PMID: 36252655

Dixon, A., Shim, M. S., Nettesheim, A., Coyne, A., Su, C.-C., Gong, H., & Liton, P. B. (2023). Autophagy deficiency protects against ocular hypertension and neurodegeneration in experimental and spontaneous glaucoma mouse models. *Cell Death & Disease*, *14*(8), 554. DOI: 10.1038/s41419-023-06086-3 PMID: 37620383

Eisenberg, T., Knauer, H., Schauer, A., Büttner, S., Ruckenstuhl, C., Carmona-Gutierrez, D., Ring, J., Schroeder, S., Magnes, C., Antonacci, L., Fussi, H., Deszcz, L., Hartl, R., Schraml, E., Criollo, A., Megalou, E., Weiskopf, D., Laun, P., Heeren, G., & Madeo, F. (2009). Induction of autophagy by spermidine promotes longevity. *Nature Cell Biology*, *11*(11), 1305–1314. DOI: 10.1038/ncb1975 PMID: 19801973

Esteban-Martínez, L., Sierra-Filardi, E., McGreal, R. S., Salazar-Roa, M., Mariño, G., Seco, E., Durand, S., Enot, D., Graña, O., Malumbres, M., Cvekl, A., Cuervo, A. M., Kroemer, G., & Boya, P. (2017). Programmed mitophagy is essential for the glycolytic switch during cell differentiation. *The EMBO Journal*, *36*(12), 1688–1706. DOI: 10.15252/embj.201695916 PMID: 28465321

Gao, Y., Tan, L., Yu, J.-T., & Tan, L. (2018). Tau in Alzheimer's disease: Mechanisms and therapeutic strategies. *Current Alzheimer Research*, *15*(3), 283–300. DOI: 10.2174/1567205014666170417111859 PMID: 28413986

Ghosh, I., Sankhe, R., Mudgal, J., Arora, D., & Nampoothiri, M. (2020). Spermidine, an autophagy inducer, as a therapeutic strategy in neurological disorders. *Neuropeptides*, *83*, 102083. DOI: 10.1016/j.npep.2020.102083 PMID: 32873420

Golestaneh, N., Chu, Y., Xiao, Y.-Y., Stoleru, G. L., & Theos, A. C. (2018). Dysfunctional autophagy in RPE, a contributing factor in age-related macular degeneration. *Cell Death & Disease*, *8*(1), e2537–e2537. DOI: 10.1038/cddis.2016.453 PMID: 28055007

Gómez-Virgilio, L., Silva-Lucero, M.-C., Flores-Morelos, D.-S., Gallardo-Nieto, J., Lopez-Toledo, G., Abarca-Fernandez, A.-M., Zacapala-Gómez, A.-E., Luna-Muñoz, J., Montiel-Sosa, F., Soto-Rojas, L. O., Pacheco-Herrero, M., & Cardenas-Aguayo, M.-C. (2022). Autophagy: a key regulator of homeostasis and disease: an overview of molecular mechanisms and modulators. *Cells*, *11*(15), 2262. DOI: 10.3390/cells11152262 PMID: 35892559

Green, D. R., & Levine, B. (2014). To be or not to be? How selective autophagy and cell death govern cell fate. *Cell*, *157*(1), 65–75. DOI: 10.1016/j.cell.2014.02.049 PMID: 24679527

Greene, A. W., Grenier, K., Aguileta, M. A., Muise, S., Farazifard, R., Haque, M. E., McBride, H. M., Park, D. S., & Fon, E. A. (2012). Mitochondrial processing peptidase regulates PINK1 processing, import and Parkin recruitment. *EMBO Reports*, *13*(4), 378–385. DOI: 10.1038/embor.2012.14 PMID: 22354088

Han, W., Li, H., & Chen, B. (2022). Research progress and potential applications of spermidine in ocular diseases. *Pharmaceutics*, *14*(7), 1500. DOI: 10.3390/pharmaceutics14071500 PMID: 35890394

Hara, T., Nakamura, K., Matsui, M., Yamamoto, A., Nakahara, Y., Suzuki-Migishima, R., Yokoyama, M., Mishima, K., Saito, I., Okano, H., & Mizushima, N. (2006). Suppression of basal autophagy in neural cells causes neurodegenerative disease in mice. *Nature*, *441*(7095), 885–889. DOI: 10.1038/nature04724 PMID: 16625204

He, C., Bassik, M. C., Moresi, V., Sun, K., Wei, Y., Zou, Z., An, Z., Loh, J., Fisher, J., Sun, Q., Korsmeyer, S., Packer, M., May, H. I., Hill, J. A., Virgin, H. W., Gilpin, C., Xiao, G., Bassel-Duby, R., Scherer, P. E., & Levine, B. (2012). Exercise-induced BCL2-regulated autophagy is required for muscle glucose homeostasis. *Nature*, *481*(7382), 511–515. DOI: 10.1038/nature10758 PMID: 22258505

Hyttinen, J. M., Viiri, J., Kaarniranta, K., & Błasiak, J. (2018). Mitochondrial quality control in AMD: Does mitophagy play a pivotal role? *Cellular and Molecular Life Sciences : CMLS*, *75*(16), 2991–3008. DOI: 10.1007/s00018-018-2843-7 PMID: 29777261

Illing, M. E., Rajan, R. S., Bence, N. F., & Kopito, R. R. (2002). A rhodopsin mutant linked to autosomal dominant retinitis pigmentosa is prone to aggregate and interacts with the ubiquitin proteasome system. *The Journal of Biological Chemistry*, *277*(37), 34150–34160. DOI: 10.1074/jbc.M204955200 PMID: 12091393

Jiménez-Loygorri, J. I., Benítez-Fernández, R., Viedma-Poyatos, Á., Zapata-Muñoz, J., Villarejo-Zori, B., Gómez-Sintes, R., & Boya, P. (2023). Mitophagy in the retina: Viewing mitochondrial homeostasis through a new lens. *Progress in Retinal and Eye Research, 96*, 101205. DOI: 10.1016/j.preteyeres.2023.101205 PMID: 37454969

Jindal, V. (2015). Interconnection between brain and retinal neurodegenerations. *Molecular Neurobiology, 51*(3), 885–892. DOI: 10.1007/s12035-014-8733-6 PMID: 24826919

Johansen, T., & Lamark, T. (2020). Selective autophagy: ATG8 family proteins, LIR motifs and cargo receptors. *Journal of Molecular Biology, 432*(1), 80–103. DOI: 10.1016/j.jmb.2019.07.016 PMID: 31310766

Kaarniranta, K., Blasiak, J., Liton, P., Boulton, M., Klionsky, D. J., & Sinha, D. (2023). Autophagy in age-related macular degeneration. *Autophagy, 19*(2), 388–400. DOI: 10.1080/15548627.2022.2069437 PMID: 35468037

Kaplan, N., Wang, J., Wray, B., Patel, P., Yang, W., Peng, H., & Lavker, R. M. (2019). Single-cell RNA transcriptome helps define the limbal/corneal epithelial stem/early transit amplifying cells and how autophagy affects this population. *Investigative Ophthalmology & Visual Science, 60*(10), 3570–3583. DOI: 10.1167/iovs.19-27656 PMID: 31419300

Kaufman, P. L. (2005). Changes in aqueous humor dynamics with age and glaucoma. *Progress in Retinal and Eye Research, 24*(5), 612–637. DOI: 10.1016/j.preteyeres.2004.10.003 PMID: 15919228

Kaushik, S., & Cuervo, A. M. (2018). The coming of age of chaperone-mediated autophagy. *Nature Reviews. Molecular Cell Biology, 19*(6), 365–381. DOI: 10.1038/s41580-018-0001-6 PMID: 29626215

Kim, E. C., Meng, H., & Jun, A. S. (2013). Lithium treatment increases endothelial cell survival and autophagy in a mouse model of Fuchs endothelial corneal dystrophy. *The British Journal of Ophthalmology, 97*(8), 1068–1073. DOI: 10.1136/bjophthalmol-2012-302881 PMID: 23759441

Kim, J., Kundu, M., Viollet, B., & Guan, K.-L. (2011). AMPK and mTOR regulate autophagy through direct phosphorylation of Ulk1. *Nature Cell Biology, 13*(2), 132–141. DOI: 10.1038/ncb2152 PMID: 21258367

Kim, J. H., Kim, J. H., Yu, Y. S., Mun, J. Y., & Kim, K.-W. (2010). Autophagy-induced regression of hyaloid vessels in early ocular development. *Autophagy, 6*(7), 922–928. DOI: 10.4161/auto.6.7.13306 PMID: 20818164

Knöferle, J., Koch, J. C., Ostendorf, T., Michel, U., Planchamp, V., Vutova, P., Tönges, L., Stadelmann, C., Brück, W., Bähr, M., & Lingor, P. (2010). Mechanisms of acute axonal degeneration in the optic nerve in vivo. *Proceedings of the National Academy of Sciences of the United States of America, 107*(13), 6064–6069. DOI: 10.1073/pnas.0909794107 PMID: 20231460

Kroken, A. R., Gajenthra Kumar, N., Yahr, T. L., Smith, B. E., Nieto, V., Horneman, H., Evans, D. J., & Fleiszig, S. M. (2022). Exotoxin S secreted by internalized Pseudomonas aeruginosa delays lytic host cell death. *PLoS Pathogens, 18*(2), e1010306. DOI: 10.1371/journal.ppat.1010306 PMID: 35130333

Kuma, A., Hatano, M., Matsui, M., Yamamoto, A., Nakaya, H., Yoshimori, T., Ohsumi, Y., Tokuhisa, T., & Mizushima, N. (2004). The role of autophagy during the early neonatal starvation period. *Nature, 432*(7020), 1032–1036. DOI: 10.1038/nature03029 PMID: 15525940

Lakkaraju, A., Umapathy, A., Tan, L. X., Daniele, L., Philp, N. J., Boesze-Battaglia, K., & Williams, D. S. (2020). The cell biology of the retinal pigment epithelium. *Progress in Retinal and Eye Research, 78*, 100846. DOI: 10.1016/j.preteyeres.2020.100846 PMID: 32105772

Lamark, T., & Johansen, T. (2021). Mechanisms of selective autophagy. *Annual Review of Cell and Developmental Biology, 37*(1), 143–169. DOI: 10.1146/annurev-cellbio-120219-035530 PMID: 34152791

Lamark, T., Kirkin, V., Dikic, I., & Johansen, T. (2009). NBR1 and p62 as cargo receptors for selective autophagy of ubiquitinated targets. *Cell Cycle (Georgetown, Tex.), 8*(13), 1986–1990. DOI: 10.4161/cc.8.13.8892 PMID: 19502794

Levine, B., & Kroemer, G. (2019). Biological functions of autophagy genes: A disease perspective. *Cell, 176*(1), 11–42. DOI: 10.1016/j.cell.2018.09.048 PMID: 30633901

Li, M., Gao, Z.-L., Zhang, Q.-P., Luo, A.-X., Xu, W.-Y., Duan, T.-Q., Wen, X.-P., Zhang, R.-Q., Zeng, R., & Huang, J.-F. (2022). Autophagy in glaucoma pathogenesis: Therapeutic potential and future perspectives. *Frontiers in Cell and Developmental Biology, 10*, 1068213. DOI: 10.3389/fcell.2022.1068213 PMID: 36589756

Lin, W., & Kuang, H.-y. (2014). Oxidative stress induces autophagy in response to multiple noxious stimuli in retinal ganglion cells. *Autophagy, 10*(10), 1692–1701. DOI: 10.4161/auto.36076 PMID: 25207555

Lin, W., & Xu, G. (2019). Autophagy: A role in the apoptosis, survival, inflammation, and development of the retina. *Ophthalmic Research, 61*(2), 65–72. DOI: 10.1159/000487486 PMID: 29694961

Liton, P. B., Boesze-Battaglia, K., Boulton, M. E., Boya, P., Ferguson, T. A., Ganley, I. G., Kauppinnen, A., Laurie, G. W., Mizushima, N., Morishita, H., Russo, R., Sadda, J., Shyam, R., Sinha, D., Thompson, D. A., & Zacks, D. N. (2023). Autophagy in the eye: From physiology to pathophysiology. *Autophagy Reports*, *2*(1), 2178996. DOI: 10.1080/27694127.2023.2178996 PMID: 37034386

Liu, J., Bassal, M., Schlichting, S., Braren, I., Di Spiezio, A., Saftig, P., & Bartsch, U. (2022). Intravitreal gene therapy restores the autophagy-lysosomal pathway and attenuates retinal degeneration in cathepsin D-deficient mice. *Neurobiology of Disease*, *164*, 105628. DOI: 10.1016/j.nbd.2022.105628 PMID: 35033660

Liu, W., Liu, S., Li, P., & Yao, K. (2022). Retinitis pigmentosa: Progress in molecular pathology and biotherapeutical strategies. *International Journal of Molecular Sciences*, *23*(9), 4883. DOI: 10.3390/ijms23094883 PMID: 35563274

Madeo, F., Eisenberg, T., Büttner, S., Ruckenstuhl, C., & Kroemer, G. (2010). Spermidine: A novel autophagy inducer and longevity elixir. *Autophagy*, *6*(1), 160–162. DOI: 10.4161/auto.6.1.10600 PMID: 20110777

Manley, S., Williams, J. A., & Ding, W.-X. (2013). Role of p62/SQSTM1 in liver physiology and pathogenesis. *Experimental Biology and Medicine*, *238*(5), 525–538. DOI: 10.1177/1535370213489446 PMID: 23856904

Matsui, M., Yamamoto, A., Kuma, A., Ohsumi, Y., & Mizushima, N. (2006). Organelle degradation during the lens and erythroid differentiation is independent of autophagy. *Biochemical and Biophysical Research Communications*, *339*(2), 485–489. DOI: 10.1016/j.bbrc.2005.11.044 PMID: 16300732

McWilliams, T. G., Prescott, A. R., Villarejo-Zori, B., Ball, G., Boya, P., & Ganley, I. G. (2019). A comparative map of macroautophagy and mitophagy in the vertebrate eye. *Autophagy*, *15*(7), 1296–1308. DOI: 10.1080/15548627.2019.1580509 PMID: 30786807

Mellén, M., De la Rosa, E. J., & Boya, P. (2008). The autophagic machinery is necessary for removal of cell corpses from the developing retinal neuroepithelium. *Cell Death and Differentiation*, *15*(8), 1279–1290. DOI: 10.1038/cdd.2008.40 PMID: 18369370

Meng, H., Matthaei, M., Ramanan, N., Grebe, R., Chakravarti, S., Speck, C. L., Kimos, M., Vij, N., Eberhart, C. G., & Jun, A. S. (2013). L450W and Q455K Col8a2 knock-in mouse models of Fuchs endothelial corneal dystrophy show distinct phenotypes and evidence for altered autophagy. *Investigative Ophthalmology & Visual Science*, *54*(3), 1887–1897. DOI: 10.1167/iovs.12-11021 PMID: 23422828

Mitter, S. K., Song, C., Qi, X., Mao, H., Rao, H., Akin, D., Lewin, A., Grant, M., Dunn, W.Jr, Ding, J., Bowes Rickman, C., & Boulton, M. (2014). Dysregulated autophagy in the RPE is associated with increased susceptibility to oxidative stress and AMD. *Autophagy*, *10*(11), 1989–2005. DOI: 10.4161/auto.36184 PMID: 25484094

Mizushima, N. (2020). The ATG conjugation systems in autophagy. *Current Opinion in Cell Biology*, *63*, 1–10. DOI: 10.1016/j.ceb.2019.12.001 PMID: 31901645

Mizushima, N., & Klionsky, D. J. (2007). Protein turnover via autophagy: Implications for metabolism. *Annual Review of Nutrition*, *27*(1), 19–40. DOI: 10.1146/annurev.nutr.27.061406.093749 PMID: 17311494

Mizushima, N., & Komatsu, M. (2011). Autophagy: Renovation of cells and tissues. *Cell*, *147*(4), 728–741. DOI: 10.1016/j.cell.2011.10.026 PMID: 22078875

Mizushima, N., & Levine, B. (2010). Autophagy in mammalian development and differentiation. *Nature Cell Biology*, *12*(9), 823–830. DOI: 10.1038/ncb0910-823 PMID: 20811354

Moreau, K. L., & King, J. A. (2012). Protein misfolding and aggregation in cataract disease and prospects for prevention. *Trends in Molecular Medicine*, *18*(5), 273–282. DOI: 10.1016/j.molmed.2012.03.005 PMID: 22520268

Moreno, M.-L., Mérida, S., Bosch-Morell, F., Miranda, M., & Villar, V. M. (2018). Autophagy dysfunction and oxidative stress, two related mechanisms implicated in retinitis pigmentosa. *Frontiers in Physiology*, *9*, 1008. DOI: 10.3389/fphys.2018.01008 PMID: 30093867

Morishita, H. (2022). Role of autophagy in the eye: From physiology to disease. *Current Opinion in Physiology*, *30*, 100592. DOI: 10.1016/j.cophys.2022.100592

Morishita, H., Eguchi, S., Kimura, H., Sasaki, J., Sakamaki, Y., Robinson, M. L., Sasaki, T., & Mizushima, N. (2013). Deletion of autophagy-related 5 (Atg5) and Pik3c3 genes in the lens causes cataract independent of programmed organelle degradation. *The Journal of Biological Chemistry*, *288*(16), 11436–11447. DOI: 10.1074/jbc.M112.437103 PMID: 23479732

Morishita, H., Eguchi, T., Tsukamoto, S., Sakamaki, Y., Takahashi, S., Saito, C., Koyama-Honda, I., & Mizushima, N. (2021). Organelle degradation in the lens by PLAAT phospholipases. *Nature*, *592*(7855), 634–638. DOI: 10.1038/s41586-021-03439-w PMID: 33854238

Nakahira, K., Haspel, J. A., Rathinam, V. A., Lee, S.-J., Dolinay, T., Lam, H. C., Englert, J. A., Rabinovitch, M., Cernadas, M., Kim, H. P., Fitzgerald, K. A., Ryter, S. W., & Choi, A. M. K. (2011). Autophagy proteins regulate innate immune responses by inhibiting the release of mitochondrial DNA mediated by the NALP3 inflammasome. *Nature Immunology*, *12*(3), 222–230. DOI: 10.1038/ni.1980 PMID: 21151103

Nettesheim, A., Shim, M. S., Hirt, J., & Liton, P. B. (2019). Transcriptome analysis reveals autophagy as regulator of TGFβ/Smad-induced fibrogenesis in trabecular meshwork cells. *Scientific Reports*, *9*(1), 16092. DOI: 10.1038/s41598-019-52627-2 PMID: 31695131

Newton, F., & Megaw, R. (2020). Mechanisms of photoreceptor death in retinitis pigmentosa. *Genes*, *11*(10), 1120. DOI: 10.3390/genes11101120 PMID: 32987769

Nishida, Y., Arakawa, S., Fujitani, K., Yamaguchi, H., Mizuta, T., Kanaseki, T., Komatsu, M., Otsu, K., Tsujimoto, Y., & Shimizu, S. (2009). Discovery of Atg5/Atg7-independent alternative macroautophagy. *Nature*, *461*(7264), 654–658. DOI: 10.1038/nature08455 PMID: 19794493

Pankiv, S., Clausen, T. H., Lamark, T., Brech, A., Bruun, J.-A., Outzen, H., Øvervatn, A., Bjørkøy, G., & Johansen, T. (2007). p62/SQSTM1 binds directly to Atg8/LC3 to facilitate degradation of ubiquitinated protein aggregates by autophagy. *The Journal of Biological Chemistry*, *282*(33), 24131–24145. DOI: 10.1074/jbc.M702824200 PMID: 17580304

Park, H., Kim, J., & Park, C. (2012). Activation of autophagy induces retinal ganglion cell death in a chronic hypertensive glaucoma model. *Cell Death & Disease*, *3*(4), e290–e290. DOI: 10.1038/cddis.2012.26 PMID: 22476098

Park, H.-Y. L., Kim, J. H., & Park, C. K. (2018). Different contributions of autophagy to retinal ganglion cell death in the diabetic and glaucomatous retinas. *Scientific Reports*, *8*(1), 13321. DOI: 10.1038/s41598-018-30165-7 PMID: 30190527

Park, J. K., Peng, H., Katsnelson, J., Yang, W., Kaplan, N., Dong, Y., Rappoport, J. Z., He, C., & Lavker, R. M. (2016). MicroRNAs-103/107 coordinately regulate macropinocytosis and autophagy. *The Journal of Cell Biology*, *215*(5), 667–685. DOI: 10.1083/jcb.201604032 PMID: 27872138

Parzych, K. R., & Klionsky, D. J. (2014). An overview of autophagy: Morphology, mechanism, and regulation. *Antioxidants & Redox Signaling*, *20*(3), 460–473. DOI: 10.1089/ars.2013.5371 PMID: 23725295

Peng, H., Park, J. K., & Lavker, R. M. (2017). Eyeing autophagy and macropinocytosis in the corneal/limbal epithelia. *Autophagy*, *13*(5), 975–977. DOI: 10.1080/15548627.2017.1287658 PMID: 28402214

Perusek, L., Sahu, B., Parmar, T., Maeno, H., Arai, E., Le, Y.-Z., Subauste, C. S., Chen, Y., Palczewski, K., & Maeda, A. (2015). Di-retinoid-pyridinium-ethanolamine (A2E) accumulation and the maintenance of the visual cycle are independent of Atg7-mediated autophagy in the retinal pigmented epithelium. *The Journal of Biological Chemistry*, *290*(48), 29035–29044. DOI: 10.1074/jbc.M115.682310 PMID: 26468292

Puertollano, R., Ferguson, S. M., Brugarolas, J., & Ballabio, A. (2018). The complex relationship between TFEB transcription factor phosphorylation and subcellular localization. *The EMBO Journal*, *37*(11), e98804. DOI: 10.15252/embj.201798804 PMID: 29764979

Qi, X., Mitter, S. K., Yan, Y., Busik, J. V., Grant, M. B., & Boulton, M. E. (2020). Diurnal rhythmicity of autophagy is impaired in the diabetic retina. *Cells*, *9*(4), 905. DOI: 10.3390/cells9040905 PMID: 32272782

Qu, X., Zou, Z., Sun, Q., Luby-Phelps, K., Cheng, P., Hogan, R. N., Gilpin, C., & Levine, B. (2007). Autophagy gene-dependent clearance of apoptotic cells during embryonic development. *Cell*, *128*(5), 931–946. DOI: 10.1016/j.cell.2006.12.044 PMID: 17350577

Ravenhill, B. J., Boyle, K. B., von Muhlinen, N., Ellison, C. J., Masson, G. R., Otten, E. G., Foeglein, A., Williams, R., & Randow, F. (2019). The cargo receptor NDP52 initiates selective autophagy by recruiting the ULK complex to cytosol-invading bacteria. *Molecular Cell*, *74*(2), 320-329. e326.

Reme, C., Wirz-Justice, A., Rhyner, A., & Hofmann, S. (1986). Circadian rhythm in the light response of rat retinal disk-shedding and autophagy. *Brain Research*, *369*(1-2), 356–360. DOI: 10.1016/0006-8993(86)90550-0 PMID: 3697752

Remé, C. E., & Sulser, M. (1977). Diurnal variation of autophagy in rod visual cells in the rat. *Graefe's Archive for Clinical and Experimental Ophthalmology*, *203*, 261–270. DOI: 10.1007/BF00409832 PMID: 303474

Reyes-Reveles, J., Dhingra, A., Alexander, D., Bragin, A., Philp, N. J., & Boesze-Battaglia, K. (2017). Phagocytosis-dependent ketogenesis in retinal pigment epithelium. *The Journal of Biological Chemistry*, *292*(19), 8038–8047. DOI: 10.1074/jbc.M116.770784 PMID: 28302729

Rodríguez-Muela, N., Germain, F., Mariño, G., Fitze, P. S., & Boya, P. (2012). Autophagy promotes survival of retinal ganglion cells after optic nerve axotomy in mice. *Cell Death and Differentiation*, *19*(1), 162–169. DOI: 10.1038/cdd.2011.88 PMID: 21701497

Ross, C. A., & Pickart, C. M. (2004). The ubiquitin–proteasome pathway in Parkinson's disease and other neurodegenerative diseases. *Trends in Cell Biology*, *14*(12), 703–711. DOI: 10.1016/j.tcb.2004.10.006 PMID: 15564047

Rubinsztein, D. C., Mariño, G., & Kroemer, G. (2011). Autophagy and aging. *Cell*, *146*(5), 682–695. DOI: 10.1016/j.cell.2011.07.030 PMID: 21884931

Schlamp, C. L., Li, Y., Dietz, J. A., Janssen, K. T., & Nickells, R. W. (2006). Progressive ganglion cell loss and optic nerve degeneration in DBA/2J mice is variable and asymmetric. *BMC Neuroscience*, *7*(1), 1–14. DOI: 10.1186/1471-2202-7-66 PMID: 17018142

Schneider, J. L., & Cuervo, A. M. (2014). Autophagy and human disease: Emerging themes. *Current Opinion in Genetics & Development*, *26*, 16–23. DOI: 10.1016/j.gde.2014.04.003 PMID: 24907664

Scrivo, A., Bourdenx, M., Pampliega, O., & Cuervo, A. M. (2018). Selective autophagy as a potential therapeutic target for neurodegenerative disorders. *Lancet Neurology*, *17*(9), 802–815. DOI: 10.1016/S1474-4422(18)30238-2 PMID: 30129476

Shi, H., Zhang, Z., Wang, X., Li, R., Hou, W., Bi, W., & Zhang, X. (2015). Inhibition of autophagy induces IL-1β release from ARPE-19 cells via ROS mediated NLRP3 inflammasome activation under high glucose stress. *Biochemical and Biophysical Research Communications*, *463*(4), 1071–1076. DOI: 10.1016/j.bbrc.2015.06.060 PMID: 26102024

Shinde, V., Hu, N., Mahale, A., Maiti, G., Daoud, Y., Eberhart, C. G., Maktabi, A., Jun, A. S., Al-Swailem, S. A., & Chakravarti, S. (2020). RNA sequencing of corneas from two keratoconus patient groups identifies potential biomarkers and decreased NRF2-antioxidant responses. *Scientific Reports*, *10*(1), 9907. DOI: 10.1038/s41598-020-66735-x PMID: 32555404

Shukla, D., Rowe, C. L., Dong, Y., Racaniello, V. R., & Spear, P. G. (1999). The murine homolog (Mph) of human herpesvirus entry protein B (HveB) mediates entry of pseudorabies virus but not herpes simplex virus types 1 and 2. *Journal of Virology*, *73*(5), 4493–4497. DOI: 10.1128/JVI.73.5.4493-4497.1999 PMID: 10196354

Sinha, D., Klise, A., Sergeev, Y., Hose, S., Bhutto, I. A., Hackler, L.Jr, Malpic-Llanos, T., Samtani, S., Grebe, R., Goldberg, M. F., Hejtmancik, J. F., Nath, A., Zack, D. J., Fariss, R. N., McLeod, D. S., Sundin, O., Broman, K. W., Lutty, G. A., & Zigler, J. S.Jr. (2008). βA3/A1-crystallin in astroglial cells regulates retinal vascular remodeling during development. *Molecular and Cellular Neurosciences*, *37*(1), 85–95. DOI: 10.1016/j.mcn.2007.08.016 PMID: 17931883

Sinha, R. A., Singh, B. K., & Yen, P. M. (2017). Reciprocal crosstalk between autophagic and endocrine signaling in metabolic homeostasis. *Endocrine Reviews*, *38*(1), 69–102. DOI: 10.1210/er.2016-1103 PMID: 27901588

Skobo, T., Benato, F., Grumati, P., Meneghetti, G., Cianfanelli, V., Castagnaro, S., Chrisam, M., Di Bartolomeo, S., Bonaldo, P., Cecconi, F., & Valle, L. D. (2014). Zebrafish ambra1a and ambra1b knockdown impairs skeletal muscle development. *PLoS One*, *9*(6), e99210. DOI: 10.1371/journal.pone.0099210 PMID: 24922546

Song, Y., Wang, F., Luo, H., Hu, H., Pang, Y., Xu, K., & Zhang, X. (2025). Rapamycin protects glucocorticoid-induced glaucoma model mice against trabecular meshwork fibrosis by suppressing mTORC1/2 signaling. *European Journal of Pharmacology*, *990*, 177269. DOI: 10.1016/j.ejphar.2025.177269 PMID: 39805488

Takamura, A., Komatsu, M., Hara, T., Sakamoto, A., Kishi, C., Waguri, S., Eishi, Y., Hino, O., Tanaka, K., & Mizushima, N. (2011). Autophagy-deficient mice develop multiple liver tumors. *Genes & Development*, *25*(8), 795–800. DOI: 10.1101/gad.2016211 PMID: 21498569

Tang, P. H., Kono, M., Koutalos, Y., Ablonczy, Z., & Crouch, R. K. (2013). New insights into retinoid metabolism and cycling within the retina. *Progress in Retinal and Eye Research*, *32*, 48–63. DOI: 10.1016/j.preteyeres.2012.09.002 PMID: 23063666

Tekirdag, K., & Cuervo, A. M. (2018). Chaperone-mediated autophagy and endosomal microautophagy: Jointed by a chaperone. *The Journal of Biological Chemistry*, *293*(15), 5414–5424. DOI: 10.1074/jbc.R117.818237 PMID: 29247007

Valapala, M., Edwards, M., Hose, S., Grebe, R., Bhutto, I. A., Cano, M., Berger, T., Mak, T. W., Wawrousek, E., Handa, J. T., Lutty, G. A., Samuel Zigler, J.Jr, & Sinha, D. (2014). Increased L ipocalin-2 in the retinal pigment epithelium of Cryba1 c KO mice is associated with a chronic inflammatory response. *Aging Cell*, *13*(6), 1091–1094. DOI: 10.1111/acel.12274 PMID: 25257511

Valapala, M., Wilson, C., Hose, S., Bhutto, I. A., Grebe, R., Dong, A., Greenbaum, S., Gu, L., Sengupta, S., Cano, M., Hackett, S., Xu, G., Lutty, G. A., Dong, L., Sergeev, Y., Handa, J. T., Campochiaro, P., Wawrousek, E., Zigler, J. S.Jr, & Sinha, D. (2014). Lysosomal-mediated waste clearance in retinal pigment epithelial cells is regulated by CRYBA1/βA3/A1-crystallin via V-ATPase-MTORC1 signaling. *Autophagy*, *10*(3), 480–496. DOI: 10.4161/auto.27292 PMID: 24468901

Vargas, J. N. S., Wang, C., Bunker, E., Hao, L., Maric, D., Schiavo, G., Randow, F., & Youle, R. J. (2019). Spatiotemporal control of ULK1 activation by NDP52 and TBK1 during selective autophagy. *Molecular cell, 74*(2), 347-362. e346.

Vijmasi, T., Chen, F. Y., Balasubbu, S., Gallup, M., McKown, R. L., Laurie, G. W., & McNamara, N. A. (2014). Topical administration of lacritin is a novel therapy for aqueous-deficient dry eye disease. *Investigative Ophthalmology & Visual Science*, *55*(8), 5401–5409. DOI: 10.1167/iovs.14-13924 PMID: 25034600

Wilkinson, J. E., Burmeister, L., Brooks, S. V., Chan, C. C., Friedline, S., Harrison, D. E., Hejtmancik, J. F., Nadon, N., Strong, R., Wood, L. K., Woodward, M. A., & Miller, R. A. (2012). Rapamycin slows aging in mice. *Aging Cell*, *11*(4), 675–682. DOI: 10.1111/j.1474-9726.2012.00832.x PMID: 22587563

Xavier, S., Gilbert, V., Rastaldi, M. P., Krick, S., Kollins, D., Reddy, A., Bottinger, E., Cohen, C. D., & Schlondorff, D. (2010). BAMBI is expressed in endothelial cells and is regulated by lysosomal/autolysosomal degradation. *PLoS One*, *5*(9), e12995. DOI: 10.1371/journal.pone.0012995 PMID: 20886049

Xiao, J., Yao, J., Jia, L., Ferguson, T. A., Weber, S., Sundstrom, J. M., Wubben, T. J., Besirli, C. G., & Zacks, D. N. (2021). Autophagy activation and photoreceptor survival in retinal detachment. *Experimental Eye Research*, *205*, 108492. DOI: 10.1016/j.exer.2021.108492 PMID: 33609513

Yang, Z., & Klionsky, D. J. (2010). Eaten alive: A history of macroautophagy. *Nature Cell Biology*, *12*(9), 814–822. DOI: 10.1038/ncb0910-814 PMID: 20811353

Yao, J., Jia, L., Shelby, S. J., Ganios, A. M., Feathers, K., Thompson, D. A., & Zacks, D. N. (2014). Circadian and noncircadian modulation of autophagy in photoreceptors and retinal pigment epithelium. *Investigative Ophthalmology & Visual Science*, *55*(5), 3237–3246. DOI: 10.1167/iovs.13-13336 PMID: 24781939

Yu, L., Chen, Y., & Tooze, S. A. (2018). Autophagy pathway: Cellular and molecular mechanisms. *Autophagy*, *14*(2), 207–215. DOI: 10.1080/15548627.2017.1378838 PMID: 28933638

Zhang, Z.-Y., Bao, X.-L., Cong, Y.-Y., Fan, B., & Li, G.-Y. (2020). Autophagy in age-related macular degeneration: A regulatory mechanism of oxidative stress. *Oxidative Medicine and Cellular Longevity*, *2020*(1), 2896036. DOI: 10.1155/2020/2896036 PMID: 32831993

Zhao, M., Lv, H., Yang, N., & Peng, G.-H. (2022). Rapamycin Improved Retinal Function and Morphology in a Mouse Model of Retinal Degeneration. *Frontiers in Neuroscience*, *16*, 846584. DOI: 10.3389/fnins.2022.846584 PMID: 35295093

Zheng, T., Zhao, C., Zhao, B., Liu, H., Wang, S., Wang, L., & Liu, P. (2020). Impairment of the autophagy-lysosomal pathway and activation of pyroptosis in macular corneal dystrophy. *Cell Death Discovery*, *6*(1), 85. DOI: 10.1038/s41420-020-00320-z PMID: 32983576

Zhou, Z., Doggett, T., Sene, A., Apte, R., & Ferguson, T. (2015). Autophagy supports survival and phototransduction protein levels in rod photoreceptors. *Cell Death and Differentiation*, *22*(3), 488–498. DOI: 10.1038/cdd.2014.229 PMID: 25571975

# Chapter 7
# Polycystic Ovary Syndrome PCOS and Depression Role of Neuroinflammation

**Mohammad Reza Ramezanpour**
*Pediatric Urology and Regenerative Medicine Research Center, Gene, Cell, and Tissue Research Institute, Children's Medical Center, Tehran University of Medical Sciences, Tehran, Iran*

**Omid Bahrami**
*Pediatric Urology and Regenerative Medicine Research Center, Gene, Cell, and Tissue Research Institute, Children's Medical Center, Tehran University of Medical Sciences, Tehran, Iran*

**Muhammad Parsa Pashazadeh**
*Pediatric Urology and Regenerative Medicine Research Center, Gene, Cell, and Tissue Research Institute, Children's Medical Center, Tehran University of Medical Sciences, Tehran, Iran*

**Rojin Sarallah**
*Islamic Azad University Medical Branch of Tehran, Tehran, Iran*

**Reza Nikfar**
*Pediatric Urology and Regenerative Medicine Research Center, Gene, Cell, and Tissue Research Institute, Children's Medical Center, Tehran University of Medical Sciences, Tehran, Iran*

**Masoumeh Majidi Zolbin**
https://orcid.org/0000-0002-1351-8955
*Iranian Tissue Bank and Research Center, Tehran University of Medical Sciences, Tehran, Iran*

DOI: 10.4018/979-8-3693-5908-2.ch007

## ABSTRACT

*Polycystic ovary syndrome (PCOS) is a multidimensional condition with metabolic, hormonal, and neuropsychiatric effects. Depression is common among PCOS patients, and emerging research shows neuroinflammation as a probable reason. Chronic low-grade inflammation, insulin resistance, hyperandrogenism, and gut dysbiosis all contribute to neuroinflammatory processes, which result in hypothalamic dysfunction, neurotransmitter imbalances, and hypothalamic-pituitary-adrenal (HPA) axis dysregulation. Abnormal microglial activation and oxidative stress may further damage neuroplasticity, interfering with proper mood regulation. This chapter investigates how neuroinflammatory pathways influence the relationship between PCOS and depression, focusing on their effects on brain function and mood regulation. This chapter addresses neuroinflammation as a mechanism contributing to psychiatric symptoms in PCOS by identifying key inflammatory drivers, as well as prospective treatment targets.*

## INTRODUCTION

Polycystic Ovary Syndrome (PCOS) is a common condition that affects between 4% and 20% of women during their reproductive years (Deswal et al., 2020). Although genetic factors, particularly genes linked to ovarian and adrenal hormone production, like those in the CYP family, contribute to PCOS, environmental influences also play a major role in how the condition develops and progresses. Environmental toxins, such as endocrine-disrupting chemicals (EDCs) like bisphenol A (BPA) and advanced glycation end-products (AGEs), may aid in the development and exacerbation of PCOS by disrupting hormonal and metabolic balance, especially when exposure occurs during critical developmental periods (Rutkowska & Diamanti-Kandarakis, 2016). When combined with genetic susceptibility, environmental factors can worsen the metabolic and hormonal imbalances commonly seen in PCOS-affected individuals.

The genetic and environmental metabolic impairments underlying PCOS primarily drive the condition through functional ovarian hyperandrogenism, characterized by dysregulated androgen secretion. This dysfunction is associated with the overexpression of steroidogenic enzymes, particularly cytochrome P450c17, in ovarian theca cells. Insulin resistance and obesity deteriorate hyperandrogenism through increasing ovarian sensitivity to luteinizing hormone (LH) (Rosenfield & Ehrmann, 2016). These disruptions cause diverse physical and psychological symptoms.

PCOS symptoms include menstruation abnormalities, weight gain, exhaustion, hirsutism, thinning hair, infertility, acne, mood changes, pelvic discomfort, and sleep difficulties, with depression being one of the most often reported psychological symptoms (Witchel et al., 2019). The link between PCOS and depression is being more recognized, with research revealing that several physiological pathways play an important role in their cohabitation (Zehra & Khursheed, 2018). Emerging evidence suggests that inflammation and mitochondrial dysfunction may be key mechanisms linking these conditions.

The relationship between PCOS and depression is well-documented, with evidence suggesting that chronic low-grade inflammation and mitochondrial dysfunction play critical roles in their coexistence. Women with PCOS often show higher levels of inflammation in their bodies, including markers like tumor necrosis factor-alpha (TNF-α), C-reactive protein (CRP), and interleukin-6 (IL-6). This ongoing inflammation may result in both metabolic problems and mental health challenges (Dabravolski et al., 2021). Research suggests that mitochondrial abnormalities may worsen insulin resistance and oxidative stress, which may further connect depression and PCOS. Given these findings, further exploration of neuroinflammatory pathways may provide deeper insights into the PCOS-depression link.

These results highlight the need for a multidisciplinary strategy that addresses both hormonal and psychological components of PCOS management. Recent studies suggest hypothalamic inflammation may integrate the metabolic, hormonal, and psychological aspects of PCOS (D. Barlampa et al., 2021).

## PCOS

In the absence of any particular diagnoses, ovarian dysfunction and androgen excess are combined to result in PCOS (Ashraf et al., 2019). PCOS is a complex illness that has significant implications for metabolic, reproductive, and mental health. A wide range of symptoms, such as hyperandrogenism, persistent anovulation, and polycystic ovarian morphology, characterize it (Su et al., 2025). Up to 105 million women of reproductive age worldwide are thought to be affected by PCOS. Significant comorbidities, including obesity, eating, sleeping, and sexual difficulties, cardiovascular illnesses, type 2 diabetes, insulin resistance, and mental health issues like depression and anxiety, are linked to PCOS (Boldis et al., 2024).

## Symptoms

Individual differences can exist in the clinical symptoms of PCOS (Escobar-Morreale, 2022). According to the National Institutes of Health (NIH), common manifestations include irregular menstruation, such as amenorrhea (absence of periods), oligomenorrhea (infrequent periods), or anovulatory bleeding (menstruation without ovulation). Women may also experience hirsutism, characterized by increased hair growth on the face, chest, abdomen, or thighs, as well as persistent or severe acne that resists standard treatments. Other common features include insulin sensitivity, weight gain or central obesity, and difficulty losing weight. Additionally, some individuals develop acanthosis nigricans, which presents as dark, thickened, velvety patches of skin, typically in body folds (National Institute of Child Health and Human Development, n.d.).

PCOS impacts the body in different ways, such as increased testosterone levels, which can cause problems with fertility because women with PCOS frequently have trouble ovulating (Louwers & Laven, 2020).

## Diagnosis

Women with at least two of these three symptoms, high testosterone levels, irregular menstruation, and ovarian cysts, are usually diagnosed with PCOS (Chaudhuri, 2023). When a woman exhibits any two of the following symptoms, PCOS is usually diagnosed: irregular or nonexistent menstrual periods, increased androgens, or numerous ovarian cysts seen on a transvaginal ultrasound (Dason et al., 2024). The well-known Rotterdam criteria state that a woman is diagnosed with PCOS if she possesses at least two of the following: Infrequent or nonexistent ovulation, indications or laboratory results of increased androgens, and ultrasound evidence of polycystic ovaries (Christ & Cedars, 2023).

## Pathophysiology

PCOS is a complicated, diverse pathophysiology that is influenced by metabolic abnormalities and reproductive dysfunction. It also includes complex connections between genetic, hormonal, and environmental variables (Harada, 2022). Alterations influence the condition's pathophysiology in adipocyte function, inflammation, neuroendocrine regulation, metabolic processes, insulin secretion, ovarian follicle development, hormone production, and sympathetic nervous system activity (Ibáñez et al., 2017).

One of PCOS's main characteristics is hyperandrogenism, which is characterized by elevated testosterone levels that cause hirsutism, acne, and hair loss. This condition's severity is frequently correlated with hyperandrogenism (Fahs et al., 2023). Disturbances in the hypothalamic-pituitary-ovarian (HPO) axis, which regulates reproductive hormones, can also result in hormonal abnormalities (Singh et al., 2023). Elevated LH levels, influenced by genetic factors and changes in cortisol metabolism, can lead to the development of numerous ovarian cysts and symptoms like acne and excessive hair growth in PCOS (Bulsara et al., 2021).

Insulin causes sex hormone-binding globulin (SHBG) levels to drop, allowing freer and more active testosterone to circulate in the body, which can cause hyperandrogenism symptoms. High insulin levels also trigger increased androgen synthesis in the adrenal glands and ovaries, which exacerbates the hormone imbalance in diseases like PCOS. The excess of androgens leads to an increase in insulin, causing a hormonal imbalance. This cycle intensifies symptoms, as insulin stimulates androgen production further (Unluhizarci et al., 2021).

Genetic studies implicate several steroidogenesis-related gene variants, such as CYP family genes, in PCOS susceptibility (Hiam et al., 2019). Women who have poor-quality diets, physical inactivity, smoke, snore, consume a high-calorie diet, use plastic dinnerware, and decorate inside are more likely to suffer PCOS (Zhang et al., 2020).

A well-known feature of PCOS is persistent low-grade systemic inflammation, which is necessary for the development of hyperandrogenism and insulin resistance. Women with PCOS have increased levels of inflammatory markers, such as IL-1β, IL-18, CRP, TNF-α, and IL-6 (Mancini et al., 2021). Follicle endoplasmic reticulum stress (ER stress) in PCOS is caused by gonadotrophins and intraovarian factors, activating granulosa cells due to local hyperandrogenism. This stress exacerbates ovarian dysfunction by stimulating factors such as TGF-β1, promoting apoptosis, altering steroid metabolism, and affecting the sizes of the cumulus-oocyte complex, potentially leading to ovulatory issues (Harada, 2022).

## Psychological Complications of PCOS

Women with PCOS are at a higher risk of fertility-related issues, cardiovascular health, metabolic issues, and psychological well-being. Compared to women without PCOS, they have a higher risk of developing endometrial cancer, type 2 diabetes, hypertension, and cardiovascular issues (Palomba et al., 2015). According to the data, this illness may result in emotional disorders, bodily issues, and a lower quality of life in addition to psychological issues. 56.9% of women with this condition have at least one mental illness, according to studies (Hasan et al., 2022). The relationship between PCOS and depression is complex, comprising both biological and psychosocial factors. Depression is one of the most common psychological diseases associated with PCOS (Cooney et al., 2017) About 40% of PCOS-afflicted women experience depression (S. Sadeeqa et al., 2018).

The bidirectional relationship between PCOS and depression complicates treatment, as depressive symptoms may reduce adherence to lifestyle interventions and exacerbate PCOS symptoms, creating a vicious cycle. Recognizing and addressing this condition in women with PCOS is essential for comprehensive care.

## PCOS and Depression

The etiopathogenesis of depression in PCOS emerges from various interplay of multisystemic physiological dysregulations, characterized by several pathophysiological mechanisms (Danai Barlampa et al., 2021; Peigné & Dewailly, 2014). Several factors may contribute to mood issues and depression in individuals, such as insulin resistance, problems with the HPA axis, high androgen levels, obesity, and struggles with infertility. Neurotransmitter imbalances, particularly reduced serotonin, norepinephrine, and dopamine in mood-related brain regions like the hippocampus and hypothalamus, further contribute to depressive symptoms (Goeldner et al., 2010).

*Insulin Resistance Association with Depression in PCOS.* Recent studies show a strong connection between insulin resistance and depression in women with PCOS. Approximately 75% of these women meet the WHO criteria for insulin resistance (Greenwood et al., 2018). Studies demonstrate a robust independent connect between insulin resistance and depression, with a 2.3-fold increased risk of depressive symptoms (Greenwood et al., 2018). Insulin plays an important role in brain function beyond its traditional metabolic understanding. Research highlights its significance in psychiatric and cognitive processes (Hamer et al., 2019).

Insulin's neurological effects are primarily transmitted through two critical molecular signaling cascades; Phosphoinositide-3 Kinase (PI3K)/Akt pathway and Ras/Mitogen-Activated Protein Kinase (MAPK) Signaling Cascade (Berry & Nedivi, 2017). Binding insulin to its receptor, initiates abovementioned pathways.

Brain insulin receptors are present in mood and reward centers, including the striatum, VTA, hippocampus, and hypothalamus (Gruber et al., 2023). Excess insulin crosses the blood-brain barrier (BBB) (Stuart et al., 2015), and insulin resistance may interfere dopamine signaling, contributing to depressive symptoms (Kullmann et al., 2016).

*HPA Axis Dysregulation Association with Depression in PCOS.* Emerging evidence suggests that chronic high insulin levels may trigger the HPA axis, leading to increased cortisol through excess corticotropin-releasing hormone (CRH) (Laura G Cooney & Anuja Dokras, 2017; Sarwar et al., 2022). This disrupts glucocorticoid feedback regulation and contributes to the development of mood disturbances (Watson et al., 2018).

*Hyperandrogenism Association with Depression in PCOS.* Hyperandrogenism is a common manifestation among women with PCOS, characterized by clinical characteristics such as hirsutism, acne, irregular menstruation, androgenic alopecia, and potential fertility complications (Diamanti-Kandarakis, 2010; Saleha Sadeeqa et al., 2018). Elevated dehydroepiandrosterone sulfate (DHEA-S) levels can serve as a biochemical indicator of hyperandrogenism, particularly in female patients (Hu et al., 2015). Previous studies have demonstrated that in dehydroepiandrosterone

(DHEA)-induced PCOS mice, there is a remarkable reduction in monoamine levels (e.g., dopamine, serotonin) compared to the control group (Lai et al., 2014; Yu et al., 2016). These findings indicate a possible link between hyperandrogenism and depression-like behavioral manifestations (Ethirajulu et al., 2021).

*Obesity Association with Depression in PCOS.* Based on many studies 40-60% of PCOS individuals have obesity (Moran et al., 2009). Hyperactivity of the HPA axis is observed in obesity. Consequently, the resulting hormonal imbalance, particularly elevated cortisol levels, leads to proinflammatory cytokine secretion, which contributes to insulin resistance (Rodriguez et al., 2015). Moreover, obesity can exert significant psychological negative effects on patients, potentially causing mood dysfunction (Ethirajulu et al., 2021).

Elucidating the inflammatory pathways and their constituent elements involved in depression associated with PCOS, can enhance our understanding of the underlying mechanisms involved. Furthermore, such insights may prove valuable in informing the design and refinement of effective therapeutic strategies.

## Neuroinflammation

The term "neuroinflammation" describes the various immune responses that take place within the central nervous system (CNS). The common causes of neuroinflammation fall into several categories, including viral infections, autoimmune conditions, peripheral organ inflammation, psychological stress, metabolic diseases, and various lifestyle-related factors (Sun et al., 2022a).

### Potential Pathways Triggering Neuroinflammation in the Hypothalamus

*Glucose Overconsumption.* Excessive intake of saturated fats and sugars synergistically promotes the formation of glycation end-products, triggering hypothalamic inflammation and microgliosis, leading to neuronal dysfunction and reduced energy availability (Mota et al., 2023). This mechanism leads to neuronal dysfunction and reduced energy availability (Gao et al., 2017). Interestingly, evidence from rodent studies proposes that female brains may be more susceptible to inflammation triggered by high-fat, high-sugar diets. This heightened sensitivity could partly explain the greater prevalence of inflammation-linked psychiatric conditions in women consuming Western-style diets (Church et al., 2022).

*Saturated Fatty Acids.* Long-chain saturated fatty acids trigger the JNK and NF-κB pathways, interfering with insulin and leptin signaling and encouraging the production of cytokines like TNF, IL-6, and IL-1β (D. Barlampa et al., 2021). Altered gut microbiota and LPS-induced systemic inflammation can disrupt BBB integrity and reduce levels of neurotrophins like BDNF and S100B, promoting neuroinflammation (Cicek et al., 2014). LPS-induced systemic inflammation and endotoxemia, as well as the subsequent immunological activation and cytokine release, can disrupt the tight connections between endothelial cells in the BBB, resulting in increased permeability (Reijerkerk et al., 2012).

*Uric Acid.* Elevated levels of uric acid in the bloodstream have been identified as a significant factor in neuroinflammation within the hypothalamus. Studies in mice have shown that diets high in uric acid activate the NF-κB signaling pathway, resulting in the production of inflammatory cytokines and reactive gliosis in the hypothalamus. Additionally, there is growing evidence indicating a crucial connection between increased uric acid levels and neuroinflammation, which could play a role in cognitive deficits and mood disorders (Shao et al., 2016).

*Stress.* Glucocorticoids are essential in the inflammation of the hypothalamus associated with stress (Duque Ede & Munhoz, 2016). Increased levels of glucocorticoids in the hypothalamus lead to a rise in the expression of ionized calcium-binding adaptor protein 1 (Iba1), a protein specific to microglia and macrophages that binds calcium. Iba1 plays a role in membrane ruffling and phagocytosis in active microglia and is used as an indicator of their activity. Stress elevates IL-1β levels in the hypothalamus, which contributes to subsequent effects. Prolonged stress results in glucocorticoid resistance, heightened IL-1β expression, and increased activity of the HPA axis (D. Barlampa et al., 2021). In relation to the behavioral impacts of chronic stress, a recent study found that female rats show stronger behavioral and neuroinflammatory reactions to chronic stress. This is marked by more significant decreases in hippocampal IL-10 and monoamine metabolites, which may help clarify why females are more vulnerable to depression triggered by stress (Xia et al., 2022).

The factors mentioned underscore the complex, multifaceted characteristics of neuroinflammation. These conditions foster a pro-inflammatory atmosphere in the brain, potentially leading to a range of neurological issues, particularly mood disorders and depression, which will be examined in the next section.

## Neuroinflammation and Depression

The association between neuroinflammation and depression is bidirectional, neuroinflammation may give rise to depression, and depression itself can exacerbate neuroinflammatory processes. Multiple studies have identified elevated levels of pro-inflammatory cytokines in individuals with major depressive disorder (MDD), with consistent increases observed in TNF-α and IL-6 levels (Cassano et al., 2017). Additional cytokines, especially IL-1β and interferon-gamma (IFN-γ), play a role in neuroinflammation, resulting in changes to neuronal function and imbalances in neurotransmitters that are associated with depression.

*Neurotransmitter's Disruption.* Neuroinflammation affects important neurotransmitters related to depression, such as serotonin, dopamine, glutamate, gamma-aminobutyric acid (GABA), and norepinephrine. One significant mechanism behind this is the disruption of tryptophan metabolism. In inflammatory states, the enzyme indoleamine 2,3-dioxygenase (IDO) redirects tryptophan from serotonin production to the kynurenine pathway. This shift leads to decreased serotonin levels and increased neurotoxic byproducts like quinolinic acid, which can harm neurons and exacerbate depressive symptoms (Correia & Vale, 2022; Schoenichen et al., 2019).

Dopaminergic signaling is also disrupted by neuroinflammation, with reduced dopamine synthesis, packaging, and release. These changes can diminish the efficacy of antidepressants and contribute to symptoms like anhedonia and impaired motivation (Grace, 2016; He et al., 2020).

During neuroinflammation, levels of glutamate—the main excitatory neurotransmitter in the brain—rise, mainly as a result of microglial activation and metabolites from the kynurenine pathway, such as quinolinic acid. This increase causes excessive glutamatergic activity, resulting in excitotoxicity and harm to the neuronal circuits that play a role in mood regulation (Haroon et al., 2017; Nayak et al., 2025; Zhang et al., 2022)

Conversely, GABA, the main inhibitory neurotransmitter, is often suppressed in inflammatory states. Impaired GABAergic signaling contributes to an imbalance between excitatory and inhibitory input, promoting neural overactivity and emotional dysregulation. This imbalance may also affect the glutamate/GABA-glutamine cycle, further contributing to neurotransmitter disruption in depression (Crowley et al., 2016; Mamelak, 2024). Altogether, these neuroinflammatory effects on neurotransmission, particularly involving serotonin, dopamine, glutamate, and GABA, highlight the complex neurochemical basis of depression and its potential relevance in PCOS-related mood disorders.

*Microglial Activation.* When activated, microglia produce pro-inflammatory cytokines such as IL-1β, TNF-α, and IL-6, which alter neuronal function and have been associated to depression symptoms. One key method by which neuroinflammation influences mood is the interruption of serotonin production. Proinflammatory cytokines stimulate IDO, an enzyme that converts tryptophan, a precursor of serotonin, to kynurenine.

The kynurenine pathway intensifies neurotoxic effects via its byproducts. Quinolinic acid acts as an NMDA receptor agonist, leading to excitotoxicity and damage to neurons. Increased levels of quinolinic acid have frequently been observed in individuals with depression, indicating a strong connection between neuroinflammation and neurodegeneration in mood disorders (Hassamal, 2023).

Chronic stress worsens this cycle by keeping microglia activated, resulting in continuous neuroinflammation. This creates a self-sustaining feedback loop where ongoing inflammation impacts serotonin and other neurotransmitter systems, worsening symptoms of depression. Recognizing these mechanisms highlights the need to focus on neuroinflammatory pathways in treating depression, particularly when standard antidepressant medications fail to provide relief (Sălcudean et al., 2025).

*HPA Axis Dysregulation.* Dysregulation of the HPA axis is a key factor in major depressive disorder (MDD), often leading to increased cortisol levels due to insufficient negative feedback. Chronic stress can worsen this imbalance by triggering neuroinflammation, as pro-inflammatory cytokines stimulate the HPA axis, resulting in mood fluctuations and cognitive difficulties. This interplay indicates a two-way relationship where neuroinflammation perpetuates hyperactivity of the HPA axis, thereby intensifying depressive symptoms (Mikulska et al., 2021; P & Vellapandian, 2024).

*Reduction in BDNF.* Brain-derived neurotrophic factor (BDNF) plays a vital role in synaptic plasticity and the well-being of neurons in the central nervous system. Its activation leads to neuroinflammation, which plays a role in the development of depression. Consequently, lower levels of BDNF not only hinder neural plasticity but also foster an inflammatory environment, both of which have been linked to depression (Colucci-D'Amato et al., 2020; Ding et al., 2020).

*P2X7-NLRP3 Inflammasome Cascade.* The P2X7-NLRP3 inflammasome pathway is critical in the link between neuroinflammation and depression. The activation of the P2X7 receptor by extracellular ATP causes potassium efflux, which activates the NLRP3 inflammasome in microglia. This cleaves pro-inflammatory cytokines including IL-1β and IL-18 into active forms, leading to neuroinflammation. Excessive production of IL-1β disturbs synaptic activity and changes monoaminergic and glutamatergic neurotransmission, both of which play a role in depression. Furthermore, NLRP3 activation in peripheral immune cells increases Th17 polarization, which weakens the blood-brain barrier and exacerbates neuroinflammation (Troubat et al., 2021).

As seen in Figure 1, neuroinflammation and depression form a bidirectional feedback loop. On the one hand, neuroinflammatory mechanisms such microglial activation and disturbed neurotransmitter networks lead to depression symptoms. On the other hand, depression can exacerbate neuroinflammation through processes such as HPA axis dysfunction and inflammasome activation, resulting in a self-reinforcing loop.

## INTERSECTION OF PCOS, DEPRESSION, AND NEUROINFLAMMATION

As previously mentioned, there is evidence linking PCOS to depression and anxiety. While this association is frequently attributed to endocrine and metabolic disruptions, such as hyperandrogenism, insulin resistance, and obesity, these factors alone fail to describe the complexity of the psychiatric burden experienced by women with PCOS (Laura G. Cooney & Anuja Dokras, 2017). Recent work suggests that neuroinflammation may serve as a missing mechanistic bridge (Wolugbom et al., 2023). A review of the literature reveals several pathways through which neuroinflammation can lead to depression in patients with PCOS.

*Stress and HPA Axis Dysregulation.* Alterations in serum FSH and LH levels, along with hyperandrogenism in PCOS, have a pivotal role in dysregulation of the HPA axis (Danai Barlampa et al., 2021). Chronic stress in PCOS patients is linked to sustained systemic glucocorticoid elevation due to HPA axis hyperactivity, resulting in resistance to glucocorticoid, particularly in the PVN of the hypothalamus, which contains CRH neurons (Merkulov et al., 2017). This condition promotes the upregulation of microglia markers contributing to the synthesis of many inflammatory factors (Frank et al., 2014; Wohleb et al., 2012). Rodent models further showed that psychological stress increases IL-1β expression in the hypothalamus, indicating stress as a direct contributor to central inflammation (Frank et al., 2014). Furthermore, fast GnRH pulse secretion, a hallmark of PCOS, increases gonadotropin abnormalities, which trigger ovarian hyperandrogenism and ovulatory dysfunction (McCartney & Campbell, 2020). In vitro findings show that elevated levels of leuprolide acetate, a GnRH agonist, raised the expression of the M1 macrophage polarization marker (CD86) which contributes to higher inflammatory factors (Wang et al., 2023). This synergistic loop between endocrine dysfunction, chronic stress, and neuroinflammation activates hypothalamic immune responses and disrupts emotion-regulating neural circuits via microglial activation, making PCOS patients more vulnerable to depression through neuroimmune dysregulation.

*Systemic Inflammation and Inflammatory Markers.* PCOS is associated with mild, chronic inflammation, shown by increased levels of certain inflammatory markers (Zhai & Pang, 2022). Inflammation in PCOS functions both as a symptom and a contributing driver of the syndrome, reinforcing a self-perpetuating cycle (Zhai & Pang, 2022). A recently published meta-analysis indicated that levels of CRP and IL-6 are significantly increased in people with PCOS (Aboeldalyl et al., 2021). A separate meta-analysis of animal studies showed that peripheral immune activation reliably triggers microglial priming and activation in the brain, providing mechanistic insight into how systemic inflammation may translate into neuroinflammation in PCOS (Hoogland et al., 2015). Systemic inflammation also raises levels of pro-inflammatory cytokines like IL-6, IL-18, TNF-α, and growth-related oncogene-α (GRO-α) (Reichmann et al., 2015). The relationship of these mediators with insulin resistance in PCOS patients was described in the previous literature (Mazloomi et al., 2021). Additionally, this is well-established that these factors can cross the BBB and trigger neuroinflammation. This occurs as circulating inflammatory cytokines lead to the breakdown of the BBB (Sun et al., 2022b). Inflammatory mediators can impair the tight junction of brain endothelial cells, resulting in BBB permeability (McKim et al., 2018). Besides, the elevation of inflammatory mediators induces microglia to phagocytose astrocytic end feet, which also leads to BBB leakage (Haruwaka et al., 2019). Vascular endothelial growth factor-A (VEGF-A) is also upregulated in astrocytes, causing higher nitric oxide production and higher permeability (Linnerbauer et al., 2020). These mechanisms promote a pro-inflammatory neural state, with microglial activation from systemic inflammation contributing to increased depression vulnerability in PCOS.

*Insulin Resistance and Oxidative Stress.* Insulin resistance, another defining disturbance of PCOS, aid in exacerbating neuroinflammation. Insulin resistance can result from androgens influencing skeletal muscle and adipose tissue insulin sensitivity, altering the secretion of adipokines, and driving the accumulation of visceral fat (Diamanti-Kandarakis & Dunaif, 2012). This condition, itself, can trigger inflammation through different molecular pathways (Shimobayashi et al., 2018). Pro-inflammatory factors like TNF-α, IL-6, and monocyte chemoattractant protein-1 (MCP-1) released by adipose tissue and macrophages perpetuate inflammation (Shoelson et al., 2006). Lipid overaccumulation in adipocytes also increases the generation of reactive oxygen species (ROS) through NADPH oxidases, amplifying oxidative stress and diminishing levels of anti-inflammatory molecules like adiponectin (Furukawa et al., 2004). These inflammatory and oxidative processes create a self-reinforcing cycle that exacerbates insulin resistance (Shoelson et al., 2006), and also extend their effect into the brain. Furthermore, higher oxidative stress, caused by insulin resistance, impairs mitochondrial function and possibly increases the risk of depression (Bansal & Kuhad, 2016). The link between metabolic and neuroimmune dysfunction suggests insulin resistance may contribute to mood disorders via neuroinflammation. Though unproven, future studies should track insulin resistance, inflammation, and mood symptoms to explore insulin-sensitizing therapies as potential treatments for both metabolic and psychiatric aspects of PCOS.

*Anti-inflammatory Effect of Estrogen.* Estrogen deficiency is a significant, though often underemphasized, component of PCOS pathophysiology, particularly in patients with anovulation and impaired follicular development (Yang & Chen, 2024). Studies have proposed that the reduction in estrogen level after oophorectomy is linked to increased level of inflammatory markers, indicating the anti-inflammatory effect of estrogen (Au et al., 2016). As mentioned above, this condition can induce neuroinflammation, thereby offering a mechanistic explanation for the heightened risk of mood disorders in this population. Besides, estrogen itself affects neurotransmitters and mood regulatory system (Au et al., 2016; Newhouse & Albert, 2015), which can be associated with depression in PCOS. Reduced estrogen, therefore, may impair neurotransmitter balance in mood-regulating brain regions like the prefrontal cortex and hippocampus, compounding the psychological burden of PCOS. Furthermore, lower estrogen levels are associated with heightened oxidative stress and elevated ROS production (Cagnacci et al., 2023) alongside insulin resistance (De Paoli et al., 2021), which shows estrogen can induce neuroinflammation through these pathways. These interconnected pathways underscore estrogen's key role in PCOS-related mood disturbances, suggesting that targeting estrogen signaling may help manage depressive symptoms.

*Kynurenine Pathway*. Nearly 95% of the tryptophan obtained from the diet undergoes metabolism via the kynurenine pathway, predominantly in the liver through the action of tryptophan 2,3-dioxygenase (TDO), with a smaller portion being processed by IDO (Badawy, 2015). Additionally, tryptophan is converted into tryptamine and indole derivatives via tryptophanase, and it also serves as a precursor for serotonin synthesis through tryptophan hydroxylase 1 (Taleb, 2019). Notably, PCOS-associated inflammation and rise in CRP upregulates IDO activity and shift tryptophan metabolism toward the kynurenine pathway (Jovanovic et al., 2022) which is known to precipitate neuroinflammation and CNS disorders, including depression (Mithaiwala et al., 2021). As discussed earlier, neuroinflammation itself can further dysregulate tryptophan metabolism favoring kynurenine pathway, thereby exacerbating mood disturbances. Tryptophan metabolism thus represents a promising intersectional biomarker, and potentially modifiable target, for future intervention.

*Gut-Brain Axis*. It is well established that gut dysbiosis influences mood and cognition disorders through different pathways, including HPA axis alteration and changes in neurotransmitters (Liu & Zhu, 2018). Microbiota imbalance can change gut permeability and increase the level of endotoxin-LPS in the blood system (Liu & Zhu, 2018). Even low-grade LPS exposure can precipitate anxiety- and depression-like behaviors by promoting neuroimmune responses (Bested et al., 2013). In addition, LPS-induced inflammation can enhance the activity of IDO, an enzyme involved in tryptophan metabolism via the kynurenine pathway, which has been positively linked to depressive symptoms (Bested et al., 2013). Notably, PCOS has been established to significantly change the gut microbiota in comparison to the normal population (Yurtdaş & Akdevelioğlu, 2020); however, it remains uncertain whether gut microbiota dysbiosis is a cause of PCOS or a consequence of it (Yurtdaş & Akdevelioğlu, 2020).

In addition to the hypothesized mechanisms discussed earlier, several experimental studies have directly explored the link between neuroinflammation and depression in PCOS. Foda et al. demonstrated that improving insulin resistance and hormonal imbalance in PCOS rat models led to better neurobehavioral outcomes. These effects were mediated through modulation of the IRS-1/PI3K/AKT/GSK-3β pathway, which reduced GSK-3β activity, tau hyperphosphorylation, amyloid-β accumulation, and NF-κB-mediated neuroinflammation (Foda et al., 2024). Similarly, Saad et al. reported heightened Notch signaling in PCOS rats, which promoted amyloid-β buildup, neuroinflammation, neuronal apoptosis, and reduced hippocampal acetylcholine levels (Saad et al., 2021).

Epigenetic factors have also been implicated. Olaniyi & Areloegbe observed that depressive-like behaviors in PCOS models were associated with elevated HDAC2 expression and DNA hypermethylation in the prefrontal cortex and hippocampus, accompanied by inflammation and disrupted acetylcholine signaling (Olaniyi & Areloegbe, 2022). Another study highlighted this pathway, showing that acetate improved depressive-like behavior in PCOS rats (Wolugbom et al., 2023). Additionally, Gao et al. showed that promoting anti-inflammatory M2 microglial polarization in the hypothalamus reduced neuroinflammation and reversed PCOS-like features induced by dihydrotestosterone (Gao et al., 2024).

Taken together, these findings underscore the multifaceted and bidirectional relationship between PCOS, neuroinflammation, and depression, an intersection shaped by metabolic, hormonal, immune, and epigenetic factors. Figure 2 illustrates key mechanisms, including HPA axis dysregulation, systemic inflammation, estrogen deficiency, altered tryptophan metabolism, insulin resistance, and gut dysbiosis.. While much of the clinical attention in PCOS has historically centered on reproductive and metabolic concerns, the growing understanding of its neuropsychiatric component necessitates a more comprehensive approach to diagnosis and therapy. This neuroimmune perspective not only provides mechanistic insights into the high prevalence of mood disturbances in PCOS but also opens new therapeutic possibilities.

## TREATMENTS OF PCOS

A multidisciplinary strategy that takes into account metabolic, reproductive, and psychological factors is necessary for the effective therapy of PCOS. Individual goals, such as those related to mental health and fertility, should guide treatment methods (Wolf et al., 2018). Growing insights into PCOS pathophysiology support more comprehensive and modern therapeutic strategies.

## Lifestyle Modifications

International guidelines highlight lifestyle modifications, including diet, exercise, and stress management, as first-line therapy for managing PCOS. These interventions play a central role in improving metabolic, reproductive, and psychological outcomes (Cowan et al., 2023). A balanced, low-glycemic diet enhances insulin sensitivity and supports weight loss, with even modest reductions (2–5%) helping to normalize androgen levels and improve symptoms (Lerchbaum et al., 2021). Furthermore, anti-inflammatory and Mediterranean diets have demonstrated potential in lowering inflammation and enhancing metabolic profiles in PCOS patients (Ajorlouie et al., 2025).

Individualized exercise regimens, particularly high-intensity interval training (HIIT), may provide superior metabolic benefits compared to traditional exercise protocols (Patten et al., 2020). Psychological support, especially cognitive-behavioral therapy (CBT), is also crucial. CBT not only improves mental health and quality of life but also enhances adherence to lifestyle changes and fertility treatment compliance, leading to improved outcomes, including higher pregnancy rates (Majidzadeh et al., 2023).

## Pharmaceutical and Clinical Treatments

Pharmaceutical interventions for PCOS remain diverse but are often associated with limitations, including side effects and adherence challenges. Emerging research continues to refine these therapies, calling for more personalized approaches (Gautam et al., 2025). The following treatment strategies may have potential relevance not only for addressing the metabolic and hormonal aspects of PCOS but also for mitigating its associated neuroinflammatory and mood-related symptoms.

*Hormonal Therapy.* In addition to offering relief from hirsutism, acne, and irregular menstruation, combined oral contraceptives (COCs) are the first-line treatment for hyperandrogenism and irregular menstruation. Newer formulations and combinations are being explored to minimize side effects while maintaining efficacy (Jia, 2023). To counter estrogen dominance in PCOS, progesterone or progestins like Prometrium and Provera help regulate cycles, reduce PMS, and support mood. Anti-androgens such as spironolactone remain key for hirsutism, though newer alternatives with fewer side effects are being explored (Pu et al., 2023).

*Insulin Sensitizers.* Insulin-sensitizing agents like metformin and thiazolidinediones may strengthen metabolic profiles and alleviate depressive symptoms by reducing systemic inflammation and restoring insulin signaling in the brain, potentially mitigating neuroimmune dysregulation (Attia et al., 2023). Newer insulin sensitizers, including GLP-1 receptor agonists, are currently being explored for their potential in managing both insulin resistance and weight in PCOS (Szczesnowicz et al., 2023).

*Psychotropic Medications.* The effect of antidepressants, especially selective serotonin reuptake inhibitors (SSRIs), in managing the mental health aspect of PCOS is well-documented, with newer studies underscoring the important role of early intervention in treating anxiety and depression. However, emerging evidence points to the need for personalized pharmacotherapy, taking into account individual patient profiles, to avoid unnecessary side effects (Masoudi et al., 2021).

*Gut microbiota.* Recent research highlights the potential link between gut microbiota and PCOS, with studies showing that women with PCOS tend to have less diverse gut flora, which may influence insulin resistance and androgen levels (Qi et al., 2019). Although the precise mechanisms are still being studied, probiotics and prebiotics are becoming more and more popular as adjuvant medicines that may have a good impact on metabolic profiles. Larger clinical trials are required to develop defined treatment procedures in this still-emerging field of study (Rau et al., 2024).

*Anti-inflammatory Agents.* Using dietary supplements like omega-3 fatty acids, vitamin E, and other antioxidants to reduce inflammation has become a growing area of interest in managing PCOS (Tefagh et al., 2022). Studies have demonstrated that combining omega-3s with vitamin E improves overall antioxidant defenses and lowers malondialdehyde (MDA) levels in the blood of women with PCOS. It also decreases the expression of lipoprotein (a), oxidized LDL-related genes, triglycerides, VLDL, LDL cholesterol, and the total cholesterol to HDL ratio (Mirmasoumi et al., 2018).

## Integrated Approaches

The most effective management of PCOS involves an integrated, individualized approach that combines lifestyle modifications, pharmacologic treatment, and psychological support. Recent studies reinforce the importance of considering comorbidities such as metabolic syndrome and mental health disorders when developing treatment plans (Lau et al., 2022). Emerging evidence also calls for more in-depth studies on the neuroinflammatory aspects of depression in PCOS, suggesting that future research should focus on the intersection of mental health and hormonal regulation to refine therapeutic strategies (Zerrouki et al., 2024).

## REFERENCES


Aboeldalyl, S., James, C., Seyam, E., Ibrahim, E. M., Shawki, H. E., & Amer, S. (2021). The Role of Chronic Inflammation in Polycystic Ovarian Syndrome—A Systematic Review and Meta-Analysis. *International Journal of Molecular Sciences*, *22*(5), 2734. DOI: 10.3390/ijms22052734 PMID: 33800490

Ajorlouie, Z., Moshkian, P., Baghdadi, G., Amiri, R., Biglari, F., & Rahimlou, M. (2025). The association between the Mediterranean Diet and the Prime Diet Quality Score and polycystic ovary syndrome: A case control study. *BMC Nutrition*, *11*(1), 80. DOI: 10.1186/s40795-025-01067-5 PMID: 40241157

Ashraf, S., Nabi, M., Rasool, S., Rashid, F., & Amin, S. (2019). Hyperandrogenism in polycystic ovarian syndrome and role of CYP gene variants: A review. *The Egyptian Journal of Medical Human Genetics*, *20*(1), 25. DOI: 10.1186/s43042-019-0031-4

Attia, G. M., Almouteri, M. M., & Alnakhli, F. T. (2023). Role of Metformin in Polycystic Ovary Syndrome (PCOS)-Related Infertility. *Cureus*, *15*(8), e44493. DOI: 10.7759/cureus.44493 PMID: 37791160

Au, A., Feher, A., McPhee, L., Jessa, A., Oh, S., & Einstein, G. (2016). Estrogens, inflammation and cognition. *Frontiers in Neuroendocrinology*, *40*, 87–100. DOI: 10.1016/j.yfrne.2016.01.002 PMID: 26774208

Badawy, A. A. B. (2015). Tryptophan metabolism, disposition and utilization in pregnancy. *Bioscience Reports*, *35*(5), e00261. DOI: 10.1042/BSR20150197 PMID: 26381576

Bansal, Y., & Kuhad, A. (2016). Mitochondrial dysfunction in depression. *Current Neuropharmacology*, *14*(6), 610–618. DOI: 10.2174/1570159X14666160229114755 PMID: 26923778

Barlampa, D., Bompoula, M. S., Bargiota, A., Kalantaridou, S., Mastorakos, G., & Valsamakis, G. (2021). Hypothalamic Inflammation as a Potential Pathophysiologic Basis for the Heterogeneity of Clinical, Hormonal, and Metabolic Presentation in PCOS. *Nutrients*, *13*(2), 520. DOI: 10.3390/nu13020520 PMID: 33562540

Barlampa, D., Bompoula, M. S., Bargiota, A., Kalantaridou, S., Mastorakos, G., & Valsamakis, G. (2021). Hypothalamic Inflammation as a Potential Pathophysiologic Basis for the Heterogeneity of Clinical, Hormonal, and Metabolic Presentation in PCOS. *Nutrients*, *13*(2), 520. Advance online publication. DOI: 10.3390/nu13020520 PMID: 33562540

Berry, K. P., & Nedivi, E. (2017). Spine dynamics: Are they all the same? *Neuron*, *96*(1), 43–55. DOI: 10.1016/j.neuron.2017.08.008 PMID: 28957675

Bested, A. C., Logan, A. C., & Selhub, E. M. (2013). Intestinal microbiota, probiotics and mental health: from Metchnikoff to modern advances: Part II–contemporary contextual research. *Gut Pathogens*, *5*(1), 1–14. DOI: 10.1186/1757-4749-5-3

Boldis, B. V., Grünberger, I., Cederström, A., Björk, J., Nilsson, A., & Helgertz, J. (2024). Comorbidities in women with polycystic ovary syndrome: A sibling study. *BMC Women's Health*, *24*(1), 221. DOI: 10.1186/s12905-024-03028-9 PMID: 38580996

Bulsara, J., Patel, P., Soni, A., & Acharya, S. (2021). A review: Brief insight into Polycystic Ovarian syndrome. *Endocrine and Metabolic Science*, *3*, 100085. DOI: 10.1016/j.endmts.2021.100085

Cagnacci, A., Gazzo, I., Stigliani, S., Paoletti, A. M., Anserini, P., Londero, A. P., & Xholli, A. (2023). Oxidative Stress: The Role of Estrogen and Progesterone. *Journal of Clinical Medicine*, *12*(23), 7304. DOI: 10.3390/jcm12237304 PMID: 38068356

Cassano, P., Bui, E., Rogers, A. H., Walton, Z. E., Ross, R., Zeng, M., Nadal-Vicens, M., Mischoulon, D., Baker, A. W., Keshaviah, A., Worthington, J., Hoge, E. A., Alpert, J., Fava, M., Wong, K. K., & Simon, N. M. (2017). Inflammatory cytokines in major depressive disorder: A case–control study. *The Australian and New Zealand Journal of Psychiatry*, *51*(1), 23–31. DOI: 10.1177/0004867416652736 PMID: 27313138

Chaudhuri, A. (2023). Polycystic ovary syndrome: Causes, symptoms, pathophysiology, and remedies. *Obesity Medicine*, *39*, 100480. DOI: 10.1016/j.obmed.2023.100480

Christ, J. P., & Cedars, M. I. (2023). Current Guidelines for Diagnosing PCOS. *Diagnostics (Basel)*, *13*(6), 1113. Advance online publication. DOI: 10.3390/diagnostics13061113 PMID: 36980421

Church, J. S., Renzelman, M. L., & Schwartzer, J. J. (2022). Ten-week high fat and high sugar diets in mice alter gut-brain axis cytokines in a sex-dependent manner. *The Journal of Nutritional Biochemistry*, *100*, 108903. DOI: 10.1016/j.jnutbio.2021.108903 PMID: 34748922

Cicek, I. E., Cicek, E., Kayhan, F., Uguz, F., Erayman, I., Kurban, S., Yerlikaya, F. H., & Kaya, N. (2014). The roles of BDNF, S100B, and oxidative stress in interferon-induced depression and the effect of antidepressant treatment in patients with chronic viral hepatitis: A prospective study. *Journal of Psychosomatic Research*, *76*(3), 227–232. DOI: 10.1016/j.jpsychores.2014.01.003 PMID: 24529042

Colucci-D'Amato, L., Speranza, L., & Volpicelli, F. (2020). Neurotrophic Factor BDNF, Physiological Functions and Therapeutic Potential in Depression, Neurodegeneration and Brain Cancer. *International Journal of Molecular Sciences*, *21*(20), 7777. Advance online publication. DOI: 10.3390/ijms21207777 PMID: 33096634

Cooney, L. G., & Dokras, A. (2017). Depression and Anxiety in Polycystic Ovary Syndrome: Etiology and Treatment. *Current Psychiatry Reports*, *19*(11), 83. DOI: 10.1007/s11920-017-0834-2 PMID: 28929349

Cooney, L. G., & Dokras, A. (2017). Depression and anxiety in polycystic ovary syndrome: Etiology and treatment. *Current Psychiatry Reports*, *19*(11), 1–10. DOI: 10.1007/s11920-017-0834-2 PMID: 28929349

Cooney, L. G., Lee, I., Sammel, M. D., & Dokras, A. (2017). High prevalence of moderate and severe depressive and anxiety symptoms in polycystic ovary syndrome: A systematic review and meta-analysis. *Human Reproduction (Oxford, England)*, *32*(5), 1075–1091. DOI: 10.1093/humrep/dex044 PMID: 28333286

Correia, A. S., & Vale, N. (2022). Tryptophan Metabolism in Depression: A Narrative Review with a Focus on Serotonin and Kynurenine Pathways. *International Journal of Molecular Sciences*, *23*(15), 8493. Advance online publication. DOI: 10.3390/ijms23158493 PMID: 35955633

Cowan, S., Lim, S., Alycia, C., Pirotta, S., Thomson, R., Gibson-Helm, M., Blackmore, R., Naderpoor, N., Bennett, C., Ee, C., Rao, V., Mousa, A., Alesi, S., & Moran, L. (2023). Lifestyle management in polycystic ovary syndrome - beyond diet and physical activity. *BMC Endocrine Disorders*, *23*(1), 14. DOI: 10.1186/s12902-022-01208-y PMID: 36647089

Crowley, T., Cryan, J. F., Downer, E. J., & O'Leary, O. F. (2016). Inhibiting neuroinflammation: The role and therapeutic potential of GABA in neuro-immune interactions. *Brain, Behavior, and Immunity*, *54*, 260–277. DOI: 10.1016/j.bbi.2016.02.001 PMID: 26851553

Dabravolski, S. A., Nikiforov, N. G., Eid, A. H., Nedosugova, L. V., Starodubova, A. V., Popkova, T. V., Bezsonov, E. E., & Orekhov, A. N. (2021). Mitochondrial Dysfunction and Chronic Inflammation in Polycystic Ovary Syndrome. *International Journal of Molecular Sciences*, *22*(8), 3923. Advance online publication. DOI: 10.3390/ijms22083923 PMID: 33920227

Dason, E. S., Koshkina, O., Chan, C., & Sobel, M. (2024). Diagnosis and management of polycystic ovarian syndrome. *Canadian Medical Association Journal*, *196*(3), E85–e94. DOI: 10.1503/cmaj.231251 PMID: 38286488

De Paoli, M., Zakharia, A., & Werstuck, G. H. (2021). The Role of Estrogen in Insulin Resistance: A Review of Clinical and Preclinical Data. *American Journal of Pathology, 191*(9), 1490–1498. DOI: 10.1016/j.ajpath.2021.05.011 PMID: 34102108

Deswal, R., Narwal, V., Dang, A., & Pundir, C. S. (2020). The Prevalence of Polycystic Ovary Syndrome: A Brief Systematic Review. *Journal of Human Reproductive Sciences, 13*(4). PMID: 33627974

Diamanti-Kandarakis, E. (2010). PCOS in adolescents. *Best Practice & Research. Clinical Obstetrics & Gynaecology, 24*(2), 173–183. DOI: 10.1016/j.bpobgyn.2009.09.005 PMID: 19932060

Diamanti-Kandarakis, E., & Dunaif, A. (2012). Insulin resistance and the polycystic ovary syndrome revisited: An update on mechanisms and implications. *Endocrine Reviews, 33*(6), 981–1030. DOI: 10.1210/er.2011-1034 PMID: 23065822

Ding, H., Chen, J., Su, M., Lin, Z., Zhan, H., Yang, F., Li, W., Xie, J., Huang, Y., Liu, X., Liu, B., & Zhou, X. (2020). BDNF promotes activation of astrocytes and microglia contributing to neuroinflammation and mechanical allodynia in cyclophosphamide-induced cystitis. *Journal of Neuroinflammation, 17*(1), 19. DOI: 10.1186/s12974-020-1704-0 PMID: 31931832

Duque Ede, A., & Munhoz, C. D. (2016). The Pro-inflammatory Effects of Glucocorticoids in the Brain. *Frontiers in Endocrinology (Lausanne), 7*, 78. DOI: 10.3389/fendo.2016.00078 PMID: 27445981

Escobar-Morreale, H. F. (2022). Defining PCOS: A syndrome with an intrinsic heterogeneous nature. In Diamanti-Kandarakis, E. (Ed.), *Polycystic Ovary Syndrome* (pp. 3–13). Elsevier., DOI: 10.1016/B978-0-12-823045-9.00012-2

Ethirajulu, A., Alkasabera, A., Onyali, C. B., Anim-Koranteng, C., Shah, H. E., Bhawnani, N., & Mostafa, J. A. (2021). Insulin resistance, hyperandrogenism, and its associated symptoms are the precipitating factors for depression in women with polycystic ovarian syndrome. *Cureus, 13*(9). Advance online publication. DOI: 10.7759/cureus.18013 PMID: 34667688

Fahs, D., Salloum, D., Nasrallah, M., & Ghazeeri, G. (2023). Polycystic Ovary Syndrome: Pathophysiology and Controversies in Diagnosis. *Diagnostics (Basel), 13*(9), 1559. Advance online publication. DOI: 10.3390/diagnostics13091559 PMID: 37174950

Foda, A. M., Ibrahim, S. S., Ibrahim, S. M., & Elbaz, E. M. (2024). Pterostilbene Ameliorates Cognitive Impairment in Polycystic Ovary Syndrome Rat Model through Improving Insulin Resistance via the IRS-1/PI3K/Akt/GSK-3β Pathway: A Comparative Study with Metformin. *ACS Chemical Neuroscience*, *15*(16), 3064–3077. DOI: 10.1021/acschemneuro.4c00352 PMID: 39119909

Frank, M. G., Hershman, S. A., Weber, M. D., Watkins, L. R., & Maier, S. F. (2014). Chronic exposure to exogenous glucocorticoids primes microglia to pro-inflammatory stimuli and induces NLRP3 mRNA in the hippocampus. *Psychoneuroendocrinology*, *40*, 191–200. DOI: 10.1016/j.psyneuen.2013.11.006 PMID: 24485491

Furukawa, S., Fujita, T., Shimabukuro, M., Iwaki, M., Yamada, Y., Nakajima, Y., Nakayama, O., Makishima, M., Matsuda, M., & Shimomura, I. (2004). Increased oxidative stress in obesity and its impact on metabolic syndrome. *The Journal of Clinical Investigation*, *114*(12), 1752–1761. DOI: 10.1172/JCI21625 PMID: 15599400

Gao, Y., Bielohuby, M., Fleming, T., Grabner, G. F., Foppen, E., Bernhard, W., Guzmán-Ruiz, M., Layritz, C., Legutko, B., Zinser, E., García-Cáceres, C., Buijs, R. M., Woods, S. C., Kalsbeek, A., Seeley, R. J., Nawroth, P. P., Bidlingmaier, M., Tschöp, M. H., & Yi, C. X. (2017). Dietary sugars, not lipids, drive hypothalamic inflammation. *Molecular Metabolism*, *6*(8), 897–908. DOI: 10.1016/j.molmet.2017.06.008 PMID: 28752053

Gao, Z., Tan, H., Song, X., Zhuang, T., Kong, R., Wang, Y., Yan, X., & Yao, R. (2024). Troxerutin dampened hypothalamic neuroinflammation via microglial IL-22/IL-22R1/IRF3 activation in dihydrotestosterone-induced polycystic ovary syndrome rats. *Phytomedicine*, *124*, 155280. DOI: 10.1016/j.phymed.2023.155280 PMID: 38183697

Gautam, R., Maan, P., Jyoti, A., Kumar, A., Malhotra, N., & Arora, T. (2025). The Role of Lifestyle Interventions in PCOS Management: A Systematic Review. *Nutrients*, *17*(2), 310. DOI: 10.3390/nu17020310 PMID: 39861440

Goeldner, C., Reiss, D., Kieffer, B. L., & Ouagazzal, A. M. (2010). Endogenous nociceptin/orphanin-FQ in the dorsal hippocampus facilitates despair-related behavior. *Hippocampus*, *20*(8), 911–916. DOI: 10.1002/hipo.20760 PMID: 20087887

Grace, A. A. (2016). Dysregulation of the dopamine system in the pathophysiology of schizophrenia and depression. *Nature Reviews. Neuroscience*, *17*(8), 524–532. DOI: 10.1038/nrn.2016.57 PMID: 27256556

Greenwood, E. A., Pasch, L. A., Cedars, M. I., Legro, R. S., Eisenberg, E., Huddleston, H. G., & Network, H. D. R. M. (2018). Insulin resistance is associated with depression risk in polycystic ovary syndrome. *Fertility and Sterility*, *110*(1), 27–34. DOI: 10.1016/j.fertnstert.2018.03.009 PMID: 29908775

Gruber, J., Hanssen, R., Qubad, M., Bouzouina, A., Schack, V., Sochor, H., Schiweck, C., Aichholzer, M., Matura, S., Slattery, D. A., Zopf, Y., Borgland, S. L., Reif, A., & Thanarajah, S. E. (2023). Impact of insulin and insulin resistance on brain dopamine signalling and reward processing–An underexplored mechanism in the pathophysiology of depression? *Neuroscience and Biobehavioral Reviews*, *149*, 105179. DOI: 10.1016/j.neubiorev.2023.105179 PMID: 37059404

Hamer, J. A., Testani, D., Mansur, R. B., Lee, Y., Subramaniapillai, M., & McIntyre, R. S. (2019). Brain insulin resistance: A treatment target for cognitive impairment and anhedonia in depression. *Experimental Neurology*, *315*, 1–8. DOI: 10.1016/j.expneurol.2019.01.016 PMID: 30695707

Harada, M. (2022). Pathophysiology of polycystic ovary syndrome revisited: Current understanding and perspectives regarding future research. *Reproductive Medicine and Biology*, *21*(1), e12487. DOI: 10.1002/rmb2.12487 PMID: 36310656

Haroon, E., Miller, A. H., & Sanacora, G. (2017). Inflammation, Glutamate, and Glia: A Trio of Trouble in Mood Disorders. *Neuropsychopharmacology : Official Publication of the American College of Neuropsychopharmacology*, *42*(1), 193–215. DOI: 10.1038/npp.2016.199 PMID: 27629368

Haruwaka, K., Ikegami, A., Tachibana, Y., Ohno, N., Konishi, H., Hashimoto, A., Matsumoto, M., Kato, D., Ono, R., Kiyama, H., Moorhouse, A. J., Nabekura, J., & Wake, H. (2019). Dual microglia effects on blood brain barrier permeability induced by systemic inflammation. *Nature Communications*, *10*(1), 5816. DOI: 10.1038/s41467-019-13812-z PMID: 31862977

Hasan, M., Sultana, S., Sohan, M., Parvin, S., Rahman, M. A., Hossain, M. J., Rahman, M. S., & Islam, M. R. (2022). Prevalence and associated risk factors for mental health problems among patients with polycystic ovary syndrome in Bangladesh: A nationwide cross-Sectional study. *PLoS One*, *17*(6), e0270102. DOI: 10.1371/journal.pone.0270102 PMID: 35731829

Hassamal, S. (2023). Chronic stress, neuroinflammation, and depression: An overview of pathophysiological mechanisms and emerging anti-inflammatories. *Frontiers in Psychiatry*, *14*, 1130989. DOI: 10.3389/fpsyt.2023.1130989 PMID: 37252156

He, J., Zhu, G., Wang, G., & Zhang, F. (2020). Oxidative Stress and Neuroinflammation Potentiate Each Other to Promote Progression of Dopamine Neurodegeneration. *Oxidative Medicine and Cellular Longevity, 6137521*, 1–12. Advance online publication. DOI: 10.1155/2020/6137521 PMID: 32714488

Hiam, D., Moreno-Asso, A., Teede, H. J., Laven, J. S. E., Stepto, N. K., Moran, L. J., & Gibson-Helm, M. (2019). The Genetics of Polycystic Ovary Syndrome: An Overview of Candidate Gene Systematic Reviews and Genome-Wide Association Studies. *Journal of Clinical Medicine*, *8*(10), 1606. Advance online publication. DOI: 10.3390/jcm8101606 PMID: 31623391

Hoogland, I. C. M., Houbolt, C., van Westerloo, D. J., van Gool, W. A., & van de Beek, D. (2015). Systemic inflammation and microglial activation: Systematic review of animal experiments. *Journal of Neuroinflammation*, *12*(1), 114. DOI: 10.1186/s12974-015-0332-6 PMID: 26048578

Hu, M., Richard, J. E., Maliqueo, M., Kokosar, M., Fornes, R., Benrick, A., Jansson, T., Ohlsson, C., Wu, X., Skibicka, K. P., & Stener-Victorin, E. (2015). Maternal testosterone exposure increases anxiety-like behavior and impacts the limbic system in the offspring. *Proceedings of the National Academy of Sciences of the United States of America*, *112*(46), 14348–14353. DOI: 10.1073/pnas.1507514112 PMID: 26578781

Ibáñez, L., Oberfield, S. E., Witchel, S., Auchus, R. J., Chang, R. J., Codner, E., Dabadghao, P., Darendeliler, F., Elbarbary, N. S., Gambineri, A., Garcia Rudaz, C., Hoeger, K. M., López-Bermejo, A., Ong, K., Peña, A. S., Reinehr, T., Santoro, N., Tena-Sempere, M., Tao, R., & Lee, P. A. (2017). An International Consortium Update: Pathophysiology, Diagnosis, and Treatment of Polycystic Ovarian Syndrome in Adolescence. *Hormone Research in Paediatrics*, *88*(6), 371–395. DOI: 10.1159/000479371 PMID: 29156452

Jia, S. (2023). Combined Oral Contraceptives (COCs) Management on symptoms of Polycystic Ovary Syndrom: A Systematic Review. *Highlights in Science. Engineering and Technology*, *74*, 1532–1539.

Jovanovic, F., Sudhakar, A., & Knezevic, N. N. (2022). The Kynurenine Pathway and Polycystic Ovary Syndrome: Inflammation as a Common Denominator. *International Journal of Tryptophan Research : IJTR*, *15*. DOI: 10.1177/11786469221099214 PMID: 35620306

Kullmann, S., Heni, M., Hallschmid, M., Fritsche, A., Preissl, H., & Häring, H.-U. (2016). Brain insulin resistance at the crossroads of metabolic and cognitive disorders in humans. *Physiological Reviews*, *96*(4), 1169–1209. DOI: 10.1152/physrev.00032.2015 PMID: 27489306

Lai, H., Jia, X., Yu, Q., Zhang, C., Qiao, J., Guan, Y., & Kang, J. (2014). High-fat diet induces significant metabolic disorders in a mouse model of polycystic ovary syndrome. *Biology of Reproduction, 91*(5), 127, 121–111.

Lau, G. M., Elghobashy, M., Thanki, M., Ibegbulam, S., Latthe, P., Gillett, C. D., O'Reilly, M. W., Arlt, W., Lindenmeyer, A., & Kempegowda, P. (2022). A systematic review of lived experiences of people with polycystic ovary syndrome highlights the need for holistic care and co-creation of educational resources. *Frontiers in Endocrinology*, *13*, 1064937. DOI: 10.3389/fendo.2022.1064937 PMID: 36531482

Linnerbauer, M., Wheeler, M. A., & Quintana, F. J. (2020). Astrocyte crosstalk in CNS inflammation. *Neuron*, *108*(4), 608–622. DOI: 10.1016/j.neuron.2020.08.012 PMID: 32898475

Liu, L., & Zhu, G. (2018). Gut–brain axis and mood disorder. *Frontiers in Psychiatry*, *9*, 223. DOI: 10.3389/fpsyt.2018.00223 PMID: 29896129

Louwers, Y. V., & Laven, J. S. E. (2020). Characteristics of polycystic ovary syndrome throughout life. *Therapeutic Advances in Reproductive Health*, *14*. DOI: 10.1177/2633494120911038 PMID: 32518918

Majidzadeh, S., Mirghafourvand, M., Farvareshi, M., & Yavarikia, P. (2023). The effect of cognitive behavioral therapy on depression and anxiety of women with polycystic ovary syndrome: A randomized controlled trial. *BMC Psychiatry*, *23*(1), 332. DOI: 10.1186/s12888-023-04814-9 PMID: 37170227

Mamelak, M. (2024). Depression and the Glutamate/GABA-Glutamine Cycle. *Current Neuropharmacology*, *23*(1), 75–84. DOI: 10.2174/1570159X22666240815120244 PMID: 39150032

Mancini, A., Bruno, C., Vergani, E., d'Abate, C., Giacchi, E., & Silvestrini, A. (2021). Oxidative Stress and Low-Grade Inflammation in Polycystic Ovary Syndrome: Controversies and New Insights. *International Journal of Molecular Sciences*, *22*(4), 1667. Advance online publication. DOI: 10.3390/ijms22041667 PMID: 33562271

Masoudi, M., Ansari, S., Kashani, L., Tavolinejad, H., Hossein Rashidi, B., Esalatmanesh, S., Ghazizadeh-Hashemi, M., Noorbala, A. A., & Akhondzadeh, S. (2021). Effect of sertraline on depression severity and prolactin levels in women with polycystic ovary syndrome: A placebo-controlled randomized trial. *International Clinical Psychopharmacology*, *36*(5), 238–243. DOI: 10.1097/YIC.0000000000000367 PMID: 34030169

Mazloomi, S., Sheikh, N., Sanoee Farimani, M., & Pilehvari, S. (2021). Association of Prx4, Total Oxidant Status, and Inflammatory Factors with Insulin Resistance in Polycystic Ovary Syndrome. *International Journal of Endocrinology*, *2021*(1), 9949753. DOI: 10.1155/2021/9949753 PMID: 34239559

McCartney, C. R., & Campbell, R. E. (2020). Abnormal GnRH pulsatility in polycystic ovary syndrome: Recent insights. *Current Opinion in Endocrine and Metabolic Research*, *12*, 78–84. DOI: 10.1016/j.coemr.2020.04.005 PMID: 32676541

McKim, D. B., Weber, M. D., Niraula, A., Sawicki, C. M., Liu, X., Jarrett, B. L., Ramirez-Chan, K., Wang, Y., Roeth, R. M., Sucaldito, A. D., Sobol, C. G., Quan, N., Sheridan, J. F., & Godbout, J. P. (2018). Microglial recruitment of IL-1β-producing monocytes to brain endothelium causes stress-induced anxiety. *Molecular Psychiatry*, *23*(6), 1421–1431. DOI: 10.1038/mp.2017.64 PMID: 28373688

Merkulov, V., Merkulova, T., & Bondar, N. (2017). Mechanisms of brain glucocorticoid resistance in stress-induced psychopathologies. *Biochemistry (Moscow)*, *82*(3), 351–365. DOI: 10.1134/S0006297917030142 PMID: 28320277

Mikulska, J., Juszczyk, G., Gawrońska-Grzywacz, M., & Herbet, M. (2021). HPA Axis in the Pathomechanism of Depression and Schizophrenia: New Therapeutic Strategies Based on Its Participation. *Brain Sciences*, *11*(10), 1298. DOI: 10.3390/brainsci11101298 PMID: 34679364

Mirmasoumi, G., Fazilati, M., Foroozanfard, F., Vahedpoor, Z., Mahmoodi, S., Taghizadeh, M., Esfeh, N. K., Mohseni, M., Karbassizadeh, H., & Asemi, Z. (2018). The Effects of Flaxseed Oil Omega-3 Fatty Acids Supplementation on Metabolic Status of Patients with Polycystic Ovary Syndrome: A Randomized, Double-Blind, Placebo-Controlled Trial. *Experimental and Clinical Endocrinology & Diabetes*, *126*(4), 222–228. DOI: 10.1055/s-0043-119751 PMID: 29117618

Mithaiwala, M. N., Santana-Coelho, D., Porter, G. A., & O'Connor, J. C. (2021). Neuroinflammation and the Kynurenine Pathway in CNS Disease: Molecular Mechanisms and Therapeutic Implications. *Cells*, *10*(6), 1548. DOI: 10.3390/cells10061548 PMID: 34205235

Moran, L. J., Pasquali, R., Teede, H. J., Hoeger, K. M., & Norman, R. J. (2009). Treatment of obesity in polycystic ovary syndrome: A position statement of the Androgen Excess and Polycystic Ovary Syndrome Society. *Fertility and Sterility*, *92*(6), 1966–1982. DOI: 10.1016/j.fertnstert.2008.09.018 PMID: 19062007

Mota, B., Ramos, M., Marques, S. I., Silva, A., Pereira, P. A., Madeira, M. D., Mateus, N., & Cardoso, A. (2023). Effects of High-Fat and High-Fat High-Sugar Diets in the Anxiety, Learning and Memory, and in the Hippocampus Neurogenesis and Neuroinflammation of Aged Rats. *Nutrients*, *15*(6), 1370. DOI: 10.3390/nu15061370 PMID: 36986100

National Institute of Child Health and Human Development. (n.d.). *What are the symptoms of PCOS?* Retrieved February 3 from https://www.nichd.nih.gov/health/topics/pcos/conditioninfo/symptoms

Nayak, U., Manikkath, J., Arora, D., & Mudgal, J. (2025). Impact of neuroinflammation on brain glutamate and dopamine signalling in schizophrenia: An update. *Metabolic Brain Disease*, *40*(2), 119. DOI: 10.1007/s11011-025-01548-3 PMID: 39907868

Newhouse, P., & Albert, K. (2015). Estrogen, Stress, and Depression: A Neurocognitive Model. *JAMA Psychiatry*, *72*(7), 727–729. DOI: 10.1001/jamapsychiatry.2015.0487 PMID: 26018188

Palomba, S., Santagni, S., Falbo, A., & La Sala, G. B. (2015). Complications and challenges associated with polycystic ovary syndrome: Current perspectives. *International Journal of Women's Health*, *7*, 745–763. DOI: 10.2147/IJWH.S70314 PMID: 26261426

Patten, R. K., Boyle, R. A., Moholdt, T., Kiel, I., Hopkins, W. G., Harrison, C. L., & Stepto, N. K. (2020). Exercise Interventions in Polycystic Ovary Syndrome: A Systematic Review and Meta-Analysis. *Frontiers in Physiology*, *11*, 606. DOI: 10.3389/fphys.2020.00606 PMID: 32733258

Qi, X., Yun, C., Sun, L., Xia, J., Wu, Q., Wang, Y., Wang, L., Zhang, Y., Liang, X., Wang, L., Gonzalez, F. J., Patterson, A. D., Liu, H., Mu, L., Zhou, Z., Zhao, Y., Li, R., Liu, P., Zhong, C., & Qiao, J. (2019). Gut microbiota–bile acid–interleukin-22 axis orchestrates polycystic ovary syndrome. *Nature Medicine*, *25*(8), 1225–1233. DOI: 10.1038/s41591-019-0509-0 PMID: 31332392

Rau, S., Gregg, A., Yaceczko, S., & Limketkai, B. (2024). Prebiotics and Probiotics for Gastrointestinal Disorders. *Nutrients*, *16*(6), 778. Advance online publication. DOI: 10.3390/nu16060778 PMID: 38542689

Reichmann, F., Hassan, A. M., Farzi, A., Jain, P., Schuligoi, R., & Holzer, P. (2015). Dextran sulfate sodium-induced colitis alters stress-associated behaviour and neuropeptide gene expression in the amygdala-hippocampus network of mice. *Scientific Reports*, *5*(1), 9970. DOI: 10.1038/srep09970 PMID: 26066467

Reijerkerk, A., Lakeman, K. A., Drexhage, J. A., van Het Hof, B., van Wijck, Y., van der Pol, S. M., Kooij, G., Geerts, D., & de Vries, H. E. (2012). Brain endothelial barrier passage by monocytes is controlled by the endothelin system. *Journal of Neurochemistry*, *121*(5), 730–737. DOI: 10.1111/j.1471-4159.2011.07393.x PMID: 21777246

Rodriguez, A. C. I., Epel, E. S., White, M. L., Standen, E. C., Seckl, J. R., & Tomiyama, A. J. (2015). Hypothalamic-pituitary-adrenal axis dysregulation and cortisol activity in obesity: A systematic review. *Psychoneuroendocrinology*, *62*, 301–318. DOI: 10.1016/j.psyneuen.2015.08.014 PMID: 26356039

Rosenfield, R. L., & Ehrmann, D. A. (2016). The Pathogenesis of Polycystic Ovary Syndrome (PCOS): The Hypothesis of PCOS as Functional Ovarian Hyperandrogenism Revisited. *Endocrine Reviews*, *37*(5), 467–520. DOI: 10.1210/er.2015-1104 PMID: 27459230

Rutkowska, A. Z., & Diamanti-Kandarakis, E. (2016). Polycystic ovary syndrome and environmental toxins. *Fertility and Sterility*, *106*(4), 948–958. DOI: 10.1016/j.fertnstert.2016.08.031 PMID: 27559705

Saad, M. A., Eltarzy, M. A., Abdel Salam, R. M., & Ahmed, M. A. E. (2021). Liraglutide mends cognitive impairment by averting Notch signaling pathway overexpression in a rat model of polycystic ovary syndrome. *Life Sciences*, *265*, 118731. DOI: 10.1016/j.lfs.2020.118731 PMID: 33160995

Sadeeqa, S., Mustafa, T., & Latif, S. (2018). Polycystic Ovarian Syndrome-Related Depression in Adolescent Girls: A Review. *Journal of Pharmacy & Bioallied Sciences*, *10*(2), 55–59. DOI: 10.4103/JPBS.JPBS_1_18 PMID: 29962792

Sadeeqa, S., Mustafa, T., & Latif, S. (2018). Polycystic ovarian syndrome–related depression in adolescent girls: A review. *Journal of Pharmacy & Bioallied Sciences*, *10*(2), 55–59. DOI: 10.4103/JPBS.JPBS_1_18 PMID: 29962792

Sălcudean, A., Bodo, C. R., Popovici, R. A., Cozma, M. M., Păcurar, M., Crăciun, R. E., Crisan, A. I., Enatescu, V. R., Marinescu, I., Cimpian, D. M., Nan, A. G., Sasu, A. B., Anculia, R. C., & Strete, E. G. (2025). Neuroinflammation-A Crucial Factor in the Pathophysiology of Depression-A Comprehensive Review. *Biomolecules*, *15*(4), 502. Advance online publication. DOI: 10.3390/biom15040502 PMID: 40305200

Sarwar, H., Rafiqi, S. I., Ahmad, S., Jinna, S., Khan, S. A., Karim, T., Qureshi, O., Zahid, Z. A., Elhai, J. D., Levine, J. C., Naqvi, S. J., Jaume, J. C., & Imam, S. (2022). Hyperinsulinemia associated depression. *Clinical Medicine Insights. Endocrinology and Diabetes, 15*. DOI: 10.1177/11795514221090244 PMID: 35494421

Schoenichen, C., Bode, C., & Duerschmied, D. (2019). Role of platelet serotonin in innate immune cell recruitment. *Frontiers in Bioscience, 24*(3), 514–526. DOI: 10.2741/4732 PMID: 30468670

Shao, X., Lu, W., Gao, F., Li, D., Hu, J., Li, Y., Zuo, Z., Jie, H., Zhao, Y., & Cen, X. (2016). Uric Acid Induces Cognitive Dysfunction through Hippocampal Inflammation in Rodents and Humans. *The Journal of Neuroscience : The Official Journal of the Society for Neuroscience, 36*(43), 10990–11005. DOI: 10.1523/JNEUROSCI.1480-16.2016 PMID: 27798180

Shimobayashi, M., Albert, V., Woelnerhanssen, B., Frei, I. C., Weissenberger, D., Meyer-Gerspach, A. C., Clement, N., Moes, S., Colombi, M., Meier, J. A., Swierczynska, M. M., Jenö, P., Beglinger, C., Peterli, R., & Hall, M. N. (2018). Insulin resistance causes inflammation in adipose tissue. *The Journal of Clinical Investigation, 128*(4), 1538–1550. DOI: 10.1172/JCI96139 PMID: 29528335

Shoelson, S. E., Lee, J., & Goldfine, A. B. (2006). Inflammation and insulin resistance. *The Journal of Clinical Investigation, 116*(7), 1793–1801. DOI: 10.1172/JCI29069 PMID: 16823477

Singh, S., Pal, N., Shubham, S., Sarma, D. K., Verma, V., Marotta, F., & Kumar, M. (2023). Polycystic Ovary Syndrome: Etiology, Current Management, and Future Therapeutics. *Journal of Clinical Medicine, 12*(4), 1454. Advance online publication. DOI: 10.3390/jcm12041454 PMID: 36835989

Stuart, M. J., Singhal, G., & Baune, B. T. (2015). Systematic review of the neurobiological relevance of chemokines to psychiatric disorders. *Frontiers in Cellular Neuroscience, 9*, 357. DOI: 10.3389/fncel.2015.00357 PMID: 26441528

Su, P., Chen, C., & Sun, Y. (2025). Physiopathology of polycystic ovary syndrome in endocrinology, metabolism and inflammation. *Journal of Ovarian Research, 18*(1), 34. DOI: 10.1186/s13048-025-01621-6 PMID: 39980043

Sun, Y., Koyama, Y., & Shimada, S. (2022a). Inflammation From Peripheral Organs to the Brain: How Does Systemic Inflammation Cause Neuroinflammation? *Frontiers in Aging Neuroscience, 14*, 903455. DOI: 10.3389/fnagi.2022.903455 PMID: 35783147

Sun, Y., Koyama, Y., & Shimada, S. (2022b). Inflammation from peripheral organs to the brain: How does systemic inflammation cause neuroinflammation? [In.]. *Frontiers in Aging Neuroscience*, *14*, 903455. DOI: 10.3389/fnagi.2022.903455 PMID: 35783147

Szczesnowicz, A., Szeliga, A., Niwczyk, O., Bala, G., & Meczekalski, B. (2023). Do GLP-1 Analogs Have a Place in the Treatment of PCOS? New Insights and Promising Therapies. *Journal of Clinical Medicine*, *12*(18), 5915. Advance online publication. DOI: 10.3390/jcm12185915 PMID: 37762856

Taleb, S. (2019). Tryptophan dietary impacts gut barrier and metabolic diseases. *Frontiers in Immunology*, *10*, 2113. DOI: 10.3389/fimmu.2019.02113 PMID: 31552046

Tefagh, G., Payab, M., Qorbani, M., Sharifi, F., Sharifi, Y., Ebrahimnegad Shirvani, M. S., Pourghazi, F., Atlasi, R., Shadman, Z., Rezaei, N., Mohammadi-Vajari, E., Larijani, B., & Ebrahimpur, M. (2022). Effect of vitamin E supplementation on cardiometabolic risk factors, inflammatory and oxidative markers and hormonal functions in PCOS (polycystic ovary syndrome): A systematic review and meta-analysis. *Scientific Reports*, *12*(1), 5770. DOI: 10.1038/s41598-022-09082-3 PMID: 35388031

Troubat, R., Barone, P., Leman, S., Desmidt, T., Cressant, A., Atanasova, B., Brizard, B., El Hage, W., Surget, A., Belzung, C., & Camus, V. (2021). Neuroinflammation and depression: A review. *The European Journal of Neuroscience*, *53*(1), 151–171. DOI: 10.1111/ejn.14720 PMID: 32150310

Unluhizarci, K., Karaca, Z., & Kelestimur, F. (2021). Role of insulin and insulin resistance in androgen excess disorders. *World Journal of Diabetes*, *12*(5), 616–629. DOI: 10.4239/wjd.v12.i5.616 PMID: 33995849

Wang, F., Zhang, Z., Han, J., Zheng, J., Wang, X., & Wang, Z. (2023). Discovery of microglia gonadotropin-releasing hormone receptor and its potential role in polycystic ovarian syndrome. *Molecular Medicine Reports*, *27*(4), 1–9. DOI: 10.3892/mmr.2023.12964 PMID: 36799164

Watson, K., Nasca, C., Aasly, L., McEwen, B., & Rasgon, N. (2018). Insulin resistance, an unmasked culprit in depressive disorders: Promises for interventions. *Neuropharmacology*, *136*, 327–334. DOI: 10.1016/j.neuropharm.2017.11.038 PMID: 29180223

Witchel, S. F., Burghard, A. C., Tao, R. H., & Oberfield, S. E. (2019). The diagnosis and treatment of PCOS in adolescents: An update. *Current Opinion in Pediatrics*, *31*(4), 562–569. DOI: 10.1097/MOP.0000000000000778 PMID: 31299022

Wohleb, E. S., Fenn, A. M., Pacenta, A. M., Powell, N. D., Sheridan, J. F., & Godbout, J. P. (2012). Peripheral innate immune challenge exaggerated microglia activation, increased the number of inflammatory CNS macrophages, and prolonged social withdrawal in socially defeated mice. *Psychoneuroendocrinology, 37*(9), 1491–1505. DOI: 10.1016/j.psyneuen.2012.02.003 PMID: 22386198

Wolf, W. M., Wattick, R. A., Kinkade, O. N., & Olfert, M. D. (2018). The Current Description and Future Need for Multidisciplinary PCOS Clinics. *Journal of Clinical Medicine, 7*(11), 395. Advance online publication. DOI: 10.3390/jcm7110395 PMID: 30373317

Wolugbom, J. A.Jr, Areloegbe, S. E., & Olaniyi, K. S. (2023). Protective Role of Acetate Against Depressive-Like Behaviour Associated with Letrozole-Induced PCOS Rat Model: Involvement of HDAC2 and DNA Methylation. *Molecular Neurobiology, 60*(1), 355–368. DOI: 10.1007/s12035-022-03074-0 PMID: 36269541

Xia, J., Wang, H., Zhang, C., Liu, B., Li, Y., Li, K., Li, P., & Song, C. (2022). The comparison of sex differences in depression-like behaviors and neuroinflammatory changes in a rat model of depression induced by chronic stress. *Frontiers in Behavioral Neuroscience, 16*, 1059594. DOI: 10.3389/fnbeh.2022.1059594 PMID: 36703721

Yang, J., & Chen, C. (2024). Hormonal changes in PCOS. *The Journal of Endocrinology, 261*(1). Advance online publication. DOI: 10.1530/JOE-23-0342 PMID: 38285626

Yu, Q., Hao, S., Wang, H., Song, X., Shen, Q., & Kang, J. (2016). Depression-like behavior in a dehydroepiandrosterone-induced mouse model of polycystic ovary syndrome. *Biology of Reproduction, 95*(4), 79, 71–10.

Yurtdaş, G., & Akdevelioğlu, Y. (2020). A new approach to polycystic ovary syndrome: The gut microbiota. *Journal of the American College of Nutrition, 39*(4), 371–382. DOI: 10.1080/07315724.2019.1657515 PMID: 31513473

Zehra, B., & Khursheed, A. (2018). Polycystic ovarian syndrome: symptoms, treatment and diagnosis: a review. *Journal of Pharmacognosy and Phytochemistry, 7*(6), 875–880.

Zerrouki, D., Rami, I., Assarrar, I., Bouichrat, N., Rouf, S., & Latrech, H. (2024). Is there any association between vitamin D status and PCOS disease? *Gynecological Endocrinology, 40*(1), 2381501. DOI: 10.1080/09513590.2024.2381501 PMID: 39481002

Zhai, Y., & Pang, Y. (2022). Systemic and ovarian inflammation in women with polycystic ovary syndrome. *Journal of Reproductive Immunology*, *151*, 103628. DOI: 10.1016/j.jri.2022.103628 PMID: 35472833

Zhang, B., Zhou, W., Shi, Y., Zhang, J., Cui, L., & Chen, Z.-J. (2020). Lifestyle and environmental contributions to ovulatory dysfunction in women of polycystic ovary syndrome. *BMC Endocrine Disorders*, *20*(1), 19. DOI: 10.1186/s12902-020-0497-6 PMID: 32000752

Zhang, Y., Chu, J. M., & Wong, G. T. (2022). Cerebral Glutamate Regulation and Receptor Changes in Perioperative Neuroinflammation and Cognitive Dysfunction. *Biomolecules*, *12*(4), 597. Advance online publication. DOI: 10.3390/biom12040597 PMID: 35454185

## KEY TERMS AND DEFINITIONS

**Gut-Brain Axis:** The communication network between the gut and brain. Dysbiosis in PCOS may drive systemic and neuroinflammation, worsening depressive symptoms.

**Hypothalamic-Pituitary-Adrenal (HPA) Axis Dysregulation:** A stress-response imbalance common in PCOS and depression, leading to elevated cortisol levels and contributing to mood disturbances.

**Kynurenine Pathway:** A major route of tryptophan metabolism that is upregulated in inflammation. Increased kynurenine pathway activity, often observed in PCOS and depression, reduces serotonin availability and leads to the accumulation of neurotoxic metabolites associated with neuroinflammation.

**Neuroinflammation:** Activation of immune cells (mainly microglia and astrocytes) in the central nervous system. Chronic neuroinflammation is associated with depression and is increasingly recognized in PCOS.

**Oxidative Stress:** An imbalance between reactive oxygen species (ROS) and antioxidants, contributing to neuroinflammation, insulin resistance, and mood disorders in PCOS.

**Polycystic Ovary Syndrome (PCOS):** A common endocrine disorder in women of reproductive age, marked by hyperandrogenism, ovulatory dysfunction, and polycystic ovaries. It is linked to metabolic disturbances, insulin resistance, and increased risk of depression.

## APPENDIX

*Figure 1. Bidirectional association between neuroinflammation and depression*

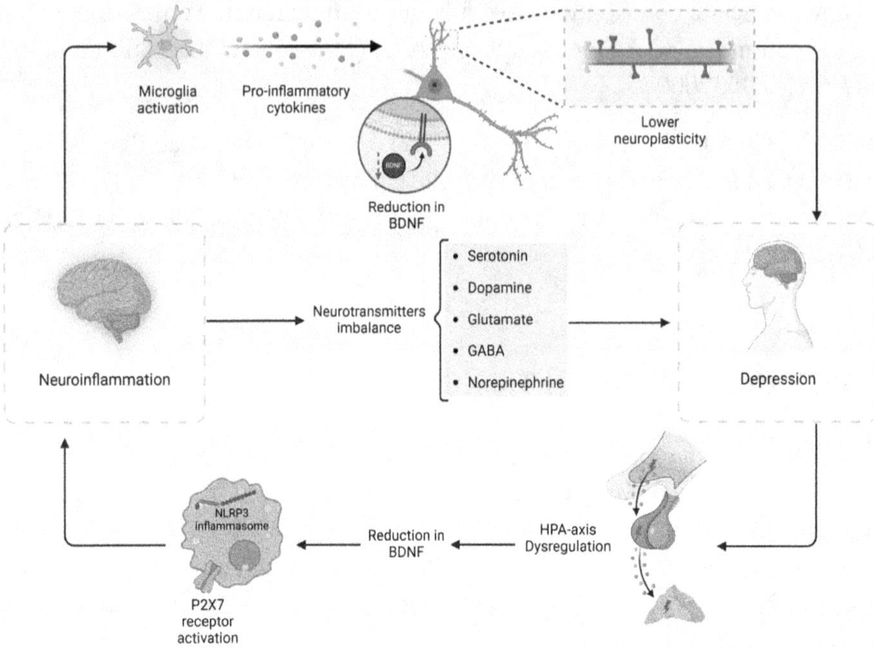


Neuroinflammation contributes to depression through two key pathways: (1) microglial activation and reduced neuroplasticity, and (2) neurotransmitter imbalances. Conversely, depression can further induce neuroinflammation via HPA axis dysregulation and activation of the P2X7-NLRP3 inflammasome cascade. This figure was created by the authors specifically for this chapter.

*Figure 2. Suggested pathways through which PCOS may contribute to neuroinflammation*

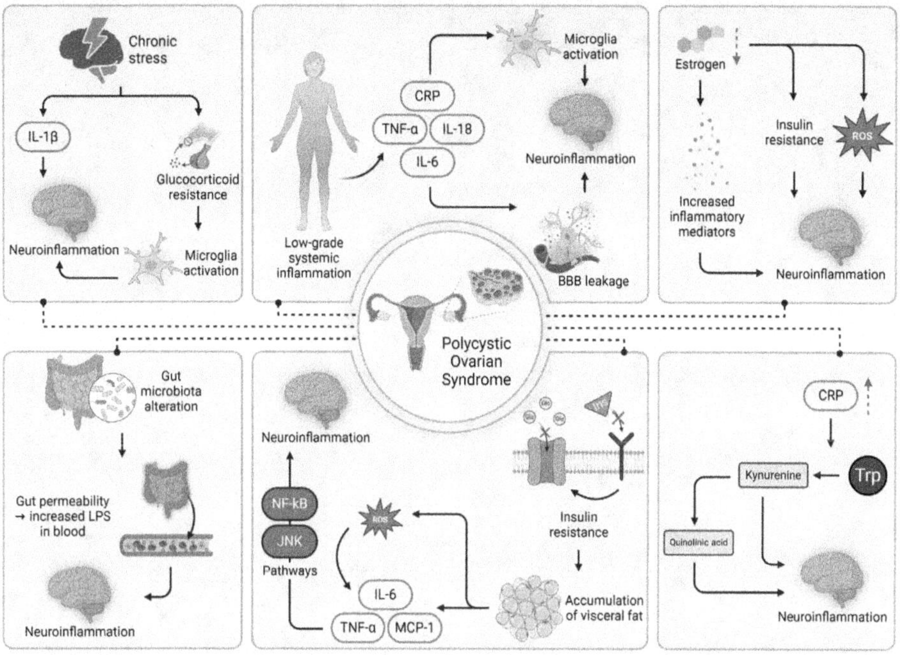


This figure illustrates six primary mechanisms: (1) stress-induced dysregulation of the HPA axis, (2) low-grade systemic inflammation and the impact of inflammatory markers on the CNS, (3) loss of estrogen's anti-inflammatory effects, (4) altered tryptophan metabolism shifting toward the kynurenine pathway, (5) insulin resistance and oxidative stress, and (6) gut microbiota dysbiosis. This figure was created by the authors specifically for this chapter.

# Chapter 8
# Unveiling Nano-Specific Toxicity: Empowering Medical Literacy and Exploring Its Connection to Autophagy

**Kadir Uludag**
https://orcid.org/0000-0003-3713-4670
*Shanghai Jiao Tong University, China*

**Amjad Islam Amjad**
https://orcid.org/0000-0002-4250-7526
*School Education Department, Government of Punjab, Kasur, Pakistan*

## ABSTRACT

*This chapter explores the misconception surrounding nano-specific toxicity and emphasises the critical role of improved medical literacy in addressing this issue. By delving into the realm of nanotechnology and its potential health implications, this study highlights the importance of accurate information dissemination and education to dispel unfounded fears. Through enhanced medical literacy, stakeholders can navigate the complexities of nano-toxicity concerns more effectively, paving the way for informed decision-making and fostering a deeper understanding of nanomaterials' true impact on health and well-being. Furthermore, this chapter delves into the fascinating relationship between nano-specific toxicity and autophagy, unveiling insights that could have profound implications for both medicine and nanotechnology. The authors have also discussed the potential connection between fasting and medical understanding of autophagy.*

DOI: 10.4018/979-8-3693-5908-2.ch008

## INTRODUCTION

Nanotechnology refers to the scientific and technological study of objects that are small in size (Sahu et al., 2023). In reality, nanotechnology refers to any technology operating at the nanoscale, which boasts numerous practical applications across various real - world domains (Nasrollahzadeh, Sajadi, Sajjadi, & Issaabadi, 2019) such as cognitive health (Bi, Po, & Uludag, 2025). Nanotechnology's rapid advancement necessitates rigorous nanotoxicology research to assess and mitigate the potential health and environmental risks of engineered nanomaterials. Nanotoxicology is a rapidly evolving field, yet it is often surrounded by myths and misconceptions that can obscure the true nature of the risks and benefits associated with nanomaterials (Rahman, Rahman, Parvez, & Kim, 2023). These misunderstandings can mislead public perception of nanomaterials and hinder the rational development and application of this emerging field. Nanomaterials derived from nanotechnology have initiated an exciting era of technology that will influence every facet of human existence (Patel et al., 2021). They frequently demonstrate distinctive characteristics and are intriguing across a range of scientific disciplines, such as biomedicine (Erisen & Uludag, 2024). These objects possess unique physical and chemical characteristics. Nanotechnology encompasses advancements and progress in the atomic, molecular, and macromolecular domains, which have the potential to influence a nation's competitive edge significantly (Ibrohim et al., 2023). Its advancements in atomic, molecular, and macromolecular domains serve as a multiplier of national power in economics, technology, and security. By learning from history's successes and failures, nations can position nanotechnology as a driver of inclusive, sustainable competitiveness—avoiding the pitfalls of unaccountable expertise while harnessing the power of evidence-driven progress. Previously, medical professionals would make decisions about medical diagnosis, prevention, and treatment based on the opinions of authoritative experts. These experts would rely on their own experiences, intuition, and reputation rather than conducting a thorough evaluation of causal relationships and utilizing evidence-based criteria. This approach was prevalent in the medical field during that time (Guzelian et al., 2005). This approach had both advantages and disadvantages. On one hand, it allowed experienced practitioners to rely on their own judgment and expertise, which could be valuable in complex and unique cases. However, it also meant that medical decision-making lacked a standardized and rigorous approach, and there was limited accountability for the accuracy and effectiveness of treatments. Drawing lessons from historical shifts in medical decision-making—where expertise evolved from authority-based to evidence-driven—nations must guide nanotechnology's advancement through rigorous, evidence-based frameworks that balance innovation with accountability, ensuring its role as a multiplier of sustainable national power while avoiding the pitfalls of unregulated expertise.

## PUBLIC PERCEPTION, MISCONCEPTIONS OF NANOTECHNOLOGY, AND MEDICAL LITERACY

Public perception and misconceptions of nanotechnology play a significant role in shaping public attitudes and acceptance of this emerging field, The toxicity of nanomaterials must be understood for the sake of human and ecological health (Luty-Błocho et al., 2023). By grounding discussions in evidence and ethics, society can harness nanotechnology's transformative power. The growing discourse over the use of nano-enabled items draws on this particular focus, which is essential for understanding nanomaterial toxicity (Domingues et al., 2022; Ostrowski et al., 2009). Public perception of nanotechnology is a dynamic ecosystem where scientific evidence, cultural values, and communication strategies intersect.

The future of medicine is thought to be centered around nanotechnology, which would allow for the creation of personalized medicines based on individual needs (Haleem et al., 2023). Nanomaterials have gained significant recognition for their potential and are now being utilized not just in treatments but also in other consumer products (Foulkes et al., 2020; Kumar et al., 2020). Nanomaterials are ubiquitous (Lebre et al., 2020). Due to the ubiquitous nature of these nanoparticles, human people are openly and unavoidably exposed to them (Jose et al., 2022). However, our understanding of the health and environmental hazards posed by these manufactured nanomaterials is currently restricted and primarily insufficient. Several studies demonstrate the harmful effects of these substances in animal models, which can result in teratogenicity, genotoxicity, and organ abnormalities such as liver, lung, kidney, and immune system dysfunction (El-Zawi & Elyasseri, 2021; Li et al., 2022). The hazardous reactions are contingent upon numerous factors, such as shape, dimension, charge, structure, and surface area (Jamil et al., 2018). These findings highlight the urgent need for systematic, mechanistic, and translational studies to better characterize exposure pathways, dose-response relationships, and long-term health outcomes in humans, particularly vulnerable populations.

The risk perceived by the general population was considerably higher than that of other stakeholders. The public trusts academics and government personnel more than some other trusted individuals, with scientists being the most trusted. Greater levels of public trust were often correlated with reduced perceptions of danger (Kamarulzaman et al., 2020). All stakeholders perceived the application of nanomaterials in food products and cosmetics/sunscreens as riskier, regardless of their involvement. However, those who were knowledgeable about nanotechnology had a lower perception of risk (Capon et al., 2015). This pattern highlights the influence of scientific literacy on risk perception and underscores the importance of targeted education and outreach strategies to bridge knowledge gaps and foster informed public engagement with emerging technologies. While nanotechnology holds immense promise for advancing medicine, energy, and environmental sustainability, a vocal segment of the public remains staunchly opposed to its development, often rooted in misconceptions or fear of the unknown.

To sum up, the way the public perceives and misunderstands nanotechnology greatly influences their attitudes and acceptance of this developing area. It is essential to counter these misunderstandings through education and open communication, so that people can grasp the potential advantages and risks linked with nanotechnology and make informed choices.

## MEDICAL LITERACY AND HEALTH RISK COMMUNICATION

The global community has researched the general public's perception of nanotechnology (Kamarulzaman et al., 2020; Pokrajac et al., 2021). A study at Yale indicated that demographic and cultural characteristics had an impact on the public's risk perception. This influence was so strong that it was simpler to erode public trust than to acquire it, mainly when leaders failed to take responsibility for any negative consequences that may have arisen (Aluya, 2015).

Nanotechnology encompasses the creation, manipulation, and regulation of things at the atomic scale, which may introduce new uncertainties (Renn & Roco, 2020). Given the potential similarities with other contentious technologies, it is necessary to cultivate a thorough comprehension of the ways in which the public perceives uncertainties, risks, and rewards associated with nanotechnology. This understanding should also encompass communication challenges that are relevant to this field (Sahu et al., 2023). Examining perceptions at such an early level in the technological development process is likely unprecedented in the area of risk perception and interaction. Consequently, it also presents novel methodological and conceptual difficulties (Halamoda-Kenzaoui et al., 2021). These challenges encompass managing the inherent variety within the nanotechnology area, grappling with the new and abstract nature of the notion, which lacks clear comparisons to guide understanding or risk assessment, and addressing the ethical issues that underlie numerous disputes surrounding nanotechnology (Pidgeon et al., 2011).

Nanotechnology-enabled items possess an extended lifespan and exhibit enhanced performance (Gidiagba et al., 2023; Sriram & Suttee, 2020). However, the rise of nanotechnology not only brings advantages but also risks to the general population, which is linked to unpredictable risks in this field (Malik et al., 2023). Although mutations in genes and atomic power have been extensively demonstrated to enhance human well-being, the public tends to reject these technologies because their early development fails to consider public benefits and risk perceptions adequately. Therefore, it is crucial to assess the public's perception of the benefits and risks associated with the development of nanotechnology to enhance their comprehension and promote the acceptance of this developing technology (Joubert et al., 2020). Effectively conveying the advantages and potential dangers of nanotechnology is a strategic method for establishing public confidence and equipping the public with accurate knowledge. This will empower individuals to make well-informed choices and prevent them from succumbing to apprehension caused by a lack of information. The report centres on Malaysia's efforts to position nanotechnology as a key driver of the country's economy (Arumugam & Thambiratnam, 2023). Therefore, it is necessary to comprehend the general opinion of the public on nanotechnology to enhance communication with the public and to design ethical policies for the advancement of nanotechnology. Previous study mentioned the public's opinions of the benefits and risks associated with the growth of nanotechnology, considering demographic, psychological, and sociological elements (Kamarulzaman et al., 2020). These elements create nuanced patterns of acceptance or skepticism, underscoring the need for tailored communication strategies and policy-making.

## RISK PERCEPTIONS AND COMMUNICATION

The advancements in nanotechnology provide both new possibilities for improvements and novel complexitics (Sahu et al., 2023; Usman et al., 2020). Nanoscale fabrication of typical substances may alter or improve their chemical and electrical characteristics, which could have unintended consequences for individuals and their surroundings. The literature supports these issues and concludes that nanomaterials should be regulated as new substances and that more in-depth investigation into novel toxicologist risks is vital (Zielińska et al., 2020). Considering the possible similarities to other contentious technologies like biotech or nuclear energy, the study emphasised the significance of comprehending the general acceptance and opinion of nanotechnologies (Rathore & Mahesh, 2021). Ecological and consumer groups have stressed the need for strict regulations to deal with potential hazards before developing or using nanotechnology on a large scale, while academics and industrial supporters have voiced concern that a new industry might face stigmatisation and increased concern among the public in an arrangement that is strangely similar to previous controversies. Nanotechnology was poised to become the following major example of social reinforcement of risk (Renn & Roco, 2020). However, as we will see in the next section, history has taken a slightly different turn. Because the social amplifying of risk perceptions necessitates the coming together over time of a complicated mixture of intervening factors including, but not limited to, risk-related incidents with high signal value, widespread coverage in the press, regulatory or corporate failure, mistrust, and blame nanotechnologies did not inevitably face the same levels of controversy as biotechnology (Macnaghten & Guivant, 2020). The question of whether the rise of nanotechnology fears is characterised by amplification or reduction is an important one that needs further investigation (Pidgeon et al., 2011).

In conclusion, nanotechnology presents a dual-edged sword with its advancements offering promising improvements yet introducing novel complexities, as underscored by various studies; the need for regulation, in - depth toxicological research, and understanding public acceptance is evident, and while it was initially expected to face significant social reinforcement of risk similar to other controversial technologies, the actual trajectory has been influenced by a complex interplay of factors, leaving the question of whether risk perceptions of nanotechnology will be amplified or reduced as a crucial area for further exploration.

## ETHICS AND NANOTECHNOLOGY RISKS

A discourse on the ethical implications of nanotechnology has emerged alongside the public discourse on its advantages and disadvantages (Archibald et al., 2020). There is a proposition that suggests the need for a new set of ethical principles specifically for nanotechnology, known as "nano-ethics" (Li, 2020). The main focus of this argument is on the positive and destructive visionary and speculative breakthroughs associated with nanotechnology. Several studies explored new ethical dimensions of nanotechnology beyond previous approaches. There are few if any, novel ethical considerations that arise from nanotechnology (Sparrow, 2020). The issue primarily involves subtle changes in focus and significance of topics that are already recognised, leading to ethical concerns on nanotechnology. Structurally fresh ethical aspects emerge from the significant influence of visions in discourse in society. The convergence of previously distinct ethical lines of thought in the area of nanotechnology also raises new inquiries (Grunwald, 2020).

In November 2020, delegates from prominent nanotechnology academic institutions convened for the online international workshop on Nanotechnology for a Sustainable Tomorrow. The discourse aimed to explore the potential of nanotechnology and its uses in achieving sustainable objectives. One of the main results was the realisation of the necessity for an International Network for Sustainable Nanotechnology to provide a worldwide platform for addressing this goal and to welcome individuals with innovative perspectives and concepts. The main focus of this network is to facilitate communication and collaboration between scientists, engineers, authorities, and policy-makers worldwide. The aim is to promote the swift implementation of technology-driven solutions to address our pressing issues (Pokrajac et al., 2021). In conclusion, the development of nanotechnology has sparked a rich and complex conversation. On one hand, the ethical implications of nanotechnology have become a central part of public discourse, with the concept of "nano - ethics" emerging as a call for new ethical principles. While some argue that nanotechnology brings few truly novel ethical considerations, others highlight the structural and converging ethical aspects that demand our attention. On the other hand, the global community is actively working towards harnessing the potential of nanotechnology for a sustainable future. The establishment of the International Network for Sustainable Nanotechnology is a testament to this effort, aiming to bridge the gap between different stakeholders and accelerate the adoption of nano-technology - based solutions to our most pressing challenges. As we move forward, it is crucial that we continue to engage in these discussions, balancing the pursuit of innovation with a steadfast commitment to ethical and sustainable practices in the field of nanotechnology.

## AUTOPHAGY, NANOTECHNOLOGY AND TOXICOLOGY

Autophagy serves intracellular antiviral functions and contributes to the activation of innate and adaptive immune responses against infections (Tanida, 2011). Autophagy plays various roles in both innate and adaptive immunity, including providing resistance to pathogen invasion (Yang & Klionsky, 2009). Autophagy is a non-specific response of the body to starvation (Cao, Li, Yang, & Cao, 2021). Due to the swift advancement of nanotechnology, inorganic nanomaterials have found extensive use in contemporary society. Given the unavoidable human contact with these inorganic nanomaterials, a thorough evaluation of their safety is essential. It is common knowledge that autophagy serves a dual function in cellular processes, contributing to both cell survival and cell demise (Guo, He, Zhao, Liu, & Deng, 2020). Autophagy process has been preserved throughout evolution (Vijayakumar & Cho, 2019; Wei & Duan, 2019). This intersection of nanotechnology and cellular biology presents a vital area for future investigation, aiming to ensure the safe application of advancements in nanotechnology.

The integration of autophagy research, nanotechnology advancements, and toxicological studies holds immense promise for developing innovative approaches to disease treatment, drug delivery, and environmental protection. Collaborative efforts across these disciplines are vital for harnessing the full potential of nanotechnology while safeguarding human health and the environment.

## AUTOPHAGY AND MEDICAL LITERACY

Understanding autophagy is vital for medical literacy as it provides insights into how cells maintain their health and how dysregulation of this process can contribute to disease development. Researchers are actively studying autophagy pathways to develop new treatments for conditions where autophagy is impaired. Autophagy represents a powerful, natural mechanism for maintaining health and preventing disease. However, the public's ability to understand and apply this knowledge depends heavily on medical literacy. Bridging this gap through targeted education and responsible communication can promote healthier lifestyles, reduce chronic disease burden, and can create a more scientifically informed society.

## METHODS

The authors adopted a narrative review approach for this chapter and it should be noted that it is not a systematic review. Moreover, we have emphasized importance of medical education. We also relate it to toxicology and autophagy. We have used PubMed and Google Scholar to search for articles. Chatbots are employed to improve the language of manuscripts.

## RESULTS

The intersection of autophagy, nanotechnology, and toxicology is a promising area for future research. Autophagy, with its roles in immunity and response to starvation, has a dual function in cell survival and death. As inorganic nanomaterials are widely used, evaluating their safety is essential. Integrating autophagy research, nanotechnology advancements, and toxicological studies holds great potential for innovative medical and environmental solutions. Additionally, understanding autophagy is fundamental to medical literacy, as it is closely related to cell health and disease development, and ongoing research aims to develop new treatments for autophagy - impaired conditions.

## DISCUSSION

This section investigates the misunderstandings related to nano-specific toxicity and underscores the crucial importance of enhanced healthcare knowledge in tackling this concern. By examining the field of nanotechnology and its possible health effects, this research underscores the significance of spreading precise information and providing education to counter baseless anxieties.

## MEDICAL EDUCATION AND TOXICOLOGY

Medical education plays a pivotal role in shaping healthcare professionals who are equipped to make informed decisions, ensuring patient safety and public health. One critical aspect of this education is toxicology literacy, which involves understanding the effects of chemicals, drugs, and environmental toxins on human health.

## TOXICOLOGY AND AUTOPHAGY

Autophagy is a vital cellular mechanism that coordinates the breakdown of cellular elements within lysosomes to uphold cellular balance and react to cellular stress. Extensive research in the past decade has demonstrated the indispensable nature of autophagy for continual protein and organelle renewal, as well as for the survival of embryos/neonates and cells under environmental stress conditions (Bolt & Klimecki, 2012). Autophagy has been observed as the primary reaction of a cell to a toxic metal in a manner influenced by both the concentration of the metal and the duration of exposure. Different signaling pathways are activated by cellular proteins and/or protein kinases, which can induce autophagy, apoptosis (or necroptosis), and necrosis (Chatterjee, Sarkar, & Bhattacharya, 2014). Together, these findings indicate that autophagy serves as a safeguard against cell toxicity induced by manganese. The antioxidants NAC and resveratrol primarily shield against manganese toxicity by preserving mitochondrial dynamics and function, rather than by directly influencing the flow of autophagy (Zhou et al., 2018). Recent studies have shown that autophagy is not just a fundamental consequence of genotoxic substances but also a crucial factor in the cellular response to DNA damage induced by genotoxic stress (Zhang, Shang, & Zhou, 2015).

Cadmium is a harmful pollutant linked to severe human illnesses. It can be readily absorbed in significant amounts from sources like air pollution, industrial emissions, cigarette smoke, food, and water, predominantly impacting the liver, kidneys, and lungs. Furthermore, cadmium disrupts hepatic autophagy, a natural breakdown process that eliminates damaged cell components or dysfunctional proteins through vacuole-mediated containment and lysosomal breakdown (Niture et al., 2021). Cadmium's toxicity is a multifaceted threat, with its disruption of hepatic autophagy serving as a key mechanism driving liver damage and systemic health effects. By impairing this essential cellular cleanup process, cadmium undermines the liver's ability to maintain homeostasis, leading to chronic disease and long-term health consequences.

In conclusion, the intricate interplay between toxicology and autophagy underscores the pivotal role of cellular self-cleaning mechanisms in mitigating the impact of toxic substances on biological systems.

## TOXICOLOGY, AUTOPHAGY AND FASTING

Toxicology is the scientific study of the adverse effects that chemicals, physical agents, and biological substances can have on living organisms (Pope, Schlenk, & Baud, 2020). It involves understanding the mechanisms by which toxins exert their effects, as well as how these effects can be prevented or mitigated. Autophagy, on the other hand, is a cellular process that involves the degradation and recycling of cellular components.

Fasting has been linked to the stimulation of autophagy in cells, which is thought to be a protective mechanism that helps cells survive times of nutrient deprivation by recycling damaged organelles and proteins. This process is believed to play a role in the body's response to stress, including exposure to toxins.

The possible contribution of increased autophagic lysosomal clearance of impaired proteins in the evolutionary development of stress resilience in intertidal mollusks is examined, drawing comparisons with the expanding indications of autophagy's role in hormesis (Moore, Shaw, Ferrar Adams, & Viarengo, 2015).

Out of various stress triggers, fasting stand out as the most powerful non-genetic activators of autophagy (Kendler, 2005). The extent to which autophagy contributes to the overall toxicity of these compounds remains uncertain, and additional studies are required to better understand the toxicological relevance of this mechanism (Orrenius, Kaminskyy, & Zhivotovsky, 2013). Autophagy plays a vital role in maintaining normal cellular homeostasis, and its complex interplay with cellular stress, along with the discovery of certain xenobiotics that can influence this process, highlights the significance of autophagy in the field of toxicology (Bolt & Klimecki, 2012).

Autophagy, a term derived from the Greek words "auto" (self) and "phagy" (eating), is a cellular process that involves the breakdown and recycling of damaged or dysfunctional parts of cells (Hamad, 2017). This plays a crucial role in maintaining cellular health and promoting longevity (Madeo, Tavernarakis, & Kroemer, 2010).

Fasting, on the other hand, is the voluntary abstention from food and, in some cases, drink for a set period. Fasting has been practiced for centuries for various reasons, including religious observances, weight management, and potential health benefits (e.g., metabolism). One of the intriguing connections between autophagy and fasting is that fasting can stimulate autophagy in cells. When the body is deprived of nutrients, especially glucose, it switches to alternative energy sources and initiates various cellular processes to maintain energy balance. Autophagy is one of these processes that are upregulated during fasting.

This phrase could be interpreted as the study of how toxins impact the body in relation to autophagy and fasting. Toxicology involves the study of harmful substances, while autophagy fasting refers to the process of cellular self-cleansing that occurs during fasting. A previous study showed that intermittent fasting may increase autophagy in mouse liver cells (Chaudhary et al., 2022). Intermittent fasting may help shed body fat as well (Yoshii, Higashida, & Nakai, 2023). These findings underscore the potential of intermittent fasting as a non-pharmacological intervention for obesity and metabolic disorders. However, further clinical studies are warranted to determine its long-term efficacy, optimal fasting protocols for diverse patient populations, and its safety profile in individuals with comorbidities such as type 2 diabetes, cardiovascular disease, or endocrine disorders. Future research should also explore the molecular pathways underlying these metabolic effects to better inform patient-specific treatment strategies.

## CONCLUSION

In conclusion, by increasing medical literacy and dispelling the m yth of nano-specific toxicity, we can foster a more informed understanding of nanotechnology and its potential benefits in the field of medicine. Moreover, understanding the medical aspects of fasting can support the process of autophagy. This will ultimately lead to greater acceptance and support for further research and development in this promising area.

Further studies should bridge the gaps between medical education, nanotechnology innovation, and autophagy research to create a holistic, evidence-driven ecosystem. By enhancing public understanding of both the nuanced risks of nanomaterials and the cellular benefits of fasting-induced autophagy, these efforts can foster trust, drive ethical advancements, and unlock the full potential of nanotech in improving human health—while ensuring that progress is inclusive, safe, and aligned with societal values. Further studies should explore repurposing apparently safe nanomaterials to intentionally trigger beneficial autophagy.

## ACKNOWLEDGMENT

**Conflict of interest:** There is no conflict of interest to state.
**Author Statement:** K.U. and A.I.A. wrote the manuscript.
**Data Availability Statement:** There is no data associated with the study.
**Ethical Permission:** Not required.
**Limitations:** It is not a systematic review. We didn't use a systematic approach to compile the articles.

## REFERENCES

Aluya, J. (2015). Nanotechnology implications and global leadership perspectives. *Energy Sources. Part B, Economics, Planning, and Policy*, *10*(1), 31–37. DOI: 10.1080/15567249.2010.506472

Archibald, J., Garzone, G., & Catenaccio, P. (2020). Debating evolutions in science, technology and society: Ethical and ideological perspectives. *Lingue e Linguaggi, 34*(speciale).

Bi, Y. H., Po, J., & Uludag, K. (2025). Enhancing Offspring Cognitive Health: Addressing Prenatal Stress and Problems Through Nanotechnology. In Nanomaterials and the Nervous System (pp. 341-358). IGI Global.

Bolt, A. M., & Klimecki, W. T. (2012). Autophagy in toxicology: Self-consumption in times of stress and plenty. *Journal of Applied Toxicology*, *32*(7), 465–479. DOI: 10.1002/jat.1787 PMID: 22334383

Cao, W., Li, J., Yang, K., & Cao, D. (2021). An overview of autophagy: Mechanism, regulation and research progress. *Bulletin du Cancer*, *108*(3), 304–322. DOI: 10.1016/j.bulcan.2020.11.004 PMID: 33423775

Capon, A., Gillespie, J., Rolfe, M., & Smith, W. (2015). Perceptions of risk from nanotechnologies and trust in stakeholders: A cross sectional study of public, academic, government and business attitudes. *BMC Public Health*, *15*(1), 1–13. DOI: 10.1186/s12889-015-1795-1 PMID: 25928741

Chatterjee, S., Sarkar, S., & Bhattacharya, S. (2014). Toxic metals and autophagy. *Chemical Research in Toxicology*, *27*(11), 1887–1900. DOI: 10.1021/tx500264s PMID: 25310621

Chaudhary, R., Liu, B., Bensalem, J., Sargeant, T. J., Page, A. J., Wittert, G. A., Hutchison, A. T., & Heilbronn, L. K. (2022). Intermittent fasting activates markers of autophagy in mouse liver, but not muscle from mouse or humans. *Nutrition (Burbank, Los Angeles County, Calif.)*, *101*, 111662. DOI: 10.1016/j.nut.2022.111662 PMID: 35660501

Domingues, C., Santos, A., Alvarez-Lorenzo, C., Concheiro, A., Jarak, I., Veiga, F., Barbosa, I., Dourado, M., & Figueiras, A. (2022). Where is nano today and where is it headed? A review of nanomedicine and the dilemma of nanotoxicology. *ACS Nano*, *16*(7), 9994–10041. DOI: 10.1021/acsnano.2c00128 PMID: 35729778

El-Zawi, N. R., & Elyasseri, A. M. A. (2021). A review of toxicity, genotoxicity and teratogenicity of formaldehyde. *Cancer*, *15*, 43.

Erisen, D. E., & Uludag, K. (2024). Estimating Biosafety of Biodegradable Biomedical Materials from In Vitro Ion Tolerance Parameters and Toxicity of Nanomaterials in Brain. In *Transformative Approaches to Patient Literacy and Healthcare Innovation* (pp. 201–221). IGI Global. DOI: 10.4018/979-8-3693-3661-8.ch010

Foulkes, R., Man, E., Thind, J., Yeung, S., Joy, A., & Hoskins, C. (2020). The regulation of nanomaterials and nanomedicines for clinical application: Current and future perspectives. *Biomaterials Science*, *8*(17), 4653–4664. DOI: 10.1039/D0BM00558D PMID: 32672255

Gidiagba, J. O., Daraojimba, C., Ofonagoro, K. A., Eyo-Udo, N. L., Egbokhaebho, B. A., Ogunjobi, O. A., & Banso, A. A. (2023). Economic impacts and innovations in materials science: A holistic exploration of nanotechnology and advanced materials. *Engineering Science & Technology Journal*, *4*(3), 84–100. DOI: 10.51594/estj.v4i3.553

Grunwald, A. (2020). Nanotechnology-A new field of ethical inquiry? In *The Ethics of Nanotechnology, Geoengineering, and Clean Energy* (pp. 17-31). Routledge. https://doi.org/DOI: 10.4324/9781003075028-3

Guo, L., He, N., Zhao, Y., Liu, T., & Deng, Y. (2020). Autophagy modulated by inorganic nanomaterials. *Theranostics*, *10*(7), 3206–3222. DOI: 10.7150/thno.40414 PMID: 32194863

Guzelian, P. S., Victoroff, M. S., Halmes, N. C., James, R. C., & Guzelian, C. P. (2005). Evidence-based toxicology: A comprehensive framework for causation. *Human and Experimental Toxicology*, *24*(4), 161–201. DOI: 10.1191/0960327105ht517oa PMID: 15957536

Halamoda-Kenzaoui, B., Vandebriel, R. J., Howarth, A., Siccardi, M., David, C. A. W., Liptrott, N. J., Santin, M., Borgos, S. E., Bremer-Hoffmann, S., & Caputo, F. (2021). Methodological needs in the quality and safety characterisation of nanotechnology-based health products: Priorities for method development and standardisation. *Journal of Controlled Release, 336*, 192–206. DOI: 10.1016/j.jconrel.2021.06.016 PMID: 34126169

Haleem, A., Javaid, M., Singh, R. P., Rab, S., & Suman, R. (2023). Applications of nanotechnology in medical field: A brief review. *Global Health Journal (Amsterdam, Netherlands)*, *7*(2), 70–77. DOI: 10.1016/j.glohj.2023.02.008

Ibrohim, I., Kusumaningrum, I. K., Hamimi, E., Putra, W. E., Utomo, J., Kundariati, M., & Malek, N. A. N. B. N. (2023). Science students' literacy and faculty members' perspective toward nanotechnology: Is it needed in 21st century education? *Eurasia Journal of Mathematics, Science and Technology Education*, *19*(12), em2362. Advance online publication. DOI: 10.29333/ejmste/13835

Jamil, B., Javed, R., Qazi, A. S., & Syed, M. A. (2018). Nanomaterials: Toxicity, risk management, and public perception. *Nanomaterials: Ecotoxicity, safety, and public perception*, 283-304. DOI: 10.1007/978-3-030-05144-0_14

Jose, J. P. A., Younus, L. A., Reddy, K. B., Sana, S. S., Gangadhar, L., Hou, T., & Bhardwaj, P. (2022). Environmental impact on toxicity of nanomaterials. In *Biogenic Sustainable Nanotechnology* (pp. 315–355). Elsevier. DOI: 10.1016/B978-0-323-88535-5.00011-1

Joubert, I. A., Geppert, M., Ess, S., Nestelbacher, R., Gadermaier, G., Duschl, A., Bathke, A. C., & Himly, M. (2020). Public perception and knowledge on nanotechnology: A study based on a citizen science approach. *NanoImpact*, *17*, 100201. DOI: 10.1016/j.impact.2019.100201

Kamarulzaman, N. A., Lee, K. E., Siow, K. S., & Mokhtar, M. (2020). Public benefit and risk perceptions of nanotechnology development: Psychological and sociological aspects. *Technology in Society*, *62*, 101329. DOI: 10.1016/j.techsoc.2020.101329

Kendler, K. S. (2005). Toward a philosophical structure for psychiatry. *The American Journal of Psychiatry*, *162*(3), 433–440. DOI: 10.1176/appi.ajp.162.3.433 PMID: 15741457

Kumar, P., Mahajan, P., Kaur, R., & Gautam, S. (2020). Nanotechnology and its challenges in the food sector: A review. *Materials Today. Chemistry*, *17*, 100332. DOI: 10.1016/j.mtchem.2020.100332 PMID: 32835156

Li, M., Gong, J., Gao, L., Zou, T., Kang, J., & Xu, H. (2022). Advanced human developmental toxicity and teratogenicity assessment using human organoid models. *Ecotoxicology and Environmental Safety, 235*, 113429. DOI: 10.1016/j.ecoenv.2022.113429 PMID: 35325609

Li, S. (2020). Three Dimensions of Nano-ethics. *Chinese Philosophy of Technology: Classical Readings and Contemporary Work*, 165-181. DOI: 10.1007/978-981-15-1952-9_11

Luty-Błocho, M., Hoyos, M. V., & Hessel, V. (2023). Implications of the Use of Functional Nanomaterials for the Environment and Health: Risk Assessment and Challenges. In *Functional Nanomaterials for Sensors* (pp. 289–319). CRC Press. DOI: 10.1201/9781003263852-14

Macnaghten, P., & Guivant, J. S. (2020). Narrative as a resource for inclusive governance: a UK–Brazil comparison of public responses to nanotechnology. *Journal of Responsible Innovation, 7*(sup1), 13-33. DOI: 10.1080/23299460.2020.1842643

Madeo, F., Tavernarakis, N., & Kroemer, G. (2010). Can autophagy promote longevity? *Nature Cell Biology, 12*(9), 842–846. DOI: 10.1038/ncb0910-842 PMID: 20811357

Malik, S., Muhammad, K., & Waheed, Y. (2023). Nanotechnology: A revolution in modern industry. *Molecules (Basel, Switzerland), 28*(2), 661. DOI: 10.3390/molecules28020661 PMID: 36677717

Moore, M. N., Shaw, J. P., Ferrar Adams, D. R., & Viarengo, A. (2015). Anti-oxidative cellular protection effect of fasting-induced autophagy as a mechanism for hormesis. *Marine Environmental Research, 107*, 35–44. DOI: 10.1016/j.marenvres.2015.04.001 PMID: 25881010

Nasrollahzadeh, M., Sajadi, S. M., Sajjadi, M., & Issaabadi, Z. (2019). An Introduction to Nanotechnology. In Nasrollahzadeh, M., Sajadi, S. M., Sajjadi, M., Issaabadi, Z., & Atarod, M. (Eds.), *Interface Science and Technology* (Vol. 28, pp. 1–27). Elsevier.

Niture, S., Lin, M., Qi, Q., Moore, J. T., Levine, K. E., Fernando, R. A., & Kumar, D. (2021). Role of Autophagy in Cadmium-Induced Hepatotoxicity and Liver Diseases. *Journal of Toxicology, 2021*(1), 9564297. DOI: 10.1155/2021/9564297 PMID: 34422041

Orrenius, S., Kaminskyy, V. O., & Zhivotovsky, B. (2013). Autophagy in toxicology: Cause or consequence? *Annual Review of Pharmacology and Toxicology, 53*(1), 275–297. DOI: 10.1146/annurev-pharmtox-011112-140210 PMID: 23072380

Ostrowski, A. D., Martin, T., Conti, J., Hurt, I., & Harthorn, B. H. (2009). Nanotoxicology: Characterizing the scientific literature, 2000–2007. *Journal of Nanoparticle Research, 11*(2), 251–257. DOI: 10.1007/s11051-008-9579-5 PMID: 21170129

Patel, J. K., Patel, A., & Bhatia, D. (2021). Introduction to nanomaterials and nanotechnology. In *Emerging technologies for nanoparticle manufacturing* (pp. 3–23). Springer International Publishing. DOI: 10.1007/978-3-030-50703-9_1

Pidgeon, N., Harthorn, B., & Satterfield, T. (2011). Nanotechnology risk perceptions and communication: Emerging technologies, emerging challenges. *Risk Analysis, 31*(11), 1694–1700. DOI: 10.1111/j.1539-6924.2011.01738.x PMID: 22084861

Pokrajac, L., Abbas, A., Chrzanowski, W., Dias, G. M., Eggleton, B. J., Maguire, S., ... Mitra, S. (2021). Nanotechnology for a sustainable future: Addressing global challenges with the international network4sustainable nanotechnology. DOI: 10.1021/acsnano.1c10919

Pope, C. N., Schlenk, D., & Baud, F. J. (2020). History and basic concepts of toxicology. In *An Introduction to Interdisciplinary Toxicology* (pp. 3–15). Elsevier. DOI: 10.1016/B978-0-12-813602-7.00001-6

Rahman, M. A., Rahman, M. S., Parvez, M. A. K., & Kim, B. (2023). The Emerging Role of Autophagy as a Target of Environmental Pollutants: An Update on Mechanisms. *Toxics, 11*(2), 135. Advance online publication. DOI: 10.3390/toxics11020135 PMID: 36851010

Rathore, A., & Mahesh, G. (2021). Public perception of nanotechnology: A contrast between developed and developing countries. *Technology in Society, 67*, 101751. DOI: 10.1016/j.techsoc.2021.101751

Sahu, M. K., Yadav, R., & Tiwari, S. P. (2023). Recent advances in nanotechnology. *International Journal of Nanomaterials, Nanotechnology and Nanomedicine, 9*(1), 15-23. DOI: 10.17352/2455-3492.000053

Sriram, P., & Suttee, A. (2020). Nanotechnology advances, benefits, and applications in daily life. In *Nanotechnology* (pp. 23–44). CRC Press. DOI: 10.1201/9781003082859-3

Tanida, I. (2011). Autophagy basics. *Microbiology and Immunology, 55*(1), 1–11. DOI: 10.1111/j.1348-0421.2010.00271.x PMID: 21175768

Usman, M., Farooq, M., Wakeel, A., Nawaz, A., Cheema, S. A., Rehman, H., Ashraf, I., & Sanaullah, M. (2020). Nanotechnology in agriculture: Current status, challenges and future opportunities. *The Science of the Total Environment, 721*, 137778. DOI: 10.1016/j.scitotenv.2020.137778 PMID: 32179352

Vijayakumar, K., & Cho, G. W. (2019). Autophagy: An evolutionarily conserved process in the maintenance of stem cells and aging. *Cell Biochemistry and Function*, *37*(6), 452–458. DOI: 10.1002/cbf.3427 PMID: 31318072

Wei, F., & Duan, Y. (2019). Crosstalk between Autophagy and Nanomaterials: Internalization, Activation, Termination. *Advanced Biosystems*, *3*(1), 1800259. DOI: 10.1002/adbi.201800259 PMID: 32627344

Yang, Z., & Klionsky, D. J. (2009). An overview of the molecular mechanism of autophagy. *Autophagy in infection and immunity*, 1-32.

Yoshii, R., Higashida, K., & Nakai, N. (2023). Intermittent fasting reduces mouse body fat while maintaining muscle mass by regulating protein synthesis and autophagy. *Nutrition (Burbank, Los Angeles County, Calif.)*, *115*, 112130. DOI: 10.1016/j.nut.2023.112130 PMID: 37454541

Zhang, S.-M., Shang, Z.-F., & Zhou, P.-K. (2015). Autophagy as the effector and player in DNA damage response of cells to genotoxicants. *Toxicology Research*, *4*(3), 613–622. DOI: 10.1039/C5TX00043B

Zhou, Q., Fu, X., Wang, X., Wu, Q., Lu, Y., Shi, J., Klaunig, J. E., & Zhou, S. (2018). Autophagy plays a protective role in Mn-induced toxicity in PC12 cells. *Toxicology*, *394*, 45–53. DOI: 10.1016/j.tox.2017.12.001 PMID: 29222055

Zielińska, A., Costa, B., Ferreira, M. V., Miguéis, D., Louros, J. M., Durazzo, A., & Souto, E. B. (2020). Nanotoxicology and nanosafety: Safety-by-design and testing at a glance. *International Journal of Environmental Research and Public Health*, *17*(13), 4657. DOI: 10.3390/ijerph17134657 PMID: 32605255

# Compilation of References


Abbaszadeh, F., Fakhri, S., & Khan, H. (2020). Targeting apoptosis and autophagy following spinal cord injury: Therapeutic approaches to polyphenols and candidate phytochemicals. *Pharmacological Research*, *160*, 105069. DOI: 10.1016/j.phrs.2020.105069 PMID: 32652198

Abdallah, H. M., El-Gohary, R. M., Khattab, H., Farghal, E. E., Ghabrial, M. M., & Hashad, A. F. (2024). Potential Synergestic effect of quercetin and exercise on depressive like behavior in male albino rats: Targeting oxidative stress, inflammation, neural apoptosis and Mitophagy. *Bulletin of Egyptian Society for Physiological Sciences.*, *44*(2), 54–68. DOI: 10.21608/besps.2023.243730.1157

Abdoullah, B., Durand, G. A., Basco, L. K., El Bara, A., Bollahi, M. A., Bosio, L., Geulen, M., Briolant, S., & Boukhary, A. O. M. S. (2023). Seroprevalence of Alphaviruses (Togaviridae) among Urban Population in Nouakchott, Mauritania, West Africa. *Viruses*, *15*(7), 1588. DOI: 10.3390/v15071588 PMID: 37515274

Abelaira, H. M., Réus, G. Z., Neotti, M. V., & Quevedo, J. (2014). The role of mTOR in depression and antidepressant responses. *Life Sciences*, *101*(1-2), 10–14. DOI: 10.1016/j.lfs.2014.02.014 PMID: 24582593

Abeliovich, A., & Gitler, A. D. (2016). Defects in trafficking bridge Parkinson's disease pathology and genetics. *Nature*, *539*(7628), 207–216. DOI: 10.1038/nature20414 PMID: 27830778

Aber, E. R., Griffey, C. J., Davies, T., Li, A. M., Yang, Y. J., Croce, K. R., Goldman, J. E., Grutzendler, J., Canman, J. C., & Yamamoto, A. (2022). Oligodendroglial macroautophagy is essential for myelin sheath turnover to prevent neurodegeneration and death. *Cell Reports*, *41*(3), 111480. DOI: 10.1016/j.celrep.2022.111480 PMID: 36261002

Aboeldalyl, S., James, C., Seyam, E., Ibrahim, E. M., Shawki, H. E., & Amer, S. (2021). The Role of Chronic Inflammation in Polycystic Ovarian Syndrome—A Systematic Review and Meta-Analysis. *International Journal of Molecular Sciences*, *22*(5), 2734. DOI: 10.3390/ijms22052734 PMID: 33800490

Abou-Saleh, H., Younes, N., Rasool, K., Younis, M. H., Prieto, R. M., Yassine, H. M., Mahmoud, K. A., Pintus, G., & Nasrallah, G. K. (2019). Impaired liver size and compromised neurobehavioral activity are elicited by chitosan nanoparticles in the zebrafish embryo model. *Nanomaterials (Basel, Switzerland)*, *9*(1), 122. DOI: 10.3390/nano9010122 PMID: 30669437

Abrignani, V., Salvo, A., Pacinella, G., & Tuttolomondo, A. (2024). The Mediterranean Diet, Its Microbiome Connections, and Cardiovascular Health: A Narrative Review. *International Journal of Molecular Sciences*, *25*(9), 4942. DOI: 10.3390/ijms25094942 PMID: 38732161

Absalyamova, M., Traktirov, D., Burdinskaya, V., Artemova, V., Muruzheva, Z., & Karpenko, M. (2025). Molecular basis of the development of Parkinson's disease. *Neuroscience*, *565*, 292–300. DOI: 10.1016/j.neuroscience.2024.12.009 PMID: 39653246

Adamu, A., Li, S., Gao, F., & Xue, G. (2024). The role of neuroinflammation in neurodegenerative diseases: Current understanding and future therapeutic targets. *Frontiers in Aging Neuroscience*, *16*, 1347987. DOI: 10.3389/fnagi.2024.1347987 PMID: 38681666

Ahmad, I., Teotia, P., Erickson, H., & Xia, X. (2020). Recapitulating developmental mechanisms for retinal regeneration. *Progress in Retinal and Eye Research*, *76*, 100824. DOI: 10.1016/j.preteyeres.2019.100824 PMID: 31843569

Ahuja, C. S., Wilson, J. R., Nori, S., Kotter, M., Druschel, C., Curt, A., & Fehlings, M. G. (2017). Traumatic spinal cord injury. *Nature Reviews. Disease Primers*, *3*(1), 1–21. DOI: 10.1038/nrdp.2017.18 PMID: 28447605

Ajorlouie, Z., Moshkian, P., Baghdadi, G., Amiri, R., Biglari, F., & Rahimlou, M. (2025). The association between the Mediterranean Diet and the Prime Diet Quality Score and polycystic ovary syndrome: A case control study. *BMC Nutrition*, *11*(1), 80. DOI: 10.1186/s40795-025-01067-5 PMID: 40241157

Alcocer-Gómez, E., Casas-Barquero, N., Núñez-Vasco, J., Navarro-Pando, J. M., & Bullón, P. (2017). Psychological status in depressive patients correlates with metabolic gene expression. *CNS Neuroscience & Therapeutics*, *23*(10), 843–845. DOI: 10.1111/cns.12755 PMID: 28879683

Alcocer-Gómez, E., Casas-Barquero, N., Williams, M. R., Romero-Guillena, S. L., Cañadas-Lozano, D., Bullón, P., Sánchez-Alcazar, J. A., Navarro-Pando, J. M., & Cordero, M. D. (2017). Antidepressants induce autophagy dependent-NLRP3-inflammasome inhibition in Major depressive disorder. *Pharmacological Research, 121*, 114–121. DOI: 10.1016/j.phrs.2017.04.028 PMID: 28465217

Alers, S., Löffler, A. S., Wesselborg, S., & Stork, B. (2012). Role of AMPK-mTOR-Ulk1/2 in the regulation of autophagy: Cross talk, shortcuts, and feedbacks. *Molecular and Cellular Biology, 32*(1), 2–11. DOI: 10.1128/MCB.06159-11 PMID: 22025673

Ali, T., Rahman, S. U., Hao, Q., Li, W., Liu, Z., Ali Shah, F., Murtaza, I., Zhang, Z., Yang, X., Liu, G., & Li, S. (2020). Melatonin prevents neuroinflammation and relieves depression by attenuating autophagy impairment through FOXO3a regulation. *Journal of Pineal Research, 69*(2), e12667. DOI: 10.1111/jpi.12667 PMID: 32375205

Almannai, M., Marafi, D., & El-Hattab, A. W. (2022). WIPI proteins: Biological functions and related syndromes. *Frontiers in Molecular Neuroscience, 15*, 1011918. Advance online publication. DOI: 10.3389/fnmol.2022.1011918 PMID: 36157071

Almeida, S. S., Zizzi, F. B., Cattaneo, A., Comandini, A., Di Dato, G., Lubrano, E., Pellicano, C., Spallone, V., Tongiani, S., & Torta, R. (2020). Management and treatment of patients with major depressive disorder and chronic diseases: A multidisciplinary approach. *Frontiers in Psychology, 11*, 542444. DOI: 10.3389/fpsyg.2020.542444 PMID: 33101117

Alonaizan, R. (2024). Molecular regulation of NLRP3 inflammasome activation during parasitic infection. *Bioscience Reports, 44*(5), BSR20231918. Advance online publication. DOI: 10.1042/BSR20231918 PMID: 38623843

Alpuche Lazcano, S. P., Scarborough, R. J., & Gatignol, A. (2024). MicroRNAs and long noncoding RNAs during transcriptional regulation and latency of HIV and HTLV. *Retrovirology, 21*(1), 5. DOI: 10.1186/s12977-024-00637-y PMID: 38424561

Alqahtani, S., Xia, L., & Shannahan, J. H. (2022). Enhanced silver nanoparticle-induced pulmonary inflammation in a metabolic syndrome mouse model and resolvin D1 treatment. *Particle and Fibre Toxicology, 19*(1), 54. DOI: 10.1186/s12989-022-00495-6 PMID: 35933425

Aluya, J. (2015). Nanotechnology implications and global leadership perspectives. *Energy Sources. Part B, Economics, Planning, and Policy, 10*(1), 31–37. DOI: 10.1080/15567249.2010.506472

Amartumur, S., Nguyen, H., Huynh, T., Kim, T. S., Woo, R. S., Oh, E., Kim, K. K., Lee, L. P., & Heo, C. (2024). Neuropathogenesis-on-chips for neurodegenerative diseases. *Nature Communications*, *15*(1), 2219. DOI: 10.1038/s41467-024-46554-8 PMID: 38472255

Amiri, S., Alijanpour, S., Tirgar, F., Haj-Mirzaian, A., Amini-Khoei, H., Rahimi-Balaei, M., Rastegar, M., Ghaderi, M., Ghazi-Khansari, M., & Zarrindast, M.-R. (2016). NMDA receptors are involved in the antidepressant-like effects of capsaicin following amphetamine withdrawal in male mice. *Neuroscience*, *329*, 122–133. DOI: 10.1016/j.neuroscience.2016.05.003 PMID: 27167081

Amor, S., Peferoen, L. A., Vogel, D. Y., Breur, M., van der Valk, P., Baker, D., & van Noort, J. M. (2014). Inflammation in neurodegenerative diseases—An update. *Immunology*, *142*(2), 151–166. DOI: 10.1111/imm.12233 PMID: 24329535

Andreatta Maduro, P., Guimarães, M. P., de Sousa Rodrigues, M., Pereira Rolim Coimbra Pinto, A. P., da Mota, A. A.Junior, Lima Rocha, A. S., Matoso, J. M. D., Bavaresco Gambassi, B., & Schwingel, P. A. (2024). Comparing the Efficacy of Two Cognitive Screening Tools in Identifying Gray and White Matter Brain Damage among Older Adults. *Journal of Aging Research*, *5527225*, 1–14. Advance online publication. DOI: 10.1155/2024/5527225 PMID: 38690079

Andrus, B., Blizinsky, K., Vedell, P., Dennis, K., Shukla, P., Schaffer, D., Radulovic, J., Churchill, G. A., & Redei, E. E. (2012). Gene expression patterns in the hippocampus and amygdala of endogenous depression and chronic stress models. *Molecular Psychiatry*, *17*(1), 49–61. DOI: 10.1038/mp.2010.119 PMID: 21079605

Angelopoulou, E., Pyrgelis, E. S., & Piperi, C. (2022). Emerging Potential of the Phosphodiesterase (PDE) Inhibitor Ibudilast for Neurodegenerative Diseases: An Update on Preclinical and Clinical Evidence. *Molecules (Basel, Switzerland)*, *27*(23), 8448. Advance online publication. DOI: 10.3390/molecules27238448 PMID: 36500540

Anjomshoa, M., Boroujeni, S. N., Ghasemi, S., Lorigooini, Z., Amiri, A., Balali-Dehkordi, S., & Amini-khoei, H. (2020). Rutin via increase in the CA3 diameter of the hippocampus exerted antidepressant-like effect in mouse model of maternal separation stress: Possible involvement of NMDA receptors. *Behavioural Neurology*, *2020*(1), 4813616. DOI: 10.1155/2020/4813616 PMID: 32587637

Anjum, A., Yazid, M. D., Fauzi Daud, M., Idris, J., Ng, A. M. H., Selvi Naicker, A., Ismail, O. H. R., Athi Kumar, R. K., & Lokanathan, Y. (2020). Spinal cord injury: Pathophysiology, multimolecular interactions, and underlying recovery mechanisms. *International Journal of Molecular Sciences*, *21*(20), 7533. DOI: 10.3390/ijms21207533 PMID: 33066029

Archibald, J., Garzone, G., & Catenaccio, P. (2020). Debating evolutions in science, technology and society: Ethical and ideological perspectives. *Lingue e Linguaggi, 34*(speciale).

Ardianto, C., Budiatin, A. S., Sumartha, I. N. B., Nurrahmi, N., Rahmadi, M., & Khotib, J. (2021). Resveratrol ameliorates physical and psychological stress-induced depressive-like behavior. *Journal of Basic and Clinical Physiology and Pharmacology, 32*(4), 335–340. DOI: 10.1515/jbcpp-2020-0437 PMID: 34214312

Arning, L., & Nguyen, H. P. (2022). Huntington disease update: New insights into the role of repeat instability in disease pathogenesis. *Medizinische Genetik : Mitteilungsblatt des Berufsverbandes Medizinische Genetik e.V, 33*(4), 293–300. DOI: 10.1515/medgen-2021-2101 PMID: 38835439

Arya, P., Munshi, M., & Kumar, P. (2023). Diosgenin: Chemistry, extraction, quantification and health benefits. *Food Chemistry Advances, 2*, 100170. DOI: 10.1016/j.focha.2022.100170

Ashraf, S., Nabi, M., Rasool, S., Rashid, F., & Amin, S. (2019). Hyperandrogenism in polycystic ovarian syndrome and role of CYP gene variants: A review. *The Egyptian Journal of Medical Human Genetics, 20*(1), 25. DOI: 10.1186/s43042-019-0031-4

Assaye, M. A., & Gizaw, S. T. (2022). Chaperone-mediated autophagy and its implications for neurodegeneration and cancer. *International Journal of General Medicine, 15*, 5635–5649. DOI: 10.2147/IJGM.S368364 PMID: 35734200

Attia, G. M., Almouteri, M. M., & Alnakhli, F. T. (2023). Role of Metformin in Polycystic Ovary Syndrome (PCOS)-Related Infertility. *Cureus, 15*(8), e44493. DOI: 10.7759/cureus.44493 PMID: 37791160

Au, A., Feher, A., McPhee, L., Jessa, A., Oh, S., & Einstein, G. (2016). Estrogens, inflammation and cognition. *Frontiers in Neuroendocrinology, 40*, 87–100. DOI: 10.1016/j.yfrne.2016.01.002 PMID: 26774208

Auti, S. T., & Kulkarni, Y. A. (2019). Neuroprotective Effect of Cardamom Oil Against Aluminum Induced Neurotoxicity in Rats. *Frontiers in Neurology, 10*, 399. DOI: 10.3389/fneur.2019.00399 PMID: 31114535

Azzini, E., Peña-Corona, S. I., Hernández-Parra, H., Chandran, D., Saleena, L. A. K., Sawikr, Y., Peluso, I., Dhumal, S., Kumar, M., Leyva-Gómez, G., Martorell, M., Sharifi-Rad, J., & Calina, D. (2024). Neuroprotective and anti-inflammatory effects of curcumin in Alzheimer's disease: Targeting neuroinflammation strategies. *Phytotherapy Research, 38*(6), 3169–3189. DOI: 10.1002/ptr.8200 PMID: 38616356

Badawy, A. A. B. (2015). Tryptophan metabolism, disposition and utilization in pregnancy. *Bioscience Reports*, *35*(5), e00261. DOI: 10.1042/BSR20150197 PMID: 26381576

Bagherniya, M., Butler, A. E., Barreto, G. E., & Sahebkar, A. (2018). The effect of fasting or calorie restriction on autophagy induction: A review of the literature. *Ageing Research Reviews*, *47*, 183–197. DOI: 10.1016/j.arr.2018.08.004 PMID: 30172870

Bai, B., Wang, X., Li, Y., Chen, P.-C., Yu, K., Dey, K. K., Yarbro, J. M., Han, X., Lutz, B. M., Rao, S., Jiao, Y., Sifford, J. M., Han, J., Wang, M., Tan, H., Shaw, T. I., Cho, J.-H., Zhou, S., Wang, H., & Peng, J. (2020). Deep Multilayer Brain Proteomics Identifies Molecular Networks in Alzheimer's Disease Progression. *Neuron*, *106*(4), 700. DOI: 10.1016/j.neuron.2020.04.031 PMID: 32437656

Balestri, W., Sharma, R., da Silva, V. A., Bobotis, B. C., Curle, A. J., Kothakota, V., Kalantarnia, F., Hangad, M. V., Hoorfar, M., Jones, J. L., Tremblay, M.-È., El-Jawhari, J. J., Willerth, S. M., & Reinwald, Y. (2024). Modeling the neuroimmune system in Alzheimer's and Parkinson's diseases. *Journal of Neuroinflammation*, *21*(1), 32. DOI: 10.1186/s12974-024-03024-8 PMID: 38263227

Bansal, Y., & Kuhad, A. (2016). Mitochondrial dysfunction in depression. *Current Neuropharmacology*, *14*(6), 610–618. DOI: 10.2174/1570159X14666160229114755 PMID: 26923778

Barlampa, D., Bompoula, M. S., Bargiota, A., Kalantaridou, S., Mastorakos, G., & Valsamakis, G. (2021). Hypothalamic Inflammation as a Potential Pathophysiologic Basis for the Heterogeneity of Clinical, Hormonal, and Metabolic Presentation in PCOS. *Nutrients*, *13*(2), 520. DOI: 10.3390/nu13020520 PMID: 33562540

Barturen, G., Carnero-Montoro, E., Martínez-Bueno, M., Rojo-Rello, S., Sobrino, B., Porras-Perales, Ó., Alcántara-Domínguez, C., Bernardo, D., & Alarcón-Riquelme, M. E. (2022). Whole blood DNA methylation analysis reveals respiratory environmental traits involved in COVID-19 severity following SARS-CoV-2 infection. *Nature Communications*, *13*(1), 4597. DOI: 10.1038/s41467-022-32357-2 PMID: 35933486

Bayly-Jones, C., Lupton, C. J., D'Andrea, L., Chang, Y.-G., Jones, G. D., Steele, J. R., Venugopal, H., Schittenhelm, R. B., Halls, M. L., & Ellisdon, A. M. (2024). Structure of the human TSC:WIPI3 lysosomal recruitment complex. *Science Advances*, *10*(47), eadr5807. Advance online publication. DOI: 10.1126/sciadv.adr5807 PMID: 39565846

Belgrad, J., & Khvorova, A. (2024). More than 185 CAG repeats: A point of no return in Huntington's disease biology. *Brain*, *147*(5), 1601–1603. DOI: 10.1093/brain/awae105 PMID: 38588052

Belleau, E. L., Treadway, M. T., & Pizzagalli, D. A. (2019). The impact of stress and major depressive disorder on hippocampal and medial prefrontal cortex morphology. *Biological Psychiatry*, *85*(6), 443–453. DOI: 10.1016/j.biopsych.2018.09.031 PMID: 30470559

Bennani-Baiti, I. M., Machado, I., Llombart-Bosch, A., & Kovar, H. (2012). Lysine-specific demethylase 1 (LSD1/KDM1A/AOF2/BHC110) is expressed and is an epigenetic drug target in chondrosarcoma, Ewing's sarcoma, osteosarcoma, and rhabdomyosarcoma. *Human Pathology*, *43*(8), 1300–1307. DOI: 10.1016/j.humpath.2011.10.010 PMID: 22245111

Ben-Shlomo, Y., Darweesh, S., Llibre-Guerra, J., Marras, C., San Luciano, M., & Tanner, C. (2024). The epidemiology of Parkinson's disease. *Lancet*, *403*(10423), 283–292. DOI: 10.1016/S0140-6736(23)01419-8 PMID: 38245248

Berger, Z., Ravikumar, B., Menzies, F. M., Oroz, L. G., Underwood, B. R., Pangalos, M. N., Schmitt, I., Wullner, U., Evert, B. O., O'Kane, C. J., & Rubinsztein, D. C. (2006). Rapamycin alleviates toxicity of different aggregate-prone proteins. *Human Molecular Genetics*, *15*(3), 433–442. DOI: 10.1093/hmg/ddi458 PMID: 16368705

Berry, K. P., & Nedivi, E. (2017). Spine dynamics: Are they all the same? *Neuron*, *96*(1), 43–55. DOI: 10.1016/j.neuron.2017.08.008 PMID: 28957675

Besirli, C. G., Chinskey, N. D., Zheng, Q.-D., & Zacks, D. N. (2011). Autophagy activation in the injured photoreceptor inhibits fas-mediated apoptosis. *Investigative Ophthalmology & Visual Science*, *52*(7), 4193–4199. DOI: 10.1167/iovs.10-7090 PMID: 21421874

Bested, A. C., Logan, A. C., & Selhub, E. M. (2013). Intestinal microbiota, probiotics and mental health: from Metchnikoff to modern advances: Part II–contemporary contextual research. *Gut Pathogens*, *5*(1), 1–14. DOI: 10.1186/1757-4749-5-3

Bhattacharya, A., Torggler, R., Reiter, W., Romanov, N., Licheva, M., Ciftci, A., Mari, M., Kolb, L., Kaiser, D., Reggiori, F., Ammerer, G., Hollenstein, D. M., & Kraft, C. (2024). Decoding the function of Atg13 phosphorylation reveals a role of Atg11 in bulk autophagy initiation. *EMBO Reports*, *25*(2), 813–831. DOI: 10.1038/s44319-023-00055-9 PMID: 38233718

Bhatt, S., Nagappa, A. N., & Patil, C. R. (2020). Role of oxidative stress in depression. *Drug Discovery Today*, *25*(7), 1270–1276. DOI: 10.1016/j.drudis.2020.05.001 PMID: 32404275

Bi, Y. H., Po, J., & Uludag, K. (2025). Enhancing Offspring Cognitive Health: Addressing Prenatal Stress and Problems Through Nanotechnology. In Nanomaterials and the Nervous System (pp. 341-358). IGI Global.

Boga, J. A., Caballero, B., Potes, Y., Perez-Martinez, Z., Reiter, R. J., Vega-Naredo, I., & Coto-Montes, A. (2019). Therapeutic potential of melatonin related to its role as an autophagy regulator: A review. *Journal of Pineal Research*, *66*(1), e12534. DOI: 10.1111/jpi.12534 PMID: 30329173

Boldis, B. V., Grünberger, I., Cederström, A., Björk, J., Nilsson, A., & Helgertz, J. (2024). Comorbidities in women with polycystic ovary syndrome: A sibling study. *BMC Women's Health*, *24*(1), 221. DOI: 10.1186/s12905-024-03028-9 PMID: 38580996

Bolt, A. M., & Klimecki, W. T. (2012). Autophagy in toxicology: Self-consumption in times of stress and plenty. *Journal of Applied Toxicology*, *32*(7), 465–479. DOI: 10.1002/jat.1787 PMID: 22334383

Booth, L. A., Tavallai, S., Hamed, H. A., Cruickshanks, N., & Dent, P. (2014). The role of cell signalling in the crosstalk between autophagy and apoptosis. *Cellular Signalling*, *26*(3), 549–555. DOI: 10.1016/j.cellsig.2013.11.028 PMID: 24308968

Borghammer, P., Okkels, N., & Weintraub, D. (2024). Parkinson's Disease and Dementia with Lewy Bodies: One and the Same. *Journal of Parkinson's Disease*, *14*(3), 383–397. DOI: 10.3233/JPD-240002 PMID: 38640172

Bose, J. K., Huang, C. C., & Shen, C. K. J. (2011). Regulation of autophagy by neuropathological protein TDP-43. *The Journal of Biological Chemistry*, *286*(52), 44441–44448. DOI: 10.1074/jbc.M111.237115 PMID: 22052911

Boukhalfa, A., Nascimbeni, A. C., Ramel, D., Dupont, N., Hirsch, E., Gayral, S., Laffargue, M., Codogno, P., & Morel, E. (2020). PI3KC2α-dependent and VPS34-independent generation of PI3P controls primary cilium-mediated autophagy in response to shear stress. *Nature Communications*, *11*(1), 294. DOI: 10.1038/s41467-019-14086-1 PMID: 31941925

Bové, J., Martínez-Vicente, M., & Vila, M. (2011). Fighting neurodegeneration with rapamycin: Mechanistic insights. *Nature Reviews. Neuroscience*, *12*(8), 437–452. DOI: 10.1038/nrn3068 PMID: 21772323

Bradke, F., Fawcett, J. W., & Spira, M. E. (2012). Assembly of a new growth cone after axotomy: The precursor to axon regeneration. *Nature Reviews. Neuroscience*, *13*(3), 183–193. DOI: 10.1038/nrn3176 PMID: 22334213

Brogi, S., Tabanelli, R., Puca, S., & Calderone, V. (2024). Intermittent Fasting: Myths, Fakes and Truth on This Dietary Regimen Approach. *Foods, 13*(13), 1960. DOI: 10.3390/foods13131960 PMID: 38998465

Brosse, A. L., Sheets, E. S., Lett, H. S., & Blumenthal, J. A. (2002). Exercise and the treatment of clinical depression in adults: Recent findings and future directions. *Sports Medicine (Auckland, N.Z.), 32*(12), 741–760. DOI: 10.2165/00007256-200232120-00001 PMID: 12238939

Bruno, A., Dovizio, M., Milillo, C., Aruffo, E., Pesce, M., Gatta, M., Chiacchiaretta, P., Di Carlo, P., & Ballerini, P. (2024). Orally Ingested Micro- and Nano-Plastics: A Hidden Driver of Inflammatory Bowel Disease and Colorectal Cancer. *Cancers (Basel), 16*(17), 3079. DOI: 10.3390/cancers16173079 PMID: 39272937

Bulsara, J., Patel, P., Soni, A., & Acharya, S. (2021). A review: Brief insight into Polycystic Ovarian syndrome. *Endocrine and Metabolic Science, 3*, 100085. DOI: 10.1016/j.endmts.2021.100085

Bunting, E. L., Hamilton, J., & Tabrizi, S. J. (2022). Polyglutamine diseases. *Current Opinion in Neurobiology, 72*, 39–47. DOI: 10.1016/j.conb.2021.07.001 PMID: 34488036

Burgos, R. A., Alarcón, P., Quiroga, J., Manosalva, C., & Hancke, J. (2020). Andrographolide, an anti-inflammatory multitarget drug: All roads lead to cellular metabolism. *Molecules (Basel, Switzerland), 26*(1), 5. DOI: 10.3390/molecules26010005 PMID: 33374961

Burman, C., & Ktistakis, N. T. (2010). Regulation of autophagy by phosphatidylinositol 3-phosphate. *FEBS Letters, 584*(7), 1302–1312. DOI: 10.1016/j.febslet.2010.01.011 PMID: 20074568

Bustos, V., Pulina, M. V., Bispo, A., Lam, A., Flajolet, M., Gorelick, F. S., & Greengard, P. (2017). Phosphorylated Presenilin 1 decreases β-amyloid by facilitating autophagosome-lysosome fusion. *Proceedings of the National Academy of Sciences of the United States of America, 114*(27), 7148–7153. DOI: 10.1073/pnas.1705240114 PMID: 28533369

Bustos, V., Pulina, M. V., Kelahmetoglu, Y., Sinha, S. C., Gorelick, F. S., Flajolet, M., & Greengard, P. (2017). Bidirectional regulation of Aβ levels by Presenilin 1. *Proceedings of the National Academy of Sciences of the United States of America, 114*(27), 7142–7147. DOI: 10.1073/pnas.1705235114 PMID: 28533411

Caballero, B., Bourdenx, M., Luengo, E., Diaz, A., Sohn, P. D., Chen, X., Wang, C., Juste, Y. R., Wegmann, S., Patel, B., Young, Z. T., Kuo, S. Y., Rodriguez-Navarro, J. A., Shao, H., Lopez, M. G., Karch, C. M., Goate, A. M., Gestwicki, J. E., Hyman, B. T., & Cuervo, A. M. (2021). Acetylated tau inhibits chaperone-mediated autophagy and promotes tau pathology propagation in mice. *Nature Communications*, *12*(1), 2238. DOI: 10.1038/s41467-021-22501-9 PMID: 33854069

Cagnacci, A., Gazzo, I., Stigliani, S., Paoletti, A. M., Anserini, P., Londero, A. P., & Xholli, A. (2023). Oxidative Stress: The Role of Estrogen and Progesterone. *Journal of Clinical Medicine*, *12*(23), 7304. DOI: 10.3390/jcm12237304 PMID: 38068356

Cai, Q., & Ganesan, D. (2022). Regulation of neuronal autophagy and the implications in neurodegenerative diseases. *Neurobiology of Disease*, *162*(105582), 1–20. DOI: 10.1016/j.nbd.2021.105582 PMID: 34890791

Campagnoli, L. I. M., Marchesi, N., Varesi, A., Morozzi, M., Mascione, L., Ricevuti, G., Esposito, C., Galeotti, N., & Pascale, A. (2024). New therapeutic avenues in multiple sclerosis: Is there a place for gut microbiota-based treatments? *Pharmacological Research*, *209*, 107456. DOI: 10.1016/j.phrs.2024.107456 PMID: 39389400

Cantoni, C., Smirnov, R. A., Firulyova, M., Andhey, P. S., Bradstreet, T. R., Esaulova, E., Terekhova, M., Schwarzkopf, E. A., Abdalla, N. M., Kleverov, M., Sabatino, J. J.Jr, Liu, K., Schwab, N., Meyer Zu Hörste, G., Cross, A. H., Artyomov, M. N., Edelson, B. T., & Wu, G. F. (2025). A single-cell compendium of human cerebrospinal fluid identifies disease-associated immune cell populations. *The Journal of Clinical Investigation*, *135*(1), e177793. DOI: 10.1172/JCI177793 PMID: 39744938

Cao, W., Li, J., Yang, K., & Cao, D. (2021). An overview of autophagy: Mechanism, regulation and research progress. *Bulletin du Cancer*, *108*(3), 304–322. DOI: 10.1016/j.bulcan.2020.11.004 PMID: 33423775

Cao, Z., Wang, Y., Long, Z., & He, G. (2019). Interaction between autophagy and the NLRP3 inflammasome. *Acta Biochimica et Biophysica Sinica*, *51*(11), 1087–1095. DOI: 10.1093/abbs/gmz098 PMID: 31609412

Capon, A., Gillespie, J., Rolfe, M., & Smith, W. (2015). Perceptions of risk from nanotechnologies and trust in stakeholders: A cross sectional study of public, academic, government and business attitudes. *BMC Public Health*, *15*(1), 1–13. DOI: 10.1186/s12889-015-1795-1 PMID: 25928741

Caporali, S., De Stefano, A., Calabrese, C., Giovannelli, A., Pieri, M., Savini, I., Tesauro, M., Bernardini, S., Minieri, M., & Terrinoni, A. (2022). Anti-inflammatory and active biological properties of the plant-derived bioactive compounds luteolin and luteolin 7-glucoside. *Nutrients*, *14*(6), 1155. DOI: 10.3390/nu14061155 PMID: 35334812

Carosi, J. M., Denton, D., Kumar, S., & Sargeant, T. J. (2021). Retromer dysfunction at the nexus of tauopathies. *Cell Death and Differentiation*, *28*(3), 884–899. DOI: 10.1038/s41418-020-00727-2 PMID: 33473181

Carrillo-Martinez, E. J., Flores-Hernández, F. Y., Salazar-Montes, A. M., Nario-Chaidez, H. F., & Hernández-Ortega, L. D. (2024). Quercetin, a flavonoid with great pharmacological capacity. *Molecules (Basel, Switzerland)*, *29*(5), 1000. DOI: 10.3390/molecules29051000 PMID: 38474512

Casiraghi, V., Sorce, M. N., Santangelo, S., Invernizzi, S., Bossolasco, P., Lattuada, C., Battaglia, C., Venturin, M., Silani, V., Colombrita, C., & Ratti, A. (2024). Modeling of TDP-43 proteinopathy by chronic oxidative stress identifies rapamycin as beneficial in ALS patient-derived 2D and 3D iPSC models. *Experimental Neurology*, *383*(115057), 1–17. DOI: 10.1016/j.expneurol.2024.115057 PMID: 39536963

Cassano, P., Bui, E., Rogers, A. H., Walton, Z. E., Ross, R., Zeng, M., Nadal-Vicens, M., Mischoulon, D., Baker, A. W., Keshaviah, A., Worthington, J., Hoge, E. A., Alpert, J., Fava, M., Wong, K. K., & Simon, N. M. (2017). Inflammatory cytokines in major depressive disorder: A case–control study. *The Australian and New Zealand Journal of Psychiatry*, *51*(1), 23–31. DOI: 10.1177/0004867416652736 PMID: 27313138

Castedo, M., Perfettini, J. L., Roumier, T., & Kroemer, G. (2002). Cyclin-dependent kinase-1: Linking apoptosis to cell cycle and mitotic catastrophe. *Cell Death and Differentiation*, *9*(12), 1287–1293. DOI: 10.1038/sj.cdd.4401130 PMID: 12478465

Chandrasekaran, P., Weiskirchen, S., & Weiskirchen, R. (2024). Effects of Probiotics on Gut Microbiota: An Overview. *International Journal of Molecular Sciences*, *25*(11), 6022. DOI: 10.3390/ijms25116022 PMID: 38892208

Chatterjee, S., Sarkar, S., & Bhattacharya, S. (2014). Toxic metals and autophagy. *Chemical Research in Toxicology*, *27*(11), 1887–1900. DOI: 10.1021/tx500264s PMID: 25310621

Chaudhary, R., Liu, B., Bensalem, J., Sargeant, T. J., Page, A. J., Wittert, G. A., Hutchison, A. T., & Heilbronn, L. K. (2022). Intermittent fasting activates markers of autophagy in mouse liver, but not muscle from mouse or humans. *Nutrition (Burbank, Los Angeles County, Calif.)*, *101*, 111662. DOI: 10.1016/j.nut.2022.111662 PMID: 35660501

Chaudhuri, A. (2023). Polycystic ovary syndrome: Causes, symptoms, pathophysiology, and remedies. *Obesity Medicine*, *39*, 100480. DOI: 10.1016/j.obmed.2023.100480

Chen, A., Xiong, L.-J., Tong, Y., & Mao, M. (2013). Neuroprotective effect of brain-derived neurotrophic factor mediated by autophagy through the PI3K/Akt/mTOR pathway. *Molecular Medicine Reports*, *8*(4), 1011–1016. DOI: 10.3892/mmr.2013.1628 PMID: 23942837

Chen, H.-C., Fong, T.-H., Lee, A.-W., & Chiu, W.-T. (2012). Autophagy is activated in injured neurons and inhibited by methylprednisolone after experimental spinal cord injury. *Spine*, *37*(6), 470–475. DOI: 10.1097/BRS.0b013e318221e859 PMID: 21587101

Chen, J., He, H.-J., Ye, Q., Feng, F., Wang, W.-W., Gu, Y., Han, R., & Xie, C. (2021). Defective Autophagy and Mitophagy in Alzheimer's Disease: Mechanisms and Translational Implications. *Molecular Neurobiology*, *58*(10), 5289–5302. DOI: 10.1007/s12035-021-02487-7 PMID: 34279771

Chen, J., Ma, Z., Jiao, X., Fariss, R., Kantorow, W. L., Kantorow, M., Pras, E., Frydman, M., Pras, E., Riazuddin, S., Riazuddin, S. A., & Hejtmancik, J. F. (2011). Mutations in FYCO1 cause autosomal-recessive congenital cataracts. *American Journal of Human Genetics*, *88*(6), 827–838. DOI: 10.1016/j.ajhg.2011.05.008 PMID: 21636066

Chen, L., Gao, T., Zhou, P., Xia, W., Yao, H., Xu, S., & Xu, J. (2024). Recent advances of vacuolar protein-sorting 34 inhibitors targeting autophagy. *Bioorganic Chemistry*, *143*, 107039. DOI: 10.1016/j.bioorg.2023.107039 PMID: 38134519

Chen, M., Peng, L., Gong, P., Zheng, X., Sun, T., Zhang, X., & Huo, J. (2022). Baicalein Induces Mitochondrial Autophagy to Prevent Parkinson's Disease in Rats via miR-30b and the SIRT1/AMPK/mTOR Pathway. *Frontiers in Neurology*, *12*(646817), 1–11. DOI: 10.3389/fneur.2021.646817 PMID: 35237220

Chen, P., Huo, X., Liu, W., Li, K., Sun, Z., & Tian, J. (2020). Apigenin exhibits anti-inflammatory effects in LPS-stimulated BV2 microglia through activating GSK3β/Nrf2 signaling pathway. *Immunopharmacology and Immunotoxicology*, *42*(1), 9–16. DOI: 10.1080/08923973.2019.1688345 PMID: 31760890

Chen, S., Zhou, Q., Ni, Y., & Le, W. (2020). Autophagy and Alzheimer's Disease. *Advances in Experimental Medicine and Biology*, *1207*, 3–19. DOI: 10.1007/978-981-15-4272-5_1 PMID: 32671736

Chen, X.-B., Wang, Z.-L., Yang, Q.-Y., Zhao, F.-Y., Qin, X.-L., Tang, X.-E., Du, J.-L., Chen, Z.-H., Zhang, K., & Huang, F.-J. (2018). Diosgenin glucoside protects against spinal cord injury by regulating autophagy and alleviating apoptosis. *International Journal of Molecular Sciences*, *19*(8), 2274. DOI: 10.3390/ijms19082274 PMID: 30072674

Chen, X., Wang, Z., Zheng, P., Dongol, A., Xie, Y., Ge, X., Zheng, M., Dang, X., Seyhan, Z. B., Nagaratnam, N., Yu, Y., & Huang, X. F. (2023). Impaired mitophagosome–lysosome fusion mediates olanzapine-induced aging. *Aging Cell*, *22*(11), 1–19. DOI: 10.1111/acel.14003 PMID: 37828862

Chen, Y., Sawada, O., Kohno, H., Le, Y.-Z., Subauste, C., Maeda, T., & Maeda, A. (2013). Autophagy protects the retina from light-induced degeneration. *The Journal of Biological Chemistry*, *288*(11), 7506–7518. DOI: 10.1074/jbc.M112.439935 PMID: 23341467

Chen, Y., Wang, H., Ying, Z., & Gao, Q. (2020). Ibudilast enhances the clearance of SOD1 and TDP-43 aggregates through TFEB-mediated autophagy and lysosomal biogenesis: The new molecular mechanism of ibudilast and its implication for neuroprotective therapy. *Biochemical and Biophysical Research Communications*, *526*(1), 231–238. DOI: 10.1016/j.bbrc.2020.03.051 PMID: 32204915

Chen, Z., Fu, Q., Shen, B., Huang, X., Wang, K., He, P., Li, F., Zhang, F., & Shen, H. (2014). Enhanced p62 expression triggers concomitant autophagy and apoptosis in a rat chronic spinal cord compression model. *Molecular Medicine Reports*, *9*(6), 2091–2096. DOI: 10.3892/mmr.2014.2124 PMID: 24715058

Chinskey, N. D., Zheng, Q.-D., & Zacks, D. N. (2014). Control of photoreceptor autophagy after retinal detachment: The switch from survival to death. *Investigative Ophthalmology & Visual Science*, *55*(2), 688–695. DOI: 10.1167/iovs.13-12951 PMID: 24408986

Choi, S.-I., Kim, B.-Y., Dadakhujaev, S., Oh, J.-Y., Kim, T.-I., Kim, J. Y., & Kim, E. K. (2012). Impaired autophagy and delayed autophagic clearance of transforming growth factor β-induced protein (TGFBI) in granular corneal dystrophy type 2. *Autophagy*, *8*(12), 1782–1797. DOI: 10.4161/auto.22067 PMID: 22995918

Chopra, G., Shabir, S., Yousuf, S., Kauts, S., Bhat, S. A., Mir, A. H., & Singh, M. P. (2022). Proteinopathies: Deciphering Physiology and Mechanisms to Develop Effective Therapies for Neurodegenerative Diseases. *Molecular Neurobiology*, *59*(12), 7513–7540. DOI: 10.1007/s12035-022-03042-8 PMID: 36205914

Chouksey, A., & Pandey, S. (2020). Phenotypic variability in huntington's disease. *Annals of Indian Academy of Neurology*, *23*(2), 153. DOI: 10.4103/aian.AIAN_543_19 PMID: 32189853

Christ, J. P., & Cedars, M. I. (2023). Current Guidelines for Diagnosing PCOS. *Diagnostics (Basel)*, *13*(6), 1113. Advance online publication. DOI: 10.3390/diagnostics13061113 PMID: 36980421

Christopher, M. A., Myrick, D. A., Barwick, B. G., Engstrom, A. K., Porter-Stransky, K. A., Boss, J. M., Weinshenker, D., Levey, A. I., & Katz, D. J. (2017). LSD1 protects against hippocampal and cortical neurodegeneration. *Nature Communications*, *8*(1), 805. DOI: 10.1038/s41467-017-00922-9 PMID: 28993646

Chua, J. P., De Calbiac, H., Kabashi, E., & Barmada, S. J. (2022). Autophagy and ALS: Mechanistic insights and therapeutic implications. *Autophagy*, *18*(2), 254–282. DOI: 10.1080/15548627.2021.1926656 PMID: 34057020

Church, J. S., Renzelman, M. L., & Schwartzer, J. J. (2022). Ten-week high fat and high sugar diets in mice alter gut-brain axis cytokines in a sex-dependent manner. *The Journal of Nutritional Biochemistry*, *100*, 108903. DOI: 10.1016/j.jnutbio.2021.108903 PMID: 34748922

Cicek, I. E., Cicek, E., Kayhan, F., Uguz, F., Erayman, I., Kurban, S., Yerlikaya, F. H., & Kaya, N. (2014). The roles of BDNF, S100B, and oxidative stress in interferon-induced depression and the effect of antidepressant treatment in patients with chronic viral hepatitis: A prospective study. *Journal of Psychosomatic Research*, *76*(3), 227–232. DOI: 10.1016/j.jpsychores.2014.01.003 PMID: 24529042

Cichalewska-Studzinska, M., Szymanski, J., Stec-Martyna, E., Perdas, E., Studzinska, M., Jerczynska, H., Kulczycka-Wojdala, D., Stawski, R., & Mycko, M. P. (2024). The Role of miR-155 in Modulating Gene Expression in CD4+ T Cells: Insights into Alternative Immune Pathways in Autoimmune Encephalomyelitis. *International Journal of Molecular Sciences*, *25*(21), 11355. DOI: 10.3390/ijms252111355 PMID: 39518908

Clarke, P. G. (1990). Developmental cell death: Morphological diversity and multiple mechanisms. *Anatomy and Embryology*, *181*(3), 195–213. DOI: 10.1007/BF00174615 PMID: 2186664

Cleary, C., Linde, J., Hiscock, K., Hadas, I., Belmaker, R., Agam, G., Flaisher-Grinberg, S., & Einat, H. (2008). Antidepressive-like effects of rapamycin in animal models: Implications for mTOR inhibition as a new target for treatment of affective disorders. *Brain Research Bulletin, 76*(5), 469–473. DOI: 10.1016/j.brainresbull.2008.03.005 PMID: 18534253

Cloonan, S. M., & Williams, D. C. (2011). The antidepressants maprotiline and fluoxetine induce Type II autophagic cell death in drug-resistant Burkitt's lymphoma. *International Journal of Cancer, 128*(7), 1712–1723. DOI: 10.1002/ijc.25477 PMID: 20503272

Cohen, J. S., Phillips, J., Das, S. R., Olm, C. A., Radhakrishnan, H., Rhodes, E., Cousins, K. A. Q., Xie, S. X., Nasrallah, I. M., Yushkevich, P. A., Wolk, D. A., Lee, E. B., Weintraub, D., Irwin, D. J., & McMillan, C. T. (2025). Posterior hippocampal sparing in Lewy body disorders with Alzheimer's copathology: An in vivo MRI study. *NeuroImage. Clinical, 45*, 103714. DOI: 10.1016/j.nicl.2024.103714 PMID: 39675237

Colacurcio, D. J., & Nixon, R. A. (2016). Disorders of lysosomal acidification—The emerging role of v-ATPase in aging and neurodegenerative disease. *Ageing Research Reviews, 32*, 75–88. DOI: 10.1016/j.arr.2016.05.004 PMID: 27197071

Colman, R. J., Beasley, T. M., Kemnitz, J. W., Johnson, S. C., Weindruch, R., & Anderson, R. M. (2014). Caloric restriction reduces age-related and all-cause mortality in rhesus monkeys. *Nature Communications, 5*(1), 3557. DOI: 10.1038/ncomms4557 PMID: 24691430

Colucci-D'Amato, L., Speranza, L., & Volpicelli, F. (2020). Neurotrophic Factor BDNF, Physiological Functions and Therapeutic Potential in Depression, Neurodegeneration and Brain Cancer. *International Journal of Molecular Sciences, 21*(20), 7777. Advance online publication. DOI: 10.3390/ijms21207777 PMID: 33096634

Cooney, L. G., & Dokras, A. (2017). Depression and Anxiety in Polycystic Ovary Syndrome: Etiology and Treatment. *Current Psychiatry Reports, 19*(11), 83. DOI: 10.1007/s11920-017-0834-2 PMID: 28929349

Cooney, L. G., Lee, I., Sammel, M. D., & Dokras, A. (2017). High prevalence of moderate and severe depressive and anxiety symptoms in polycystic ovary syndrome: A systematic review and meta-analysis. *Human Reproduction (Oxford, England), 32*(5), 1075–1091. DOI: 10.1093/humrep/dex044 PMID: 28333286

Correia, A. S., & Vale, N. (2022). Tryptophan Metabolism in Depression: A Narrative Review with a Focus on Serotonin and Kynurenine Pathways. *International Journal of Molecular Sciences*, *23*(15), 8493. Advance online publication. DOI: 10.3390/ijms23158493 PMID: 35955633

Corti, O., Blomgren, K., Poletti, A., & Beart, P. M. (2020). Autophagy in neurodegeneration: New insights underpinning therapy for neurological diseases. *Journal of Neurochemistry*, *154*(4), 354–371. DOI: 10.1111/jnc.15002 PMID: 32149395

Cowan, S., Lim, S., Alycia, C., Pirotta, S., Thomson, R., Gibson-Helm, M., Blackmore, R., Naderpoor, N., Bennett, C., Ee, C., Rao, V., Mousa, A., Alesi, S., & Moran, L. (2023). Lifestyle management in polycystic ovary syndrome - beyond diet and physical activity. *BMC Endocrine Disorders*, *23*(1), 14. DOI: 10.1186/s12902-022-01208-y PMID: 36647089

Cozzi, M., & Ferrari, V. (2022). Autophagy Dysfunction in ALS: From Transport to Protein Degradation. *Journal of Molecular Neuroscience*, *72*(7), 1456–1481. DOI: 10.1007/s12031-022-02029-3 PMID: 35708843

Crocetta, A., Liloia, D., Costa, T., Duca, S., Cauda, F., & Manuello, J. (2024). From gut to brain: Unveiling probiotic effects through a neuroimaging perspective-A systematic review of randomized controlled trials. *Frontiers in Nutrition*, *11*, 1446854. DOI: 10.3389/fnut.2024.1446854 PMID: 39360283

Crowley, T., Cryan, J. F., Downer, E. J., & O'Leary, O. F. (2016). Inhibiting neuroinflammation: The role and therapeutic potential of GABA in neuro-immune interactions. *Brain, Behavior, and Immunity*, *54*, 260–277. DOI: 10.1016/j.bbi.2016.02.001 PMID: 26851553

Cruz-Pereira, J. S., Rea, K., Nolan, Y. M., O'Leary, O. F., Dinan, T. G., & Cryan, J. F. (2020). Depression's unholy trinity: Dysregulated stress, immunity, and the microbiome. *Annual Review of Psychology*, *71*(1), 49–78. DOI: 10.1146/annurev-psych-122216-011613 PMID: 31567042

Cui, X., Lin, Q., & Liang, Y. (2020). Plant-derived antioxidants protect the nervous system from aging by inhibiting oxidative stress. *Frontiers in Aging Neuroscience*, *12*, 209. DOI: 10.3389/fnagi.2020.00209 PMID: 32760268

Cui, X., Li, X., Zheng, H., Su, Y., Zhang, S., Li, M., Hao, X., Zhang, S., Hu, Z., Xia, Z., Shi, C., Xu, Y., & Mao, C. (2024). Human midbrain organoids: A powerful tool for advanced Parkinson's disease modeling and therapy exploration. *NPJ Parkinson's Disease*, *10*(1), 189. DOI: 10.1038/s41531-024-00799-8 PMID: 39428415

Cunha, N. S. C., Malvea, A., Sadat, S., Ibrahim, G. M., & Fehlings, M. G. (2023). Pediatric Spinal Cord Injury: A Review. *Children (Basel, Switzerland), 10*(9), 1456. DOI: 10.3390/children10091456 PMID: 37761417

Dabravolski, S. A., Nikiforov, N. G., Eid, A. H., Nedosugova, L. V., Starodubova, A. V., Popkova, T. V., Bezsonov, E. E., & Orekhov, A. N. (2021). Mitochondrial Dysfunction and Chronic Inflammation in Polycystic Ovary Syndrome. *International Journal of Molecular Sciences, 22*(8), 3923. Advance online publication. DOI: 10.3390/ijms22083923 PMID: 33920227

Dai, W., Feng, K., Sun, X., Xu, L., Wu, S., Rahmand, K., Jia, D., & Han, T. (2022). Natural products for the treatment of stress-induced depression: Pharmacology, mechanism and traditional use. *Journal of Ethnopharmacology, 285*, 114692. DOI: 10.1016/j.jep.2021.114692 PMID: 34742864

Dai, W., Qiao, X., Fang, Y., Guo, R., Bai, P., Liu, S., Li, T., Jiang, Y., Wei, S., Na, Z., Xiao, X., & Li, D. (2024). Epigenetics-targeted drugs: Current paradigms and future challenges. *Signal Transduction and Targeted Therapy, 9*(332), 332. Advance online publication. DOI: 10.1038/s41392-024-02039-0 PMID: 39592582

Daley, E., Wilkie, D., Loesch, A., Hargreaves, I., Kendall, D., Pilkington, G., & Bates, T. E. (2005). Chlorimipramine: A novel anticancer agent with a mitochondrial target. *Biochemical and Biophysical Research Communications, 328*(2), 623–632. DOI: 10.1016/j.bbrc.2005.01.028 PMID: 15694394

Dason, E. S., Koshkina, O., Chan, C., & Sobel, M. (2024). Diagnosis and management of polycystic ovarian syndrome. *Canadian Medical Association Journal, 196*(3), E85–e94. DOI: 10.1503/cmaj.231251 PMID: 38286488

Datta, S., Rahman, M. A., Koka, S., & Boini, K. M. (2024). High Mobility Group Box 1 (HMGB1): Molecular Signaling and Potential Therapeutic Strategies. *Cells, 13*(23), 1946. DOI: 10.3390/cells13231946 PMID: 39682695

David, M. D., Cantí, C., & Herreros, J. (2010). Wnt-3a and Wnt-3 differently stimulate proliferation and neurogenesis of spinal neural precursors and promote neurite outgrowth by canonical signaling. *Journal of Neuroscience Research, 88*(14), 3011–3023. DOI: 10.1002/jnr.22464 PMID: 20722074

Davoody, S., Asgari Taei, A., Khodabakhsh, P., & Dargahi, L. (2024). mTOR signaling and Alzheimer's disease: What we know and where we are? *CNS Neuroscience & Therapeutics, 30*(4), 1–17. DOI: 10.1111/cns.14463 PMID: 37721413

De Duve, C., & Wattiaux, R. (1966). Functions of lysosomes. *Annual Review of Physiology, 28*(1), 435–492. DOI: 10.1146/annurev.ph.28.030166.002251 PMID: 5322983

de Oliveira Souza, R., Duarte Júnior, J. W. B., Della Casa, V. S., Santoro Rosa, D., Renia, L., & Claser, C. (2024). Unraveling the complex interplay: Immunopathology and immune evasion strategies of alphaviruses with emphasis on neurological implications. *Frontiers in Cellular and Infection Microbiology, 14*, 1421571. DOI: 10.3389/fcimb.2024.1421571 PMID: 39211797

De Paoli, M., Zakharia, A., & Werstuck, G. H. (2021). The Role of Estrogen in Insulin Resistance: A Review of Clinical and Preclinical Data. *American Journal of Pathology, 191*(9), 1490–1498. DOI: 10.1016/j.ajpath.2021.05.011 PMID: 34102108

Decuypere, J.-P., Bultynck, G., & Parys, J. B. (2011). A dual role for Ca2+ in autophagy regulation. *Cell Calcium, 50*(3), 242–250. DOI: 10.1016/j.ceca.2011.04.001 PMID: 21571367

Dedoni, S., Scherma, M., Camoglio, C., Siddi, C., Dazzi, L., Puliga, R., Frau, J., Cocco, E., & Fadda, P. (2023). Overall view of the most common experimental models for multiple sclerosis. *Neurobiology of Disease, 184*, 106230. DOI: 10.1016/j.nbd.2023.106230 PMID: 37453561

Del Bello, B., Gamberucci, A., Marcolongo, P., & Maellaro, E. (2022). The autophagy inducer trehalose stimulates macropinocytosis in NF1-deficient glioblastoma cells. *Cancer Cell International, 22*(1), 1–17. DOI: 10.1186/s12935-022-02652-5 PMID: 35864494

Delgadillo, D. R., Borelli, J. L., Mayer, E. A., Labus, J. S., Cross, M. P., & Pressman, S. D. (2025). Biological, environmental, and psychological stress and the human gut microbiome in healthy adults. *Scientific Reports, 15*(1), 362. DOI: 10.1038/s41598-024-77473-9 PMID: 39747287

Demeter, A., Romero-Mulero, M. C., Csabai, L., Ölbei, M., Sudhakar, P., Haerty, W., & Korcsmáros, T. (2020). ULK1 and ULK2 are less redundant than previously thought: Computational analysis uncovers distinct regulation and functions of these autophagy induction proteins. *Scientific Reports, 10*(1), 10940. DOI: 10.1038/s41598-020-67780-2 PMID: 32616830

Deswal, R., Narwal, V., Dang, A., & Pundir, C. S. (2020). The Prevalence of Polycystic Ovary Syndrome: A Brief Systematic Review. *Journal of Human Reproductive Sciences, 13*(4). PMID: 33627974

Dhapola, R., Beura, S. K., Sharma, P., Singh, S. K., & HariKrishnaReddy, D. (2024). Oxidative stress in Alzheimer's disease: Current knowledge of signaling pathways and therapeutics. *Molecular Biology Reports*, *51*(1), 1–18. DOI: 10.1007/s11033-023-09021-z PMID: 38165499

Dhillon, S. K., Simoens, C., Cuypers, L., Bode, J., Bonde, J., Corbisier, P., Cocuzza, C. E., Van Ranst, M., & Arbyn, M. (2023). Assessment of the clinical and analytical performance of the Aptima SARS-CoV-2 assay using the VALCOR protocol. *Virology Journal*, *20*(1), 35. DOI: 10.1186/s12985-023-01986-4 PMID: 36829164

Dhingra, A., Bell, B. A., Peachey, N. S., Daniele, L. L., Reyes-Reveles, J., Sharp, R. C., Jun, B., Bazan, N. G., Sparrow, J. R., Kim, H. J., Philp, N. J., & Boesze-Battaglia, K. (2018). Microtubule-associated protein 1 light chain 3B,(LC3B) is necessary to maintain lipid-mediated homeostasis in the retinal pigment epithelium. *Frontiers in Cellular Neuroscience*, *12*, 351. DOI: 10.3389/fncel.2018.00351 PMID: 30349463

Di Petrillo, A., Orrù, G., Fais, A., & Fantini, M. C. (2022). Quercetin and its derivates as antiviral potentials: A comprehensive review. *Phytotherapy Research*, *36*(1), 266–278. DOI: 10.1002/ptr.7309 PMID: 34709675

Diab, R., Dimachkie, L., Zein, O., Dakroub, A., & Eid, A. H. (2024). Intermittent fasting regulates metabolic homeostasis and improves cardiovascular health. *Cell Biochemistry and Biophysics*, *82*(3), 1583–1597. DOI: 10.1007/s12013-024-01314-9 PMID: 38847940

Diamanti-Kandarakis, E. (2010). PCOS in adolescents. *Best Practice & Research. Clinical Obstetrics & Gynaecology*, *24*(2), 173–183. DOI: 10.1016/j.bpobgyn.2009.09.005 PMID: 19932060

Diamanti-Kandarakis, E., & Dunaif, A. (2012). Insulin resistance and the polycystic ovary syndrome revisited: An update on mechanisms and implications. *Endocrine Reviews*, *33*(6), 981–1030. DOI: 10.1210/er.2011-1034 PMID: 23065822

Dias-Teixeira, K. L., Gh, M. S., Romano, J., Norouzi, F., & Laurie, G. W. (2022). Autophagy in the normal and diseased cornea. *Experimental Eye Research*, *225*, 109274. DOI: 10.1016/j.exer.2022.109274 PMID: 36252655

Ding, H., Chen, J., Su, M., Lin, Z., Zhan, H., Yang, F., Li, W., Xie, J., Huang, Y., Liu, X., Liu, B., & Zhou, X. (2020). BDNF promotes activation of astrocytes and microglia contributing to neuroinflammation and mechanical allodynia in cyclophosphamide-induced cystitis. *Journal of Neuroinflammation*, *17*(1), 19. DOI: 10.1186/s12974-020-1704-0 PMID: 31931832

Diskin, T., Tal-Or, P., Erlich, S., Mizrachy, L., Alexandrovich, A., Shohami, E., & Pinkas-Kramarski, R. (2005). Closed head injury induces upregulation of Beclin 1 at the cortical site of injury. *Journal of Neurotrauma*, *22*(7), 750–762. DOI: 10.1089/neu.2005.22.750 PMID: 16004578

Dixon, A., Shim, M. S., Nettesheim, A., Coyne, A., Su, C.-C., Gong, H., & Liton, P. B. (2023). Autophagy deficiency protects against ocular hypertension and neurodegeneration in experimental and spontaneous glaucoma mouse models. *Cell Death & Disease*, *14*(8), 554. DOI: 10.1038/s41419-023-06086-3 PMID: 37620383

Djajadikerta, A., Keshri, S., Pavel, M., Prestil, R., Ryan, L., & Rubinsztein, D. C. (2020). Autophagy Induction as a Therapeutic Strategy for Neurodegenerative Diseases. *Journal of Molecular Biology*, *432*(8), 2799–2821. DOI: 10.1016/j.jmb.2019.12.035 PMID: 31887286

Domingues, C., Santos, A., Alvarez-Lorenzo, C., Concheiro, A., Jarak, I., Veiga, F., Barbosa, I., Dourado, M., & Figueiras, A. (2022). Where is nano today and where is it headed? A review of nanomedicine and the dilemma of nanotoxicology. *ACS Nano*, *16*(7), 9994–10041. DOI: 10.1021/acsnano.2c00128 PMID: 35729778

Dong, W., Peng, Q., Liu, Z., Xie, Z., Guo, X., Li, Y., & Chen, C. (2023). Estrogen plays an important role by influencing the NLRP3 inflammasome. *Biomedicine and Pharmacotherapy*, *167*, 115554. DOI: 10.1016/j.biopha.2023.115554 PMID: 37738797

Dong, Y., He, Y., Geng, Y., Wei, M., Zhou, X., Lian, J., & Hallajzadeh, J. (2024). Autophagy-related lncRNAs and exosomal lncRNAs in colorectal cancer: Focusing on lncRNA-targeted strategies. *Cancer Cell International*, *24*(1), 328. DOI: 10.1186/s12935-024-03503-1 PMID: 39342235

Dudek, K. A., Dion-Albert, L., Kaufmann, F. N., Tuck, E., Lebel, M., & Menard, C. (2021). Neurobiology of resilience in depression: Immune and vascular insights from human and animal studies. *The European Journal of Neuroscience*, *53*(1), 183–221. DOI: 10.1111/ejn.14547 PMID: 31421056

Duman, R. S., & Li, N. (2012). A neurotrophic hypothesis of depression: Role of synaptogenesis in the actions of NMDA receptor antagonists. *Philosophical Transactions of the Royal Society of London. Series B, Biological Sciences*, *367*(1601), 2475–2484. DOI: 10.1098/rstb.2011.0357 PMID: 22826346

Dupont, N., Jiang, S., Pilli, M., Ornatowski, W., Bhattacharya, D., & Deretic, V. (2011). Autophagy-based unconventional secretory pathway for extracellular delivery of IL-1β. *The EMBO Journal*, *30*(23), 4701–4711. DOI: 10.1038/emboj.2011.398 PMID: 22068051

Duque Ede, A., & Munhoz, C. D. (2016). The Pro-inflammatory Effects of Glucocorticoids in the Brain. *Frontiers in Endocrinology (Lausanne)*, *7*, 78. DOI: 10.3389/fendo.2016.00078 PMID: 27445981

Ebrahimi, B., Mokhtari, T., Ghaffari, N., Adabi, M., & Hassanzadeh, G. (2024). Acellular spinal cord scaffold containing quercetin-encapsulated nanoparticles plays an anti-inflammatory role in functional recovery from spinal cord injury in rats. *Inflammopharmacology*, *32*(4), 1–20. DOI: 10.1007/s10787-024-01478-z PMID: 38702577

Eisenberg, T., Knauer, H., Schauer, A., Büttner, S., Ruckenstuhl, C., Carmona-Gutierrez, D., Ring, J., Schroeder, S., Magnes, C., Antonacci, L., Fussi, H., Deszcz, L., Hartl, R., Schraml, E., Criollo, A., Megalou, E., Weiskopf, D., Laun, P., Heeren, G., & Madeo, F. (2009). Induction of autophagy by spermidine promotes longevity. *Nature Cell Biology*, *11*(11), 1305–1314. DOI: 10.1038/ncb1975 PMID: 19801973

El-Zawi, N. R., & Elyasseri, A. M. A. (2021). A review of toxicity, genotoxicity and teratogenicity of formaldehyde. *Cancer*, *15*, 43.

Erisen, D. E., & Uludag, K. (2024). Estimating Biosafety of Biodegradable Biomedical Materials from In Vitro Ion Tolerance Parameters and Toxicity of Nanomaterials in Brain. In *Transformative Approaches to Patient Literacy and Healthcare Innovation* (pp. 201–221). IGI Global. DOI: 10.4018/979-8-3693-3661-8.ch010

Escobar-Morreale, H. F. (2022). Defining PCOS: A syndrome with an intrinsic heterogeneous nature. In Diamanti-Kandarakis, E. (Ed.), *Polycystic Ovary Syndrome* (pp. 3–13). Elsevier., DOI: 10.1016/B978-0-12-823045-9.00012-2

Esteban-Martínez, L., Sierra-Filardi, E., McGreal, R. S., Salazar-Roa, M., Mariño, G., Seco, E., Durand, S., Enot, D., Graña, O., Malumbres, M., Cvekl, A., Cuervo, A. M., Kroemer, G., & Boya, P. (2017). Programmed mitophagy is essential for the glycolytic switch during cell differentiation. *The EMBO Journal*, *36*(12), 1688–1706. DOI: 10.15252/embj.201695916 PMID: 28465321

Ethirajulu, A., Alkasabera, A., Onyali, C. B., Anim-Koranteng, C., Shah, H. E., Bhawnani, N., & Mostafa, J. A. (2021). Insulin resistance, hyperandrogenism, and its associated symptoms are the precipitating factors for depression in women with polycystic ovarian syndrome. *Cureus*, *13*(9). Advance online publication. DOI: 10.7759/cureus.18013 PMID: 34667688

Fabiani, C., & Antollini, S. S. (2019). Alzheimer's Disease as a Membrane Disorder: Spatial Cross-Talk Among Beta-Amyloid Peptides, Nicotinic Acetylcholine Receptors and Lipid Rafts. *Frontiers in Cellular Neuroscience*, *13*, 309. DOI: 10.3389/fncel.2019.00309 PMID: 31379503

Fabisiak, A., Caban, M., Dudek, P., Strigáč, A., Małecka-Wojciesko, E., & Talar-Wojnarowska, R. (2025). Advancements in dual biologic therapy for inflammatory bowel diseases: Efficacy, safety, and future directions. *Therapeutic Advances in Gastroenterology*, *18*, 17562848241309871. DOI: 10.1177/17562848241309871 PMID: 39758970

Fabrizi, C., De Vito, S., Somma, F., Pompili, E., Catizone, A., Leone, S., Lenzi, P., Fornai, F., & Fumagalli, L. (2014). Lithium improves survival of PC12 pheochromocytoma cells in high-density cultures and after exposure to toxic compounds. *International Journal of Cell Biology*, *2014*(1), 135908. DOI: 10.1155/2014/135908 PMID: 24563652

Fahs, D., Salloum, D., Nasrallah, M., & Ghazeeri, G. (2023). Polycystic Ovary Syndrome: Pathophysiology and Controversies in Diagnosis. *Diagnostics (Basel)*, *13*(9), 1559. Advance online publication. DOI: 10.3390/diagnostics13091559 PMID: 37174950

Farsi, Z., & Sheng, M. (2023). Molecular mechanisms of schizophrenia: Insights from human genetics. *Current Opinion in Neurobiology*, *81*, 102731. DOI: 10.1016/j.conb.2023.102731 PMID: 37245257

Feldman, E. L., Goutman, S. A., Petri, S., Mazzini, L., Savelieff, M. G., Shaw, P. J., & Sobue, G. (2022). Amyotrophic lateral sclerosis. *Lancet*, *400*(10360), 1363–1380. DOI: 10.1016/S0140-6736(22)01272-7 PMID: 36116464

Fernandes, S. M., Mayer, J., Nilsson, P., & Shimozawa, M. (2025). How close is autophagy-targeting therapy for Alzheimer's disease to clinical use? A summary of autophagy modulators in clinical studies. *Frontiers in Cell and Developmental Biology*, *12*, 1520949. Advance online publication. DOI: 10.3389/fcell.2024.1520949 PMID: 39845082

Ferrari, A. J., Somerville, A. J., Baxter, A. J., Norman, R., Patten, S. B., Vos, T., & Whiteford, H. A. (2013). Global variation in the prevalence and incidence of major depressive disorder: A systematic review of the epidemiological literature. *Psychological Medicine*, *43*(3), 471–481. DOI: 10.1017/S0033291712001511 PMID: 22831756

Findlay Black, H., Kay, C., Dawson, J., Bortnick, S., Javier, K., Xia, Q., Chau, C. H., Leavitt, T., Arning, L., Nguyen, H. P., & Hayden, M. R. (2024). Ascertainment of uninterrupted CAG repeat length and disease-modifying variants in fragment-based genetic testing for Huntington Disease. *Genetics in Medicine Open*, *2*, 101882. DOI: 10.1016/j.gimo.2024.101882 PMID: 39669608

Fleming, A., Bourdenx, M., Fujimaki, M., Karabiyik, C., Krause, G. J., Lopez, A., Martín-Segura, A., Puri, C., Scrivo, A., Skidmore, J., Son, S. M., Stamatakou, E., Wrobel, L., Zhu, Y., Cuervo, A. M., & Rubinsztein, D. C. (2022). The different autophagy degradation pathways and neurodegeneration. *Neuron, 110*(6), 935–966. DOI: 10.1016/j.neuron.2022.01.017 PMID: 35134347

Foda, A. M., Ibrahim, S. S., Ibrahim, S. M., & Elbaz, E. M. (2024). Pterostilbene Ameliorates Cognitive Impairment in Polycystic Ovary Syndrome Rat Model through Improving Insulin Resistance via the IRS-1/PI3K/Akt/GSK-3β Pathway: A Comparative Study with Metformin. *ACS Chemical Neuroscience*, *15*(16), 3064–3077. DOI: 10.1021/acschemneuro.4c00352 PMID: 39119909

Foulkes, R., Man, E., Thind, J., Yeung, S., Joy, A., & Hoskins, C. (2020). The regulation of nanomaterials and nanomedicines for clinical application: Current and future perspectives. *Biomaterials Science*, *8*(17), 4653–4664. DOI: 10.1039/D0BM00558D PMID: 32672255

Fracchiolla, D., Chang, C., Hurley, J. H., & Martens, S. (2020). A PI3K-WIPI2 positive feedback loop allosterically activates LC3 lipidation in autophagy. *The Journal of Cell Biology*, *219*(7), e201912098. Advance online publication. DOI: 10.1083/jcb.201912098 PMID: 32437499

Frank, M. G., Hershman, S. A., Weber, M. D., Watkins, L. R., & Maier, S. F. (2014). Chronic exposure to exogenous glucocorticoids primes microglia to pro-inflammatory stimuli and induces NLRP3 mRNA in the hippocampus. *Psychoneuroendocrinology*, *40*, 191–200. DOI: 10.1016/j.psyneuen.2013.11.006 PMID: 24485491

Fujioka, Y., Alam, J. M., Noshiro, D., Mouri, K., Ando, T., Okada, Y., May, A. I., Knorr, R. L., Suzuki, K., Ohsumi, Y., & Noda, N. N. (2020). Phase separation organizes the site of autophagosome formation. *Nature*, *578*(7794), 301–305. DOI: 10.1038/s41586-020-1977-6 PMID: 32025038

Fu, Q., Qiu, R., Liang, J., Wu, S., Huang, D., Qin, Y., Li, Q., Shi, X., Xiong, X., Jiang, Z., Chen, Y., & Cheng, Y. (2025). Sugemule-7 alleviates oxidative stress, neuroinflammation, and cell death, promoting synaptic plasticity recovery in mice with postpartum depression. *Scientific Reports*, *15*(1), 1426. DOI: 10.1038/s41598-025-85276-9 PMID: 39789071

Furukawa, S., Fujita, T., Shimabukuro, M., Iwaki, M., Yamada, Y., Nakajima, Y., Nakayama, O., Makishima, M., Matsuda, M., & Shimomura, I. (2004). Increased oxidative stress in obesity and its impact on metabolic syndrome. *The Journal of Clinical Investigation*, *114*(12), 1752–1761. DOI: 10.1172/JCI21625 PMID: 15599400

Fu, X., Han, H., Yang, H., Xu, B., Dai, W., Liu, L., He, T., Du, X., & Pei, X. (2024). Nrf2-mediated ferroptosis of spermatogenic cells involved in male reproductive toxicity induced by polystyrene nanoplastics in mice. *Journal of Zhejiang University. Science. B., 25*(4), 307–323. DOI: 10.1631/jzus.B2300138 PMID: 38584093

Gadhave, D. G., Sugandhi, V. V., Jha, S. K., Nangare, S. N., Gupta, G., Singh, S. K., Dua, K., Cho, H., Hansbro, P. M., & Paudel, K. R. (2024). Neurodegenerative disorders: Mechanisms of degeneration and therapeutic approaches with their clinical relevance. *Ageing Research Reviews, 99*, 102357. DOI: 10.1016/j.arr.2024.102357 PMID: 38830548

Galluzzi, L., Bravo-San Pedro, J. M., Levine, B., Green, D. R., & Kroemer, G. (2017). Pharmacological modulation of autophagy: Therapeutic potential and persisting obstacles. *Nature Reviews. Drug Discovery, 16*(7), 487–511. DOI: 10.1038/nrd.2017.22 PMID: 28529316

Gan, H., Ma, Q., Hao, W., Yang, N., Chen, Z.-S., Deng, L., & Chen, J. (2024). Targeting autophagy to counteract neuroinflammation: A novel antidepressant strategy. *Pharmacological Research, 202*, 107112. DOI: 10.1016/j.phrs.2024.107112 PMID: 38403256

Gao, K., Niu, J., & Dang, X. (2020). Neuroprotection of melatonin on spinal cord injury by activating autophagy and inhibiting apoptosis via SIRT1/AMPK signaling pathway. *Biotechnology Letters, 42*(10), 2059–2069. DOI: 10.1007/s10529-020-02939-5 PMID: 32514788

Gao, K., Niu, J., & Dang, X. (2020). Wnt-3a improves functional recovery through autophagy activation via inhibiting the mTOR signaling pathway after spinal cord injury. *Neuroscience Letters, 737*, 135305. DOI: 10.1016/j.neulet.2020.135305 PMID: 32818590

Gao, L., Peng, L., Wang, J., Zhang, J. H., & Xia, Y. (2024). Mitochondrial stress: A key role of neuroinflammation in stroke. *Journal of Neuroinflammation, 21*(1), 44. DOI: 10.1186/s12974-024-03033-7 PMID: 38321473

Gao, Y., Bielohuby, M., Fleming, T., Grabner, G. F., Foppen, E., Bernhard, W., Guzmán-Ruiz, M., Layritz, C., Legutko, B., Zinser, E., García-Cáceres, C., Buijs, R. M., Woods, S. C., Kalsbeek, A., Seeley, R. J., Nawroth, P. P., Bidlingmaier, M., Tschöp, M. H., & Yi, C. X. (2017). Dietary sugars, not lipids, drive hypothalamic inflammation. *Molecular Metabolism, 6*(8), 897–908. DOI: 10.1016/j.molmet.2017.06.008 PMID: 28752053

Gao, Y., Tan, L., Yu, J.-T., & Tan, L. (2018). Tau in Alzheimer's disease: Mechanisms and therapeutic strategies. *Current Alzheimer Research, 15*(3), 283–300. DOI: 10.2174/1567205014666170417111859 PMID: 28413986

Gao, Z., Tan, H., Song, X., Zhuang, T., Kong, R., Wang, Y., Yan, X., & Yao, R. (2024). Troxerutin dampened hypothalamic neuroinflammation via microglial IL-22/IL-22R1/IRF3 activation in dihydrotestosterone-induced polycystic ovary syndrome rats. *Phytomedicine, 124*, 155280. DOI: 10.1016/j.phymed.2023.155280 PMID: 38183697

Gassen, N. C., Hartmann, J., Schmidt, M. V., & Rein, T. (2015). FKBP5/FKBP51 enhances autophagy to synergize with antidepressant action. *Autophagy, 11*(3), 578–580. DOI: 10.1080/15548627.2015.1017224 PMID: 25714272

Gassen, N. C., Hartmann, J., Zschocke, J., Stepan, J., Hafner, K., Zellner, A., Kirmeier, T., Kollmannsberger, L., Wagner, K. V., Dedic, N., Balsevich, G., Deussing, J. M., Kloiber, S., Lucae, S., Holsboer, F., Eder, M., Uhr, M., Ising, M., Schmidt, M. V., & Rein, T. (2014). Association of FKBP51 with priming of autophagy pathways and mediation of antidepressant treatment response: Evidence in cells, mice, and humans. *PLoS Medicine, 11*(11), e1001755. DOI: 10.1371/journal.pmed.1001755 PMID: 25386878

Gassen, N. C., & Rein, T. (2019). Is There a Role of Autophagy in Depression and Antidepressant Action? *Frontiers in Psychiatry, 10*, 337. DOI: 10.3389/fpsyt.2019.00337 PMID: 31156481

Gautam, R., Maan, P., Jyoti, A., Kumar, A., Malhotra, N., & Arora, T. (2025). The Role of Lifestyle Interventions in PCOS Management: A Systematic Review. *Nutrients, 17*(2), 310. DOI: 10.3390/nu17020310 PMID: 39861440

Geng, J., Liu, J., Yuan, X., Liu, W., & Guo, W. (2019). Andrographolide triggers autophagy-mediated inflammation inhibition and attenuates chronic unpredictable mild stress (CUMS)-induced depressive-like behavior in mice. *Toxicology and Applied Pharmacology, 379*, 114688. DOI: 10.1016/j.taap.2019.114688 PMID: 31340160

Georgin-Lavialle, S., Moura, D. S., Salvador, A., Chauvet-Gélinier, J.-C., Launay, J.-M., Damaj, G., Côté, F., Soucié, E., Chandesris, M.-O., Barète, S., Grandpeix-Guyodo, C., Bachmeyer, C., Alyanakian, M.-A., Aouba, A., Lortholary, O., Dubreuil, P., Teyssier, J.-R., Trojak, B., Haffen, E., & Gaillard, R. (2016). Mast cells' involvement in inflammation pathways linked to depression: Evidence in mastocytosis. *Molecular Psychiatry, 21*(11), 1511–1516. DOI: 10.1038/mp.2015.216 PMID: 26809839

Ge, P., Dawson, V. L., & Dawson, T. M. (2020). PINK1 and Parkin mitochondrial quality control: A source of regional vulnerability in Parkinson's disease. *Molecular Neurodegeneration, 15*(1), 1–18. DOI: 10.1186/s13024-020-00367-7 PMID: 32169097

Gerson, J. E., Safren, N., Fischer, S., Patel, R., Crowley, E. V., Welday, J. P., Windle, A. K., Barmada, S., Paulson, H. L., & Sharkey, L. M. (2020). Ubiquilin-2 differentially regulates polyglutamine disease proteins. *Human Molecular Genetics, 29*(15), 2596–2610. DOI: 10.1093/hmg/ddaa152 PMID: 32681165

Ghaffari, N., Mokhtari, T., Adabi, M., Ebrahimi, B., Kamali, M., Gholaminejhad, M., & Hassanzadeh, G. (2024). Neurological recovery and neurogenesis by curcumin sustained-release system cross-linked with an acellular spinal cord scaffold in rat spinal cord injury: Targeting NLRP3 inflammasome pathway. *Phytotherapy Research, 38*(6), 2669–2686. DOI: 10.1002/ptr.8179 PMID: 38500263

Ghosh, I., Sankhe, R., Mudgal, J., Arora, D., & Nampoothiri, M. (2020). Spermidine, an autophagy inducer, as a therapeutic strategy in neurological disorders. *Neuropeptides, 83*, 102083. DOI: 10.1016/j.npep.2020.102083 PMID: 32873420

Gidiagba, J. O., Daraojimba, C., Ofonagoro, K. A., Eyo-Udo, N. L., Egbokhaebho, B. A., Ogunjobi, O. A., & Banso, A. A. (2023). Economic impacts and innovations in materials science: A holistic exploration of nanotechnology and advanced materials. *Engineering Science & Technology Journal, 4*(3), 84–100. DOI: 10.51594/estj.v4i3.553

Giorgi, C., Lombardozzi, G., Ammannito, F., Scenna, M. S., Maceroni, E., Quintiliani, M., d'Angelo, M., Cimini, A., & Castelli, V. (2024). Brain Organoids: A Game-Changer for Drug Testing. *Pharmaceutics, 16*(4), 443. DOI: 10.3390/pharmaceutics16040443 PMID: 38675104

Goeldner, C., Reiss, D., Kieffer, B. L., & Ouagazzal, A. M. (2010). Endogenous nociceptin/orphanin-FQ in the dorsal hippocampus facilitates despair-related behavior. *Hippocampus, 20*(8), 911–916. DOI: 10.1002/hipo.20760 PMID: 20087887

Golestaneh, N., Chu, Y., Xiao, Y.-Y., Stoleru, G. L., & Theos, A. C. (2018). Dysfunctional autophagy in RPE, a contributing factor in age-related macular degeneration. *Cell Death & Disease, 8*(1), e2537–e2537. DOI: 10.1038/cddis.2016.453 PMID: 28055007

Gómez-Virgilio, L., Silva-Lucero, M.-D.-C., Flores-Morelos, D.-S., Gallardo-Nieto, J., Lopez-Toledo, G., Abarca-Fernandez, A.-M., Zacapala-Gómez, A.-E., Luna-Muñoz, J., Montiel-Sosa, F., Soto-Rojas, L. O., Pacheco-Herrero, M., & Cardenas-Aguayo, M.-D.-C. (2022). Autophagy: A Key Regulator of Homeostasis and Disease: An Overview of Molecular Mechanisms and Modulators. *Cells, 11*(15), 2262. Advance online publication. DOI: 10.3390/cells11152262 PMID: 35892559

Gong, X., Wang, Y., Zhou, Y., & Pan, L. (2024). Structure of the WIPI3/ATG16L1 Complex Reveals the Molecular Basis for the Recruitment of the ATG12~ATG5-ATG16L1 Complex by WIPI3. *Cells, 13*(24), 2113. DOI: 10.3390/cells13242113 PMID: 39768203

González-Fernández, C., Fernández-Martos, C. M., Shields, S. D., Arenas, E., & Javier Rodriguez, F. (2014). Wnts are expressed in the spinal cord of adult mice and are differentially induced after injury. *Journal of Neurotrauma, 31*(6), 565–581. DOI: 10.1089/neu.2013.3067 PMID: 24367909

González-Rodríguez, P., Delorme-Axford, E., Bernard, A., Keane, L., Stratoulias, V., Grabert, K., Engskog-Vlachos, P., Füllgrabe, J., Klionsky, D. J., & Joseph, B. (2022). SETD2 transcriptional control of ATG14L/S isoforms regulates autophagosome–lysosome fusion. *Cell Death & Disease, 13*(11), 953. DOI: 10.1038/s41419-022-05381-9 PMID: 36371383

González-Rodríguez, P., Klionsky, D. J., & Joseph, B. (2022). Autophagy regulation by RNA alternative splicing and implications in human diseases. *Nature Communications, 13*(1), 2735. DOI: 10.1038/s41467-022-30433-1 PMID: 35585060

Goode, A., Butler, K., Long, J., Cavey, J., Scott, D., Shaw, B., Sollenberger, J., Gell, C., Johansen, T., Oldham, N. J., Searle, M. S., & Layfield, R. (2016). Defective recognition of LC3B by mutant SQSTM1/p62 implicates impairment of autophagy as a pathogenic mechanism in ALS-FTLD. *Autophagy, 12*(7), 1094–1104. DOI: 10.1080/15548627.2016.1170257 PMID: 27158844

Grace, A. A. (2016). Dysregulation of the dopamine system in the pathophysiology of schizophrenia and depression. *Nature Reviews. Neuroscience, 17*(8), 524–532. DOI: 10.1038/nrn.2016.57 PMID: 27256556

Grassner, L., Wutte, C., Klein, B., Mach, O., Riesner, S., Panzer, S., Vogel, M., Bühren, V., Strowitzki, M., Vastmans, J., & Maier, D. (2016). Early decompression (< 8 h) after traumatic cervical spinal cord injury improves functional outcome as assessed by spinal cord independence measure after one year. *Journal of Neurotrauma, 33*(18), 1658–1666. DOI: 10.1089/neu.2015.4325 PMID: 27050499

Green, D. R., & Levine, B. (2014). To be or not to be? How selective autophagy and cell death govern cell fate. *Cell*, *157*(1), 65–75. DOI: 10.1016/j.cell.2014.02.049 PMID: 24679527

Greene, A. W., Grenier, K., Aguileta, M. A., Muise, S., Farazifard, R., Haque, M. E., McBride, H. M., Park, D. S., & Fon, E. A. (2012). Mitochondrial processing peptidase regulates PINK1 processing, import and Parkin recruitment. *EMBO Reports*, *13*(4), 378–385. DOI: 10.1038/embor.2012.14 PMID: 22354088

Greenwood, E. A., Pasch, L. A., Cedars, M. I., Legro, R. S., Eisenberg, E., Huddleston, H. G., & Network, H. D. R. M. (2018). Insulin resistance is associated with depression risk in polycystic ovary syndrome. *Fertility and Sterility*, *110*(1), 27–34. DOI: 10.1016/j.fertnstert.2018.03.009 PMID: 29908775

Gruber, J., Hanssen, R., Qubad, M., Bouzouina, A., Schack, V., Sochor, H., Schiweck, C., Aichholzer, M., Matura, S., Slattery, D. A., Zopf, Y., Borgland, S. L., Reif, A., & Thanarajah, S. E. (2023). Impact of insulin and insulin resistance on brain dopamine signalling and reward processing–An underexplored mechanism in the pathophysiology of depression? *Neuroscience and Biobehavioral Reviews*, *149*, 105179. DOI: 10.1016/j.neubiorev.2023.105179 PMID: 37059404

Grunwald, A. (2020). Nanotechnology-A new field of ethical inquiry? In *The Ethics of Nanotechnology, Geoengineering, and Clean Energy* (pp. 17-31). Routledge. https://doi.org/DOI: 10.4324/9781003075028-3

Guan, Y., Wang, J., Wu, X., Song, L., Wang, Y., Gong, M., & Li, B. (2021). Quercetin reverses chronic unpredictable mild stress-induced depression-like behavior in vivo by involving nuclear factor-E2-related factor 2. *Brain Research*, *1772*, 147661. DOI: 10.1016/j.brainres.2021.147661 PMID: 34529966

Gulbins, A., Schumacher, F., Becker, K. A., Wilker, B., Soddemann, M., Boldrin, F., Müller, C. P., Edwards, M. J., Goodman, M., Caldwell, C. C., Kleuser, B., Kornhuber, J., Szabo, I., & Gulbins, E. (2018). Antidepressants act by inducing autophagy controlled by sphingomyelin–ceramide. *Molecular Psychiatry*, *23*(12), 2324–2346. DOI: 10.1038/s41380-018-0090-9 PMID: 30038230

Guo, F., Liu, X., Cai, H., & Le, W. (2018). Autophagy in neurodegenerative diseases: Pathogenesis and therapy. *Brain Pathology (Zurich, Switzerland)*, *28*(1), 3–13. DOI: 10.1111/bpa.12545 PMID: 28703923

Guo, L., He, N., Zhao, Y., Liu, T., & Deng, Y. (2020). Autophagy modulated by inorganic nanomaterials. *Theranostics*, *10*(7), 3206–3222. DOI: 10.7150/thno.40414 PMID: 32194863

Gu, Y., Chen, D., Zhou, L., Zhao, X., Lin, J., Lin, B., Lin, T., Chen, Z., Chen, Z., Wang, Z., & Liu, W. (2021). Lysine-specific demethylase 1 inhibition enhances autophagy and attenuates early-stage post-spinal cord injury apoptosis. *Cell Death Discovery*, *7*(1), 69. DOI: 10.1038/s41420-021-00455-7 PMID: 33824301

Guzelian, P. S., Victoroff, M. S., Halmes, N. C., James, R. C., & Guzelian, C. P. (2005). Evidence-based toxicology: A comprehensive framework for causation. *Human and Experimental Toxicology*, *24*(4), 161–201. DOI: 10.1191/0960327105ht517oa PMID: 15957536

Hagberg, H., Mallard, C., Rousset, C. I., & Thornton, C. (2014). Mitochondria: Hub of injury responses in the developing brain. *Lancet Neurology*, *13*(2), 217–232. DOI: 10.1016/S1474-4422(13)70261-8 PMID: 24457191

Ha, J., & Kim, J. (2016). Novel pharmacological modulators of autophagy: An updated patent review (2012-2015). *Expert Opinion on Therapeutic Patents*, *26*(11), 1273–1289. DOI: 10.1080/13543776.2016.1217996 PMID: 27476990

Halamoda-Kenzaoui, B., Vandebriel, R. J., Howarth, A., Siccardi, M., David, C. A. W., Liptrott, N. J., Santin, M., Borgos, S. E., Bremer-Hoffmann, S., & Caputo, F. (2021). Methodological needs in the quality and safety characterisation of nanotechnology-based health products: Priorities for method development and standardisation. *Journal of Controlled Release*, *336*, 192–206. DOI: 10.1016/j.jconrel.2021.06.016 PMID: 34126169

Haleem, A., Javaid, M., Singh, R. P., Rab, S., & Suman, R. (2023). Applications of nanotechnology in medical field: A brief review. *Global Health Journal (Amsterdam, Netherlands)*, *7*(2), 70–77. DOI: 10.1016/j.glohj.2023.02.008

Hamer, J. A., Testani, D., Mansur, R. B., Lee, Y., Subramaniapillai, M., & McIntyre, R. S. (2019). Brain insulin resistance: A treatment target for cognitive impairment and anhedonia in depression. *Experimental Neurology*, *315*, 1–8. DOI: 10.1016/j.expneurol.2019.01.016 PMID: 30695707

Hamilton, A. M., Krout, I. N., White, A. C., & Sampson, T. R. (2024). Microbiome-based therapeutics for Parkinson's disease. *Neurotherapeutics; the Journal of the American Society for Experimental NeuroTherapeutics*, *21*(6), e00462. DOI: 10.1016/j.neurot.2024.e00462 PMID: 39393983

Hands, S. L., Proud, C. G., & Wyttenbach, A. (2009). mTOR's role in ageing: Protein synthesis or autophagy? *Aging (Albany NY)*, *1*(7), 586–597. DOI: 10.18632/aging.100070 PMID: 20157541

Han, W., Li, H., & Chen, B. (2022). Research progress and potential applications of spermidine in ocular diseases. *Pharmaceutics*, *14*(7), 1500. DOI: 10.3390/pharmaceutics14071500 PMID: 35890394

Han, X., Cai, C., Deng, W., Shi, Y., Li, L., Wang, C., Zhang, J., Rong, M., Liu, J., Fang, B., He, H., Liu, X., Deng, C., He, X., & Cao, X. (2024). Landscape of human organoids: Ideal model in clinics and research. *Innovation (Cambridge (Mass.))*, *5*(3), 100620. DOI: 10.1016/j.xinn.2024.100620 PMID: 38706954

Han, X., Xu, T., Fang, Q., Zhang, H., Yue, L., Hu, G., & Sun, L. (2021). Quercetin hinders microglial activation to alleviate neurotoxicity via the interplay between NLRP3 inflammasome and mitophagy. *Redox Biology*, *44*, 102010. DOI: 10.1016/j.redox.2021.102010 PMID: 34082381

Hao, W., Ma, Q., Wang, L., Yuan, N., Gan, H., He, L., Li, X., Huang, J., & Chen, J. (2024). Gut dysbiosis induces the development of depression-like behavior through abnormal synapse pruning in microglia, which is mediated by complement C3. *Microbiome*, *12*(34), 34. Advance online publication. DOI: 10.1186/s40168-024-01756-6 PMID: 38378622

Harada, M. (2022). Pathophysiology of polycystic ovary syndrome revisited: Current understanding and perspectives regarding future research. *Reproductive Medicine and Biology*, *21*(1), e12487. DOI: 10.1002/rmb2.12487 PMID: 36310656

Hara, T., Nakamura, K., Matsui, M., Yamamoto, A., Nakahara, Y., Suzuki-Migishima, R., Yokoyama, M., Mishima, K., Saito, I., Okano, H., & Mizushima, N. (2006). Suppression of basal autophagy in neural cells causes neurodegenerative disease in mice. *Nature*, *441*(7095), 885–889. DOI: 10.1038/nature04724 PMID: 16625204

Haroon, E., Miller, A. H., & Sanacora, G. (2017). Inflammation, Glutamate, and Glia: A Trio of Trouble in Mood Disorders. *Neuropsychopharmacology : Official Publication of the American College of Neuropsychopharmacology*, *42*(1), 193–215. DOI: 10.1038/npp.2016.199 PMID: 27629368

Haruwaka, K., Ikegami, A., Tachibana, Y., Ohno, N., Konishi, H., Hashimoto, A., Matsumoto, M., Kato, D., Ono, R., Kiyama, H., Moorhouse, A. J., Nabekura, J., & Wake, H. (2019). Dual microglia effects on blood brain barrier permeability induced by systemic inflammation. *Nature Communications*, *10*(1), 5816. DOI: 10.1038/s41467-019-13812-z PMID: 31862977

Hasan, M., Sultana, S., Sohan, M., Parvin, S., Rahman, M. A., Hossain, M. J., Rahman, M. S., & Islam, M. R. (2022). Prevalence and associated risk factors for mental health problems among patients with polycystic ovary syndrome in Bangladesh: A nationwide cross-Sectional study. *PLoS One*, *17*(6), e0270102. DOI: 10.1371/journal.pone.0270102 PMID: 35731829

Hassamal, S. (2023). Chronic stress, neuroinflammation, and depression: An overview of pathophysiological mechanisms and emerging anti-inflammatories. *Frontiers in Psychiatry*, *14*, 1130989. DOI: 10.3389/fpsyt.2023.1130989 PMID: 37252156

He, C., Bassik, M. C., Moresi, V., Sun, K., Wei, Y., Zou, Z., An, Z., Loh, J., Fisher, J., Sun, Q., Korsmeyer, S., Packer, M., May, H. I., Hill, J. A., Virgin, H. W., Gilpin, C., Xiao, G., Bassel-Duby, R., Scherer, P. E., & Levine, B. (2012). Exercise-induced BCL2-regulated autophagy is required for muscle glucose homeostasis. *Nature*, *481*(7382), 511–515. DOI: 10.1038/nature10758 PMID: 22258505

Heinrich, M. A., Huynh, N. T., Heinrich, L., & Prakash, J. (2024). Understanding glioblastoma stromal barriers against NK cell attack via a three-culture 3D spheroid model. *Heliyon*, *10*(3), e24808. DOI: 10.1016/j.heliyon.2024.e24808 PMID: 38317968

Heiseke, A., Aguib, Y., Riemer, C., Baier, M., & Schätzl, H. M. (2009). Lithium induces clearance of protease resistant prion protein in prion-infected cells by induction of autophagy. *Journal of Neurochemistry*, *109*(1), 25–34. DOI: 10.1111/j.1471-4159.2009.05906.x PMID: 19183256

He, J., Zhu, G., Wang, G., & Zhang, F. (2020). Oxidative Stress and Neuroinflammation Potentiate Each Other to Promote Progression of Dopamine Neurodegeneration. *Oxidative Medicine and Cellular Longevity*, *6137521*, 1–12. Advance online publication. DOI: 10.1155/2020/6137521 PMID: 32714488

Hellal, F., Hurtado, A., Ruschel, J., Flynn, K. C., Laskowski, C. J., Umlauf, M., Kapitein, L. C., Strikis, D., Lemmon, V., Bixby, J., Hoogenraad, C. C., & Bradke, F. (2011). Microtubule stabilization reduces scarring and causes axon regeneration after spinal cord injury. *Science*, *331*(6019), 928–931. DOI: 10.1126/science.1201148 PMID: 21273450

He, M., Ding, Y., Chu, C., Tang, J., Xiao, Q., & Luo, Z.-G. (2016). Autophagy induction stabilizes microtubules and promotes axon regeneration after spinal cord injury. *Proceedings of the National Academy of Sciences of the United States of America*, *113*(40), 11324–11329. DOI: 10.1073/pnas.1611282113 PMID: 27638205

Heneka, M. T., Carson, M. J., El Khoury, J., Landreth, G. E., Brosseron, F., & Feinstein, D. L.. (2015). Neuroinflammation in Alzheimer's disease. *Lancet Neurology*, *14*(4), 388–405. DOI: 10.1016/S1474-4422(15)70016-5 PMID: 25792098

He, R., Liu, Y., Fu, W., He, X., Liu, S., Xiao, D., & Tao, Y. (2024). Mechanisms and cross-talk of regulated cell death and their epigenetic modifications in tumor progression. *Molecular Cancer*, *23*(1), 267. DOI: 10.1186/s12943-024-02172-y PMID: 39614268

He, Y., She, H., Zhang, T., Xu, H., Cheng, L., Yepes, M., Zhao, Y., & Mao, Z. (2018). p38 MAPK inhibits autophagy and promotes microglial inflammatory responses by phosphorylating ULK1. *The Journal of Cell Biology*, *217*(1), 315–328. DOI: 10.1083/jcb.201701049 PMID: 29196462

Hiam, D., Moreno-Asso, A., Teede, H. J., Laven, J. S. E., Stepto, N. K., Moran, L. J., & Gibson-Helm, M. (2019). The Genetics of Polycystic Ovary Syndrome: An Overview of Candidate Gene Systematic Reviews and Genome-Wide Association Studies. *Journal of Clinical Medicine*, *8*(10), 1606. Advance online publication. DOI: 10.3390/jcm8101606 PMID: 31623391

Hiebert, J. B., Shen, Q., Thimmesch, A. R., & Pierce, J. D. (2015). Traumatic brain injury and mitochondrial dysfunction. *The American Journal of the Medical Sciences*, *350*(2), 132–138. DOI: 10.1097/MAJ.0000000000000506 PMID: 26083647

Hirata, E., Shirai, K., Kawaoka, T., Sato, K., Kodama, F., & Suzuki, K. (2021). Atg15 in Saccharomyces cerevisiae consists of two functionally distinct domains. *Molecular Biology of the Cell*, *32*(8), 645–663. DOI: 10.1091/mbc.E20-07-0500 PMID: 33625870

Hoeffer, C. A., & Klann, E. (2010). mTOR signaling: At the crossroads of plasticity, memory and disease. *Trends in Neurosciences*, *33*(2), 67–75. DOI: 10.1016/j.tins.2009.11.003 PMID: 19963289

Holzer, E., Martens, S., & Tulli, S. (2024). The Role of ATG9 Vesicles in Autophagosome Biogenesis. *Journal of Molecular Biology*, *436*(15), 168489. DOI: 10.1016/j.jmb.2024.168489 PMID: 38342428

Hoogland, I. C. M., Houbolt, C., van Westerloo, D. J., van Gool, W. A., & van de Beek, D. (2015). Systemic inflammation and microglial activation: Systematic review of animal experiments. *Journal of Neuroinflammation*, *12*(1), 114. DOI: 10.1186/s12974-015-0332-6 PMID: 26048578

Hormazabal, J., Saavedra, F., Espinoza-Arratia, C., Martinez, N. W., Cruces, T., Alfaro, I. E., & Loyola, A. (2022). Chaperone mediated autophagy contributes to the newly synthesized histones H3 and H4 quality control. *Nucleic Acids Research*, *50*(4), 1875–1887. DOI: 10.1093/nar/gkab1296 PMID: 35037039

Hou, H., Zhang, L., Zhang, L., & Tang, P. (2014). Acute spinal cord injury in rats should target activated autophagy. *Journal of Neurosurgery. Spine*, *20*(5), 568–577. DOI: 10.3171/2014.1.SPINE13237 PMID: 24559459

Hou, X., Watzlawik, J. O., Fiesel, F. C., & Springer, W. (2020). Autophagy in Parkinson's Disease. *Journal of Molecular Biology*, *432*(8), 2651–2672. DOI: 10.1016/j.jmb.2020.01.037 PMID: 32061929

Hsu, S. S., Chen, W. C., Lo, Y. K., Cheng, J. S., Yeh, J. H., Cheng, H. H., Chen, J.-S., Chang, H.-T., Jiann, B.-P., Huang, J.-K., & Jan, C.-R. (2004). Effect of the antidepressant maprotiline on Ca2+ movement and proliferation in human prostate cancer cells. *Clinical and Experimental Pharmacology & Physiology*, *31*(7), 444–449. DOI: 10.1111/j.1440-1681.2004.04024.x PMID: 15236632

Huang, Q., Ye, X., Wang, L., & Pan, J. (2019). Salvianolic acid B abolished chronic mild stress-induced depression through suppressing oxidative stress and neuro-inflammation via regulating NLRP3 inflammasome activation. *Journal of Food Biochemistry*, *43*(3), e12742. PMID: 31353549

Huang, W., Huang, M., Ouyang, H., Peng, J., & Liang, J. (2018). Oridonin inhibits vascular inflammation by blocking NF-κB and MAPK activation. *European Journal of Pharmacology*, *826*, 133–139. DOI: 10.1016/j.ejphar.2018.02.044 PMID: 29518395

Huang, X., Wu, H., Jiang, R., Sun, G., Shen, J., Ma, M., Ma, C., Zhang, S., Huang, Z., Wu, Q., Chen, G., & Tao, W. (2018). The antidepressant effects of α-tocopherol are related to activation of autophagy via the AMPK/mTOR pathway. *European Journal of Pharmacology*, *833*, 1–7. DOI: 10.1016/j.ejphar.2018.05.020 PMID: 29782858

Huang, Y., Meng, S., Wu, B., Shi, H., Wang, Y., Xiang, J., Li, J., Shi, Z., Wu, G., Lyu, Y., Jia, X., Hu, J., Xu, Z.-X., & Gao, Y. (2023). HSPB2 facilitates neural regeneration through autophagy for sensorimotor recovery after traumatic brain injury. *JCI Insight*, *8*(16), e168919. DOI: 10.1172/jci.insight.168919 PMID: 37606039

Hu, H., Cao, B., Huang, D., Lin, Y., Zhou, B., Ying, J., Huang, L., & Zhang, L. (2024). Withaferin a modulation of microglia autophagy mitigates neuroinflammation and enhances cognitive function in POCD. *Scientific Reports*, *14*(1), 26112. DOI: 10.1038/s41598-024-75284-6 PMID: 39478022

Hu, M., Richard, J. E., Maliqueo, M., Kokosar, M., Fornes, R., Benrick, A., Jansson, T., Ohlsson, C., Wu, X., Skibicka, K. P., & Stener-Victorin, E. (2015). Maternal testosterone exposure increases anxiety-like behavior and impacts the limbic system in the offspring. *Proceedings of the National Academy of Sciences of the United States of America, 112*(46), 14348–14353. DOI: 10.1073/pnas.1507514112 PMID: 26578781

Hur, E.-M., & Zhou, F.-Q. (2012). Growing the growth cone: Remodeling the cytoskeleton to promote axon regeneration. *Trends in Neurosciences, 35*(3), 164–174. DOI: 10.1016/j.tins.2011.11.002 PMID: 22154154

Hurley, L. L., Akinfiresoye, L., Kalejaiye, O., & Tizabi, Y. (2014). Antidepressant effects of resveratrol in an animal model of depression. *Behavioural Brain Research, 268*, 1–7. DOI: 10.1016/j.bbr.2014.03.052 PMID: 24717328

Hussain, M. S., Moglad, E., Afzal, M., Sharma, S., Gupta, G., Sivaprasad, G. V., Deorari, M., Almalki, W. H., Kazmi, I., Alzarea, S. I., Shahwan, M., Pant, K., Ali, H., Singh, S. K., Dua, K., & Subramaniyan, V. (2024). Autophagy-associated noncoding RNAs: Unraveling their impact on Parkinson's disease pathogenesis. *CNS Neuroscience & Therapeutics, 30*(5), e14763. DOI: 10.1111/cns.14763 PMID: 38790149

Hyttinen, J. M., Viiri, J., Kaarniranta, K., & Błasiak, J. (2018). Mitochondrial quality control in AMD: Does mitophagy play a pivotal role? *Cellular and Molecular Life Sciences : CMLS, 75*(16), 2991–3008. DOI: 10.1007/s00018-018-2843-7 PMID: 29777261

Hyung, S., Lee, S. R., Kim, J., Kim, Y., Kim, S., Kim, H. N., & Jeon, N. L. (2021). A 3D disease and regeneration model of peripheral nervous system-on-a-chip. *Science Advances, 7*(5), eabd9749. Advance online publication. DOI: 10.1126/sciadv.abd9749 PMID: 33514550

Ibáñez, L., Oberfield, S. E., Witchel, S., Auchus, R. J., Chang, R. J., Codner, E., Dabadghao, P., Darendeliler, F., Elbarbary, N. S., Gambineri, A., Garcia Rudaz, C., Hoeger, K. M., López-Bermejo, A., Ong, K., Peña, A. S., Reinehr, T., Santoro, N., Tena-Sempere, M., Tao, R., & Lee, P. A. (2017). An International Consortium Update: Pathophysiology, Diagnosis, and Treatment of Polycystic Ovarian Syndrome in Adolescence. *Hormone Research in Paediatrics, 88*(6), 371–395. DOI: 10.1159/000479371 PMID: 29156452

Ibrohim, I., Kusumaningrum, I. K., Hamimi, E., Putra, W. E., Utomo, J., Kundariati, M., & Malek, N. A. N. B. N. (2023). Science students' literacy and faculty members' perspective toward nanotechnology: Is it needed in 21st century education? *Eurasia Journal of Mathematics, Science and Technology Education*, *19*(12), em2362. Advance online publication. DOI: 10.29333/ejmste/13835

Illing, M. E., Rajan, R. S., Bence, N. F., & Kopito, R. R. (2002). A rhodopsin mutant linked to autosomal dominant retinitis pigmentosa is prone to aggregate and interacts with the ubiquitin proteasome system. *The Journal of Biological Chemistry*, *277*(37), 34150–34160. DOI: 10.1074/jbc.M204955200 PMID: 12091393

Imran, M., Aslam Gondal, T., Atif, M., Shahbaz, M., Batool Qaisarani, T., Hanif Mughal, M., Salehi, B., Martorell, M., & Sharifi-Rad, J. (2020). Apigenin as an anticancer agent. *Phytotherapy Research*, *34*(8), 1812–1828. DOI: 10.1002/ptr.6647 PMID: 32059077

Incebacak Eltemur, R. D., Nguyen, H. P., & Weber, J. J. (2022). Calpain-mediated proteolysis as driver and modulator of polyglutamine toxicity. *Frontiers in Molecular Neuroscience*, *15*, 1–17. DOI: 10.3389/fnmol.2022.1020104 PMID: 36385755

Iriondo, M. N., Etxaniz, A., Varela, Y. R., Ballesteros, U., Lázaro, M., Valle, M., Fracchiolla, D., Martens, S., Montes, L. R., Goñi, F. M., & Alonso, A. (2023). Effect of ATG12–ATG5-ATG16L1 autophagy E3-like complex on the ability of LC3/GABARAP proteins to induce vesicle tethering and fusion. *Cellular and Molecular Life Sciences : CMLS*, *80*(2), 56. DOI: 10.1007/s00018-023-04704-z PMID: 36729310

Itakura, E., Kishi, C., Inoue, K., & Mizushima, N. (2008). Beclin 1 forms two distinct phosphatidylinositol 3-kinase complexes with mammalian Atg14 and UVRAG. *Molecular Biology of the Cell*, *19*(12), 5360–5372. DOI: 10.1091/mbc.e08-01-0080 PMID: 18843052

Jamil, B., Javed, R., Qazi, A. S., & Syed, M. A. (2018). Nanomaterials: Toxicity, risk management, and public perception. *Nanomaterials: Ecotoxicity, safety, and public perception*, 283-304. DOI: 10.1007/978-3-030-05144-0_14

Jazieh, C., Arabi, T. Z., Asim, Z., Sabbah, B. N., Alsaud, A. W., Alkattan, K., & Yaqinuddin, A. (2024). Unraveling the epigenetic fabric of type 2 diabetes mellitus: Pathogenic mechanisms and therapeutic implications. *Frontiers in Endocrinology*, *15*, 1295967. DOI: 10.3389/fendo.2024.1295967 PMID: 38323108

Jeong, S. J., Stitham, J., Evans, T. D., Zhang, X., Rodriguez-Velez, A., Yeh, Y. S., Tao, J., Takabatake, K., Epelman, S., Lodhi, I. J., Schilling, J. D., DeBosch, B. J., Diwan, A., & Razani, B. (2021). Trehalose causes low-grade lysosomal stress to activate TFEB and the autophagy-lysosome biogenesis response. *Autophagy*, *17*(11), 3740–3752. DOI: 10.1080/15548627.2021.1896906 PMID: 33706671

Jeon, S.-H., Kim, S. H., Kim, Y., Kim, Y. S., Lim, Y., Lee, Y. H., & Shin, S. Y. (2011). The tricyclic antidepressant imipramine induces autophagic cell death in U-87MG glioma cells. *Biochemical and Biophysical Research Communications*, *413*(2), 311–317. DOI: 10.1016/j.bbrc.2011.08.093 PMID: 21889492

Jernigan, C. S., Goswami, D. B., Austin, M. C., Iyo, A. H., Chandran, A., Stockmeier, C. A., & Karolewicz, B. (2011). The mTOR signaling pathway in the prefrontal cortex is compromised in major depressive disorder. *Progress in Neuro-Psychopharmacology & Biological Psychiatry*, *35*(7), 1774–1779. DOI: 10.1016/j.pnpbp.2011.05.010 PMID: 21635931

Jiang, F., Xia, M., Zhang, Y., Chang, J., Cao, J., Zhang, Z., Qian, Z., & Yang, L. (2022). Cannabinoid receptor-2 attenuates neuroinflammation by promoting autophagy-mediated degradation of the NLRP3 inflammasome post spinal cord injury. *Frontiers in Immunology*, *13*, 993168. DOI: 10.3389/fimmu.2022.993168 PMID: 36238284

Jiang, P., Guo, Y., Dang, R., Yang, M., Liao, D., Li, H., Sun, Z., Feng, Q., & Xu, P. (2017). Salvianolic acid B protects against lipopolysaccharide-induced behavioral deficits and neuroinflammatory response: Involvement of autophagy and NLRP3 inflammasome. *Journal of Neuroinflammation*, *14*(1), 1–10. DOI: 10.1186/s12974-017-1013-4 PMID: 29212498

Jia, S. (2023). Combined Oral Contraceptives (COCs) Management on symptoms of Polycystic Ovary Syndrom: A Systematic Review. *Highlights in Science. Engineering and Technology*, *74*, 1532–1539.

Jiménez-Loygorri, J. I., Benítez-Fernández, R., Viedma-Poyatos, Á., Zapata-Muñoz, J., Villarejo-Zori, B., Gómez-Sintes, R., & Boya, P. (2023). Mitophagy in the retina: Viewing mitochondrial homeostasis through a new lens. *Progress in Retinal and Eye Research*, *96*, 101205. DOI: 10.1016/j.preteyeres.2023.101205 PMID: 37454969

Jindal, V. (2015). Interconnection between brain and retinal neurodegenerations. *Molecular Neurobiology*, *51*(3), 885–892. DOI: 10.1007/s12035-014-8733-6 PMID: 24826919

Jin, G.-N., Lu, J.-M., Lan, H.-W., Lu, Y.-N., Shen, X.-Y., Xu, X., & Piao, L.-X. (2022). Protective effect of ginsenoside Rh2 against Toxoplasma gondii infection-induced neuronal injury through binding TgCDPK1 and NLRP3 to inhibit microglial NLRP3 inflammasome signaling pathway. *International Immunopharmacology*, *112*, 109176. DOI: 10.1016/j.intimp.2022.109176 PMID: 36067653

Johansen, T., & Lamark, T. (2020). Selective autophagy: ATG8 family proteins, LIR motifs and cargo receptors. *Journal of Molecular Biology*, *432*(1), 80–103. DOI: 10.1016/j.jmb.2019.07.016 PMID: 31310766

Jose, J. P. A., Younus, L. A., Reddy, K. B., Sana, S. S., Gangadhar, L., Hou, T., & Bhardwaj, P. (2022). Environmental impact on toxicity of nanomaterials. In *Biogenic Sustainable Nanotechnology* (pp. 315–355). Elsevier. DOI: 10.1016/B978-0-323-88535-5.00011-1

Joubert, I. A., Geppert, M., Ess, S., Nestelbacher, R., Gadermaier, G., Duschl, A., Bathke, A. C., & Himly, M. (2020). Public perception and knowledge on nanotechnology: A study based on a citizen science approach. *NanoImpact*, *17*, 100201. DOI: 10.1016/j.impact.2019.100201

Jovanovic, F., Sudhakar, A., & Knezevic, N. N. (2022). The Kynurenine Pathway and Polycystic Ovary Syndrome: Inflammation as a Common Denominator. *International Journal of Tryptophan Research : IJTR*, *15*. DOI: 10.1177/11786469221099214 PMID: 35620306

Jusop, A. S., Thanaskody, K., Tye, G. J., Dass, S. A., Wan Kamarul Zaman, W. S., & Nordin, F. (2023). Development of brain organoid technology derived from iPSC for the neurodegenerative disease modeling: A glance through. *Frontiers in Molecular Neuroscience*, *16*, 1173433. DOI: 10.3389/fnmol.2023.1173433 PMID: 37602192

Kaarniranta, K., Blasiak, J., Liton, P., Boulton, M., Klionsky, D. J., & Sinha, D. (2023). Autophagy in age-related macular degeneration. *Autophagy*, *19*(2), 388–400. DOI: 10.1080/15548627.2022.2069437 PMID: 35468037

Kacal, M., Zhang, B., Hao, Y., Norberg, E., & Vakifahmetoglu-Norberg, H. (2021). Quantitative proteomic analysis of temporal lysosomal proteome and the impact of the KFERQ-like motif and LAMP2A in lysosomal targeting. *Autophagy*, *17*(11), 3865–3874. DOI: 10.1080/15548627.2021.1876343 PMID: 33446043

Kagohashi, Y., Sasaki, M., May, A. I., Kawamata, T., & Ohsumi, Y. (2023). The mechanism of Atg15-mediated membrane disruption in autophagy. *The Journal of Cell Biology*, *222*(12), e202306120. Advance online publication. DOI: 10.1083/jcb.202306120 PMID: 37917025

Kallergi, E., Daskalaki, A.-D., Kolaxi, A., Camus, C., Ioannou, E., Mercaldo, V., Haberkant, P., Stein, F., Sidiropoulou, K., Dalezios, Y., Savitski, M. M., Bagni, C., Choquet, D., Hosy, E., & Nikoletopoulou, V. (2022). Dendritic autophagy degrades postsynaptic proteins and is required for long-term synaptic depression in mice. *Nature Communications*, *13*(1), 680. DOI: 10.1038/s41467-022-28301-z PMID: 35115539

Kamarulzaman, N. A., Lee, K. E., Siow, K. S., & Mokhtar, M. (2020). Public benefit and risk perceptions of nanotechnology development: Psychological and sociological aspects. *Technology in Society*, *62*, 101329. DOI: 10.1016/j.techsoc.2020.101329

Kanno, H., Ozawa, H., Sekiguchi, A., Yamaya, S., & Itoi, E. (2011). Induction of autophagy and autophagic cell death in damaged neural tissue after acute spinal cord injury in mice. *Spine*, *36*(22), E1427–E34. DOI: 10.1097/BRS.0b013e3182028c3a PMID: 21304420

Kaplan, N., Wang, J., Wray, B., Patel, P., Yang, W., Peng, H., & Lavker, R. M. (2019). Single-cell RNA transcriptome helps define the limbal/corneal epithelial stem/early transit amplifying cells and how autophagy affects this population. *Investigative Ophthalmology & Visual Science*, *60*(10), 3570–3583. DOI: 10.1167/iovs.19-27656 PMID: 31419300

Kara, N. Z., Flaisher-Grinberg, S., Anderson, G. W., Agam, G., & Einat, H. (2018). Mood-stabilizing effects of rapamycin and its analog temsirolimus: Relevance to autophagy. *Behavioural Pharmacology*, *29*(4), 379–384. DOI: 10.1097/FBP.0000000000000334 PMID: 28777104

Kara, N., Toker, L., Agam, G., Anderson, G. W., Belmaker, R., & Einat, H. (2013). Trehalose induced antidepressant-like effects and autophagy enhancement in mice. *Psychopharmacology*, *229*(2), 367–375. DOI: 10.1007/s00213-013-3119-4 PMID: 23644913

Kaufman, P. L. (2005). Changes in aqueous humor dynamics with age and glaucoma. *Progress in Retinal and Eye Research*, *24*(5), 612–637. DOI: 10.1016/j.preteyeres.2004.10.003 PMID: 15919228

Kaushik, S., & Cuervo, A. M. (2018). The coming of age of chaperone-mediated autophagy. *Nature Reviews. Molecular Cell Biology*, *19*(6), 365–381. DOI: 10.1038/s41580-018-0001-6 PMID: 29626215

Kempuraj, D., Thangavel, R., Kempuraj, D. D., Ahmed, M. E., Selvakumar, G. P., Raikwar, S. P., Zaheer, S. A., Iyer, S. S., Govindarajan, R., Chandrasekaran, P. N., & Zaheer, A. (2021). Neuroprotective effects of flavone luteolin in neuroinflammation and neurotrauma. *BioFactors (Oxford, England)*, *47*(2), 190–197. DOI: 10.1002/biof.1687 PMID: 33098588

Kendler, K. S. (2005). Toward a philosophical structure for psychiatry. *The American Journal of Psychiatry*, *162*(3), 433–440. DOI: 10.1176/appi.ajp.162.3.433 PMID: 15741457

Ke, P.-Y. (2024). Molecular Mechanism of Autophagosome–Lysosome Fusion in Mammalian Cells. *Cells*, *13*(6), 500. DOI: 10.3390/cells13060500 PMID: 38534345

Khalifeh, M., Barreto, G., & Sahebkar, A. (2021). Therapeutic potential of trehalose in neurodegenerative diseases: The knowns and unknowns. *Neural Regeneration Research*, *16*(10), 2026–2027. DOI: 10.4103/1673-5374.308085 PMID: 33642389

Khawagi, W. Y., Al-kuraishy, H. M., Hussein, N. R., Al-Gareeb, A. I., Atef, E., Elhussieny, O., Alexiou, A., Papadakis, M., Jabir, M. S., Alshehri, A. A., Saad, H. M., & Batiha, G. E.-S. (2024). Depression and type 2 diabetes: A causal relationship and mechanistic pathway. *Diabetes, Obesity & Metabolism*, *26*(8), 3031–3044. DOI: 10.1111/dom.15630 PMID: 38802993

Khayachi, A., Abuzgaya, M., Liu, Y., Jiao, C., Dejgaard, K., Schorova, L., Kamesh, A., He, Q., Cousineau, Y., Pietrantonio, A., Farhangdoost, N., Castonguay, C.-E., Chaumette, B., Alda, M., Rouleau, G. A., & Milnerwood, A. J. (2023). Molecular signatures of hyperexcitability and lithium responsiveness in bipolar disorder patient neurons provide alternative therapeutic strategies. *BioRxiv*, 1-30. DOI: 10.1101/2023.07.21.550088

Kim, A. S., & Diamond, M. S. (2023). A molecular understanding of alphavirus entry and antibody protection. *Nature Reviews. Microbiology*, *21*(5), 396–407. DOI: 10.1038/s41579-022-00825-7 PMID: 36474012

Kim, A., Lalonde, K., Truesdell, A., Welter, P. G., Brocardo, P. S., Rosenstock, T. R., & Gil-mohapel, J. (2021). New avenues for the treatment of huntington's disease. *International Journal of Molecular Sciences*, *22*(16), 1–50. DOI: 10.3390/ijms22168363 PMID: 34445070

Kim, E. C., Meng, H., & Jun, A. S. (2013). Lithium treatment increases endothelial cell survival and autophagy in a mouse model of Fuchs endothelial corneal dystrophy. *The British Journal of Ophthalmology*, *97*(8), 1068–1073. DOI: 10.1136/bjophthalmol-2012-302881 PMID: 23759441

Kim, H. R., Kim, J. C., Kang, S. Y., Kim, H. O., Park, C. W., & Chung, B. Y. (2021). Rapamycin Alleviates 2,3,7,8-Tetrachlorodibenzo-p-dioxin-Induced Aggravated Dermatitis in Mice with Imiquimod-Induced Psoriasis-Like Dermatitis by Inducing Autophagy. *International Journal of Molecular Sciences*, *22*(8), 3968. DOI: 10.3390/ijms22083968 PMID: 33921372

Kim, J. H., Kim, J. H., Yu, Y. S., Mun, J. Y., & Kim, K.-W. (2010). Autophagy-induced regression of hyaloid vessels in early ocular development. *Autophagy*, *6*(7), 922–928. DOI: 10.4161/auto.6.7.13306 PMID: 20818164

Kim, J., Kundu, M., Viollet, B., & Guan, K.-L. (2011). AMPK and mTOR regulate autophagy through direct phosphorylation of Ulk1. *Nature Cell Biology*, *13*(2), 132–141. DOI: 10.1038/ncb2152 PMID: 21258367

Kirkin, V., & Rogov, V. V. (2019). A Diversity of Selective Autophagy Receptors Determines the Specificity of the Autophagy Pathway. *Molecular Cell*, *76*(2), 268–285. DOI: 10.1016/j.molcel.2019.09.005 PMID: 31585693

Kitagishi, Y., Kobayashi, M., Kikuta, K., & Matsuda, S. (2012). Roles of PI3K/AKT/GSK3/mTOR pathway in cell signaling of mental illnesses. *Depression Research and Treatment*, *2012*(1), 752563. DOI: 10.1155/2012/752563 PMID: 23320155

Klionsky, D. J., Petroni, G., Amaravadi, R. K., Baehrecke, E. H., Ballabio, A., Boya, P., Bravo-San Pedro, J. M., Cadwell, K., Cecconi, F., Choi, A. M. K., Choi, M. E., Chu, C. T., Codogno, P., Colombo, M. I., Cuervo, A. M., Deretic, V., Dikic, I., Elazar, Z., Eskelinen, E., & Pietrocola, F. (2021). Autophagy in major human diseases. *The EMBO Journal*, *40*(19), e108863. Advance online publication. DOI: 10.15252/embj.2021108863 PMID: 34459017

Knöferle, J., Koch, J. C., Ostendorf, T., Michel, U., Planchamp, V., Vutova, P., Tönges, L., Stadelmann, C., Brück, W., Bähr, M., & Lingor, P. (2010). Mechanisms of acute axonal degeneration in the optic nerve in vivo. *Proceedings of the National Academy of Sciences of the United States of America*, *107*(13), 6064–6069. DOI: 10.1073/pnas.0909794107 PMID: 20231460

Kocaturk, N., & Gozuacik, D. (2017). Otofaji ve Nörodejeneratif Hastalıklar. Turkiye Klinikleri J Pharmacol-Special Topics, 5(1):11-20. https://www.researchgate.net/publication/316168897

Kodi, T., Sankhe, R., Gopinathan, A., Nandakumar, K., & Kishore, A. (2024). New Insights on NLRP3 Inflammasome: Mechanisms of Activation, Inhibition, and Epigenetic Regulation. *Journal of Neuroimmune Pharmacology*, *19*(1), 7. DOI: 10.1007/s11481-024-10101-5 PMID: 38421496

Kohler, O., Krogh, J., Mors, O., & Benros, M. E. (2016). Inflammation in Depression and the Potential for Anti-Inflammatory Treatment. *Current Neuropharmacology*, *14*(7), 732–742. DOI: 10.2174/1570159X14666151208113700 PMID: 27640518

Komatsu, M., Qing, J. W., Holstein, G. R., Friedrich, V. L., Iwata, J. I., Kominami, E., Chait, B. T., Tanaka, K., & Yue, Z. (2007). Essential role for autophagy protein Atg7 in the maintenance of axonal homeostasis and the prevention of axonal degeneration. *Proceedings of the National Academy of Sciences of the United States of America, 104*(36), 14489–14494. DOI: 10.1073/pnas.0701311104 PMID: 17726112

Komatsu, M., Waguri, S., Koike, M., Sou, Y., Ueno, T., Hara, T., Mizushima, N., Iwata, J., Ezaki, J., Murata, S., Hamazaki, J., Nishito, Y., Iemura, S., Natsume, T., Yanagawa, T., Uwayama, J., Warabi, E., Yoshida, H., Ishii, T., & Tanaka, K. (2007). Homeostatic Levels of p62 Control Cytoplasmic Inclusion Body Formation in Autophagy-Deficient Mice. *Cell, 131*(6), 1149–1163. DOI: 10.1016/j.cell.2007.10.035 PMID: 18083104

Komorowska, J., Wątroba, M., Bednarzak, M., Grabowska, A. D., & Szukiewicz, D. (2023). The Role of Glucose Concentration and Resveratrol in Modulating Neuroinflammatory Cytokines: Insights from an In Vitro Blood-Brain Barrier Model. *Medical Science Monitor, 29*, e941044. DOI: 10.12659/MSM.941044 PMID: 37817396

Kong, J., Zhang, Q., Zheng, H., Tang, D., Fang, L., An, S., Li, J., & Fan, Z. (2024). TGN-020 ameliorates motor dysfunction post-spinal cord injury via enhancing astrocyte autophagy and mitigating inflammation by activating AQP4/PPAR-γ/mTOR pathway. *Experimental Neurology, 382*, 114975. DOI: 10.1016/j.expneurol.2024.114975 PMID: 39326822

Kozlov, A. V., Javadov, S., & Sommer, N. (2024). Cellular ROS and Antioxidants: Physiological and Pathological Role. *Antioxidants, 13*(5), 602. DOI: 10.3390/antiox13050602 PMID: 38790707

Krause, G. J., Kirchner, P., Stiller, B., Morozova, K., Diaz, A., Chen, K.-H., Krogan, N. J., Agullo-Pascual, E., Clement, C. C., Lindenau, K., Swaney, D. L., Dilipkumar, S., Bravo-Cordero, J. J., Santambrogio, L., & Cuervo, A. M. (2023). Molecular determinants of the crosstalk between endosomal microautophagy and chaperone-mediated autophagy. *Cell Reports, 42*(12), 113529. DOI: 10.1016/j.celrep.2023.113529 PMID: 38060380

Kroemer, G., Mariño, G., & Levine, B. (2010). Autophagy and the integrated stress response. *Molecular Cell, 40*(2), 280–293. DOI: 10.1016/j.molcel.2010.09.023 PMID: 20965422

Kroken, A. R., Gajenthra Kumar, N., Yahr, T. L., Smith, B. E., Nieto, V., Horneman, H., Evans, D. J., & Fleiszig, S. M. (2022). Exotoxin S secreted by internalized Pseudomonas aeruginosa delays lytic host cell death. *PLoS Pathogens, 18*(2), e1010306. DOI: 10.1371/journal.ppat.1010306 PMID: 35130333

Kullmann, S., Heni, M., Hallschmid, M., Fritsche, A., Preissl, H., & Häring, H.-U. (2016). Brain insulin resistance at the crossroads of metabolic and cognitive disorders in humans. *Physiological Reviews*, *96*(4), 1169–1209. DOI: 10.1152/physrev.00032.2015 PMID: 27489306

Kuma, A., Hatano, M., Matsui, M., Yamamoto, A., Nakaya, H., Yoshimori, T., Ohsumi, Y., Tokuhisa, T., & Mizushima, N. (2004). The role of autophagy during the early neonatal starvation period. *Nature*, *432*(7020), 1032–1036. DOI: 10.1038/nature03029 PMID: 15525940

Kumar, A., Yap, K. C., BharathwajChetty, B., Lyu, J., Hegde, M., Abbas, M., Alqahtani, M. S., Khadlikar, S., Zarrabi, A., Khosravi, A., Kumar, A. P., & Kunnumakkara, A. B. (2024). Regulating the regulators: Long noncoding RNAs as autophagic controllers in chronic disease management. *Journal of Biomedical Science*, *31*(1), 105. DOI: 10.1186/s12929-024-01092-9 PMID: 39716252

Kumar, D., Bishnoi, M., Kondepudi, K. K., & Sharma, S. S. (2025). Gut microbiota-based interventions for Parkinson's disease: Neuroprotective mechanisms and current perspective. *Probiotics and Antimicrobial Proteins*. Advance online publication. DOI: 10.1007/s12602-024-10433-x PMID: 39809955

Kumar, P., Mahajan, P., Kaur, R., & Gautam, S. (2020). Nanotechnology and its challenges in the food sector: A review. *Materials Today. Chemistry*, *17*, 100332. DOI: 10.1016/j.mtchem.2020.100332 PMID: 32835156

Lai, H., Jia, X., Yu, Q., Zhang, C., Qiao, J., Guan, Y., & Kang, J. (2014). High-fat diet induces significant metabolic disorders in a mouse model of polycystic ovary syndrome. *Biology of Reproduction, 91*(5), 127, 121–111.

Lakkaraju, A., Umapathy, A., Tan, L. X., Daniele, L., Philp, N. J., Boesze-Battaglia, K., & Williams, D. S. (2020). The cell biology of the retinal pigment epithelium. *Progress in Retinal and Eye Research*, *78*, 100846. DOI: 10.1016/j.preteyeres.2020.100846 PMID: 32105772

Lamark, T., & Johansen, T. (2021). Mechanisms of selective autophagy. *Annual Review of Cell and Developmental Biology*, *37*(1), 143–169. DOI: 10.1146/annurev-cellbio-120219-035530 PMID: 34152791

Lamark, T., Kirkin, V., Dikic, I., & Johansen, T. (2009). NBR1 and p62 as cargo receptors for selective autophagy of ubiquitinated targets. *Cell Cycle (Georgetown, Tex.)*, *8*(13), 1986–1990. DOI: 10.4161/cc.8.13.8892 PMID: 19502794

Lam, X. J., Xu, B., Yeo, P. L., Cheah, P. S., & Ling, K. H. (2023). Mitochondria dysfunction and bipolar disorder: From pathology to therapy. *IBRO Neuroscience Reports*, *14*, 407–418. DOI: 10.1016/j.ibneur.2023.04.002 PMID: 37388495

Langley, M. R., Triplet, E. M., & Scarisbrick, I. A. (2020). Dietary influence on central nervous system myelin production, injury, and regeneration. Biochimica et Biophysica Acta (BBA) -. *Biochimica et Biophysica Acta. Molecular Basis of Disease*, *1866*(7), 165779. DOI: 10.1016/j.bbadis.2020.165779 PMID: 32224154

Lau, G. M., Elghobashy, M., Thanki, M., Ibegbulam, S., Latthe, P., Gillett, C. D., O'Reilly, M. W., Arlt, W., Lindenmeyer, A., & Kempegowda, P. (2022). A systematic review of lived experiences of people with polycystic ovary syndrome highlights the need for holistic care and co-creation of educational resources. *Frontiers in Endocrinology*, *13*, 1064937. DOI: 10.3389/fendo.2022.1064937 PMID: 36531482

Le Bras, A. (2022). New insights into the origin of amyloid plaques. *Lab Animal*, *51*(7), 187–187. DOI: 10.1038/s41684-022-01007-x

Le Guerroué, F., & Youle, R. J. (2021). Ubiquitin signaling in neurodegenerative diseases: An autophagy and proteasome perspective. *Cell Death and Differentiation*, *28*(2), 439–454. DOI: 10.1038/s41418-020-00667-x PMID: 33208890

Lee, E.-J., Kim, H.-J., Choi, M. S., & Chang, J.-E. (2021). Crosstalk between Autophagy and Inflammatory Processes in Cancer. *Life (Chicago, Ill.)*, *11*(9), 903. DOI: 10.3390/life11090903 PMID: 34575052

Lee, J. A. (2012). Neuronal autophagy: A housekeeper or a fighter in neuronal cell survival? *Experimental Neurobiology*, *21*(1), 1–8. DOI: 10.5607/en.2012.21.1.1 PMID: 22438673

Lee, J. M., Wagner, M., Xiao, R., Kim, K. H., Feng, D., Lazar, M. A., & Moore, D. D. (2014). Nutrient-sensing nuclear receptors coordinate autophagy. *Nature*, *516*(7529), 112–115. DOI: 10.1038/nature13961 PMID: 25383539

Lee, J., Sung, K. W., Bae, E. J., Yoon, D., Kim, D., Lee, J. S., Park, D., Park, D. Y., Mun, S. R., Kwon, S. C., Kim, H. Y., Min, J.-O., Lee, S.-J., Suh, Y. H., & Kwon, Y. T. (2023). Targeted degradation of α-synuclein aggregates in Parkinson's disease using the AUTOTAC technology. *Molecular Neurodegeneration*, *18*(1), 1–21. DOI: 10.1186/s13024-023-00630-7 PMID: 36597122

Lee, J., Yoon, D., Sung, K. W., Bae, E. J., Park, D. H., Suh, Y. H., & Kwon, Y. T. (2024). Targeted degradation of SNCA/α-synuclein aggregates in neurodegeneration using the AUTOTAC chemical platform. *Autophagy*, *20*(2), 463–465. DOI: 10.1080/15548627.2023.2274711 PMID: 37915239

Lee, K.-M., Hwang, S.-K., & Lee, J.-A. (2013). Neuronal autophagy and neurodevelopmental disorders. *Experimental Neurobiology*, *22*(3), 133–142. DOI: 10.5607/en.2013.22.3.133 PMID: 24167408

Lee, S. W., Oh, Y. M., Victor, M. B., Yang, Y., Chen, S., Strunilin, I., Dahiya, S., Dolle, R. E., Pak, S. C., Silverman, G. A., Perlmutter, D. H., & Yoo, A. S. (2024). Longitudinal modeling of human neuronal aging reveals the contribution of the RCAN1–TFEB pathway to Huntington's disease neurodegeneration. *Nature Aging*, *4*(1), 95–109. DOI: 10.1038/s43587-023-00538-3 PMID: 38066314

Lei, Y., & Klionsky, D. J. (2023). Transcriptional regulation of autophagy and its implications in human disease. *Cell Death and Differentiation*, *30*(6), 1416–1429. DOI: 10.1038/s41418-023-01162-9 PMID: 37045910

Lepinay, E., & Cicchetti, F. (2023). Tau: A biomarker of Huntington's disease. *Molecular Psychiatry*, *28*(10), 4070–4083. DOI: 10.1038/s41380-023-02230-9 PMID: 37749233

Levine, B., & Kroemer, G. (2019). Biological functions of autophagy genes: A disease perspective. *Cell*, *176*(1), 11–42. DOI: 10.1016/j.cell.2018.09.048 PMID: 30633901

Levine, B., Liu, R., Dong, X., & Zhong, Q. (2015). Beclin orthologs: Integrative hubs of cell signaling, membrane trafficking, and physiology. *Trends in Cell Biology*, *25*(9), 533–544. DOI: 10.1016/j.tcb.2015.05.004 PMID: 26071895

Li, S. (2020). Three Dimensions of Nano-ethics. *Chinese Philosophy of Technology: Classical Readings and Contemporary Work*, 165-181. DOI: 10.1007/978-981-15-1952-9_11

Liang, C. (2010). Negative regulation of autophagy. *Cell Death and Differentiation*, *17*(12), 1807–1815. DOI: 10.1038/cdd.2010.115 PMID: 20865012

Liang, L., Wang, H., Hu, Y., Bian, H., Xiao, L., & Wang, G. (2022). Oridonin relieves depressive-like behaviors by inhibiting neuroinflammation and autophagy impairment in rats subjected to chronic unpredictable mild stress. *Phytotherapy Research*, *36*(8), 3335–3351. DOI: 10.1002/ptr.7518 PMID: 35686337

Liang, T., Zhang, Y., Wu, S., Chen, Q., & Wang, L. (2022). The Role of NLRP3 Inflammasome in Alzheimer's Disease and Potential Therapeutic Targets. *Frontiers in Pharmacology*, *13*, 845185. DOI: 10.3389/fphar.2022.845185 PMID: 35250595

Liao, D., Shangguan, D., Wu, Y., Chen, Y., Liu, N., Tang, J., Yao, D., & Shi, Y. (2023). Curcumin protects against doxorubicin induced oxidative stress by regulating the Keap1-Nrf2-ARE and autophagy signaling pathways. *Psychopharmacology*, *240*(5), 1179–1190. DOI: 10.1007/s00213-023-06357-z PMID: 36949340

Liao, H.-Y., Wang, Z.-Q., Ran, R., Zhou, K.-S., Ma, C.-W., & Zhang, H.-H. (2021). Biological functions and therapeutic potential of autophagy in spinal cord injury. *Frontiers in Cell and Developmental Biology*, *9*, 761273. DOI: 10.3389/fcell.2021.761273 PMID: 34988074

Liao, W., Shi, Y., Li, Z., & Yin, X. (2025). Advances in 3D printing combined with tissue engineering for nerve regeneration and repair. *Journal of Nanobiotechnology*, *23*(1), 5. DOI: 10.1186/s12951-024-03052-9 PMID: 39754257

Li, B., Li, P., Weng, R., Wu, Z., Qin, B., Fang, J., Wang, Y., Qiu, S., Yang, J., & Gu, L. (2021). Trehalose protects motorneuron after brachial plexus root avulsion by activating autophagy and inhibiting apoptosis mediated by the AMPK signaling pathway. *Gene*, *768*(145307), 1–7. DOI: 10.1016/j.gene.2020.145307 PMID: 33197516

Li, C., Liu, B., Wang, Z.-y., Xie, F., Qiao, W., Cheng, J., Kuang, J., Wang, Y., Zhang, M., & Liu, D. (2020). Salvianolic acid B improves myocardial function in diabetic cardiomyopathy by suppressing IGFBP3. *Journal of Molecular and Cellular Cardiology*, *139*, 98–112. DOI: 10.1016/j.yjmcc.2020.01.009 PMID: 31982427

Li, D., Liu, J., Guo, B., Liang, C., Dang, L., Lu, C., He, X., Cheung, H. Y., Xu, L., Lu, C., He, B., Liu, B., Shaikh, A. B., Li, F., Wang, L., Yang, Z., Au, D. W., Peng, S., Zhang, Z., & Zhang, G. (2016). Osteoclast-derived exosomal miR-214-3p inhibits osteoblastic bone formation. *Nature Communications*, *7*(1), 10872. DOI: 10.1038/ncomms10872 PMID: 26947250

Liénard, C., Pintart, A., & Bomont, P. (2024). Neuronal Autophagy: Regulations and Implications in Health and Disease. *Cells*, *13*(1), 1–33. DOI: 10.3390/cells13010103 PMID: 38201307

Li, J., Cao, Y., Li, L.-N., Chu, X., Wang, Y.-S., Cai, J.-J., Zhao, J., Ma, S., Li, G., & Fan, Z.-K. (2023). Neuroprotective effects of oxymatrine via triggering autophagy and inhibiting apoptosis following spinal cord injury in rats. *Molecular Neurobiology*, *60*(8), 4450–4471. DOI: 10.1007/s12035-023-03364-1 PMID: 37115405

Li, J., Zhou, Y., Liu, B. B., Liu, Q., Geng, D., Weng, L. J., & Yi, L.-T. (2013). Nobiletin Ameliorates the Deficits in Hippocampal BDNF, TrkB, and Synapsin I Induced by Chronic Unpredictable Mild Stress. *Evidence-Based Complementary and Alternative Medicine : eCAM*, *2013*, 359682. DOI: 10.1155/2013/359682 PMID: 23573124

Li, M., Gao, Z.-L., Zhang, Q.-P., Luo, A.-X., Xu, W.-Y., Duan, T.-Q., Wen, X.-P., Zhang, R.-Q., Zeng, R., & Huang, J.-F. (2022). Autophagy in glaucoma pathogenesis: Therapeutic potential and future perspectives. *Frontiers in Cell and Developmental Biology*, *10*, 1068213. DOI: 10.3389/fcell.2022.1068213 PMID: 36589756

Li, M., Gong, J., Gao, L., Zou, T., Kang, J., & Xu, H. (2022). Advanced human developmental toxicity and teratogenicity assessment using human organoid models. *Ecotoxicology and Environmental Safety*, *235*, 113429. DOI: 10.1016/j.ecoenv.2022.113429 PMID: 35325609

Limanaqi, F., Biagioni, F., Busceti, C. L., Ryskalin, L., Polzella, M., Frati, A., & Fornai, F. (2019). Phytochemicals Bridging Autophagy Induction and Alpha-Synuclein Degradation in Parkinsonism. *International Journal of Molecular Sciences*, *20*(13), 3274. DOI: 10.3390/ijms20133274 PMID: 31277285

Li, N., Lee, B., Liu, R.-J., Banasr, M., Dwyer, J. M., Iwata, M., Li, X.-Y., Aghajanian, G., & Duman, R. S. (2010). mTOR-dependent synapse formation underlies the rapid antidepressant effects of NMDA antagonists. *Science*, *329*(5994), 959–964. DOI: 10.1126/science.1190287 PMID: 20724638

Lin, C.-C., & Lin, M.-S. (2016). New insight into curcumin-based therapy in spinal cord injuries: CISD2 regulation. *Neural Regeneration Research*, *11*(2), 222–223. DOI: 10.4103/1673-5374.177718 PMID: 27073364

Ling, C., Lei, C., Zou, M., Cai, X., Xiang, Y., Xie, Y., Li, X., Huang, D., & Wang, Y. (2020). Neuroprotective effect of apigenin against cerebral ischemia/reperfusion injury. *The Journal of International Medical Research*, *48*(9). DOI: 10.1177/0300060520945859 PMID: 32993408

Lin, J., Huo, X., & Liu, X. (2017). "mTOR signaling pathway": A potential target of curcumin in the treatment of spinal cord injury. *BioMed Research International*, *2017*(1), 1634801. DOI: 10.1155/2017/1634801 PMID: 28691015

Lin, K.-H., Li, C.-Y., Hsu, Y.-M., Tsai, C.-H., Tsai, F.-J., Tang, C.-H., Yang, J.-S., Wang, Z.-H., & Yin, M.-C. (2019). Oridonin, A natural diterpenoid, protected NGF-differentiated PC12 cells against MPP+-and kainic acid-induced injury. *Food and Chemical Toxicology*, *133*, 110765. DOI: 10.1016/j.fct.2019.110765 PMID: 31430510

Linnerbauer, M., Wheeler, M. A., & Quintana, F. J. (2020). Astrocyte crosstalk in CNS inflammation. *Neuron*, *108*(4), 608–622. DOI: 10.1016/j.neuron.2020.08.012 PMID: 32898475

Lin, S., Tian, H., Lin, J., Xu, C., Yuan, Y., Gao, S., Song, C., Lv, P., & Mei, X. (2020). Zinc promotes autophagy and inhibits apoptosis through AMPK/mTOR signaling pathway after spinal cord injury. *Neuroscience Letters*, *736*, 135263. DOI: 10.1016/j.neulet.2020.135263 PMID: 32682846

Lin, T.-H., Chen, W.-L., Hsu, S.-F., Chen, I.-C., Lin, C.-H., Chang, K.-H., Wu, Y.-R., Chen, Y.-R., Yao, C.-F., Lin, W., Lee-Chen, G.-J., & Chen, C.-M. (2024). Small Molecules Inducing Autophagic Degradation of Expanded Polyglutamine Protein through Interaction with Both Mutant ATXN3 and LC3. *International Journal of Molecular Sciences*, *25*(19), 10707. DOI: 10.3390/ijms251910707 PMID: 39409036

Lin, W., & Kuang, H.-y. (2014). Oxidative stress induces autophagy in response to multiple noxious stimuli in retinal ganglion cells. *Autophagy*, *10*(10), 1692–1701. DOI: 10.4161/auto.36076 PMID: 25207555

Lin, W., & Xu, G. (2019). Autophagy: A role in the apoptosis, survival, inflammation, and development of the retina. *Ophthalmic Research*, *61*(2), 65–72. DOI: 10.1159/000487486 PMID: 29694961

Lipinski, M. M., Wu, J., Faden, A. I., & Sarkar, C. (2015). Function and mechanisms of autophagy in brain and spinal cord trauma. *Antioxidants & Redox Signaling*, *23*(6), 565–577. DOI: 10.1089/ars.2015.6306 PMID: 25808205

Li, R., Zhao, D., Qu, R., Fu, Q., & Ma, S. (2015). The effects of apigenin on lipopolysaccharide-induced depressive-like behavior in mice. *Neuroscience Letters*, *594*, 17–22. DOI: 10.1016/j.neulet.2015.03.040 PMID: 25800110

Liton, P. B., Boesze-Battaglia, K., Boulton, M. E., Boya, P., Ferguson, T. A., Ganley, I. G., Kauppinnen, A., Laurie, G. W., Mizushima, N., Morishita, H., Russo, R., Sadda, J., Shyam, R., Sinha, D., Thompson, D. A., & Zacks, D. N. (2023). Autophagy in the eye: From physiology to pathophysiology. *Autophagy Reports*, *2*(1), 2178996. DOI: 10.1080/27694127.2023.2178996 PMID: 37034386

Litwiniuk, A., Juszczak, G. R., Stankiewicz, A. M., & Urbańska, K. (2023). The role of glial autophagy in Alzheimer's disease. *Molecular Psychiatry*, *28*(11), 4528–4539. DOI: 10.1038/s41380-023-02242-5 PMID: 37679471

Liu, C., Hao, S., Zhu, M., Wang, Y., Zhang, T., & Yang, Z. (2018). Maternal separation induces different autophagic responses in the hippocampus and prefrontal cortex of adult rats. *Neuroscience*, *374*, 287–294. DOI: 10.1016/j.neuroscience.2018.01.043 PMID: 29391188

Liu, J., Bassal, M., Schlichting, S., Braren, I., Di Spiezio, A., Saftig, P., & Bartsch, U. (2022). Intravitreal gene therapy restores the autophagy-lysosomal pathway and attenuates retinal degeneration in cathepsin D-deficient mice. *Neurobiology of Disease*, *164*, 105628. DOI: 10.1016/j.nbd.2022.105628 PMID: 35033660

Liu, J., Xiao, Y., Cao, L., Lu, S., Zhang, S., Yang, R., Wang, Y., Zhang, N., Yu, Y., Wang, X., Guo, W., Wang, Z., Xu, H., Xing, C., Song, X., & Cao, L. (2024). Insights on E1-like enzyme ATG7: Functional regulation and relationships with aging-related diseases. *Communications Biology*, *7*(1), 382. DOI: 10.1038/s42003-024-06080-1 PMID: 38553562

Liu, L., Liu, C., Wang, Y., Wang, P., Li, Y., & Li, B. (2015). Herbal medicine for anxiety, depression and insomnia. *Current Neuropharmacology*, *13*(4), 481–493. DOI: 10.2174/1570159X13041508311122734 PMID: 26412068

Liu, L., Wang, H., Chen, X., Zhang, Y., Zhang, H., & Xie, P. (2023). Gut microbiota and its metabolites in depression: From pathogenesis to treatment. *EBioMedicine*, *90*, 90. DOI: 10.1016/j.ebiom.2023.104527 PMID: 36963238

Liu, L., Zhang, Q., Cai, Y., Sun, D., He, X., Wang, L., Yu, D., Li, X., Xiong, X., Xu, H., Yang, Q., & Fan, X. (2016). Resveratrol counteracts lipopolysaccharide-induced depressive-like behaviors via enhanced hippocampal neurogenesis. *Oncotarget*, *7*(35), 56045–56059. DOI: 10.18632/oncotarget.11178 PMID: 27517628

Liu, L., & Zhu, G. (2018). Gut–brain axis and mood disorder. *Frontiers in Psychiatry*, *9*, 223. DOI: 10.3389/fpsyt.2018.00223 PMID: 29896129

Liu, P., & Du, J. (2020). Oridonin is an antidepressant molecule working through the PPAR-γ/AMPA receptor signaling pathway. *Biochemical Pharmacology*, *180*, 114136. DOI: 10.1016/j.bcp.2020.114136 PMID: 32628930

Liu, Q.-R., Aseer, K. R., Yao, Q., Zhong, X., Ghosh, P., O'Connell, J. F., & Egan, J. M. (2022). Anti-inflammatory and pro-autophagy effects of the cannabinoid receptor CB2R: Possibility of modulation in type 1 diabetes. *Frontiers in Pharmacology*, *12*, 809965. DOI: 10.3389/fphar.2021.809965 PMID: 35115945

Liu, R., Jiang, L., Chen, Y., Shao, J., Chen, K., Li, X., Lv, J., Cai, W., Cai, H., Zhu, Z., Wang, C., Zhou, K., Huang, J., Xiao, J., Ni, W., & Wu, C. (2024). Ginsenoside-Rh2 promotes functional recovery after spinal cord Injury by enhancing TFEB-Mediated autophagy. *Journal of Agricultural and Food Chemistry*, *72*(26), 14727–14746. DOI: 10.1021/acs.jafc.4c02379 PMID: 38907713

Liu, R., Wu, J., Guo, H., Yao, W., Li, S., Lu, Y., Jia, Y., Liang, X., Tang, J., & Zhang, H. (2023). Posttranslational modifications of histones: Mechanisms, biological functions, and therapeutic targets. *MedComm*, *4*(3), e292. DOI: 10.1002/mco2.292 PMID: 37220590

Liu, S., Li, Y., Choi, H. M., Sarkar, C., Koh, E. Y., Wu, J., & Lipinski, M. M. (2018). Lysosomal damage after spinal cord injury causes accumulation of RIPK1 and RIPK3 proteins and potentiation of necroptosis. *Cell Death & Disease*, *9*(5), 476. DOI: 10.1038/s41419-018-0469-1 PMID: 29686269

Liu, S., Yao, S., Yang, H., Liu, S., & Wang, Y. (2023). Autophagy: Regulator of cell death. *Cell Death & Disease*, *14*(10), 648. DOI: 10.1038/s41419-023-06154-8 PMID: 37794028

Liu, T., Ma, Y., Zhang, R., Zhong, H., Wang, L., Zhao, J., Yang, L., & Fan, X. (2019). Resveratrol ameliorates estrogen deficiency-induced depression-and anxiety-like behaviors and hippocampal inflammation in mice. *Psychopharmacology*, *236*(4), 1385–1399. DOI: 10.1007/s00213-018-5148-5 PMID: 30607478

Liu, W., Liu, S., Li, P., & Yao, K. (2022). Retinitis pigmentosa: Progress in molecular pathology and biotherapeutical strategies. *International Journal of Molecular Sciences*, *23*(9), 4883. DOI: 10.3390/ijms23094883 PMID: 35563274

Liu, Y., Chu, W., Ma, H., Peng, W., Li, Q., Han, L., Wang, H., Wang, L., Zhang, B., Yang, J., & Lu, X. (2024). Fisetin orchestrates neuroinflammation resolution and facilitates spinal cord injury recovery through enhanced autophagy in pro-inflammatory glial cells. *International Immunopharmacology*, *130*, 111738. DOI: 10.1016/j.intimp.2024.111738 PMID: 38428149

Liu, Y., Pan, R., Ouyang, Y., Gu, W., Xiao, T., Yang, H., Tang, L., Wang, H., Xiang, B., & Chen, P. (2024). Pyroptosis in health and disease: Mechanisms, regulation, and clinical perspective. *Signal Transduction and Targeted Therapy*, *9*(1), 245. DOI: 10.1038/s41392-024-01958-2 PMID: 39300122

Liu, Y., Xu, X., Xu, R., & Zhang, S. (2019). Renoprotective effects of isoliquiritin against cationic bovine serum albumin-induced membranous glomerulonephritis in experimental rat model through its anti-oxidative and anti-inflammatory properties. *Drug Design, Development and Therapy*, *13*, 3735–3751. DOI: 10.2147/DDDT.S213088 PMID: 31802848

Li, W., Chen, Q., Peng, C., Yang, D., Liu, S., Lv, Y., Jiang, L., Xu, S., & Huang, L. (2025). Roles of the Receptor for Advanced Glycation End Products and Its Ligands in the Pathogenesis of Alzheimer's Disease. *International Journal of Molecular Sciences*, *26*(1), 403. DOI: 10.3390/ijms26010403 PMID: 39796257

Li, W., He, P., Huang, Y., Li, Y.-F., Lu, J., Li, M., Kurihara, H., Luo, Z., Meng, T., Onishi, M., Ma, C., Jiang, L., Hu, Y., Gong, Q., Zhu, D., Xu, Y., Liu, R., Liu, L., Yi, C., & Feng, D. (2021). Selective autophagy of intracellular organelles: Recent research advances. *Theranostics*, *11*(1), 222–256. DOI: 10.7150/thno.49860 PMID: 33391472

Li, W., Yao, S., Li, H., Meng, Z., & Sun, X. (2021). Curcumin promotes functional recovery and inhibits neuronal apoptosis after spinal cord injury through the modulation of autophagy. *The Journal of Spinal Cord Medicine*, *44*(1), 37–45. DOI: 10.1080/10790268.2019.1616147 PMID: 31162984

Li, Y. Y., Qin, Z. H., & Sheng, R. (2024). The Multiple Roles of Autophagy in Neural Function and Diseases. *Neuroscience Bulletin*, *40*(3), 363–382. DOI: 10.1007/s12264-023-01120-y PMID: 37856037

Li, Y., Guo, Y., Fan, Y., Tian, H., Li, K., & Mei, X. (2019). Melatonin enhances autophagy and reduces apoptosis to promote locomotor recovery in spinal cord injury via the PI3K/AKT/mTOR signaling pathway. *Neurochemical Research*, *44*(8), 2007–2019. DOI: 10.1007/s11064-019-02838-w PMID: 31325156

Li, Y., Lei, Z., Ritzel, R. M., He, J., Li, H., Choi, H. M. C., Lipinski, M. M., & Wu, J. (2022). Impairment of autophagy after spinal cord injury potentiates neuroinflammation and motor function deficit in mice. *Theranostics*, *12*(12), 5364–5388. DOI: 10.7150/thno.72713 PMID: 35910787

Li, Y., Li, S., & Wu, H. (2022). Ubiquitination-Proteasome System (UPS) and Autophagy Two Main Protein Degradation Machineries in Response to Cell Stress. *Cells*, *11*(5), 851. DOI: 10.3390/cells11050851 PMID: 35269473

Li, Y., Song, W., Tong, Y., Zhang, X., Zhao, J., Gao, X., Yong, J., & Wang, H. (2021). Isoliquiritin ameliorates depression by suppressing NLRP3-mediated pyroptosis via miRNA-27a/SYK/NF-κB axis. *Journal of Neuroinflammation*, *18*(1), 1. DOI: 10.1186/s12974-020-02040-8 PMID: 33402173

Li, Y., Zhang, W., Zhang, Q., Li, Y., Xin, C., Tu, R., & Yan, H. (2024). Oxidative stress of mitophagy in neurodegenerative diseases: Mechanism and potential therapeutic targets. *Archives of Biochemistry and Biophysics, 110283*. Advance online publication. DOI: 10.1016/j.abb.2024.110283 PMID: 39743032

Lizama, B. N., & Chu, C. T. (2021). Neuronal autophagy and mitophagy in Parkinson's disease. *Molecular Aspects of Medicine*, *82*, 100972. DOI: 10.1016/j.mam.2021.100972 PMID: 34130867

Lockshin, R. A., & Zakeri, Z. (2004). Apoptosis, autophagy, and more. *The International Journal of Biochemistry & Cell Biology, 36*(12), 2405–2419. DOI: 10.1016/j.biocel.2004.04.011 PMID: 15325581

Logan, S., Arzua, T., Canfield, S. G., Seminary, E. R., Sison, S. L., Ebert, A. D., & Bai, X. (2019). Studying Human Neurological Disorders Using Induced Pluripotent Stem Cells: From 2D Monolayer to 3D Organoid and Blood Brain Barrier Models. *Comprehensive Physiology, 9*(2), 565–611. DOI: 10.1002/j.2040-4603.2019.tb00076.x PMID: 30873582

Logeshwaran, A., Elsen, R., & Nayak, S. (2024). Artificial intelligence-based 3D printing strategies for bone scaffold fabrication and their application in preclinical and clinical investigations. *ACS Biomaterials Science & Engineering, 10*(2), 677–696. DOI: 10.1021/acsbiomaterials.3c01368 PMID: 38252807

Loh, J. S., Mak, W. Q., Tan, L. K. S., Ng, C. X., Chan, H. H., Yeow, S. H., Foo, J. B., Ong, Y. S., How, C. W., & Khaw, K. Y. (2024). Microbiota–gut–brain axis and its therapeutic applications in neurodegenerative diseases. *Signal Transduction and Targeted Therapy, 9*(37), 37. Advance online publication. DOI: 10.1038/s41392-024-01743-1 PMID: 38360862

Longo, V. D., Di Tano, M., Mattson, M. P., & Guidi, N. (2021). Intermittent and periodic fasting, longevity and disease. *Nature Aging, 1*(1), 47–59. DOI: 10.1038/s43587-020-00013-3 PMID: 35310455

Louwers, Y. V., & Laven, J. S. E. (2020). Characteristics of polycystic ovary syndrome throughout life. *Therapeutic Advances in Reproductive Health, 14*. DOI: 10.1177/2633494120911038 PMID: 32518918

Lu, H. C., & Mackie, K. (2021). Review of the Endocannabinoid System. *Biological Psychiatry: Cognitive Neuroscience and Neuroimaging, 6*(6), 607–615. DOI: 10.1016/j.bpsc.2020.07.016 PMID: 32980261

Luis-Ravelo, D., Estévez-Silva, H., Barroso-Chinea, P., Afonso-Oramas, D., Salas-Hernández, J., Rodríguez-Núñez, J., Acevedo-Arozena, A., Marcellino, D., & González-Hernández, T. (2018). Pramipexole reduces soluble mutant huntingtin and protects striatal neurons through dopamine D3 receptors in a genetic model of Huntington's disease. *Experimental Neurology, 299*, 137–147. DOI: 10.1016/j.expneurol.2017.10.019 PMID: 29056363

Lu, J., Wu, M., & Yue, Z. (2020). Autophagy and Parkinson's Disease. *Advances in Experimental Medicine and Biology, 1207*, 21–51. DOI: 10.1007/978-981-15-4272-5_2 PMID: 32671737

Luo, C., & Tao, L. (2020). The function and mechanisms of autophagy in spinal cord injury. Autophagy: Biology and Diseases. *Clinical Science*, 649–654.

Luo, Y., Shang, P., & Li, D. (2017). Luteolin: A flavonoid that has multiple cardio-protective effects and its molecular mechanisms. *Frontiers in Pharmacology, 8*, 692. DOI: 10.3389/fphar.2017.00692 PMID: 29056912

Luty-Błocho, M., Hoyos, M. V., & Hessel, V. (2023). Implications of the Use of Functional Nanomaterials for the Environment and Health: Risk Assessment and Challenges. In *Functional Nanomaterials for Sensors* (pp. 289–319). CRC Press. DOI: 10.1201/9781003263852-14

Ma, R. Y., Li, L., Yang, H., Zou, B., Ma, R. X., Zhang, Y., Wu, M. M., Chen, P., Yao, Y. & Li, J. (2024). Therapeutic effect of nicotinamide mononucleotide on Alzheimer's disease through activating autophagy and anti-oxidative stress. *Biomedicine and Pharmacotherapy, 178*(July). DOI: 10.1016/j.biopha.2024.117199

Machado-Vieira, R., Zanetti, M. V., Teixeira, A. L., Uno, M., Valiengo, L. L., Soeiro-de-Souza, M. G., Oba-Shinjo, S. M., de Sousa, R. T., Zarate, C. A.Jr, Gattaz, W. F., & Marie, S. K. N. (2015). Decreased AKT1/mTOR pathway mRNA expression in short-term bipolar disorder. *European Neuropsychopharmacology, 25*(4), 468–473. DOI: 10.1016/j.euroneuro.2015.02.002 PMID: 25726893

Macnaghten, P., & Guivant, J. S. (2020). Narrative as a resource for inclusive governance: a UK–Brazil comparison of public responses to nanotechnology. *Journal of Responsible Innovation, 7*(sup1), 13-33. DOI: 10.1080/23299460.2020.1842643

MacQueen, G. M., Ramakrishnan, K., Ratnasingan, R., Chen, B., & Young, L. T. (2003). Desipramine treatment reduces the long-term behavioural and neurochemical sequelae of early-life maternal separation. *The International Journal of Neuropsychopharmacology, 6*(4), 391–396. DOI: 10.1017/S1461145703003729 PMID: 14641986

Ma, D., Panda, S., & Lin, J. D. (2011). Temporal orchestration of circadian autophagy rhythm by C/EBPβ. *The EMBO Journal, 30*(22), 4642–4651. DOI: 10.1038/emboj.2011.322 PMID: 21897364

Madeo, F., Eisenberg, T., Büttner, S., Ruckenstuhl, C., & Kroemer, G. (2010). Spermidine: A novel autophagy inducer and longevity elixir. *Autophagy, 6*(1), 160–162. DOI: 10.4161/auto.6.1.10600 PMID: 20110777

Madeo, F., Tavernarakis, N., & Kroemer, G. (2010). Can autophagy promote longevity? *Nature Cell Biology, 12*(9), 842–846. DOI: 10.1038/ncb0910-842 PMID: 20811357

Maeda, S., Yamamoto, H., Kinch, L. N., Garza, C. M., Takahashi, S., Otomo, C., Grishin, N. V., Forli, S., Mizushima, N., & Otomo, T. (2020). Structure, lipid scrambling activity and role in autophagosome formation of ATG9A. *Nature Structural & Molecular Biology*, *27*(12), 1194–1201. DOI: 10.1038/s41594-020-00520-2 PMID: 33106659

Maes, M., Kubera, M., Obuchowiczwa, E., Goehler, L., & Brzeszcz, J. (2011). Depression's multiple comorbidities explained by (neuro) inflammatory and oxidative & nitrosative stress pathways. *Neuroendocrinology Letters*, *32*(1), 7–24. PMID: 21407167

Maher, B. A. (2019). Airborne magnetite- and iron-rich pollution nanoparticles: Potential neurotoxicants and environmental risk factors for neurodegenerative disease, including Alzheimer's disease. *Journal of Alzheimer's Disease*, *71*(2), 361–375. DOI: 10.3233/JAD-190204 PMID: 31381513

Maiuri, M. C., Zalckvar, E., Kimchi, A., & Kroemer, G. (2007). Self-eating and self-killing: Crosstalk between autophagy and apoptosis. *Nature Reviews. Molecular Cell Biology*, *8*(9), 741–752. DOI: 10.1038/nrm2239 PMID: 17717517

Majidzadeh, S., Mirghafourvand, M., Farvareshi, M., & Yavarikia, P. (2023). The effect of cognitive behavioral therapy on depression and anxiety of women with polycystic ovary syndrome: A randomized controlled trial. *BMC Psychiatry*, *23*(1), 332. DOI: 10.1186/s12888-023-04814-9 PMID: 37170227

Maksour, S., Finol-Urdaneta, R. K., Hulme, A. J., Cabral-da-Silva, M. E. C., Targa Dias Anastacio, H., Balez, R., Berg, T., Turner, C., Sanz Muñoz, S., Engel, M., Kalajdzic, P., Lisowski, L., Sidhu, K., Sachdev, P. S., Dottori, M., & Ooi, L. (2024). Alzheimer's disease induced neurons bearing PSEN1 mutations exhibit reduced excitability. *Frontiers in Cellular Neuroscience*, *18*, 1406970. DOI: 10.3389/fncel.2024.1406970 PMID: 39444394

Malik, S., Muhammad, K., & Waheed, Y. (2023). Nanotechnology: A revolution in modern industry. *Molecules (Basel, Switzerland)*, *28*(2), 661. DOI: 10.3390/molecules28020661 PMID: 36677717

Mallick, K., Paul, S., Banerjee, S., & Banerjee, S. (2024). Lipid Droplets and Neurodegeneration. *Neuroscience*, *549*, 13–23. DOI: 10.1016/j.neuroscience.2024.04.014 PMID: 38718916

Mamelak, M. (2024). Depression and the Glutamate/GABA-Glutamine Cycle. *Current Neuropharmacology*, *23*(1), 75–84. DOI: 10.2174/1570159X22666240815120244 PMID: 39150032

Mamun, A. A., Shao, C., Geng, P., Wang, S., & Xiao, J. (2024). Polyphenols Targeting NF-κB Pathway in Neurological Disorders: What We Know So Far? *International Journal of Biological Sciences*, *20*(4), 1332–1355. DOI: 10.7150/ijbs.90982 PMID: 38385077

Mancini, A., Bruno, C., Vergani, E., d'Abate, C., Giacchi, E., & Silvestrini, A. (2021). Oxidative Stress and Low-Grade Inflammation in Polycystic Ovary Syndrome: Controversies and New Insights. *International Journal of Molecular Sciences*, *22*(4), 1667. Advance online publication. DOI: 10.3390/ijms22041667 PMID: 33562271

Mangan, D. (2015). *Autophagy activation as a mechanism of action of psychoactive drugs, revised and expanded version*. The Winnower., DOI: 10.15200/winn.143268.86120

Manley, S., Williams, J. A., & Ding, W.-X. (2013). Role of p62/SQSTM1 in liver physiology and pathogenesis. *Experimental Biology and Medicine*, *238*(5), 525–538. DOI: 10.1177/1535370213489446 PMID: 23856904

Mannack, L. V. J. C., & Lane, J. D. (2015). The autophagosome: Current understanding of formation and maturation. *Research and Reports in Biochemistry*, 39–58.

Ma, R. (2024). A detective story of intermittent fasting effect on immunity. *Immunology*, *173*(2), 227–247. DOI: 10.1111/imm.13829 PMID: 38922825

Marino, G., Niso-Santano, M., Baehrecke, E. H., & Kroemer, G. (2014). Self-consumption: The interplay of autophagy and apoptosis. *Nature Reviews. Molecular Cell Biology*, *15*(2), 81–94. DOI: 10.1038/nrm3735 PMID: 24401948

Martinelli, S., Anderzhanova, E. A., Bajaj, T., Wiechmann, S., Dethloff, F., Weckmann, K., Heinz, D. E., Ebert, T., Hartmann, J., Geiger, T. M., Döngi, M., Hafner, K., Pöhlmann, M. L., Jollans, L., Philipsen, A., Schmidt, S. V., Schmidt, U., Maccarrone, G., Stein, V., & Gassen, N. C. (2021). Stress-primed secretory autophagy promotes extracellular BDNF maturation by enhancing MMP9 secretion. *Nature Communications*, *12*(1), 4643. DOI: 10.1038/s41467-021-24810-5 PMID: 34330919

Martinez–Vicente, M., Talloczy, Z., Wong, E., Tang, G., Koga, H., Kaushik, S., de Vries, R., Arias, E., Harris, S., Sulzer, D., & Cuervo, A. M. (2010). Cargo recognition failure is responsible for inefficient autophagy in Huntington's disease m. *Physiology & Behavior*, *13*(1), 567–576. DOI: 10.1038/nn.2528.CARGO PMID: 20383138

Masaldan, S., Callegari, S., & Dewson, G. (2022). Therapeutic targeting of mitophagy in Parkinson's disease. *Biochemical Society Transactions*, *50*(2), 783–797. DOI: 10.1042/BST20211107 PMID: 35311891

Masoudi, M., Ansari, S., Kashani, L., Tavolinejad, H., Hossein Rashidi, B., Esalatmanesh, S., Ghazizadeh-Hashemi, M., Noorbala, A. A., & Akhondzadeh, S. (2021). Effect of sertraline on depression severity and prolactin levels in women with polycystic ovary syndrome: A placebo-controlled randomized trial. *International Clinical Psychopharmacology, 36*(5), 238–243. DOI: 10.1097/YIC.0000000000000367 PMID: 34030169

Masrori, P., & Van Damme, P. (2020). Amyotrophic lateral sclerosis: A clinical review. *European Journal of Neurology, 27*(10), 1918–1929. DOI: 10.1111/ene.14393 PMID: 32526057

Massacci, G., Perfetto, L., & Sacco, F. (2023). The Cyclin-dependent kinase 1: More than a cell cycle regulator. *British Journal of Cancer, 129*(11), 1707–1716. DOI: 10.1038/s41416-023-02468-8 PMID: 37898722

Matias, J. N., Achete, G., Campanari, G. S. S., Guiguer, É. L., Araújo, A. C., Buglio, D. S., & Barbalho, S. M. (2021). A systematic review of the antidepressant effects of curcumin: Beyond monoamines theory. *The Australian and New Zealand Journal of Psychiatry, 55*(5), 451–462. DOI: 10.1177/0004867421998795 PMID: 33673739

Matsui, M., Yamamoto, A., Kuma, A., Ohsumi, Y., & Mizushima, N. (2006). Organelle degradation during the lens and erythroid differentiation is independent of autophagy. *Biochemical and Biophysical Research Communications, 339*(2), 485–489. DOI: 10.1016/j.bbrc.2005.11.044 PMID: 16300732

Mazloomi, S., Sheikh, N., Sanoee Farimani, M., & Pilehvari, S. (2021). Association of Prx4, Total Oxidant Status, and Inflammatory Factors with Insulin Resistance in Polycystic Ovary Syndrome. *International Journal of Endocrinology, 2021*(1), 9949753. DOI: 10.1155/2021/9949753 PMID: 34239559

McCartney, C. R., & Campbell, R. E. (2020). Abnormal GnRH pulsatility in polycystic ovary syndrome: Recent insights. *Current Opinion in Endocrine and Metabolic Research, 12*, 78–84. DOI: 10.1016/j.coemr.2020.04.005 PMID: 32676541

McKim, D. B., Weber, M. D., Niraula, A., Sawicki, C. M., Liu, X., Jarrett, B. L., Ramirez-Chan, K., Wang, Y., Roeth, R. M., Sucaldito, A. D., Sobol, C. G., Quan, N., Sheridan, J. F., & Godbout, J. P. (2018). Microglial recruitment of IL-1β-producing monocytes to brain endothelium causes stress-induced anxiety. *Molecular Psychiatry, 23*(6), 1421–1431. DOI: 10.1038/mp.2017.64 PMID: 28373688

McWilliams, T. G., Prescott, A. R., Villarejo-Zori, B., Ball, G., Boya, P., & Ganley, I. G. (2019). A comparative map of macroautophagy and mitophagy in the vertebrate eye. *Autophagy, 15*(7), 1296–1308. DOI: 10.1080/15548627.2019.1580509 PMID: 30786807

Mellén, M., De la Rosa, E. J., & Boya, P. (2008). The autophagic machinery is necessary for removal of cell corpses from the developing retinal neuroepithelium. *Cell Death and Differentiation, 15*(8), 1279–1290. DOI: 10.1038/cdd.2008.40 PMID: 18369370

Meng, H. Y., Shao, D. C., Li, H., Huang, X. D., Yang, G., Xu, B., & Niu, H.-Y. (2018). Resveratrol improves neurological outcome and neuroinflammation following spinal cord injury through enhancing autophagy involving the AMPK/mTOR pathway. *Molecular Medicine Reports, 18*(2), 2237–2244. DOI: 10.3892/mmr.2018.9194 PMID: 29956767

Meng, H., Matthaei, M., Ramanan, N., Grebe, R., Chakravarti, S., Speck, C. L., Kimos, M., Vij, N., Eberhart, C. G., & Jun, A. S. (2013). L450W and Q455K Col8a2 knock-in mouse models of Fuchs endothelial corneal dystrophy show distinct phenotypes and evidence for altered autophagy. *Investigative Ophthalmology & Visual Science, 54*(3), 1887–1897. DOI: 10.1167/iovs.12-11021 PMID: 23422828

Menšíková, K., Matěj, R., Colosimo, C., Rosales, R., Tučková, L., Ehrmann, J., Hraboš, D., Kolaříková, K., Vodička, R., Vrtěl, R., Procházka, M., Nevrlý, M., Kaiserová, M., Kurčová, S., Otruba, P., & Kaňovský, P. (2022). Lewy body disease or diseases with Lewy bodies? *NPJ Parkinson's Disease, 8*(1), 3. DOI: 10.1038/s41531-021-00273-9 PMID: 35013341

Menzies, F. M., Fleming, A., Caricasole, A., Bento, C. F., Andrews, S. P., Ashkenazi, A., Füllgrabe, J., Jackson, A., Jimenez Sanchez, M., Karabiyik, C., Licitra, F., Lopez Ramirez, A., Pavel, M., Puri, C., Renna, M., Ricketts, T., Schlotawa, L., Vicinanza, M., Won, H., & Rubinsztein, D. C. (2017). Autophagy and Neurodegeneration: Pathogenic Mechanisms and Therapeutic Opportunities. *Neuron, 93*(5), 1015–1034. DOI: 10.1016/j.neuron.2017.01.022 PMID: 28279350

Merenlender-Wagner, A., Malishkevich, A., Shemer, Z., Udawela, M., Gibbons, A., Scarr, E., Dean, B., Levine, J., Agam, G., & Gozes, I. (2015). Autophagy has a key role in the pathophysiology of schizophrenia. *Molecular Psychiatry, 20*(1), 126–132. DOI: 10.1038/mp.2013.174 PMID: 24365867

Merenlender-Wagner, A., Shemer, Z., Touloumi, O., Lagoudaki, R., Giladi, E., Andrieux, A., Grigoriadis, N. C., & Gozes, I. (2014). New horizons in schizophrenia treatment: Autophagy protection is coupled with behavioral improvements in a mouse model of schizophrenia. *Autophagy, 10*(12), 2324–2332. DOI: 10.4161/15548627.2014.984274 PMID: 25484074

Merkulov, V., Merkulova, T., & Bondar, N. (2017). Mechanisms of brain glucocorticoid resistance in stress-induced psychopathologies. *Biochemistry (Moscow), 82*(3), 351–365. DOI: 10.1134/S0006297917030142 PMID: 28320277

Merzoug, S., Toumi, M. L., & Tahraoui, A. (2014). Quercetin mitigates Adriamycin-induced anxiety- and depression-like behaviors, immune dysfunction, and brain oxidative stress in rats. *Naunyn-Schmiedeberg's Archives of Pharmacology, 387*(10), 921–933. DOI: 10.1007/s00210-014-1008-y PMID: 24947870

Metzger, S., Saukko, M., Van Che, H., Tong, L., Puder, Y., Riess, O., & Nguyen, H. P. (2010). Age at onset in Huntington's disease is modified by the autophagy pathway: Implication of the V471A polymorphism in Atg7. *Human Genetics, 128*(4), 453–459. DOI: 10.1007/s00439-010-0873-9 PMID: 20697744

Meza-Menchaca, T., Albores-Medina, A., Heredia-Mendez, A. J., Ruíz-May, E., Ricaño-Rodríguez, J., Gallegos-García, V., Esquivel, A., Vettoretti-Maldonado, G., & Campos-Parra, A. D. (2024). Revisiting Epigenetics Fundamentals and Its Biomedical Implications. *International Journal of Molecular Sciences, 25*(14), 7927. DOI: 10.3390/ijms25147927 PMID: 39063168

Miao, J., Ma, H., Yang, Y., Liao, Y., Lin, C., Zheng, J., Yu, M., & Lan, J. (2023). Microglia in Alzheimer's disease: Pathogenesis, mechanisms, and therapeutic potentials. *Frontiers in Aging Neuroscience, 15*, 1201982. DOI: 10.3389/fnagi.2023.1201982 PMID: 37396657

Mikulska, J., Juszczyk, G., Gawrońska-Grzywacz, M., & Herbet, M. (2021). HPA Axis in the Pathomechanism of Depression and Schizophrenia: New Therapeutic Strategies Based on Its Participation. *Brain Sciences, 11*(10), 1298. DOI: 10.3390/brainsci11101298 PMID: 34679364

Milán-Tomás, Á., Fernández-Matarrubia, M., & Rodríguez-Oroz, M. C. (2021). Lewy Body Dementias: A Coin with Two Sides? *Behavioral Sciences (Basel, Switzerland), 11*(7), 94. DOI: 10.3390/bs11070094 PMID: 34206456

Minchev, D., Kazakova, M., & Sarafian, V. (2022). Neuroinflammation and Autophagy in Parkinson's Disease—Novel Perspectives. *International Journal of Molecular Sciences, 23*(23), 14997. DOI: 10.3390/ijms232314997 PMID: 36499325

Minciuna, I., van Kleef, L. A., Stefanescu, H., & Procopet, B. (2022). Is Fasting Good When One Is at Risk of Liver Cancer? *Cancers (Basel), 14*(20), 5084. DOI: 10.3390/cancers14205084 PMID: 36291868

Mirmasoumi, G., Fazilati, M., Foroozanfard, F., Vahedpoor, Z., Mahmoodi, S., Taghizadeh, M., Esfeh, N. K., Mohseni, M., Karbassizadeh, H., & Asemi, Z. (2018). The Effects of Flaxseed Oil Omega-3 Fatty Acids Supplementation on Metabolic Status of Patients with Polycystic Ovary Syndrome: A Randomized, Double-Blind, Placebo-Controlled Trial. *Experimental and Clinical Endocrinology & Diabetes*, *126*(4), 222–228. DOI: 10.1055/s-0043-119751 PMID: 29117618

Mirrazavi, Z. S., & Behrouz, V. (2024). Various types of fasting diet and possible benefits in nonalcoholic fatty liver: Mechanism of actions and literature update. *Clinical Nutrition (Edinburgh, Lothian)*, *43*(2), 519–533. DOI: 10.1016/j.clnu.2024.01.005 PMID: 38219703

Mithaiwala, M. N., Santana-Coelho, D., Porter, G. A., & O'Connor, J. C. (2021). Neuroinflammation and the Kynurenine Pathway in CNS Disease: Molecular Mechanisms and Therapeutic Implications. *Cells*, *10*(6), 1548. DOI: 10.3390/cells10061548 PMID: 34205235

Mitter, S. K., Song, C., Qi, X., Mao, H., Rao, H., Akin, D., Lewin, A., Grant, M., Dunn, W.Jr, Ding, J., Bowes Rickman, C., & Boulton, M. (2014). Dysregulated autophagy in the RPE is associated with increased susceptibility to oxidative stress and AMD. *Autophagy*, *10*(11), 1989–2005. DOI: 10.4161/auto.36184 PMID: 25484094

Mizushima, N. (2020). The ATG conjugation systems in autophagy. *Current Opinion in Cell Biology*, *63*, 1–10. DOI: 10.1016/j.ceb.2019.12.001 PMID: 31901645

Mizushima, N., & Klionsky, D. J. (2007). Protein turnover via autophagy: Implications for metabolism. *Annual Review of Nutrition*, *27*(1), 19–40. DOI: 10.1146/annurev.nutr.27.061406.093749 PMID: 17311494

Mizushima, N., & Komatsu, M. (2011). Autophagy: Renovation of cells and tissues. *Cell*, *147*(4), 728–741. DOI: 10.1016/j.cell.2011.10.026 PMID: 22078875

Mizushima, N., & Levine, B. (2010). Autophagy in mammalian development and differentiation. *Nature Cell Biology*, *12*(9), 823–830. DOI: 10.1038/ncb0910-823 PMID: 20811354

Mokhtari, T. (2022). Targeting autophagy and neuroinflammation pathways with plant-derived natural compounds as potential antidepressant agents. *Phytotherapy Research*, *36*(9), 3470–3489. DOI: 10.1002/ptr.7551 PMID: 35794794

Mokhtari, T., Tu, Y., & Hu, L. (2019). Involvement of the hippocampus in chronic pain and depression. *Brain Science Advances.*, *5*(4), 288–298. DOI: 10.26599/BSA.2019.9050025

Mokhtari, T., & Uludag, K. (2023). Role of NLRP3 inflammasome in post-spinal-cord-injury anxiety and depression: Molecular mechanisms and therapeutic implications. *ACS Chemical Neuroscience*, *15*(1), 56–70. DOI: 10.1021/acschemneuro.3c00596 PMID: 38109051

Monzani, R., Gagliardi, M., Clemente, N., Saverio, V., Pańczyszyn, E., Santoro, C., Yissachar, N., Visciglia, A., Pane, M., Amoruso, A., & Corazzari, M. (2022). The Gut-Ex-Vivo System (GEVS) Is a Dynamic and Versatile Tool for the Study of DNBS-Induced IBD in BALB/C and C57BL/6 Mice, Highlighting the Protective Role of Probiotics. *Biology (Basel)*, *11*(11), 1574. DOI: 10.3390/biology11111574 PMID: 36358275

Moore, M. N., Shaw, J. P., Ferrar Adams, D. R., & Viarengo, A. (2015). Anti-oxidative cellular protection effect of fasting-induced autophagy as a mechanism for hormesis. *Marine Environmental Research*, *107*, 35–44. DOI: 10.1016/j.marenvres.2015.04.001 PMID: 25881010

Moran, L. J., Pasquali, R., Teede, H. J., Hoeger, K. M., & Norman, R. J. (2009). Treatment of obesity in polycystic ovary syndrome: A position statement of the Androgen Excess and Polycystic Ovary Syndrome Society. *Fertility and Sterility*, *92*(6), 1966–1982. DOI: 10.1016/j.fertnstert.2008.09.018 PMID: 19062007

Moratilla-Rivera, I., Sánchez, M., Valdés-González, J. A., & Gómez-Serranillos, M. P. (2023). Natural products as modulators of Nrf2 signaling pathway in neuroprotection. *International Journal of Molecular Sciences*, *24*(4), 3748. DOI: 10.3390/ijms24043748 PMID: 36835155

Moreau, K. L., & King, J. A. (2012). Protein misfolding and aggregation in cataract disease and prospects for prevention. *Trends in Molecular Medicine*, *18*(5), 273–282. DOI: 10.1016/j.molmed.2012.03.005 PMID: 22520268

Moreno, M.-L., Mérida, S., Bosch-Morell, F., Miranda, M., & Villar, V. M. (2018). Autophagy dysfunction and oxidative stress, two related mechanisms implicated in retinitis pigmentosa. *Frontiers in Physiology*, *9*, 1008. DOI: 10.3389/fphys.2018.01008 PMID: 30093867

More, S., Kaleem, M., Kharwade, R., Almutairy, A. F., Shahzad, N., & Mujtaba, M. A.. (2024). Depression unveiled: Insights into etiology and animal models for behavioral assessment, exploring the multifactorial nature and treatment of depression. *Brain Research*, 149313. PMID: 39515744

Morimoto, M., Bhambhani, V., Gazzaz, N., Davids, M., Sathiyaseelan, P., Macnamara, E. F., Lange, J., Lehman, A., Zerfas, P. M., Murphy, J. L., Acosta, M. T., Wang, C., Alderman, E., Adam, M., Alvarez, R. L., Alvey, J., Amendola, L., Andrews, A., Ashley, E. A., & Malicdan, M. C. V. (2023). Bi-allelic ATG4D variants are associated with a neurodevelopmental disorder characterized by speech and motor impairment. *NPJ Genomic Medicine*, *8*(1), 4. DOI: 10.1038/s41525-022-00343-8 PMID: 36765070

Morishita, H. (2022). Role of autophagy in the eye: From physiology to disease. *Current Opinion in Physiology*, *30*, 100592. DOI: 10.1016/j.cophys.2022.100592

Morishita, H., Eguchi, S., Kimura, H., Sasaki, J., Sakamaki, Y., Robinson, M. L., Sasaki, T., & Mizushima, N. (2013). Deletion of autophagy-related 5 (Atg5) and Pik3c3 genes in the lens causes cataract independent of programmed organelle degradation. *The Journal of Biological Chemistry*, *288*(16), 11436–11447. DOI: 10.1074/jbc.M112.437103 PMID: 23479732

Morishita, H., Eguchi, T., Tsukamoto, S., Sakamaki, Y., Takahashi, S., Saito, C., Koyama-Honda, I., & Mizushima, N. (2021). Organelle degradation in the lens by PLAAT phospholipases. *Nature*, *592*(7855), 634–638. DOI: 10.1038/s41586-021-03439-w PMID: 33854238

Morvaridzadeh, M., Agah, S., Dulce Estêvão, M., Hosseini, A. S., Heydari, H., Toupchian, O., Abdollahi, S., Persad, E., Abu-Zaid, A., Rezamand, G., & Heshmati, J. (2021). Effect of saffron supplementation on oxidative stress parameters: A systematic review and meta-analysis of randomized placebo-controlled trials. *Food Science & Nutrition*, *9*(10), 5809–5819. DOI: 10.1002/fsn3.2463 PMID: 34646548

Mota, B., Ramos, M., Marques, S. I., Silva, A., Pereira, P. A., Madeira, M. D., Mateus, N., & Cardoso, A. (2023). Effects of High-Fat and High-Fat High-Sugar Diets in the Anxiety, Learning and Memory, and in the Hippocampus Neurogenesis and Neuroinflammation of Aged Rats. *Nutrients*, *15*(6), 1370. DOI: 10.3390/nu15061370 PMID: 36986100

Motoi, Y., Shimada, K., Ishiguro, K., & Hattori, N. (2014). Lithium and autophagy. *ACS Chemical Neuroscience*, *5*(6), 434–442. DOI: 10.1021/cn500056q PMID: 24738557

Mozdarani, H., Ezzatizadeh, V., & Rahbar Parvaneh, R. (2020). The emerging role of the long noncoding RNA HOTAIR in breast cancer development and treatment. *Journal of Translational Medicine*, *18*(1), 152. DOI: 10.1186/s12967-020-02320-0 PMID: 32245498

Mrza, M. A., He, J., & Wang, Y. (2024). Integration of iPSC-Derived Microglia into Brain Organoids for Neurological Research. *International Journal of Molecular Sciences*, *25*(6), 3148. DOI: 10.3390/ijms25063148 PMID: 38542121

Mu, C., Gao, M., Xu, W., Sun, X., Chen, T., Xu, H., & Qiu, H. (2024). Mechanisms of microRNA-132 in central neurodegenerative diseases: A comprehensive review. *Biomedicine and Pharmacotherapy*, *170*, 116029. DOI: 10.1016/j.biopha.2023.116029 PMID: 38128185

Mueller, A. L., Payandeh, Z., Mohammadkhani, N., Mubarak, S. M. H., Zakeri, A., Alagheband Bahrami, A., Brockmueller, A., & Shakibaei, M. (2021). Recent Advances in Understanding the Pathogenesis of Rheumatoid Arthritis: New Treatment Strategies. *Cells*, *10*(11), 3017. DOI: 10.3390/cells10113017 PMID: 34831240

Muñoz-Galdeano, T., Reigada, D., Del Águila, Á., Velez, I., Caballero-López, M. J., Maza, R. M., & Nieto-Díaz, M. (2018). Cell Specific Changes of Autophagy in a Mouse Model of Contusive Spinal Cord Injury. *Frontiers in Cellular Neuroscience*, *12*, 164. DOI: 10.3389/fncel.2018.00164 PMID: 29946241

Nagayach, A., & Wang, C. (2024). Autophagy in neural stem cells and glia for brain health and diseases. *Neural Regeneration Research*, *19*(4), 729–736. DOI: 10.4103/1673-5374.382227 PMID: 37843206

Nakahira, K., Haspel, J. A., Rathinam, V. A., Lee, S.-J., Dolinay, T., Lam, H. C., Englert, J. A., Rabinovitch, M., Cernadas, M., Kim, H. P., Fitzgerald, K. A., Ryter, S. W., & Choi, A. M. K. (2011). Autophagy proteins regulate innate immune responses by inhibiting the release of mitochondrial DNA mediated by the NALP3 inflammasome. *Nature Immunology*, *12*(3), 222–230. DOI: 10.1038/ni.1980 PMID: 21151103

Nasrollahzadeh, M., Sajadi, S. M., Sajjadi, M., & Issaabadi, Z. (2019). An Introduction to Nanotechnology. In Nasrollahzadeh, M., Sajadi, S. M., Sajjadi, M., Issaabadi, Z., & Atarod, M. (Eds.), *Interface Science and Technology* (Vol. 28, pp. 1–27). Elsevier.

National Institute of Child Health and Human Development. (n.d.). *What are the symptoms of PCOS?* Retrieved February 3 from https://www.nichd.nih.gov/health/topics/pcos/conditioninfo/symptoms

Navabi, S. P., Badreh, F., Shooshtari, M. K., Hajipour, S., Vastegani, S. M., & Khoshnam, S. E. (2024). Microglia-induced neuroinflammation in hippocampal neurogenesis following traumatic brain injury. *Heliyon*, *10*(16), e35869. DOI: 10.1016/j.heliyon.2024.e35869 PMID: 39220913

Nayak, U., Manikkath, J., Arora, D., & Mudgal, J. (2025). Impact of neuroinflammation on brain glutamate and dopamine signalling in schizophrenia: An update. *Metabolic Brain Disease*, *40*(2), 119. DOI: 10.1007/s11011-025-01548-3 PMID: 39907868

Nechushtai, L., Frenkel, D., & Pinkas-Kramarski, R. (2023). Autophagy in Parkinson's Disease. *Biomolecules*, *13*(10), 1435. DOI: 10.3390/biom13101435 PMID: 37892117

Nedelsky, N. B., Todd, P. K., & Taylor, J. P. (2008). Autophagy and the ubiquitin-proteasome system: Collaborators in neuroprotection. Biochimica et Biophysica Acta (BBA)-. *Biochimica et Biophysica Acta. Molecular Basis of Disease*, *1782*(12), 691–699. DOI: 10.1016/j.bbadis.2008.10.002 PMID: 18930136

Nestler, E. J., & Hyman, S. E. (2010). Animal models of neuropsychiatric disorders. *Nature Neuroscience*, *13*(10), 1161–1169. DOI: 10.1038/nn.2647 PMID: 20877280

Netea-Maier, R. T., Plantinga, T. S., van de Veerdonk, F. L., Smit, J. W., & Netea, M. G. (2016). Modulation of inflammation by autophagy: Consequences for human disease. *Autophagy*, *12*(2), 245–260. DOI: 10.1080/15548627.2015.1071759 PMID: 26222012

Nettesheim, A., Shim, M. S., Hirt, J., & Liton, P. B. (2019). Transcriptome analysis reveals autophagy as regulator of TGFβ/Smad-induced fibrogenesis in trabecular meshwork cells. *Scientific Reports*, *9*(1), 16092. DOI: 10.1038/s41598-019-52627-2 PMID: 31695131

Newhouse, P., & Albert, K. (2015). Estrogen, Stress, and Depression: A Neurocognitive Model. *JAMA Psychiatry*, *72*(7), 727–729. DOI: 10.1001/jamapsychiatry.2015.0487 PMID: 26018188

Newton, F., & Megaw, R. (2020). Mechanisms of photoreceptor death in retinitis pigmentosa. *Genes*, *11*(10), 1120. DOI: 10.3390/genes11101120 PMID: 32987769

Nie, B.-X., Zhao, G., Yuan, X.-F., Yu, L.-X., Zhang, J., Yuan, Y., Liu, Y., Hu, J., Song, E., Zhou, Y.-C., & Shu, J. (2022). Inhibition of CDK1 attenuates neuronal apoptosis and autophagy and confers neuroprotection after chronic spinal cord injury in vivo. *Journal of Chemical Neuroanatomy*, *119*, 102053. DOI: 10.1016/j.jchemneu.2021.102053 PMID: 34839004

Nikvarz, N., Sedighi, B., Ansari, M., Shahdizade, S., Shojaei, R., & Sharififar, F. (2024). Medicinal plants used in multiple sclerosis patients, prevalence and associated factors: A descriptive cross-sectional study. *BMC Complementary Medicine and Therapies*, *24*(1), 278. DOI: 10.1186/s12906-024-04587-y PMID: 39039480

Ning, S., Chen, Y., Shao, J., Zhu, H., Zhang, Z., & Miao, J. (2024). The effects of acteoside on locomotor recovery after spinal cord injury – The role of autophagy and apoptosis signaling pathway. *Biomedicine and Pharmacotherapy*, *175*, 116607. DOI: 10.1016/j.biopha.2024.116607 PMID: 38692056

Nishida, Y., Arakawa, S., Fujitani, K., Yamaguchi, H., Mizuta, T., Kanaseki, T., Komatsu, M., Otsu, K., Tsujimoto, Y., & Shimizu, S. (2009). Discovery of Atg5/Atg7-independent alternative macroautophagy. *Nature*, *461*(7264), 654–658. DOI: 10.1038/nature08455 PMID: 19794493

Niture, S., Lin, M., Qi, Q., Moore, J. T., Levine, K. E., Fernando, R. A., & Kumar, D. (2021). Role of Autophagy in Cadmium-Induced Hepatotoxicity and Liver Diseases. *Journal of Toxicology*, *2021*(1), 9564297. DOI: 10.1155/2021/9564297 PMID: 34422041

Nixon, R. A. (2024). Autophagy–lysosomal-associated neuronal death in neurodegenerative disease. *Acta Neuropathologica*, *148*(1), 42. DOI: 10.1007/s00401-024-02799-7 PMID: 39259382

Nobili, A., D'Amelio, M., & Viscomi, M. T. (2023). Nilotinib: From animal-based studies to clinical investigation in Alzheimer's disease patients. *Neural Regeneration Research*, *18*(4), 803–804. DOI: 10.4103/1673-5374.350700 PMID: 36204843

Noce, B., Di Bello, E., Fioravanti, R., & Mai, A. (2023). LSD1 inhibitors for cancer treatment: Focus on multi-target agents and compounds in clinical trials. *Frontiers in Pharmacology*, *14*, 1120911. DOI: 10.3389/fphar.2023.1120911 PMID: 36817147

Numakawa, T., Richards, M., Nakajima, S., Adachi, N., Furuta, M., Odaka, H., & Kunugi, H. (2014). The role of brain-derived neurotrophic factor in comorbid depression: Possible linkage with steroid hormones, cytokines, and nutrition. *Frontiers in Psychiatry*, *5*, 136. DOI: 10.3389/fpsyt.2014.00136 PMID: 25309465

Ogłodek, E., Szota, A., Just, M., Moś, D., & Araszkiewicz, A. (2014). The role of the neuroendocrine and immune systems in the pathogenesis of depression. *Pharmacological Reports*, *66*(5), 776–781. DOI: 10.1016/j.pharep.2014.04.009 PMID: 25149980

Öksüz, N., Öztürk, Ş., & Doğu, O. (2022). Future Prospects in Parkinson's Disease Diagnosis and Treatment. *Nöro Psikiyatri Arşivi*, *59*, 36–41. DOI: 10.29399/npa.28169 PMID: 36578989

Oku, M., & Sakai, Y. (2018). Three Distinct Types of Microautophagy Based on Membrane Dynamics and Molecular Machineries. *BioEssays*, *40*(6), 1–6. DOI: 10.1002/bies.201800008 PMID: 29708272

Orcheston-Findlay, L., Bax, S., Utama, R., Engel, M., Govender, D., & O'Neill, G. (2021). Advanced Spheroid, Tumouroid and 3D Bioprinted In-Vitro Models of Adult and Pediatric Glioblastoma. *International Journal of Molecular Sciences, 22*(6), 2962. DOI: 10.3390/ijms22062962 PMID: 33803967

Orrenius, S., Kaminskyy, V. O., & Zhivotovsky, B. (2013). Autophagy in toxicology: Cause or consequence? *Annual Review of Pharmacology and Toxicology, 53*(1), 275–297. DOI: 10.1146/annurev-pharmtox-011112-140210 PMID: 23072380

Ostrowski, A. D., Martin, T., Conti, J., Hurt, I., & Harthorn, B. H. (2009). Nanotoxicology: Characterizing the scientific literature, 2000–2007. *Journal of Nanoparticle Research, 11*(2), 251–257. DOI: 10.1007/s11051-008-9579-5 PMID: 21170129

Otabe, H., Nibuya, M., Shimazaki, K., Toda, H., Suzuki, G., Nomura, S., & Shimizu, K. (2014). Electroconvulsive seizures enhance autophagy signaling in rat hippocampus. *Progress in Neuro-Psychopharmacology & Biological Psychiatry, 50*, 37–43. DOI: 10.1016/j.pnpbp.2013.11.012 PMID: 24316174

Otte, C., Gold, S. M., Penninx, B. W., Pariante, C. M., Etkin, A., Fava, M., Mohr, D. C., & Schatzberg, A. F. (2016). Major depressive disorder. *Nature Reviews. Disease Primers, 2*(1), 1–20. DOI: 10.1038/nrdp.2016.65 PMID: 27629598

Ou, A. H., Rosenthal, S. B., Adli, M., Akiyama, K., Akula, N., Alda, M., Amare, A. T., Ardau, R., Arias, B., Aubry, J. M., Backlund, L., Bauer, M., Baune, B. T., Bellivier, F., Benabarre, A., Bengesser, S., Bhattacharjee, A. K., Biernacka, J. M., Cervantes, P., & Kelsoe, J. R. (2024). Lithium response in bipolar disorder is associated with focal adhesion and PI3K-Akt networks: A multi-omics replication study. *Translational Psychiatry, 14*(1), 109. Advance online publication. DOI: 10.1038/s41398-024-02811-4 PMID: 38395906

Paganos, P., Ullmann, C. V., Gaglio, D., Bonanomi, M., Salmistraro, N., Arnone, M. I., & Jimenez-Guri, E. (2023). Plastic leachate-induced toxicity during sea urchin embryonic development: Insights into the molecular pathways affected by PVC. *The Science of the Total Environment, 864*, 160901. DOI: 10.1016/j.scitotenv.2022.160901 PMID: 36526210

Pagnin, D., de Queiroz, V., Pini, S., & Cassano, G. B. (2008). Efficacy of ECT in depression: A meta-analytic review. *Focus - American Psychiatric Publishing, 6*(1), 155–162. DOI: 10.1176/foc.6.1.foc155

Palomba, S., Santagni, S., Falbo, A., & La Sala, G. B. (2015). Complications and challenges associated with polycystic ovary syndrome: Current perspectives. *International Journal of Women's Health, 7*, 745–763. DOI: 10.2147/IJWH.S70314 PMID: 26261426

Pandya, C. D., Howell, K. R., & Pillai, A. (2013). Antioxidants as potential therapeutics for neuropsychiatric disorders. *Progress in Neuro-Psychopharmacology & Biological Psychiatry*, *46*, 214–223. DOI: 10.1016/j.pnpbp.2012.10.017 PMID: 23123357

Pankiv, S., Clausen, T. H., Lamark, T., Brech, A., Bruun, J.-A., Outzen, H., Øvervatn, A., Bjørkøy, G., & Johansen, T. (2007). p62/SQSTM1 binds directly to Atg8/LC3 to facilitate degradation of ubiquitinated protein aggregates by autophagy. *The Journal of Biological Chemistry*, *282*(33), 24131–24145. DOI: 10.1074/jbc.M702824200 PMID: 17580304

Panwar, V., Singh, A., Bhatt, M., Tonk, R. K., Azizov, S., Raza, A. S., Sengupta, S., Kumar, D., & Garg, M. (2023). Multifaceted role of mTOR (mammalian target of rapamycin) signaling pathway in human health and disease. *Signal Transduction and Targeted Therapy*, *8*(1), 375. DOI: 10.1038/s41392-023-01608-z PMID: 37779156

Papadopoulou, P., Polissidis, A., Kythreoti, G., Sagnou, M., Stefanatou, A., & Theoharides, T. C. (2024). Anti-Inflammatory and Neuroprotective Polyphenols Derived from the European Olive Tree, Olea europaea L., in Long COVID and Other Conditions Involving Cognitive Impairment. *International Journal of Molecular Sciences*, *25*(20), 11040. DOI: 10.3390/ijms252011040 PMID: 39456822

Park, H., Kang, J. H., & Lee, S. (2020). Autophagy in neurodegenerative diseases: A hunter for aggregates. *International Journal of Molecular Sciences*, *21*(9), 1–24. DOI: 10.3390/ijms21093369 PMID: 32397599

Park, H., Kim, J., & Park, C. (2012). Activation of autophagy induces retinal ganglion cell death in a chronic hypertensive glaucoma model. *Cell Death & Disease*, *3*(4), e290–e290. DOI: 10.1038/cddis.2012.26 PMID: 22476098

Park, H.-Y. L., Kim, J. H., & Park, C. K. (2018). Different contributions of autophagy to retinal ganglion cell death in the diabetic and glaucomatous retinas. *Scientific Reports*, *8*(1), 13321. DOI: 10.1038/s41598-018-30165-7 PMID: 30190527

Park, J. H., Min, J., Baek, S. R., Kim, S. W., Kwon, I. K., & Jeon, S. R. (2013). Enhanced neuroregenerative effects by scaffold for the treatment of a rat spinal cord injury with Wnt3a-secreting fibroblasts. *Acta Neurochirurgica*, *155*(5), 809–816. DOI: 10.1007/s00701-013-1663-7 PMID: 23456240

Park, J. K., Peng, H., Katsnelson, J., Yang, W., Kaplan, N., Dong, Y., Rappoport, J. Z., He, C., & Lavker, R. M. (2016). MicroRNAs-103/107 coordinately regulate macropinocytosis and autophagy. *The Journal of Cell Biology*, *215*(5), 667–685. DOI: 10.1083/jcb.201604032 PMID: 27872138

Park, S. W., Lee, J. G., Seo, M. K., Lee, C. H., Cho, H. Y., Lee, B. J., Seol, W., & Kim, Y. H. (2014). Differential effects of antidepressant drugs on mTOR signalling in rat hippocampal neurons. *The International Journal of Neuropsychopharmacology, 17*(11), 1831–1846. DOI: 10.1017/S1461145714000534 PMID: 24901414

Parzych, K. R., & Klionsky, D. J. (2014). An overview of autophagy: Morphology, mechanism, and regulation. *Antioxidants & Redox Signaling, 20*(3), 460–473. DOI: 10.1089/ars.2013.5371 PMID: 23725295

Patel, J. K., Patel, A., & Bhatia, D. (2021). Introduction to nanomaterials and nanotechnology. In *Emerging technologies for nanoparticle manufacturing* (pp. 3–23). Springer International Publishing. DOI: 10.1007/978-3-030-50703-9_1

Patten, R. K., Boyle, R. A., Moholdt, T., Kiel, I., Hopkins, W. G., Harrison, C. L., & Stepto, N. K. (2020). Exercise Interventions in Polycystic Ovary Syndrome: A Systematic Review and Meta-Analysis. *Frontiers in Physiology, 11*, 606. DOI: 10.3389/fphys.2020.00606 PMID: 32733258

Pattingre, S., Espert, L., Biard-Piechaczyk, M., & Codogno, P. (2008). Regulation of macroautophagy by mTOR and Beclin 1 complexes. *Biochimie, 90*(2), 313–323. DOI: 10.1016/j.biochi.2007.08.014 PMID: 17928127

Pavlinov, I., Salkovski, M., & Aldrich, L. N. (2020). Beclin 1–ATG14L Protein–Protein Interaction Inhibitor Selectively Inhibits Autophagy through Disruption of VPS34 Complex I. *Journal of the American Chemical Society, 142*(18), 8174–8182. DOI: 10.1021/jacs.9b12705 PMID: 32320221

Peng, H., Park, J. K., & Lavker, R. M. (2017). Eyeing autophagy and macropinocytosis in the corneal/limbal epithelia. *Autophagy, 13*(5), 975–977. DOI: 10.1080/15548627.2017.1287658 PMID: 28402214

Peng, X., Ju, J., Li, Z., Liu, J., Jia, X., Wang, J., Ren, J., & Gao, F. (2025). C3/C3aR Bridges Spinal Astrocyte-Microglia Crosstalk and Accelerates Neuroinflammation in Morphine-Tolerant Rats. *CNS Neuroscience & Therapeutics, 31*(1), e70216. DOI: 10.1111/cns.70216 PMID: 39801259

Peng, Y., Wang, Y., Tang, N., Sun, D., Lan, Y., Yu, Z., Zhao, X., Feng, L., Zhang, B., Jin, L., Yu, F., Ma, X., & Lv, C. (2018). Andrographolide inhibits breast cancer through suppressing COX-2 expression and angiogenesis via inactivation of p300 signaling and VEGF pathway. *Journal of Experimental & Clinical Cancer Research : CR, 37*(1), 1–14. DOI: 10.1186/s13046-018-0926-9 PMID: 30314513

Perusek, L., Sahu, B., Parmar, T., Maeno, H., Arai, E., Le, Y.-Z., Subauste, C. S., Chen, Y., Palczewski, K., & Maeda, A. (2015). Di-retinoid-pyridinium-ethanolamine (A2E) accumulation and the maintenance of the visual cycle are independent of Atg7-mediated autophagy in the retinal pigmented epithelium. *The Journal of Biological Chemistry*, *290*(48), 29035–29044. DOI: 10.1074/jbc.M115.682310 PMID: 26468292

Pickford, F., Masliah, E., Britschgi, M., Lucin, K., Narasimhan, R., Jaeger, P. A., Small, S., Spencer, B., Rockenstein, E., Levine, B., & Wyss-Coray, T. (2008). The autophagy-related protein beclin 1 shows reduced expression in early Alzheimer disease and regulates amyloid β accumulation in mice. *The Journal of Clinical Investigation*, *118*(6), 2190–2199. DOI: 10.1172/JCI33585 PMID: 18497889

Pidgeon, N., Harthorn, B., & Satterfield, T. (2011). Nanotechnology risk perceptions and communication: Emerging technologies, emerging challenges. *Risk Analysis*, *31*(11), 1694–1700. DOI: 10.1111/j.1539-6924.2011.01738.x PMID: 22084861

Pierone, B. C., Pereira, C. A., Garcez, M. L., & Kaster, M. P. (2020). Stress and signaling pathways regulating autophagy: From behavioral models to psychiatric disorders. *Experimental Neurology*, *334*(113485), 1–13. DOI: 10.1016/j.expneurol.2020.113485 PMID: 32987001

Pokrajac, L., Abbas, A., Chrzanowski, W., Dias, G. M., Eggleton, B. J., Maguire, S., ... Mitra, S. (2021). Nanotechnology for a sustainable future: Addressing global challenges with the international network4sustainable nanotechnology. DOI: 10.1021/acsnano.1c10919

Polajnar, M., & Žerovnik, E. (2014). Impaired autophagy: A link between neurodegenerative and neuropsychiatric diseases. *Journal of Cellular and Molecular Medicine*, *18*(9), 1705–1711. DOI: 10.1111/jcmm.12349 PMID: 25139375

Pope, C. N., Schlenk, D., & Baud, F. J. (2020). History and basic concepts of toxicology. In *An Introduction to Interdisciplinary Toxicology* (pp. 3–15). Elsevier. DOI: 10.1016/B978-0-12-813602-7.00001-6

Puertollano, R., Ferguson, S. M., Brugarolas, J., & Ballabio, A. (2018). The complex relationship between TFEB transcription factor phosphorylation and subcellular localization. *The EMBO Journal*, *37*(11), e98804. DOI: 10.15252/embj.201798804 PMID: 29764979

Puglisi-Allegra, S., Lazzeri, G., Busceti, C. L., Giorgi, F. S., Biagioni, F., & Fornai, F. (2023). Lithium engages autophagy for neuroprotection and neuroplasticity: Translational evidence for therapy. *Neuroscience and Biobehavioral Reviews*, *148*(105148), 1–19. DOI: 10.1016/j.neubiorev.2023.105148 PMID: 36996994

Qi, X., Mitter, S. K., Yan, Y., Busik, J. V., Grant, M. B., & Boulton, M. E. (2020). Diurnal rhythmicity of autophagy is impaired in the diabetic retina. *Cells*, *9*(4), 905. DOI: 10.3390/cells9040905 PMID: 32272782

Qi, X., Yun, C., Sun, L., Xia, J., Wu, Q., Wang, Y., Wang, L., Zhang, Y., Liang, X., Wang, L., Gonzalez, F. J., Patterson, A. D., Liu, H., Mu, L., Zhou, Z., Zhao, Y., Li, R., Liu, P., Zhong, C., & Qiao, J. (2019). Gut microbiota–bile acid–interleukin-22 axis orchestrates polycystic ovary syndrome. *Nature Medicine*, *25*(8), 1225–1233. DOI: 10.1038/s41591-019-0509-0 PMID: 31332392

Qu, X., Zou, Z., Sun, Q., Luby-Phelps, K., Cheng, P., Hogan, R. N., Gilpin, C., & Levine, B. (2007). Autophagy gene-dependent clearance of apoptotic cells during embryonic development. *Cell*, *128*(5), 931–946. DOI: 10.1016/j.cell.2006.12.044 PMID: 17350577

Ragupathi, A., Kim, C., & Jacinto, E. (2024). The mTORC2 signaling network: Targets and cross-talks. *The Biochemical Journal*, *481*(2), 45–91. DOI: 10.1042/BCJ20220325 PMID: 38270460

Rahman, M. A., Rahman, M. S., Parvez, M. A. K., & Kim, B. (2023). The Emerging Role of Autophagy as a Target of Environmental Pollutants: An Update on Mechanisms. *Toxics*, *11*(2), 135. Advance online publication. DOI: 10.3390/toxics11020135 PMID: 36851010

Rashwan, H. K., Mahgoub, S., Abuelezz, N. Z., & Amin, H. K. (2023). Black Cumin Seed (Nigella sativa) in Inflammatory Disorders: Therapeutic Potential and Promising Molecular Mechanisms. *Drugs and Drug Candidates*, *2*(2), 516–537. DOI: 10.3390/ddc2020027

Rathore, A., & Mahesh, G. (2021). Public perception of nanotechnology: A contrast between developed and developing countries. *Technology in Society*, *67*, 101751. DOI: 10.1016/j.techsoc.2021.101751

Rau, S., Gregg, A., Yaceczko, S., & Limketkai, B. (2024). Prebiotics and Probiotics for Gastrointestinal Disorders. *Nutrients*, *16*(6), 778. Advance online publication. DOI: 10.3390/nu16060778 PMID: 38542689

Ravenhill, B. J., Boyle, K. B., von Muhlinen, N., Ellison, C. J., Masson, G. R., Otten, E. G., Foeglein, A., Williams, R., & Randow, F. (2019). The cargo receptor NDP52 initiates selective autophagy by recruiting the ULK complex to cytosol-invading bacteria. *Molecular Cell, 74*(2), 320-329. e326.

Ray, S. K. (2020). Modulation of autophagy for neuroprotection and functional recovery in traumatic spinal cord injury. *Neural Regeneration Research*, *15*(9), 1601–1612. DOI: 10.4103/1673-5374.276322 PMID: 32209759

Razali, K., Othman, N., Mohd Nasir, M. H., Doolaanea, A. A., Kumar, J., Ibrahim, W. N., Mohamed Ibrahim, N., & Mohamed, W. M. Y. (2021). The Promise of the Zebrafish Model for Parkinson's Disease: Today's Science and Tomorrow's Treatment. *Frontiers in Genetics*, *12*, 655550. DOI: 10.3389/fgene.2021.655550 PMID: 33936174

Read, J. R., Sharpe, L., Modini, M., & Dear, B. F. (2017). Multimorbidity and depression: A systematic review and meta-analysis. *Journal of Affective Disorders*, *221*, 36–46. DOI: 10.1016/j.jad.2017.06.009 PMID: 28628766

Reichmann, F., Hassan, A. M., Farzi, A., Jain, P., Schuligoi, R., & Holzer, P. (2015). Dextran sulfate sodium-induced colitis alters stress-associated behaviour and neuropeptide gene expression in the amygdala-hippocampus network of mice. *Scientific Reports*, *5*(1), 9970. DOI: 10.1038/srep09970 PMID: 26066467

Reijerkerk, A., Lakeman, K. A., Drexhage, J. A., van Het Hof, B., van Wijck, Y., van der Pol, S. M., Kooij, G., Geerts, D., & de Vries, H. E. (2012). Brain endothelial barrier passage by monocytes is controlled by the endothelin system. *Journal of Neurochemistry*, *121*(5), 730–737. DOI: 10.1111/j.1471-4159.2011.07393.x PMID: 21777246

Remé, C. E., & Sulser, M. (1977). Diurnal variation of autophagy in rod visual cells in the rat. *Graefe's Archive for Clinical and Experimental Ophthalmology*, *203*, 261–270. DOI: 10.1007/BF00409832 PMID: 303474

Reme, C., Wirz-Justice, A., Rhyner, A., & Hofmann, S. (1986). Circadian rhythm in the light response of rat retinal disk shedding and autophagy. *Brain Research*, *369*(1-2), 356–360. DOI: 10.1016/0006-8993(86)90550-0 PMID: 3697752

Ren, L., Liang, H., Zhu, L., Yang, X., Zhang, H., Sun, N., Huang, D., Feng, J., Wu, Y., Xiong, L., Ke, X., Li, M., & Zhang, A. (2024). Dietary restriction improves perioperative neurocognitive disorders by inhibiting neuroinflammation and gut microbial dysbiosis. *Neuroscience*, *540*, 48–67. DOI: 10.1016/j.neuroscience.2024.01.012 PMID: 38272300

Reyes-Reveles, J., Dhingra, A., Alexander, D., Bragin, A., Philp, N. J., & Boesze-Battaglia, K. (2017). Phagocytosis-dependent ketogenesis in retinal pigment epithelium. *The Journal of Biological Chemistry*, *292*(19), 8038–8047. DOI: 10.1074/jbc.M116.770784 PMID: 28302729

Rikiishi, H. (2012). Novel insights into the interplay between apoptosis and autophagy. *International Journal of Cell Biology, 2012*(1), 317645. DOI: 10.1155/2012/317645 PMID: 22496691

Riva, N., Domi, T., Pozzi, L., Lunetta, C., Schito, P., Spinelli, E. G., Cabras, S., Matteoni, E., Consonni, M., Bella, E. D., Agosta, F., Filippi, M., Calvo, A., & Quattrini, A. (2024). Update on recent advances in amyotrophic lateral sclerosis. *Journal of Neurology, 271*(7), 4693–4723. DOI: 10.1007/s00415-024-12435-9 PMID: 38802624

Rodriguez, A. C. I., Epel, E. S., White, M. L., Standen, E. C., Seckl, J. R., & Tomiyama, A. J. (2015). Hypothalamic-pituitary-adrenal axis dysregulation and cortisol activity in obesity: A systematic review. *Psychoneuroendocrinology, 62*, 301–318. DOI: 10.1016/j.psyneuen.2015.08.014 PMID: 26356039

Rodríguez-Muela, N., Germain, F., Mariño, G., Fitze, P. S., & Boya, P. (2012). Autophagy promotes survival of retinal ganglion cells after optic nerve axotomy in mice. *Cell Death and Differentiation, 19*(1), 162–169. DOI: 10.1038/cdd.2011.88 PMID: 21701497

Rosenfield, R. L., & Ehrmann, D. A. (2016). The Pathogenesis of Polycystic Ovary Syndrome (PCOS): The Hypothesis of PCOS as Functional Ovarian Hyperandrogenism Revisited. *Endocrine Reviews, 37*(5), 467–520. DOI: 10.1210/er.2015-1104 PMID: 27459230

Ross, C. A., & Pickart, C. M. (2004). The ubiquitin–proteasome pathway in Parkinson's disease and other neurodegenerative diseases. *Trends in Cell Biology, 14*(12), 703–711. DOI: 10.1016/j.tcb.2004.10.006 PMID: 15564047

Roth, J. R., de Moraes, R. C. M., Xu, B. P., Crawley, S. R., Khan, M. A., & Melkani, G. C. (2023). Rapamycin reduces neuronal mutant huntingtin aggregation and ameliorates locomotor performance in Drosophila. *Frontiers in Aging Neuroscience, 15*, 1–11. DOI: 10.3389/fnagi.2023.1223911 PMID: 37823007

Rubinsztein, D. C., Mariño, G., & Kroemer, G. (2011). Autophagy and aging. *Cell, 146*(5), 682–695. DOI: 10.1016/j.cell.2011.07.030 PMID: 21884931

Rubinsztein, D. C., Shpilka, T., & Elazar, Z. (2012). Mechanisms of autophagosome biogenesis. *Current Biology, 22*(1), R29–R34. DOI: 10.1016/j.cub.2011.11.034 PMID: 22240478

Ruiz de Sabando, A., Ciosi, M., Galbete, A., Cumming, S. A., Álvarez, V., Martinez-Descals, A., Mila, M., Trujillo-Tiebas, M. J., López-Sendón, J. L., Fenollar-Cortés, M., Legarda, I., Bernal Noguera, S., Millán, J. M., Durán-Herrera, C., Ruiz-Martínez, J., Ruiz Onandi, R., Monckton, D. G., & Ramos-Arroyo, M. A. (2024). Somatic CAG repeat instability in intermediate alleles of the HTT gene and its potential association with a clinical phenotype. *European Journal of Human Genetics : EJHG*, *32*(7), 770–778. DOI: 10.1038/s41431-024-01546-6 PMID: 38433266

Ruschel, J., Hellal, F., Flynn, K. C., Dupraz, S., Elliott, D. A., Tedeschi, A., Bates, M., Sliwinski, C., Brook, G., Dobrindt, K., Peitz, M., Brüstle, O., Norenberg, M. D., Blesch, A., Weidner, N., Bunge, M. B., Bixby, J. L., & Bradke, F. (2015). Systemic administration of epothilone B promotes axon regeneration after spinal cord injury. *Science*, *348*(6232), 347–352. DOI: 10.1126/science.aaa2958 PMID: 25765066

Rutkowska, A. Z., & Diamanti-Kandarakis, E. (2016). Polycystic ovary syndrome and environmental toxins. *Fertility and Sterility*, *106*(4), 948–958. DOI: 10.1016/j.fertnstert.2016.08.031 PMID: 27559705

Ryskalin, L., Limanaqi, F., Frati, A., Busceti, C. L., & Fornai, F. (2018). mTOR-related brain dysfunctions in neuropsychiatric disorders. *International Journal of Molecular Sciences*, *19*(8), 2226. DOI: 10.3390/ijms19082226 PMID: 30061532

Saad, M. A., Eltarzy, M. A., Abdel Salam, R. M., & Ahmed, M. A. E. (2021). Liraglutide mends cognitive impairment by averting Notch signaling pathway overexpression in a rat model of polycystic ovary syndrome. *Life Sciences*, *265*, 118731. DOI: 10.1016/j.lfs.2020.118731 PMID: 33160995

Sabino, V., Cottone, P., Blasio, A., Iyer, M. R., Steardo, L., Rice, K. C., Conti, B., Koob, G. F., & Zorrilla, E. P. (2011). Activation of σ-receptors induces binge-like drinking in Sardinian alcohol-preferring rats. *Neuropsychopharmacology : Official Publication of the American College of Neuropsychopharmacology*, *36*(6), 1207–1218. DOI: 10.1038/npp.2011.5 PMID: 21346735

Sadeeqa, S., Mustafa, T., & Latif, S. (2018). Polycystic Ovarian Syndrome-Related Depression in Adolescent Girls: A Review. *Journal of Pharmacy & Bioallied Sciences*, *10*(2), 55–59. DOI: 10.4103/JPBS.JPBS_1_18 PMID: 29962792

Sadeghi, A., Niknam, M., Momeni-Moghaddam, M. A., Shabani, M., Aria, H., Bastin, A., Teimouri, M., Meshkani, R., & Akbari, H. (2023). Crosstalk between autophagy and insulin resistance: Evidence from different tissues. *European Journal of Medical Research*, *28*(1), 456. DOI: 10.1186/s40001-023-01424-9 PMID: 37876013

Saha, P., Ajgaonkar, S., Maniar, D., Sahare, S., Mehta, D., & Nair, S. (2024). Current insights into transcriptional role(s) for the nutraceutical Withania somnifera in inflammation and aging. *Frontiers in Nutrition, 11*, 1370951. DOI: 10.3389/fnut.2024.1370951 PMID: 38765810

Saha, S., Ash, P. E. A., Gowda, V., Liu, L., Shirihai, O., & Wolozin, B. (2015). Mutations in LRRK2 potentiate age-related impairment of autophagic flux. *Molecular Neurodegeneration, 10*(1), 1–13. DOI: 10.1186/s13024-015-0022-y PMID: 26159606

Sahu, M. K., Yadav, R., & Tiwari, S. P. (2023). Recent advances in nanotechnology. *International Journal of Nanomaterials, Nanotechnology and Nanomedicine, 9*(1), 15-23. DOI: 10.17352/2455-3492.000053

Sakai, K., Fukuda, T., & Iwadate, K. (2014). Immunohistochemical analysis of the ubiquitin proteasome system and autophagy lysosome system induced after traumatic intracranial injury: Association with time between the injury and death. *The American Journal of Forensic Medicine and Pathology, 35*(1), 38–44. DOI: 10.1097/PAF.0000000000000067 PMID: 24317096

Sălcudean, A., Bodo, C. R., Popovici, R. A., Cozma, M. M., Păcurar, M., Crăciun, R. E., Crisan, A. I., Enatescu, V. R., Marinescu, I., Cimpian, D. M., Nan, A. G., Sasu, A. B., Anculia, R. C., & Strete, E. G. (2025). Neuroinflammation-A Crucial Factor in the Pathophysiology of Depression-A Comprehensive Review. *Biomolecules, 15*(4), 502. Advance online publication. DOI: 10.3390/biom15040502 PMID: 40305200

Sánchez-Marzo, N., Pérez-Sánchez, A., Ruiz-Torres, V., Martínez-Tébar, A., Castillo, J., Herranz-López, M., & Barrajón-Catalán, E. (2019). Antioxidant and photoprotective activity of apigenin and its potassium salt derivative in human keratinocytes and absorption in Caco-2 cell monolayers. *International Journal of Molecular Sciences, 20*(9), 2148. DOI: 10.3390/ijms20092148 PMID: 31052292

Sanchez-Mirasierra, I., Ghimire, S., Hernandez-Diaz, S., & Soukup, S. F. (2022). Targeting Macroautophagy as a Therapeutic Opportunity to Treat Parkinson's Disease. *Frontiers in Cell and Developmental Biology, 10*, 1–12. DOI: 10.3389/fcell.2022.921314 PMID: 35874822

Sánchez-Vidaña, D. I., Li, J., Abokyi, S., Chan, J. N.-M., Ngai, S. P.-C., & Lau, B. W.-M. (2023). In vitro methods in autophagy research: Applications in neurodegenerative diseases and mood disorders. *Frontiers in Molecular Neuroscience, 16*, 1168948. DOI: 10.3389/fnmol.2023.1168948 PMID: 37122628

Sapolsky, R. M. (2015). Stress and the brain: Individual variability and the inverted-U. *Nature Neuroscience, 18*(10), 1344–1346. DOI: 10.1038/nn.4109 PMID: 26404708

Sarb, O. F., Sarb, A. D., Iacobescu, M., Vlad, I. M., Milaciu, M. V., Ciurmarnean, L., Vacaras, V., & Tantau, A. I. (2024). From Gut to Brain: Uncovering Potential Serum Biomarkers Connecting Inflammatory Bowel Diseases to Neurodegenerative Diseases. *International Journal of Molecular Sciences*, *25*(11), 5676. DOI: 10.3390/ijms25115676 PMID: 38891863

Sarwar, H., Rafiqi, S. I., Ahmad, S., Jinna, S., Khan, S. A., Karim, T., Qureshi, O., Zahid, Z. A., Elhai, J. D., Levine, J. C., Naqvi, S. J., Jaume, J. C., & Imam, S. (2022). Hyperinsulinemia associated depression. *Clinical Medicine Insights. Endocrinology and Diabetes*, *15*. DOI: 10.1177/11795514221090244 PMID: 35494421

Sawa-Makarska, J., Baumann, V., Coudevylle, N., von Bülow, S., Nogellova, V., Abert, C., Schuschnig, M., Graef, M., Hummer, G., & Martens, S. (2020). Reconstitution of autophagosome nucleation defines Atg9 vesicles as seeds for membrane formation. *Science*, *369*(6508), 1–23. DOI: 10.1126/science.aaz7714 PMID: 32883836

Scahill, R. I., Farag, M., Murphy, M. J., Hobbs, N. Z., Leocadi, M., Langley, C., Knights, H., Ciosi, M., Fayer, K., Nakajima, M., Thackeray, O., Gobom, J., Rönnholm, J., Weiner, S., Hassan, Y. R., Ponraj, N. K. P., Estevez-Fraga, C., Parker, C. S., Malone, I. B., & Tabrizi, S. J. (2025). Somatic CAG repeat expansion in blood associates with biomarkers of neurodegeneration in Huntington's disease decades before clinical motor diagnosis. *Nature Medicine*, *31*(3), 807–818. DOI: 10.1038/s41591-024-03424-6 PMID: 39825149

Scaini, G., Barichello, T., Fries, G. R., Kennon, E. A., Andrews, T., Nix, B. R., Zunta-Soares, G., Valvassori, S. S., Soares, J. C., & Quevedo, J. (2019). TSPO upregulation in bipolar disorder and concomitant downregulation of mitophagic proteins and NLRP3 inflammasome activation. *Neuropsychopharmacology : Official Publication of the American College of Neuropsychopharmacology*, *44*(7), 1291–1299. DOI: 10.1038/s41386-018-0293-4 PMID: 30575805

Scaini, G., Mason, B. L., Diaz, A. P., Jha, M. K., Soares, J. C., Trivedi, M. H., & Quevedo, J. (2022). Dysregulation of mitochondrial dynamics, mitophagy and apoptosis in major depressive disorder: Does inflammation play a role? *Molecular Psychiatry*, *27*(2), 1095–1102. DOI: 10.1038/s41380-021-01312-w PMID: 34650203

Schieber, M., & Chandel, N. S. (2014). ROS function in redox signaling and oxidative stress. *Current Biology*, *24*(10), R453–R62. DOI: 10.1016/j.cub.2014.03.034 PMID: 24845678

Schlamp, C. L., Li, Y., Dietz, J. A., Janssen, K. T., & Nickells, R. W. (2006). Progressive ganglion cell loss and optic nerve degeneration in DBA/2J mice is variable and asymmetric. *BMC Neuroscience*, *7*(1), 1–14. DOI: 10.1186/1471-2202-7-66 PMID: 17018142

Schnebert, S., Goguet, M., Vélez, E. J., Depincé, A., Beaumatin, F., Herpin, A., & Seiliez, I. (2022). Diving into the Evolutionary History of HSC70-Linked Selective Autophagy Pathways: Endosomal Microautophagy and Chaperone-Mediated Autophagy. *Cells*, *11*(12), 1945. DOI: 10.3390/cells11121945 PMID: 35741074

Schneider, J. L., & Cuervo, A. M. (2014). Autophagy and human disease: Emerging themes. *Current Opinion in Genetics & Development*, *26*, 16–23. DOI: 10.1016/j.gde.2014.04.003 PMID: 24907664

Schoenichen, C., Bode, C., & Duerschmied, D. (2019). Role of platelet serotonin in innate immune cell recruitment. *Frontiers in Bioscience*, *24*(3), 514–526. DOI: 10.2741/4732 PMID: 30468670

Scrivo, A., Bourdenx, M., Pampliega, O., & Cuervo, A. M. (2018). Selective autophagy as a potential therapeutic target for neurodegenerative disorders. *Lancet Neurology*, *17*(9), 802–815. DOI: 10.1016/S1474-4422(18)30238-2 PMID: 30129476

Sevriev, B., Dimitrova, S., Kehayova, G., & Dragomanova, S. (2024). Trehalose : Neuroprotective Effects and Mechanisms — An Updated Review. *NeuroSci*, *5*(4), 429–444. DOI: 10.3390/neurosci5040032 PMID: 39484301

Shao, X., Lu, W., Gao, F., Li, D., Hu, J., Li, Y., Zuo, Z., Jie, H., Zhao, Y., & Cen, X. (2016). Uric Acid Induces Cognitive Dysfunction through Hippocampal Inflammation in Rodents and Humans. *The Journal of Neuroscience : The Official Journal of the Society for Neuroscience*, *36*(43), 10990–11005. DOI: 10.1523/JNEUROSCI.1480-16.2016 PMID: 27798180

Shayganfard, M. (2020). Molecular and biological functions of resveratrol in psychiatric disorders: A review of recent evidence. *Cell & Bioscience*, *10*(1), 128. DOI: 10.1186/s13578-020-00491-3 PMID: 33292508

Shelton, R., Claiborne, J., Sidoryk-Wegrzynowicz, M., Reddy, R., Aschner, M., Lewis, D., & Mirnics, K. (2011). Altered expression of genes involved in inflammation and apoptosis in frontal cortex in major depression. *Molecular Psychiatry*, *16*(7), 751–762. DOI: 10.1038/mp.2010.52 PMID: 20479761

Sheppard, O., & Coleman, M. (2020). Alzheimer's Disease: Etiology, Neuropathology and Pathogenesis. In *Alzheimer's Disease: Drug Discovery* (pp. 1–22). Exon Publications. DOI: 10.36255/exonpublications.alzheimersdisease.2020.ch1

Shi, H., Zhang, Z., Wang, X., Li, R., Hou, W., Bi, W., & Zhang, X. (2015). Inhibition of autophagy induces IL-1β release from ARPE-19 cells via ROS mediated NLRP3 inflammasome activation under high glucose stress. *Biochemical and Biophysical Research Communications*, *463*(4), 1071–1076. DOI: 10.1016/j.bbrc.2015.06.060 PMID: 26102024

Shih, J.-H., Chiu, C.-H., Ma, K.-H., Huang, Y.-S., Shiue, C.-Y., Yeh, T.-Y., Kao, L.-T., Lin, Y.-Y., & Li, I.-H. (2019). Autophagy inhibition plays a protective role against 3, 4-methylenedioxymethamphetamine (MDMA)-induced loss of serotonin transporters and depressive-like behaviors in rats. *Pharmacological Research*, *142*, 283–293. DOI: 10.1016/j.phrs.2019.02.026 PMID: 30826457

Shimobayashi, M., Albert, V., Woelnerhanssen, B., Frei, I. C., Weissenberger, D., Meyer-Gerspach, A. C., Clement, N., Moes, S., Colombi, M., Meier, J. A., Swierczynska, M. M., Jenö, P., Beglinger, C., Peterli, R., & Hall, M. N. (2018). Insulin resistance causes inflammation in adipose tissue. *The Journal of Clinical Investigation*, *128*(4), 1538–1550. DOI: 10.1172/JCI96139 PMID: 29528335

Shinde, V., Hu, N., Mahale, A., Maiti, G., Daoud, Y., Eberhart, C. G., Maktabi, A., Jun, A. S., Al-Swailem, S. A., & Chakravarti, S. (2020). RNA sequencing of corneas from two keratoconus patient groups identifies potential biomarkers and decreased NRF2-antioxidant responses. *Scientific Reports*, *10*(1), 9907. DOI: 10.1038/s41598-020-66735-x PMID: 32555404

Shin, H. J., Kim, H., Oh, S., Lee, J. G., Kee, M., Ko, H. J., Kweon, M. N., Won, K. J., & Baek, S. H. (2016). AMPK-SKP2-CARM1 signaling cascade in transcriptional regulation of autophagy. *Nature*, *534*(7608), 553–557. DOI: 10.1038/nature18014 PMID: 27309807

Shoelson, S. E., Lee, J., & Goldfine, A. B. (2006). Inflammation and insulin resistance. *The Journal of Clinical Investigation*, *116*(7), 1793–1801. DOI: 10.1172/JCI29069 PMID: 16823477

Shu, F., Xiao, H., Li, Q.-N., Ren, X.-S., Liu, Z.-G., Hu, B.-W., Wang, H.-S., Wang, H., & Jiang, G.-M. (2023). Epigenetic and post-translational modifications in autophagy: Biological functions and therapeutic targets. *Signal Transduction and Targeted Therapy*, *8*(1), 32. DOI: 10.1038/s41392-022-01300-8 PMID: 36646695

Shukla, D., Rowe, C. L., Dong, Y., Racaniello, V. R., & Spear, P. G. (1999). The murine homolog (Mph) of human herpesvirus entry protein B (HveB) mediates entry of pseudorabies virus but not herpes simplex virus types 1 and 2. *Journal of Virology*, *73*(5), 4493–4497. DOI: 10.1128/JVI.73.5.4493-4497.1999 PMID: 10196354

Singh, S., Pal, N., Shubham, S., Sarma, D. K., Verma, V., Marotta, F., & Kumar, M. (2023). Polycystic Ovary Syndrome: Etiology, Current Management, and Future Therapeutics. *Journal of Clinical Medicine*, *12*(4), 1454. Advance online publication. DOI: 10.3390/jcm12041454 PMID: 36835989

Singulani, M. P., Ferreira, A. F. F., Figueroa, P. S., Cuyul-Vásquez, I., Talib, L. L., Britto, L. R., & Forlenza, O. V. (2024). Lithium and disease modification: A systematic review and meta-analysis in Alzheimer's and Parkinson's disease. *Ageing Research Reviews*, *95*, 1–20. DOI: 10.1016/j.arr.2024.102231 PMID: 38364914

Sinha, D., Klise, A., Sergeev, Y., Hose, S., Bhutto, I. A., Hackler, L.Jr, Malpic-Llanos, T., Samtani, S., Grebe, R., Goldberg, M. F., Hejtmancik, J. F., Nath, A., Zack, D. J., Fariss, R. N., McLeod, D. S., Sundin, O., Broman, K. W., Lutty, G. A., & Zigler, J. S.Jr. (2008). βA3/A1-crystallin in astroglial cells regulates retinal vascular remodeling during development. *Molecular and Cellular Neurosciences*, *37*(1), 85–95. DOI: 10.1016/j.mcn.2007.08.016 PMID: 17931883

Sinha, R. A., Singh, B. K., & Yen, P. M. (2017). Reciprocal crosstalk between autophagic and endocrine signaling in metabolic homeostasis. *Endocrine Reviews*, *38*(1), 69–102. DOI: 10.1210/er.2016-1103 PMID: 27901588

Skobo, T., Benato, F., Grumati, P., Meneghetti, G., Cianfanelli, V., Castagnaro, S., Chrisam, M., Di Bartolomeo, S., Bonaldo, P., Cecconi, F., & Valle, L. D. (2014). Zebrafish ambra1a and ambra1b knockdown impairs skeletal muscle development. *PLoS One*, *9*(6), e99210. DOI: 10.1371/journal.pone.0099210 PMID: 24922546

Slattery, D. A., & Cryan, J. F. (2017). Modelling depression in animals: At the interface of reward and stress pathways. *Psychopharmacology*, *234*(9-10), 1451–1465. DOI: 10.1007/s00213-017-4552-6 PMID: 28224183

Smith, E. D., Prieto, G. A., Tong, L., Sears-Kraxberger, I., Rice, J. D., Steward, O., & Cotman, C. W. (2014). Rapamycin and interleukin-1β impair brain-derived neurotrophic factor-dependent neuron survival by modulating autophagy. *The Journal of Biological Chemistry*, *289*(30), 20615–20629. DOI: 10.1074/jbc.M114.568659 PMID: 24917666

Sng, K. S., Li, G., Zhou, L.-y., Song, Y., Chen, X., Wang, Y., Yao, M., & Cui, X. (2022). Ginseng extract and ginsenosides improve neurological function and promote antioxidant effects in rats with spinal cord injury: A meta-analysis and systematic review. *Journal of Ginseng Research*, *46*(1), 11–22. DOI: 10.1016/j.jgr.2021.05.009 PMID: 35058723

Sobral, J., Empadinhas, N., Esteves, A. R., & Cardoso, S. M. (2025). Impact of nutrition on the gut microbiota: Implications for Parkinson's disease. *Nutrition Reviews*, *83*(4), 713–727. Advance online publication. DOI: 10.1093/nutrit/nuae208 PMID: 39812804

Song, M., Bai, Y., & Song, F. (2025). High-fat diet and neuroinflammation: The role of mitochondria. *Pharmacological Research*, *107615*, 107615. Advance online publication. DOI: 10.1016/j.phrs.2025.107615 PMID: 39842474

Song, Y., Wang, F., Luo, H., Hu, H., Pang, Y., Xu, K., & Zhang, X. (2025). Rapamycin protects glucocorticoid-induced glaucoma model mice against trabecular meshwork fibrosis by suppressing mTORC1/2 signaling. *European Journal of Pharmacology*, *990*, 177269. DOI: 10.1016/j.ejphar.2025.177269 PMID: 39805488

Son, S. M., Park, S. J., Fernandez-Estevez, M., & Rubinsztein, D. C. (2021). Autophagy regulation by acetylation—Implications for neurodegenerative diseases. *Experimental & Molecular Medicine*, *53*(1), 30–41. DOI: 10.1038/s12276-021-00556-4 PMID: 33483607

Sporbeck, K., Haas, M. L., Pastor-Maldonado, C. J., Schüssele, D. S., Hunter, C., Takacs, Z., Diogo de Oliveira, A. L., Franz-Wachtel, M., Charsou, C., Pfisterer, S. G., Gubas, A., Haller, P. K., Knorr, R. L., Kaulich, M., Macek, B., Eskelinen, E.-L., Simonsen, A., & Proikas-Cezanne, T. (2023). The ABL-MYC axis controls WIPI1-enhanced autophagy in lifespan extension. *Communications Biology*, *6*(1), 872. DOI: 10.1038/s42003-023-05236-9 PMID: 37620393

Sragovich, S., Merenlender-Wagner, A., & Gozes, I. (2017). ADNP Plays a Key Role in Autophagy: From Autism to Schizophrenia and Alzheimer's Disease. *BioEssays*, *39*(11), 1–10. DOI: 10.1002/bies.201700054 PMID: 28940660

Sriram, P., & Suttee, A. (2020). Nanotechnology advances, benefits, and applications in daily life. In *Nanotechnology* (pp. 23–44). CRC Press. DOI: 10.1201/9781003082859-3

Stuart, M. J., Singhal, G., & Baune, B. T. (2015). Systematic review of the neurobiological relevance of chemokines to psychiatric disorders. *Frontiers in Cellular Neuroscience*, *9*, 357. DOI: 10.3389/fncel.2015.00357 PMID: 26441528

Sukkriang, N., & Buranapin, S. (2024). Effect of intermittent fasting 16:8 and 14:10 compared with control-group on weight reduction and metabolic outcomes in obesity with type 2 diabetes patients: A randomized controlled trial. *Journal of Diabetes Investigation*, *15*(9), 1297–1305. DOI: 10.1111/jdi.14186 PMID: 38932663

Sun, X., Shukla, M., & Wang, W. (2024). Unlocking gut-liver-brain axis communication metabolites: Energy metabolism, immunity and barriers. npj Biofilms and Microbiomes, 10(136). DOI: 10.1038/s41522-024-00610-9

Sunhe, Y.-X., Zhang, Y.-H., Fu, R.-J., Xu, D.-Q., & Tang, Y.-P. (2024). Neuroprotective effect and preparation methods of berberine. *Frontiers in Pharmacology*, *15*, 1–23. DOI: 10.3389/fphar.2024.1429050 PMID: 39309003

Sun, Y., Koyama, Y., & Shimada, S. (2022a). Inflammation From Peripheral Organs to the Brain: How Does Systemic Inflammation Cause Neuroinflammation? *Frontiers in Aging Neuroscience*, *14*, 903455. DOI: 10.3389/fnagi.2022.903455 PMID: 35783147

Sun, Y., Zhang, C., Lei, T., Lin, F., Huang, J., Hu, Y., Wang, D., & Zhang, W. (2024). HIV-1 gp120 activates microglia via TLR2–NF-κB signaling to upregulate inflammatory cytokine expression and induce neuropathic pain. *Neuropharmacology*, *260*, 110136. DOI: 10.1016/j.neuropharm.2024.110136 PMID: 39216684

Su, P., Chen, C., & Sun, Y. (2025). Physiopathology of polycystic ovary syndrome in endocrinology, metabolism and inflammation. *Journal of Ovarian Research*, *18*(1), 34. DOI: 10.1186/s13048-025-01621-6 PMID: 39980043

Su, Y., D'Arcy, C., Li, M., O'Donnell, K., Caron, J., Meaney, M., & Meng, X. (2022). Specific and cumulative lifetime stressors in the aetiology of major depression: A longitudinal community-based population study. *Epidemiology and Psychiatric Sciences*, *31*, e3. DOI: 10.1017/S2045796021000779 PMID: 35078547

Szczesnowicz, A., Szeliga, A., Niwczyk, O., Bala, G., & Meczekalski, B. (2023). Do GLP-1 Analogs Have a Place in the Treatment of PCOS? New Insights and Promising Therapies. *Journal of Clinical Medicine*, *12*(18), 5915. Advance online publication. DOI: 10.3390/jcm12185915 PMID: 37762856

Takamura, A., Komatsu, M., Hara, T., Sakamoto, A., Kishi, C., Waguri, S., Eishi, Y., Hino, O., Tanaka, K., & Mizushima, N. (2011). Autophagy-deficient mice develop multiple liver tumors. *Genes & Development*, *25*(8), 795–800. DOI: 10.1101/gad.2016211 PMID: 21498569

Takebayashi, M., Kagaya, A., Inagaki, M., Kozuru, T., Jitsuiki, H., Kurata, K., Okamoto, Y., & Yamawaki, S. (2000). Effects of antidepressants on γ-aminobutyric acid-and N-methyl-D-aspartate-induced intracellular Ca2+ concentration increases in primary cultured rat cortical neurons. *Neuropsychobiology*, *42*(3), 120–126. DOI: 10.1159/000026681 PMID: 11015029

Taleb, S. (2019). Tryptophan dietary impacts gut barrier and metabolic diseases. *Frontiers in Immunology*, *10*, 2113. DOI: 10.3389/fimmu.2019.02113 PMID: 31552046

Talei, S., Mokhtari, T., Asadi, I., Arab, L., Hassanzadeh, N., Hosseinjani, E., & Hassanzadeh, G. (2021). Flaxseed oil (Linum usitatissimum) attenuates restraint stress-induced depressive-like behavior: Upregulation of neurotrophic factors in CA1 region of hippocampus. *Journal of Contemporary Medical Sciences.*, *7*(1), 61–65. DOI: 10.22317/jcms.v7i1.945

Tamaki, Y., Ross, J. P., Alipour, P., Castonguay, C. É., Li, B., Catoire, H., Rochefort, D., Urushitani, M., Takahashi, R., Sonnen, J. A., Stifani, S., Dion, P. A., & Rouleau, G. A. (2023). Spinal cord extracts of amyotrophic lateral sclerosis spread TDP-43 pathology in cerebral organoids. *PLOS Genetics*, *19*(2), e1010606. DOI: 10.1371/journal.pgen.1010606 PMID: 36745687

Tanabe, F., Yone, K., Kawabata, N., Sakakima, H., Matsuda, F., Ishidou, Y., Maeda, S., Abematsu, M., Komiya, S., & Setoguchi, T. (2011). Accumulation of p62 in degenerated spinal cord under chronic mechanical compression: Functional analysis of p62 and autophagy in hypoxic neuronal cells. *Autophagy*, *7*(12), 1462–1471. DOI: 10.4161/auto.7.12.17892 PMID: 22082874

Tang, P. H., Kono, M., Koutalos, Y., Ablonczy, Z., & Crouch, R. K. (2013). New insights into retinoid metabolism and cycling within the retina. *Progress in Retinal and Eye Research*, *32*, 48–63. DOI: 10.1016/j.preteyeres.2012.09.002 PMID: 23063666

Tang, P., Hou, H., Zhang, L., Lan, X., Mao, Z., Liu, D., He, C., Du, H., & Zhang, L. (2014). Autophagy reduces neuronal damage and promotes locomotor recovery via inhibition of apoptosis after spinal cord injury in rats. *Molecular Neurobiology*, *49*(1), 276–287. DOI: 10.1007/s12035-013-8518-3 PMID: 23954967

Tang, Y., Ju, W., Liu, Y., & Deng, Q. (2024). The role of SIRT1 in autophagy and drug resistance: Unveiling new targets and potential biomarkers in cancer therapy. *Frontiers in Pharmacology*, *15*, 1469830. DOI: 10.3389/fphar.2024.1469830 PMID: 39403142

Tanida, I. (2011). Autophagy basics. *Microbiology and Immunology*, *55*(1), 1–11. DOI: 10.1111/j.1348-0421.2010.00271.x PMID: 21175768

Tannous, P., Zhu, H., Nemchenko, A., Berry, J. M., Johnstone, J. L., Shelton, J. M., Miller, F. J.Jr, Rothermel, B. A., & Hill, J. A. (2008). Intracellular protein aggregation is a proximal trigger of cardiomyocyte autophagy. *Circulation*, *117*(24), 3070–3078. DOI: 10.1161/CIRCULATIONAHA.107.763870 PMID: 18541737

Tan, X., Yang, Y., Xu, J., Zhang, P., Deng, R., Mao, Y., He, J., Chen, Y., Zhang, Y., Ding, J., Li, H., Shen, H., Li, X., Dong, W., & Chen, G. (2020). Luteolin exerts neuroprotection via modulation of the p62/Keap1/Nrf2 pathway in intracerebral hemorrhage. *Frontiers in Pharmacology*, *10*, 1551. DOI: 10.3389/fphar.2019.01551 PMID: 32038239

Tan, Y., Zhu, J., & Hashimoto, K. (2024). Autophagy-related gene model as a novel risk factor for schizophrenia. *Translational Psychiatry*, *14*(1), 1–11. DOI: 10.1038/s41398-024-02767-5 PMID: 38351068

Tartt, A. N., Mariani, M. B., Hen, R., Mann, J. J., & Boldrini, M. (2022). Dysregulation of adult hippocampal neuroplasticity in major depression: Pathogenesis and therapeutic implications. *Molecular Psychiatry*, *27*(6), 2689–2699. DOI: 10.1038/s41380-022-01520-y PMID: 35354926

Taymans, J. M., Fell, M., Greenamyre, T., Hirst, W. D., Mamais, A., Padmanabhan, S., Peter, I., Rideout, H., & Thaler, A. (2023). Perspective on the current state of the LRRK2 field. *NPJ Parkinson's Disease*, *9*(1), 1–9. DOI: 10.1038/s41531-023-00544-7 PMID: 37393318

Tefagh, G., Payab, M., Qorbani, M., Sharifi, F., Sharifi, Y., Ebrahimnegad Shirvani, M. S., Pourghazi, F., Atlasi, R., Shadman, Z., Rezaei, N., Mohammadi-Vajari, E., Larijani, B., & Ebrahimpur, M. (2022). Effect of vitamin E supplementation on cardiometabolic risk factors, inflammatory and oxidative markers and hormonal functions in PCOS (polycystic ovary syndrome): A systematic review and meta-analysis. *Scientific Reports*, *12*(1), 5770. DOI: 10.1038/s41598-022-09082-3 PMID: 35388031

Tekirdag, K., & Cuervo, A. M. (2018). Chaperone-mediated autophagy and endosomal microautophagy: Jointed by a chaperone. *The Journal of Biological Chemistry*, *293*(15), 5414–5424. DOI: 10.1074/jbc.R117.818237 PMID: 29247007

Terzioglu, G., & Young-Pearse, T. L. (2023). Microglial function, INPP5D/SHIP1 signaling, and NLRP3 inflammasome activation: Implications for Alzheimer's disease. *Molecular Neurodegeneration*, *18*(89), 89. Advance online publication. DOI: 10.1186/s13024-023-00674-9 PMID: 38017562

Thapa, R., Moglad, E., Afzal, M., Gupta, G., Bhat, A. A., Hassan almalki, W., Kazmi, I., Alzarea, S. I., pant, K., Singh, T. G., Singh, S. K., & Ali, H. (2024). The role of sirtuin 1 in ageing and neurodegenerative disease: A molecular perspective. *Ageing Research Reviews*, *102*(102545), 1–14. DOI: 10.1016/j.arr.2024.102545 PMID: 39423873

Themistokleous, C., Bagnoli, E., Parulekar, R., & Muqit, M. M. K. (2023). Role of Autophagy Pathway in Parkinson's Disease and Related Genetic Neurological Disorders. *Journal of Molecular Biology*, *435*(12), 168144. DOI: 10.1016/j.jmb.2023.168144 PMID: 37182812

Thomas, G. M., & Huganir, R. L. (2004). MAPK cascade signalling and synaptic plasticity. *Nature Reviews. Neuroscience*, *5*(3), 173–183. DOI: 10.1038/nrn1346 PMID: 14976517

Thompson, L. J.-P., Genovese, J., Hong, Z., Singh, M. V., & Singh, V. B. (2024). HIV-Associated Neurocognitive Disorder: A Look into Cellular and Molecular Pathology. *International Journal of Molecular Sciences*, *25*(9), 4697. DOI: 10.3390/ijms25094697 PMID: 38731913

Tian, E., Sharma, G., & Dai, C. (2023). Neuroprotective Properties of Berberine: Molecular Mechanisms and Clinical Implications. *Antioxidants*, *12*(10), 1–17. DOI: 10.3390/antiox12101883 PMID: 37891961

Tobin, M. K., Bonds, J. A., Minshall, R. D., Pelligrino, D. A., Testai, F. D., & Lazarov, O. (2014). Neurogenesis and inflammation after ischemic stroke: What is known and where we go from here. *Journal of Cerebral Blood Flow and Metabolism*, *34*(10), 1573–1584. DOI: 10.1038/jcbfm.2014.130 PMID: 25074747

Todd, T. W., Shao, W., Zhang, Y., & Petrucelli, L. (2023). The endolysosomal pathway and ALS/FTD. *Trends in Neurosciences*, *46*(12), 1025–1041. DOI: 10.1016/j.tins.2023.09.004 PMID: 37827960

Toker, L., & Agam, G. (2015). Mitochondrial dysfunction in psychiatric morbidity: Current evidence and therapeutic prospects. *Neuropsychiatric Disease and Treatment*, *11*, 2441–2447. DOI: 10.2147/NDT.S70346 PMID: 26442764

Tosefsky, K. N., Zhu, J., Wang, Y. N., Lam, J. S. T., Cammalleri, A., & Appel-Cresswell, S. (2024). The Role of Diet in Parkinson's Disease. *Journal of Parkinson's Disease*, *14*(s1), S21–S34. DOI: 10.3233/JPD-230264 PMID: 38251061

Tramutola, A., Triplett, J. C., Di Domenico, F., Niedowicz, D. M., Murphy, M. P., Coccia, R., Perluigi, M., & Butterfield, D. A. (2015). Alteration of mTOR signaling occurs early in the progression of Alzheimer disease (AD): Analysis of brain from subjects with pre-clinical AD, amnestic mild cognitive impairment and late-stage AD. *Journal of Neurochemistry*, *133*(5), 739–749. DOI: 10.1111/jnc.13037 PMID: 25645581

Tran, S., Fairlie, W. D., & Lee, E. F. (2021). BECLIN1: Protein Structure, Function and Regulation. *Cells*, *10*(6), 1522. DOI: 10.3390/cells10061522 PMID: 34204202

Troubat, R., Barone, P., Leman, S., Desmidt, T., Cressant, A., Atanasova, B., Brizard, B., El Hage, W., Surget, A., Belzung, C., & Camus, V. (2021). Neuroinflammation and depression: A review. *The European Journal of Neuroscience*, *53*(1), 151–171. DOI: 10.1111/ejn.14720 PMID: 32150310

Tu, H., Yuan, B., Hou, X., Zhang, X., Pei, C., Ma, Y., Yang, Y., Fan, Y., Qin, Z., Liu, C., & Hu, L. (2021). α-synuclein suppresses microglial autophagy and promotes neurodegeneration in a mouse model of Parkinson's disease. *Aging Cell*, *20*(12), e13522. Advance online publication. DOI: 10.1111/acel.13522 PMID: 34811872

Turco, E., Savova, A., Gere, F., Ferrari, L., Romanov, J., Schuschnig, M., & Martens, S. (2021). Reconstitution defines the roles of p62, NBR1 and TAX1BP1 in ubiquitin condensate formation and autophagy initiation. *Nature Communications*, *12*(1), 1–16. DOI: 10.1038/s41467-021-25572-w PMID: 34471133

Tutas, J., Tolve, M., Özer-Yildiz, E., Ickert, L., Klein, I., Silverman, Q., Liebsch, F., Dethloff, F., Giavalisco, P., Endepols, H., Georgomanolis, T., Neumaier, B., Drzezga, A., Schwarz, G., Thorens, B., Gatto, G., Frezza, C., & Kononenko, N. L. (2025). Autophagy regulator ATG5 preserves cerebellar function by safeguarding its glycolytic activity. *Nature Metabolism*, *7*(2), 297–320. DOI: 10.1038/s42255-024-01196-4 PMID: 39815080

Uddin, M. S., Stachowiak, A., Al Mamun, A., Tzvetkov, N. T., Takeda, S., Atanasov, A. G., Bergantin, L. B., Abdel-Daim, M. M., & Stankiewicz, A. M. (2018). Autophagy and Alzheimer's disease: From molecular mechanisms to therapeutic implications. *Frontiers in Aging Neuroscience*, *10*, 1–18. DOI: 10.3389/fnagi.2018.00004 PMID: 29441009

Unluhizarci, K., Karaca, Z., & Kelestimur, F. (2021). Role of insulin and insulin resistance in androgen excess disorders. *World Journal of Diabetes*, *12*(5), 616–629. DOI: 10.4239/wjd.v12.i5.616 PMID: 33995849

Usenovic, M., Tresse, E., Mazzulli, J. R., Taylor, J. P., & Krainc, D. (2012). Deficiency of ATP13A2 leads to lysosomal dysfunction, α-synuclein accumulation, and neurotoxicity. *The Journal of Neuroscience : The Official Journal of the Society for Neuroscience*, *32*(12), 4240–4246. DOI: 10.1523/JNEUROSCI.5575-11.2012 PMID: 22442086

Usman, M., Farooq, M., Wakeel, A., Nawaz, A., Cheema, S. A., Rehman, H., Ashraf, I., & Sanaullah, M. (2020). Nanotechnology in agriculture: Current status, challenges and future opportunities. *The Science of the Total Environment*, *721*, 137778. DOI: 10.1016/j.scitotenv.2020.137778 PMID: 32179352

Valapala, M., Edwards, M., Hose, S., Grebe, R., Bhutto, I. A., Cano, M., Berger, T., Mak, T. W., Wawrousek, E., Handa, J. T., Lutty, G. A., Samuel Zigler, J.Jr, & Sinha, D. (2014). Increased L ipocalin-2 in the retinal pigment epithelium of Cryba1 c KO mice is associated with a chronic inflammatory response. *Aging Cell, 13*(6), 1091–1094. DOI: 10.1111/acel.12274 PMID: 25257511

Valapala, M., Wilson, C., Hose, S., Bhutto, I. A., Grebe, R., Dong, A., Greenbaum, S., Gu, L., Sengupta, S., Cano, M., Hackett, S., Xu, G., Lutty, G. A., Dong, L., Sergeev, Y., Handa, J. T., Campochiaro, P., Wawrousek, E., Zigler, J. S.Jr, & Sinha, D. (2014). Lysosomal-mediated waste clearance in retinal pigment epithelial cells is regulated by CRYBA1/βA3/A1-crystallin via V-ATPase-MTORC1 signaling. *Autophagy*, *10*(3), 480–496. DOI: 10.4161/auto.27292 PMID: 24468901

Valencia, M., Kim, S. R., Jang, Y., & Lee, S. H. (2021). Neuronal autophagy: Characteristic features and roles in neuronal pathophysiology. *Biomolecules & Therapeutics*, *29*(6), 605–614. DOI: 10.4062/biomolther.2021.012 PMID: 33875624

Valiukas, Z., Tangalakis, K., Apostolopoulos, V., & Feehan, J. (2025). Microglial activation states and their implications for Alzheimer's disease. *The Journal of Prevention of Alzheimer's Disease*, *12*(1), 100013. DOI: 10.1016/j.tjpad.2024.100013 PMID: 39800461

Valls-Pedret, C., Sala-Vila, A., Serra-Mir, M., Corella, D., de la Torre, R., Martínez-González, M. Á., Martínez-Lapiscina, E. H., Fitó, M., Pérez-Heras, A., Salas-Salvadó, J., Estruch, R., & Ros, E. (2015). Mediterranean Diet and Age-Related Cognitive Decline: A Randomized Clinical Trial. *JAMA Internal Medicine*, *175*(7), 1094–1103. DOI: 10.1001/jamainternmed.2015.1668 PMID: 25961184

Valverde, D. P., Yu, S., Boggavarapu, V., Kumar, N., Lees, J. A., Walz, T., Reinisch, K. M., & Melia, T. J. (2019). ATG2 transports lipids to promote autophagosome biogenesis. *The Journal of Cell Biology*, *218*(6), 1787–1798. DOI: 10.1083/jcb.201811139 PMID: 30952800

Vargas, J. N. S., Wang, C., Bunker, E., Hao, L., Maric, D., Schiavo, G., Randow, F., & Youle, R. J. (2019). Spatiotemporal control of ULK1 activation by NDP52 and TBK1 during selective autophagy. *Molecular cell, 74*(2), 347-362. e346.

Vargova, I., Machova Urdzikova, L., Karova, K., Smejkalova, B., Sursal, T., Cimermanova, V., Turnovcova, K., Gandhi, C. D., Jhanwar-Uniyal, M., & Jendelova, P. (2021). Involvement of mTOR pathways in recovery from spinal cord injury by modulation of autophagy and immune response. *Biomedicines*, *9*(6), 593. DOI: 10.3390/biomedicines9060593 PMID: 34073791

Vater, M., Rost, N., Eckstein, G., Sauer, S., Tontsch, A., Erhardt, A., Lucae, S., Brückl, T., Klopstock, T., Sämann, P. G., & Binder, E. B. (2024). Huntingtin CAG repeat size variations below the Huntington's disease threshold: Associations with depression, anxiety and basal ganglia structure. *European Journal of Human Genetics : EJHG*. Advance online publication. DOI: 10.1038/s41431-024-01737-1 PMID: 39572770

Verma, R., Virdi, J. K., Singh, N., & Jaggi, A. S. (2019). Animals models of spinal cord contusion injury. *The Korean Journal of Pain*, *32*(1), 12–21. DOI: 10.3344/kjp.2019.32.1.12 PMID: 30671199

Vijayakumar, K., & Cho, G. W. (2019). Autophagy: An evolutionarily conserved process in the maintenance of stem cells and aging. *Cell Biochemistry and Function*, *37*(6), 452–458. DOI: 10.1002/cbf.3427 PMID: 31318072

Vijmasi, T., Chen, F. Y., Balasubbu, S., Gallup, M., McKown, R. L., Laurie, G. W., & McNamara, N. A. (2014). Topical administration of lacritin is a novel therapy for aqueous-deficient dry eye disease. *Investigative Ophthalmology & Visual Science*, *55*(8), 5401–5409. DOI: 10.1167/iovs.14-13924 PMID: 25034600

von Zglinicki, T., Wan, T., & Miwa, S. (2021). Senescence in Post-Mitotic Cells: A Driver of Aging? *Antioxidants & Redox Signaling*, *34*(4), 308–323. DOI: 10.1089/ars.2020.8048 PMID: 32164429

Voros, V., Martin Gutierrez, D., Alvarez, F., Boda-Jorg, A., Kovacs, A., Tenyi, T., Fekete, S., & Osvath, P. (2020). The impact of depressive mood and cognitive impairment on quality of life of the elderly. *Psychogeriatrics*, *20*(3), 271–277. DOI: 10.1111/psyg.12495 PMID: 31811691

Vuic, B., Milos, T., Tudor, L., Konjevod, M., Nikolac Perkovic, M., Jazvinscak Jembrek, M., Nedic Erjavec, G., & Svob Strac, D. (2022). Cannabinoid cb2 receptors in neurodegenerative proteinopathies: New insights and therapeutic potential. *Biomedicines*, *10*(12), 3000. DOI: 10.3390/biomedicines10123000 PMID: 36551756

Wang, B., Fang, T., & Chen, H. (2023). Zinc and central nervous system disorders. *Nutrients*, *15*(9), 2140. DOI: 10.3390/nu15092140 PMID: 37432243

Wang, D., Yuan, T., Liu, J., Wen, Z., Shen, Y., Tang, J., Wang, Z., & Wu, X. (2022). ATG16L2 inhibits NLRP3 inflammasome activation through promoting ATG5-12-16L1 complex assembly and autophagy. *European Journal of Immunology*, *52*(8), 1321–1334. DOI: 10.1002/eji.202149764 PMID: 35426127

Wang, F., Zhang, Z., Han, J., Zheng, J., Wang, X., & Wang, Z. (2023). Discovery of microglia gonadotropin-releasing hormone receptor and its potential role in polycystic ovarian syndrome. *Molecular Medicine Reports*, *27*(4), 1–9. DOI: 10.3892/mmr.2023.12964 PMID: 36799164

Wang, G., Estrella, A., Hakim, O., Milazzo, P., Patel, S., Pintagro, C., Li, D., Zhao, R., Vance, D. E., & Li, W. (2022). Mini-Mental State Examination and Montreal Cognitive Assessment as Tools for Following Cognitive Changes in Alzheimer's Disease Neuroimaging Initiative Participants. *Journal of Alzheimer's Disease*, *90*(1), 263–270. DOI: 10.3233/JAD-220397 PMID: 36093696

Wang, H.-D., Lv, C.-L., Feng, L., Guo, J.-X., Zhao, S.-Y., & Jiang, P. (2024). The role of autophagy in brain health and disease: Insights into exosome and autophagy interactions. *Heliyon*, *10*(21), e38959. DOI: 10.1016/j.heliyon.2024.e38959 PMID: 39524893

Wang, H., Guo, Y., Qiao, Y., Zhang, J., & Jiang, P. (2020). Nobiletin ameliorates NLRP3 inflammasome-mediated inflammation through promoting autophagy via the AMPK pathway. *Molecular Neurobiology*, *57*(12), 5056–5068. DOI: 10.1007/s12035-020-02071-5 PMID: 32833187

Wang, H.-H., Zhang, Y., Qu, T.-Q., Sang, X.-Q., Li, Y.-X., Ren, F.-Z., Wen, P.-C., & Sun, Y.-N. (2023). Nobiletin improves D-galactose-induced aging mice skeletal muscle atrophy by regulating protein homeostasis. *Nutrients*, *15*(8), 1801. DOI: 10.3390/nu15081801 PMID: 37111020

Wang, J., Bian, S., Wang, S., Yang, S., Zhang, W., Zhao, D., Liu, M., & Bai, X. (2020). Ginsenoside Rh2 represses autophagy to promote cervical cancer cell apoptosis during starvation. *Chinese Medicine*, *15*(1), 1–14. DOI: 10.1186/s13020-020-00396-w PMID: 33292331

Wang, J., Li, J., Cao, N., Li, Z., Han, J., & Li, L. (2018). Resveratrol, an activator of SIRT1, induces protective autophagy in non-small-cell lung cancer via inhibiting Akt/mTOR and activating p38-MAPK. *OncoTargets and Therapy*, *11*, 7777–7786. DOI: 10.2147/OTT.S159095 PMID: 30464525

Wang, L., Huang, Z., Ping, J., Liu, X., Cai, Z., & Dai, C. (2021). Oridonin ameliorates depressive-like behaviors induced by chronic unpredictable mild stress in mice via TXNIP/NLRP3 signaling pathway. *Tropical Journal of Pharmaceutical Research*, *20*(3), 531–536. DOI: 10.4314/tjpr.v20i3.13

Wang, L., Klionsky, D. J., & Shen, H. M. (2023). The emerging mechanisms and functions of microautophagy. *Nature Reviews. Molecular Cell Biology*, *24*(3), 186–203. DOI: 10.1038/s41580-022-00529-z PMID: 36097284

Wang, P., Jiang, L., Zhou, N., Zhou, H., Liu, H., Zhao, W., Zhang, H., Zhang, X., & Hu, Z. (2018). Resveratrol ameliorates autophagic flux to promote functional recovery in rats after spinal cord injury. *Oncotarget*, *9*(9), 8427–8440. DOI: 10.18632/oncotarget.23877 PMID: 29492205

Wang, P., Nolan, T. M., Yin, Y., & Bassham, D. C. (2020). Identification of transcription factors that regulate ATG8 expression and autophagy in Arabidopsis. *Autophagy*, *16*(1), 123–139. DOI: 10.1080/15548627.2019.1598753 PMID: 30909785

Wang, R., Tan, J., Chen, T., Han, H., Tian, R., Tan, Y., Wu, Y., Cui, J., Chen, F., Li, J., Lv, L., Guan, X., Shang, S., Lu, J., & Zhang, Z. (2019). ATP13A2 facilitates HDAC6 recruitment to lysosome to promote autophagosome–lysosome fusion. *The Journal of Cell Biology*, *218*(1), 267–284. DOI: 10.1083/jcb.201804165 PMID: 30538141

Wang, S., Li, H., Yuan, M., Fan, H., & Cai, Z. (2022). Role of AMPK in autophagy. *Frontiers in Physiology*, *13*, 1015500. DOI: 10.3389/fphys.2022.1015500 PMID: 36505072

Wang, S., Yu, L., Yang, H., Li, C., Hui, Z., Xu, Y., & Zhu, X. (2016). Oridonin attenuates synaptic loss and cognitive deficits in an Aβ1–42-induced mouse model of Alzheimer's disease. *PLoS One*, *11*(3), e0151397. DOI: 10.1371/journal.pone.0151397 PMID: 26974541

Wang, W., Hu, X., Zhao, Z., Liu, P., Hu, Y., Zhou, J., Zhou, D., Wang, Z., Guo, D., & Guo, H. (2008). Antidepressant-like effects of liquiritin and isoliquiritin from Glycyrrhiza uralensis in the forced swimming test and tail suspension test in mice. *Progress in Neuro-Psychopharmacology & Biological Psychiatry*, *32*(5), 1179–1184. DOI: 10.1016/j.pnpbp.2007.12.021 PMID: 18289754

Wang, X., Hu, J., Zhang, H., & Zhou, P. (2025). Probiotics encapsulated via biological macromolecule for neurological therapy and functional food: A review. *Probiotics and Antimicrobial Proteins*, *17*(3), 1754–1768. Advance online publication. DOI: 10.1007/s12602-025-10453-1 PMID: 39821001

Wang, X., Li, X., Zuo, X., Liang, Z., Ding, T., Li, K., Ma, Y., Li, P., Zhu, Z., Ju, C., Zhang, Z., Song, Z., Quan, H., Zhang, J., Hu, X., & Wang, Z. (2021). Photobiomodulation inhibits the activation of neurotoxic microglia and astrocytes by inhibiting Lcn2/JAK2-STAT3 crosstalk after spinal cord injury in male rats. *Journal of Neuroinflammation*, *18*(1), 1–20. DOI: 10.1186/s12974-021-02312-x PMID: 34740378

Wāng, Y. (2025). Ambient fine particulate matter provokes multiple modalities of cell death via perturbation of subcellular structures. *Environment International*, *195*, 109193. DOI: 10.1016/j.envint.2024.109193 PMID: 39721566

Wang, Y.-T., & Lu, J.-H. (2022). Chaperone-Mediated Autophagy in Neurodegenerative Diseases: Molecular Mechanisms and Pharmacological Opportunities. *Cells*, *11*(14), 2250. DOI: 10.3390/cells11142250 PMID: 35883693

Wang, Z. Y., Liu, W. G., Muharram, A., Wu, Z. Y., & Lin, J. H. (2014). Neuroprotective effects of autophagy induced by rapamycin in rat acute spinal cord injury model. *Neuroimmunomodulation*, *21*(5), 257–267. DOI: 10.1159/000357382 PMID: 24603048

Warren, B. L., Iniguez, S. D., Alcantara, L. F., Wright, K. N., Parise, E. M., Weakley, S. K., & Bolanos-Guzman, C. A. (2011). Juvenile administration of concomitant methylphenidate and fluoxetine alters behavioral reactivity to reward-and mood-related stimuli and disrupts ventral tegmental area gene expression in adulthood. *The Journal of Neuroscience : The Official Journal of the Society for Neuroscience*, *31*(28), 10347–10358. DOI: 10.1523/JNEUROSCI.1470-11.2011 PMID: 21753012

Watanabe, Y., Iwasaki, Y., Sasaki, K., Motono, C., Imai, K., & Suzuki, K. (2023). Atg15 is a vacuolar phospholipase that disintegrates organelle membranes. *Cell Reports*, *42*(12), 113567. DOI: 10.1016/j.celrep.2023.113567 PMID: 38118441

Watson, K., Nasca, C., Aasly, L., McEwen, B., & Rasgon, N. (2018). Insulin resistance, an unmasked culprit in depressive disorders: Promises for interventions. *Neuropharmacology*, *136*, 327–334. DOI: 10.1016/j.neuropharm.2017.11.038 PMID: 29180223

Weber, S., Sitte, S., Voegele, A.-L., Sologub, L., Wilfer, A., Rath, T., Nägel, A., Zundler, S., Franchi, L., Opipari, A. W., Sonnewald, S., Reid, S., Hartmann, A., Eichhorn, P., Handtrack, C., Weber, K., Grützmann, R., Neufert, C., Schellerer, V. S., & Atreya, R. (2024). NLRP3 inhibition leads to impaired mucosal fibroblast function in patients with inflammatory bowel diseases. *Journal of Crohn's and Colitis*, *18*(3), 446–461. DOI: 10.1093/ecco-jcc/jjad164 PMID: 37748021

Wei, B., Fu, Y., Li, X., Chen, F., Zhang, Y., Chen, H., Tong, M., Li, L., Pan, Y., Zhang, S., Chen, S., Liu, X., & Zhong, Q. (2024). ANKFY1 bridges ATG2A-mediated lipid transfer from endosomes to phagophores. *Cell Discovery*, *10*(1), 43. DOI: 10.1038/s41421-024-00659-y PMID: 38622126

Wei, F., & Duan, Y. (2019). Crosstalk between Autophagy and Nanomaterials: Internalization, Activation, Termination. *Advanced Biosystems*, *3*(1), 1800259. DOI: 10.1002/adbi.201800259 PMID: 32627344

Wei, W., Lu, M., Lan, X., Liu, N., Su, W., Dushkin, A. V., & Yu, J. (2021). Neuroprotective effects of oxymatrine on PI3K/Akt/mTOR pathway after hypoxic-ischemic brain damage in neonatal rats. *Frontiers in Pharmacology*, *12*, 642415. DOI: 10.3389/fphar.2021.642415 PMID: 33927621

Wei, Y., Li, H., Li, Y., Zeng, Y., Quan, T., Leng, Y., Chang, E., Bai, Y., Bian, Y., & Hou, Y. (2024). Advances of curcumin in nervous system diseases: The effect of regulating oxidative stress and clinical studies. *Frontiers in Pharmacology*, *15*, 1496661. DOI: 10.3389/fphar.2024.1496661 PMID: 39555102

Wilkinson, J. E., Burmeister, L., Brooks, S. V., Chan, C. C., Friedline, S., Harrison, D. E., Hejtmancik, J. F., Nadon, N., Strong, R., Wood, L. K., Woodward, M. A., & Miller, R. A. (2012). Rapamycin slows aging in mice. *Aging Cell*, *11*(4), 675–682. DOI: 10.1111/j.1474-9726.2012.00832.x PMID: 22587563

Willis, D. E., & Twiss, J. L. (2006). The evolving roles of axonally synthesized proteins in regeneration. *Current Opinion in Neurobiology*, *16*(1), 111–118. DOI: 10.1016/j.conb.2006.01.002 PMID: 16418002

Winden, K. D., Ebrahimi-Fakhari, D., & Sahin, M. (2018). Abnormal mTOR Activation in Autism. *Annual Review of Neuroscience*, *41*(1), 1–23. DOI: 10.1146/annurev-neuro-080317-061747 PMID: 29490194

Witchel, S. F., Burghard, A. C., Tao, R. H., & Oberfield, S. E. (2019). The diagnosis and treatment of PCOS in adolescents: An update. *Current Opinion in Pediatrics*, *31*(4), 562–569. DOI: 10.1097/MOP.0000000000000778 PMID: 31299022

Wohleb, E. S., Fenn, A. M., Pacenta, A. M., Powell, N. D., Sheridan, J. F., & Godbout, J. P. (2012). Peripheral innate immune challenge exaggerated microglia activation, increased the number of inflammatory CNS macrophages, and prolonged social withdrawal in socially defeated mice. *Psychoneuroendocrinology*, *37*(9), 1491–1505. DOI: 10.1016/j.psyneuen.2012.02.003 PMID: 22386198

Wolf, W. M., Wattick, R. A., Kinkade, O. N., & Olfert, M. D. (2018). The Current Description and Future Need for Multidisciplinary PCOS Clinics. *Journal of Clinical Medicine*, *7*(11), 395. Advance online publication. DOI: 10.3390/jcm7110395 PMID: 30373317

Wolugbom, J. A.Jr, Areloegbe, S. E., & Olaniyi, K. S. (2023). Protective Role of Acetate Against Depressive-Like Behaviour Associated with Letrozole-Induced PCOS Rat Model: Involvement of HDAC2 and DNA Methylation. *Molecular Neurobiology*, *60*(1), 355–368. DOI: 10.1007/s12035-022-03074-0 PMID: 36269541

Wong, Y. C., & Holzbaur, E. L. (2015). Autophagosome dynamics in neurodegeneration at a glance. *Journal of Cell Science, 128*(7), 1259–1267. DOI: 10.1242/jcs.161216 PMID: 25829512

Wong, Y. C., & Holzbaur, E. L. F. (2014). The regulation of autophagosome dynamics by huntingtin and HAP1 is disrupted by expression of mutant huntingtin, leading to defective cargo degradation. *The Journal of Neuroscience : The Official Journal of the Society for Neuroscience, 34*(4), 1293–1305. DOI: 10.1523/JNEUROSCI.1870-13.2014 PMID: 24453320

Wood, C. S., Valentino, R. J., & Wood, S. K. (2017). Individual differences in the locus coeruleus-norepinephrine system: Relevance to stress-induced cardiovascular vulnerability. *Physiology & Behavior, 172*, 40–48. DOI: 10.1016/j.physbeh.2016.07.008 PMID: 27423323

Woo, H., Hong, C. J., Jung, S., Choe, S., & Yu, S.-W. (2018). Chronic restraint stress induces hippocampal memory deficits by impairing insulin signaling. *Molecular Brain, 11*(1), 1–13. DOI: 10.1186/s13041-018-0381-8 PMID: 29970188

Wu, D., Chen, Q., Chen, X., Han, F., Chen, Z., & Wang, Y. (2023). The blood–brain barrier: Structure, regulation and drug delivery. *Signal Transduction and Targeted Therapy, 8*(217), 217. Advance online publication. DOI: 10.1038/s41392-023-01481-w PMID: 37231000

Wu, F., Wei, X., Wu, Y., Kong, X., Hu, A., Tong, S., Liu, Y., Gong, F., Xie, L., Zhang, J., Xiao, J., & Zhang, H. (2018). Chloroquine promotes the recovery of acute spinal cord injury by inhibiting autophagy-associated inflammation and endoplasmic reticulum stress. *Journal of Neurotrauma, 35*(12), 1329–1344. DOI: 10.1089/neu.2017.5414 PMID: 29316847

Wu, J., & Lipinski, M. M. (2019). Autophagy in neurotrauma: Good, bad, or dysregulated. *Cells, 8*(7), 693. DOI: 10.3390/cells8070693 PMID: 31295858

Wu, J., Pajoohesh-Ganji, A., Stoica, B. A., Dinizo, M., Guanciale, K., & Faden, A. I. (2012). Delayed expression of cell cycle proteins contributes to astroglial scar formation and chronic inflammation after rat spinal cord contusion. *Journal of Neuroinflammation, 9*(1), 1–12. DOI: 10.1186/1742-2094-9-169 PMID: 22784881

Wu, M., Gao, Y., & Chen, B. (2022). Mechanism of acteoside-activated let-7g-5P attenuating Aβ-induced increased permeability and apoptosis of brain microvascular endothelial cells based on experimental and network pharmacology. *Neuroreport, 33*(16), 714–722. DOI: 10.1097/WNR.0000000000001837 PMID: 36165002

Wu, N. N., Tian, H., Chen, P., Wang, D., Ren, J., & Zhang, Y. (2019). Physical Exercise and Selective Autophagy: Benefit and Risk on Cardiovascular Health. *Cells*, *8*(11), 1436. DOI: 10.3390/cells8111436 PMID: 31739509

Wu, N., Zheng, W., Zhou, Y., Tian, Y., Tang, M., Feng, X., Ashrafizadeh, M., Wang, Y., Niu, X., Tambuwala, M., Wang, L., Tergaonkar, V., Sethi, G., Klionsky, D., Huang, L., & Gu, M. (2024). Autophagy in aging-related diseases and cancer: Principles, regulatory mechanisms and therapeutic potential. *Ageing Research Reviews*, *100*, 102428. DOI: 10.1016/j.arr.2024.102428 PMID: 39038742

Wu, S., & Zhou, W. (2023). Antimicrobial activity of oridonin. *Food Science and Technology (Campinas)*, *43*, e110222. DOI: 10.1590/fst.110222

Wu, X., Xu, H., Zeng, N., Li, H., Yao, G., Liu, K., Yan, C., & Wu, L. (2024). Luteolin alleviates depression-like behavior by modulating glycerophospholipid metabolism in the hippocampus and prefrontal cortex of LOD rats. *CNS Neuroscience & Therapeutics*, *30*(3), e14455. DOI: 10.1111/cns.14455 PMID: 37715585

Wu, Y., Li, L., Ning, Z., Li, C., Yin, Y., Chen, K., Li, L., Xu, F., & Gao, J. (2024). Autophagy-modulating biomaterials: Multifunctional weapons to promote tissue regeneration. *Cell Communication and Signaling : CCS*, *22*(1), 124. DOI: 10.1186/s12964-023-01346-3 PMID: 38360732

Xavier, S., Gilbert, V., Rastaldi, M. P., Krick, S., Kollins, D., Reddy, A., Bottinger, E., Cohen, C. D., & Schlondorff, D. (2010). BAMBI is expressed in endothelial cells and is regulated by lysosomal/autolysosomal degradation. *PLoS One*, *5*(9), e12995. DOI: 10.1371/journal.pone.0012995 PMID: 20886049

Xia, C.-Y., Wang, Z.-Z., Zhang, Z., Chen, J., Wang, Y.-Y., Lou, Y.-X., Gao, Y., Luo, P., Ren, Q., Du, G.-H., & Chen, N.-H. (2018). Corticosterone impairs gap junctions in the prefrontal cortical and hippocampal astrocytes via different mechanisms. *Neuropharmacology*, *131*, 20–30. DOI: 10.1016/j.neuropharm.2017.12.003 PMID: 29223529

Xia, J., Wang, H., Zhang, C., Liu, B., Li, Y., Li, K., Li, P., & Song, C. (2022). The comparison of sex differences in depression-like behaviors and neuroinflammatory changes in a rat model of depression induced by chronic stress. *Frontiers in Behavioral Neuroscience*, *16*, 1059594. DOI: 10.3389/fnbeh.2022.1059594 PMID: 36703721

Xiao, J., Yao, J., Jia, L., Ferguson, T. A., Weber, S., Sundstrom, J. M., Wubben, T. J., Besirli, C. G., & Zacks, D. N. (2021). Autophagy activation and photoreceptor survival in retinal detachment. *Experimental Eye Research*, *205*, 108492. DOI: 10.1016/j.exer.2021.108492 PMID: 33609513

Xiao, X., Shang, X., Zhai, B., Zhang, H., & Zhang, T. (2018). Nicotine alleviates chronic stress-induced anxiety and depressive-like behavior and hippocampal neuropathology via regulating autophagy signaling. *Neurochemistry International, 114*, 58–70. DOI: 10.1016/j.neuint.2018.01.004 PMID: 29339018

Xiao, Y., Ren, Q., & Wu, L. (2022). The pharmacokinetic property and pharmacological activity of acteoside: A review. *Biomedicine and Pharmacotherapy, 153*, 113296. DOI: 10.1016/j.biopha.2022.113296 PMID: 35724511

Xie, Z., Tian, Y., Guo, X., & Xie, N. (2024). The emerging role of CARM1 in cancer. Cellular oncology (Dordrecht, Netherlands), 47(5), 1503–1522. /DOI: 10.1007/s13402-024-00943-9

Xie, X. H., An, H. J., Kang, S., Hong, S., Choi, Y. P., Kim, Y. T., Choi, Y. D., & Cho, N. H. (2005). Loss of Cyclin B1 followed by downregulation of Cyclin A/Cdk2, apoptosis and antiproliferation in Hela cell line. *International Journal of Cancer, 116*(4), 520–525. DOI: 10.1002/ijc.21056 PMID: 15818617

Xin, Z., Xin, C., Huo, J., Liu, Q., Dong, H., Li, R., & Liu, Y. (2024). Neuroprotective effect of a multistrain probiotic mixture in SOD1G93A mice by reducing SOD1 aggregation and targeting the microbiota-gut-brain axis. *Molecular Neurobiology, 61*(11), 10051–10071. DOI: 10.1007/s12035-024-03988-x PMID: 38349516

Xu, C., Zhou, Z., Zhao, H., Lin, S., Zhang, P., Tian, H., & Mei, X. (2023). Zinc promotes spinal cord injury recovery by blocking the activation of NLRP3 inflammasome through SIRT3-mediated autophagy. *Neurochemical Research, 48*(2), 435–446. DOI: 10.1007/s11064-022-03762-2 PMID: 36152137

Xu, F., Wu, H., Zhang, K., Lv, P., Zheng, L., & Zhao, J. (2016). Pro-neurogenic effects of andrographolide on RSC96 Schwann cells in vitro. *Molecular Medicine Reports, 14*(4), 3573–3580. DOI: 10.3892/mmr.2016.5717 PMID: 27599453

Xu, J., Wang, G., Hou, Y., Sun, K., Zheng, Z., Guo, Z., Hou, L., Zhang, X., Ruan, Z., Ye, Y., & Guo, F. (2025). RICTOR-mediated activation of AKT/mTOR signaling and autophagy inhibition promote osteoarthritis. *International Immunopharmacology, 144*, 113681. DOI: 10.1016/j.intimp.2024.113681 PMID: 39591826

Xu, J., Wold, E. A., Ding, Y., Shen, Q., & Zhou, J. (2018). Therapeutic potential of oridonin and its analogs: From anticancer and antiinflammation to neuroprotection. *Molecules (Basel, Switzerland), 23*(2), 474. DOI: 10.3390/molecules23020474 PMID: 29470395

Xu, L., Botchway, B. O. A., Zhang, S., Zhou, J., & Liu, X. (2018). Inhibition of NF-κB signaling pathway by resveratrol improves spinal cord injury. *Frontiers in Neuroscience*, *12*, 690. DOI: 10.3389/fnins.2018.00690 PMID: 30337851

Xu, Y., Yuan, J., & Lipinski, M. M. (2013). Live imaging and single-cell analysis reveal differential dynamics of autophagy and apoptosis. *Autophagy*, *9*(9), 1418–1430. DOI: 10.4161/auto.25080 PMID: 23748697

Xu, Z., Yang, J., Xin, X., Liu, C., Li, L., Mei, X., & Li, M. (2023). Merits and challenges of iPSC-derived organoids for clinical applications. *Frontiers in Cell and Developmental Biology*, *11*, 1188905. DOI: 10.3389/fcell.2023.1188905 PMID: 37305682

Yalaev, B., Tyurin, A., Akhiiarova, K., & Khusainova, R. (2024). Hypomethylation of the RUNX2 Gene Is a New Potential Biomarker of Primary Osteoporosis in Men and Women. *International Journal of Molecular Sciences*, *25*(13), 7312. DOI: 10.3390/ijms25137312 PMID: 39000419

Yamaguchi, H., Honda, S., Torii, S., Shimizu, K., Katoh, K., Miyake, K., Miyake, N., Fujikake, N., Sakurai, H. T., Arakawa, S., & Shimizu, S. (2020). Wipi3 is essential for alternative autophagy and its loss causes neurodegeneration. *Nature Communications*, *11*(1), 5311. DOI: 10.1038/s41467-020-18892-w PMID: 33082312

Yamamoto, H. (2023). Molecular mechanisms of macroautophagy, microautophagy, and chaperone-mediated autophagy. *J Nippon Med Sch.*, *1302*(2003), 91–102. .DOI: 10.1272/jnms.JNMS.2024_91-102

Yamashima, T., & Oikawa, S. (2009). The role of lysosomal rupture in neuronal death. *Progress in Neurobiology*, *89*(4), 343–358. DOI: 10.1016/j.pneurobio.2009.09.003 PMID: 19772886

Yan, B., Li, Z., Su, H., Xue, H., Qiu, D., Xu, Z., & Tan, G. (2023). Regulatory mechanisms of autophagy-related ncRNAs in bone metabolic diseases. *Frontiers in Pharmacology*, *14*, 1178310. DOI: 10.3389/fphar.2023.1178310 PMID: 38146458

Yang, Z., & Klionsky, D. J. (2009). An overview of the molecular mechanism of autophagy. *Autophagy in infection and immunity*, 1-32.

Yang, C., Xiong, W., Dong, J., Zhao, X., Liang, G., & Zheng, W. (2025). Artemisinin protected human bronchial epithelial cells from amiodarone-induced oxidative damage via 5′-AMP-activated protein kinase (AMPK) activation. *Redox Report : Communications in Free Radical Research*, *30*(1), 2447721. DOI: 10.1080/13510002.2024.2447721 PMID: 39803706

Yang, J., & Chen, C. (2024). Hormonal changes in PCOS. *The Journal of Endocrinology*, *261*(1). Advance online publication. DOI: 10.1530/JOE-23-0342 PMID: 38285626

Yang, J., Xu, H., Zhang, C., Yang, X., Cai, W., & Chen, X. (2023). A prion-like domain of TFEB mediates the co-aggregation of TFEB and mHTT. *Autophagy*, *19*(2), 544–550. DOI: 10.1080/15548627.2022.2083857 PMID: 35635192

Yang, N., Li, L., Li, Z., Ni, C., Cao, Y., Liu, T., Tian, M., Chui, D., & Guo, X. (2017). Protective effect of dapsone on cognitive impairment induced by propofol involves hippocampal autophagy. *Neuroscience Letters*, *649*, 85–92. DOI: 10.1016/j.neulet.2017.04.019 PMID: 28411068

Yang, W., Qiu, X., Wu, Q., Chang, F., Zhou, T., Zhou, M., & Pei, J. (2023). Active constituents of saffron (Crocus sativus L.) and their prospects in treating neurodegenerative diseases [Review]. *Experimental and Therapeutic Medicine*, *25*(5), 235. DOI: 10.3892/etm.2023.11934 PMID: 37114174

Yang, Y., He, H., Miao, F., Yu, M., Wu, X., Liu, Y., Fu, J., Chen, J., Ma, L., Chen, X., Peng, X., You, Z., & Zhou, C. (2024). 3D-printed PCL framework assembling an ECM-inspired multilayer mineralized GO-Col-HAp microscaffold for in situ mandibular bone regeneration. *Journal of Translational Medicine*, *22*(1), 224. DOI: 10.1186/s12967-024-05020-1 PMID: 38429799

Yang, Y., Hu, Z., Du, X., Davies, H., Huo, X., & Fang, M. (2017). miR-16 and fluoxetine both reverse autophagic and apoptotic change in chronic unpredictable mild stress model rats. *Frontiers in Neuroscience*, *11*, 428. DOI: 10.3389/fnins.2017.00428 PMID: 28790887

Yang, Y., Yuan, R., Lu, Y., Zhu, C., Zhang, C., Lue, H., & Zhang, X. (2023). The engagement of autophagy in maniac disease. *CNS Neuroscience & Therapeutics*, *29*(12), 3684–3692. DOI: 10.1111/cns.14353 PMID: 37438945

Yang, Z., & Klionsky, D. J. (2010). Eaten alive: A history of macroautophagy. *Nature Cell Biology*, *12*(9), 814–822. DOI: 10.1038/ncb0910-814 PMID: 20811353

Yan, H., Xing, Q., Xiao, X., Yu, B., He, J., Mao, X., Yu, J., Zheng, P., Luo, Y., Wu, A., Pu, J., Lu, P., Wei, M., Khafipour, E., & Chen, D. (2024). Effect of *Saccharomyces cerevisiae* Postbiotics and Essential Oil on Growth Performance and Intestinal Health of Weanling Pigs During K88 ETEC Infection. *Journal of Animal Science*, *102*, skae007. Advance online publication. DOI: 10.1093/jas/skae007 PMID: 38198728

Yan, J., Luo, A., Gao, J., Tang, X., Zhao, Y., & Zhou, B. (2019). The role of SIRT1 in neuroinflammation and cognitive dysfunction in aged rats after anesthesia and surgery. *American Journal of Translational Research, 11*(3), 1555. PMID: 30972182

Yan, J., Porch, M. W., Court-Vazquez, B., Bennett, M. V. L., & Zukin, R. S. (2018). Activation of autophagy rescues synaptic and cognitive deficits in fragile X mice. *Proceedings of the National Academy of Sciences of the United States of America, 115*(41), 9707–9716. DOI: 10.1073/pnas.1808247115 PMID: 30242133

Yan, P., Bai, L., Lu, W., Gao, Y., Bi, Y., & Lv, G. (2017). Regulation of autophagy by AMP-activated protein kinase/sirtuin 1 pathway reduces spinal cord neurons damage. *Iranian Journal of Basic Medical Sciences., 20*(9), 1029–1036. PMID: 29085598

Yao, J., Jia, L., Shelby, S. J., Ganios, A. M., Feathers, K., Thompson, D. A., & Zacks, D. N. (2014). Circadian and noncircadian modulation of autophagy in photoreceptors and retinal pigment epithelium. *Investigative Ophthalmology & Visual Science, 55*(5), 3237–3246. DOI: 10.1167/iovs.13-13336 PMID: 24781939

Yao, R., & Shen, J. (2023). Chaperone-mediated autophagy: Molecular mechanisms, biological functions, and diseases. *MedComm, 4*(5), 1–27. DOI: 10.1002/mco2.347 PMID: 37655052

Ye, S., Fang, L., Xie, S., Hu, Y., Chen, S., Amin, N., Fang, M., & Hu, Z. (2023). Resveratrol alleviates postpartum depression-like behavior by activating autophagy via SIRT1 and inhibiting AKT/mTOR pathway. *Behavioural Brain Research, 438*, 114208. DOI: 10.1016/j.bbr.2022.114208 PMID: 36356720

Ye, Y., Tyndall, E. R., Bui, V., Tang, Z., Shen, Y., Jiang, X., Flanagan, J. M., Wang, H.-G., & Tian, F. (2021). An N-terminal conserved region in human Atg3 couples membrane curvature sensitivity to conjugase activity during autophagy. *Nature Communications, 12*(1), 374. DOI: 10.1038/s41467-020-20607-0 PMID: 33446636

Yi, L. T., Xu, H. L., Feng, J., Zhan, X., Zhou, L. P., & Cui, C. C. (2011). Involvement of monoaminergic systems in the antidepressant-like effect of nobiletin. *Physiology & Behavior, 102*(1), 1–6. DOI: 10.1016/j.physbeh.2010.10.008 PMID: 20951716

Yoshii, R., Higashida, K., & Nakai, N. (2023). Intermittent fasting reduces mouse body fat while maintaining muscle mass by regulating protein synthesis and autophagy. *Nutrition (Burbank, Los Angeles County, Calif.), 115*, 112130. DOI: 10.1016/j.nut.2023.112130 PMID: 37454541

Yu, Q., Hao, S., Wang, H., Song, X., Shen, Q., & Kang, J. (2016). Depression-like behavior in a dehydroepiandrosterone-induced mouse model of polycystic ovary syndrome. *Biology of Reproduction, 95*(4), 79, 71–10.

Yuan, Y., Fan, X., Guo, Z., Zhou, Z., & Gao, W. (2022). Metformin Protects against Spinal Cord Injury and Cell Pyroptosis via AMPK/NLRP3 Inflammasome Pathway. *Analytical Cellular Pathology (Amsterdam)*, *2022*, 3634908. DOI: 10.1155/2022/3634908 PMID: 35387358

Yu, L., Chen, Y., & Tooze, S. A. (2018). Autophagy pathway: Cellular and molecular mechanisms. *Autophagy*, *14*(2), 207–215. DOI: 10.1080/15548627.2017.1378838 PMID: 28933638

Yun, H. R., Jo, Y. H., Kim, J., Shin, Y., Kim, S. S., & Choi, T. G. (2020). Roles of Autophagy in Oxidative Stress. *International Journal of Molecular Sciences*, *21*(9), 3289. DOI: 10.3390/ijms21093289 PMID: 32384691

Yurtdaş, G., & Akdevelioğlu, Y. (2020). A new approach to polycystic ovary syndrome: The gut microbiota. *Journal of the American College of Nutrition*, *39*(4), 371–382. DOI: 10.1080/07315724.2019.1657515 PMID: 31513473

Yu, X., Zhao, H., Wang, R., Chen, Y., Ouyang, X., Li, W., Sun, Y., & Peng, A. (2024). Cancer epigenetics: From laboratory studies and clinical trials to precision medicine. *Cell Death Discovery*, *10*(28), 28. Advance online publication. DOI: 10.1038/s41420-024-01803-z PMID: 38225241

Yu, Y. S., Kim, H., Kim, K., & Baek, S. H. (2023). Epigenetic regulation of autophagy by histone-modifying enzymes under nutrient stress. *Cell Death and Differentiation*, *30*(6), 1430–1436. DOI: 10.1038/s41418-023-01154-9 PMID: 36997734

Zachari, M., Longo, M., & Ganley, I. G. (2020). Aberrant autophagosome formation occurs upon small molecule inhibition of ULK1 kinase activity. *Life Science Alliance*, *3*(12), e202000815. DOI: 10.26508/lsa.202000815 PMID: 33109685

Zatyka, M., Sarkar, S., & Barrett, T. (2020). Autophagy in Rare (NonLysosomal) Neurodegenerative Diseases. *Journal of Molecular Biology*, *432*(8), 2735–2753. DOI: 10.1016/j.jmb.2020.02.012 PMID: 32087199

Zehra, B., & Khursheed, A. (2018). Polycystic ovarian syndrome: symptoms, treatment and diagnosis: a review. *Journal of Pharmacognosy and Phytochemistry*, *7*(6), 875–880.

Zeng, M., & Zhou, J.-N. (2008). Roles of autophagy and mTOR signaling in neuronal differentiation of mouse neuroblastoma cells. *Cellular Signalling*, *20*(4), 659–665. DOI: 10.1016/j.cellsig.2007.11.015 PMID: 18207367

Zerrouki, D., Rami, I., Assarrar, I., Bouichrat, N., Rouf, S., & Latrech, H. (2024). Is there any association between vitamin D status and PCOS disease? *Gynecological Endocrinology*, *40*(1), 2381501. DOI: 10.1080/09513590.2024.2381501 PMID: 39481002

Zhai, Y., & Pang, Y. (2022). Systemic and ovarian inflammation in women with polycystic ovary syndrome. *Journal of Reproductive Immunology*, *151*, 103628. DOI: 10.1016/j.jri.2022.103628 PMID: 35472833

Zhang, B., Chen, Y., Yang, B., Jiang, F., Zhao, J., & Ouyang, Z. (2017). Neuroprotective effects of diosgenin in rats with experimental spinal cord injury via promotion of autophagy. *International Journal of Clinical and Experimental Medicine*, *10*(8), 11655–11663.

Zhang, B., Zhou, W., Shi, Y., Zhang, J., Cui, L., & Chen, Z.-J. (2020). Lifestyle and environmental contributions to ovulatory dysfunction in women of polycystic ovary syndrome. *BMC Endocrine Disorders*, *20*(1), 19. DOI: 10.1186/s12902-020-0497-6 PMID: 32000752

Zhang, C., & Yang, Y. (2023). Targeting toll-like receptor 4 (TLR4) and the NLRP3 inflammasome: Novel and emerging therapeutic targets for hyperuricemia nephropathy. *Biomolecules and Biomedicine*, *24*(4), 688–697. DOI: 10.17305/bb.2023.9838 PMID: 38041694

Zhang, D., Xuan, J., Zheng, B., Zhou, Y., Lin, Y., Wu, Y., Zhou, Y., Huang, Y., Wang, Q., Shen, L., Mao, C., Wu, Y., Wang, X., Tian, N., Xu, H.-Z., & Zhang, X. (2017). Metformin improves functional recovery after spinal cord injury via autophagy flux stimulation. *Molecular Neurobiology*, *54*(5), 3327–3341. DOI: 10.1007/s12035-016-9895-1 PMID: 27167128

Zhang, H., Li, S., Si, Y., & Xu, H. (2021). Andrographolide and its derivatives: Current achievements and future perspectives. *European Journal of Medicinal Chemistry*, *224*, 113710. DOI: 10.1016/j.ejmech.2021.113710 PMID: 34315039

Zhang, H., Shang, Y., Xiao, X., Yu, M., & Zhang, T. (2017). Prenatal stress-induced impairments of cognitive flexibility and bidirectional synaptic plasticity are possibly associated with autophagy in adolescent male-offspring. *Experimental Neurology*, *298*, 68–78. DOI: 10.1016/j.expneurol.2017.09.001 PMID: 28882411

Zhang, J. J., Gao, T. T., Wang, Y., Wang, J. L., Guan, W., Wang, Y. J., Wang, C.-N., Liu, J.-F., & Jiang, B. (2019). Andrographolide Exerts Significant Antidepressant-Like Effects Involving the Hippocampal BDNF System in Mice. *The International Journal of Neuropsychopharmacology*, *22*(9), 585–600. DOI: 10.1093/ijnp/pyz032 PMID: 31181145

Zhang, S.-M., Shang, Z.-F., & Zhou, P.-K. (2015). Autophagy as the effector and player in DNA damage response of cells to genotoxicants. *Toxicology Research*, *4*(3), 613–622. DOI: 10.1039/C5TX00043B

Zhang, S., Yazaki, E., Sakamoto, H., Yamamoto, H., & Mizushima, N. (2022). Evolutionary diversification of the autophagy-related ubiquitin-like conjugation systems. *Autophagy*, *18*(12), 2969–2984. DOI: 10.1080/15548627.2022.2059168 PMID: 35427200

Zhang, W., Wu, Y., Guo, Y., Wang, C., Long, X., & Jiang, P. (2022). Curcumin relieves corticosterone-induced behaviour deficits via PGC-1α-mediated mitochondrial biogenesis and mitophagy. *Clinical and Translational Discovery*, *2*(4), e161. DOI: 10.1002/ctd2.161 PMID: 36712172

Zhang, W., Xu, C., Sun, J., Shen, H. M., Wang, J., & Yang, C. (2022). Impairment of the autophagy-lysosomal pathway in Alzheimer's disease: Pathogenic mechanisms and therapeutic potential. *Acta Pharmaceutica Sinica. B*, *12*(3), 1019–1040. DOI: 10.1016/j.apsb.2022.01.008 PMID: 35530153

Zhang, W., Zhou, R., Lei, X., Wang, M., Duan, Q., Miao, Y., Zhang, T., Li, X., Zutong, Z., Wang, L., Jones, O. D., Xu, M., Bryant, J., Ma, J., Liu, Y., & Xu, X. (2025). Molecular mechanism on autophagy associated cardiovascular dysfunction in Drosophila melanogaster. *Frontiers in Cell and Developmental Biology*, *13*, 1512341. DOI: 10.3389/fcell.2025.1512341 PMID: 40099194

Zhang, X., Bu, H., Jiang, Y., Sun, G., Jiang, R., Huang, X., Duan, H., Huang, Z., & Wu, Q. (2019). The antidepressant effects of apigenin are associated with the promotion of autophagy via the mTOR/AMPK/ULK1 pathway. *Molecular Medicine Reports*, *20*(3), 2867–2874. DOI: 10.3892/mmr.2019.10491 PMID: 31322238

Zhang, X., Li, Y., Xu, H., & Zhang, Y. W. (2014). The γ-secretase complex: From structure to function. *Frontiers in Cellular Neuroscience*, *8*, 1–10. DOI: 10.3389/fncel.2014.00427 PMID: 25565961

Zhang, X.-W., Zhu, X.-X., Tang, D.-S., & Lu, J.-H. (2023). Targeting autophagy in Alzheimer's disease: Animal models and mechanisms. *Zoological Research*, *44*(6), 1132–1145. DOI: 10.24272/j.issn.2095-8137.2023.294 PMID: 37963840

Zhang, Y., Chu, J. M., & Wong, G. T. (2022). Cerebral Glutamate Regulation and Receptor Changes in Perioperative Neuroinflammation and Cognitive Dysfunction. *Biomolecules*, *12*(4), 597. Advance online publication. DOI: 10.3390/biom12040597 PMID: 35454185

Zhang, Y., Liu, C., Zhao, Y., Zhang, X., Li, B., & Cui, R. (2015). The effects of calorie restriction in depression and potential mechanisms. *Current Neuropharmacology, 13*(4), 536–542. DOI: 10.2174/1570159X13666150326003852 PMID: 26412073

Zhang, Y., Sun, S., Chen, J., Ren, P., Hu, Y., Cao, Z., Sun, H., & Ding, Y. (2014). Oxymatrine induces mitochondria dependent apoptosis in human osteosarcoma MNNG/HOS cells through inhibition of PI3K/Akt pathway. *Tumour Biology : The Journal of the International Society for Oncodevelopmental Biology and Medicine, 35*(2), 1619–1625. DOI: 10.1007/s13277-013-1223-z PMID: 24078465

Zhang, Y., Yang, H., Hou, S., Xia, Y., & Wang, Y. (2024). Influence of the brain-gut axis on neuroinflammation in cerebral ischemia-reperfusion injury [Review]. *International Journal of Molecular Medicine, 53*(3), 30. DOI: 10.3892/ijmm.2024.5354 PMID: 38299236

Zhang, Z., Cai, X., Yao, Z., Wen, F., Fu, Z., Zhang, J., Zhong, Z., Huang, Y., & Qu, S. (2020). EA ameliorated depressive behaviors in CUMS rats and was related to its suppressing autophagy in the hippocampus. *Neural Plasticity, 2020*(1), 8860968. DOI: 10.1155/2020/8860968 PMID: 33029121

Zhang, Z.-Y., Bao, X.-L., Cong, Y.-Y., Fan, B., & Li, G.-Y. (2020). Autophagy in age-related macular degeneration: A regulatory mechanism of oxidative stress. *Oxidative Medicine and Cellular Longevity, 2020*(1), 2896036. DOI: 10.1155/2020/2896036 PMID: 32831993

Zhang, Z., Yang, W., Wang, L., Zhu, C., Cui, S., Wang, T., Gu, X., Liu, Y., & Qiu, P. (2024). Unraveling the role and mechanism of mitochondria in postoperative cognitive dysfunction: A narrative review. *Journal of Neuroinflammation, 21*(1), 293. DOI: 10.1186/s12974-024-03285-3 PMID: 39533332

Zhang, Z., Yang, X., Song, Y.-Q., & Tu, J. (2021). Autophagy in Alzheimer's disease pathogenesis: Therapeutic potential and future perspectives. *Ageing Research Reviews, 72*, 101464. DOI: 10.1016/j.arr.2021.101464 PMID: 34551326

Zhao, D. Y., Bäuerlein, F. J. B., Saha, I., Hartl, F. U., Baumeister, W., & Wilfling, F. (2024). Autophagy preferentially degrades non-fibrillar polyQ aggregates. *Molecular Cell, 84*(10), 1980–1994.e8. DOI: 10.1016/j.molcel.2024.04.018 PMID: 38759629

Zhao, H., Chen, S., Gao, K., Zhou, Z., Wang, C., Shen, Z., Guo, Y., Li, Z., Wan, Z., Liu, C., & Mei, X. (2017). Resveratrol protects against spinal cord injury by activating autophagy and inhibiting apoptosis mediated by the SIRT1/AMPK signaling pathway. *Neuroscience, 348*, 241–251. DOI: 10.1016/j.neuroscience.2017.02.027 PMID: 28238848

Zhao, L., Wang, J.-L., Liu, R., Li, X.-X., Li, J.-F., & Zhang, L. (2013). Neuroprotective, anti-amyloidogenic and neurotrophic effects of apigenin in an Alzheimer's disease mouse model. *Molecules (Basel, Switzerland)*, *18*(8), 9949–9965. DOI: 10.3390/molecules18089949 PMID: 23966081

Zhao, M., Lv, H., Yang, N., & Peng, G.-H. (2022). Rapamycin Improved Retinal Function and Morphology in a Mouse Model of Retinal Degeneration. *Frontiers in Neuroscience*, *16*, 846584. DOI: 10.3389/fnins.2022.846584 PMID: 35295093

Zhao, S., Chen, R., Gao, Y., Lu, Y., Bai, X., & Zhang, J. (2023). Fundamental roles of the Optineurin gene in the molecular pathology of Amyotrophic Lateral Sclerosis. *Frontiers in Neuroscience*, *17*, 1–8. DOI: 10.3389/fnins.2023.1319706 PMID: 38178841

Zhao, X., Zhang, Q., Wang, Y., Li, S., Yu, X., Wang, B., & Wang, X. (2021). Oridonin induces autophagy-mediated cell death in pancreatic cancer by activating the c-Jun N-terminal kinase pathway and inhibiting phosphoinositide 3-kinase signaling. *Annals of Translational Medicine*, *9*(13), 1084. DOI: 10.21037/atm-21-2630 PMID: 34422996

Zhao, Y. G., Codogno, P., & Zhang, H. (2021). Machinery, regulation and pathophysiological implications of autophagosome maturation. *Nature Reviews. Molecular Cell Biology*, *22*(11), 733–750. DOI: 10.1038/s41580-021-00392-4 PMID: 34302147

Zhao, Z., Yan, J., Huang, L., & Yang, X. (2024). Phytochemicals targeting Alzheimer's disease via the AMP-activated protein kinase pathway, effects, and mechanisms of action. *Biomedicine and Pharmacotherapy*, *173*, 116373. DOI: 10.1016/j.biopha.2024.116373 PMID: 38442672

Zhao, Z., Zhang, L., Guo, X.-D., Cao, L.-L., Xue, T.-F., Zhao, X.-J., Yang, D.-D., Yang, J., Ji, J., Huang, J.-Y., & Sun, X.-L. (2017). Rosiglitazone exerts an antidepressive effect in unpredictable chronic mild-stress-induced depressive mice by maintaining essential neuron autophagy and inhibiting excessive astrocytic apoptosis. *Frontiers in Molecular Neuroscience*, *10*, 293. DOI: 10.3389/fnmol.2017.00293 PMID: 28959186

Zha, S., Liu, H., Li, H., Li, H., Wong, K.-L., & Homayoun, A. (2024). Functionalized nanomaterials capable of crossing the blood–brain barrier. *ACS Nano*, *18*(3), 1820–1845. DOI: 10.1021/acsnano.3c10674 PMID: 38193927

Zheng, Q., Chen, Y., Chen, D., Zhao, H., Feng, Y., Meng, Q., Zhao, Y., & Zhang, H. (2022). Calcium transients on the ER surface trigger liquid-liquid phase separation of FIP200 to specify autophagosome initiation sites. *Cell*, *185*(22), 4082–4098. DOI: 10.1016/j.cell.2022.09.001 PMID: 36198318

Zheng, T., Zhao, C., Zhao, B., Liu, H., Wang, S., Wang, L., & Liu, P. (2020). Impairment of the autophagy-lysosomal pathway and activation of pyroptosis in macular corneal dystrophy. *Cell Death Discovery*, *6*(1), 85. DOI: 10.1038/s41420-020-00320-z PMID: 32983576

Zhen, Y., & Zhang, H. (2019). NLRP3 Inflammasome and Inflammatory Bowel Disease. *Frontiers in Immunology*, *10*, 276. DOI: 10.3389/fimmu.2019.00276 PMID: 30873162

Zhong, W., Qian, K., Xiong, J., Ma, K., Wang, A., & Zou, Y. (2016). Curcumin alleviates lipopolysaccharide induced sepsis and liver failure by suppression of oxidative stress-related inflammation via PI3K/AKT and NF-κB related signaling. *Biomedicine and Pharmacotherapy*, *83*, 302–313. DOI: 10.1016/j.biopha.2016.06.036 PMID: 27393927

Zhou, F., Wu, Z., Zhao, M., Murtazina, R., Cai, J., Zhang, A., Li, R., Sun, D., Li, W., Zhao, L., Li, Q., Zhu, J., Cong, X., Zhou, Y., Xie, Z., Gyurkovska, V., Li, L., Huang, X., Xue, Y., & Segev, N. (2019). Rab5-dependent autophagosome closure by ESCRT. *The Journal of Cell Biology*, *218*(6), 1908–1927. DOI: 10.1083/jcb.201811173 PMID: 31010855

Zhou, Q., Fu, X., Wang, X., Wu, Q., Lu, Y., Shi, J., Klaunig, J. E., & Zhou, S. (2018). Autophagy plays a protective role in Mn-induced toxicity in PC12 cells. *Toxicology*, *394*, 45–53. DOI: 10.1016/j.tox.2017.12.001 PMID: 29222055

Zhou, Y., Li, X., Gong, W., Tian, J., Gao, X., Gao, L., Zhang, X., Du, G., & Qin, X. (2017). Protective effect of isoliquiritin against corticosterone-induced neurotoxicity in PC12 cells. *Food & Function*, *8*(3), 1235–1244. DOI: 10.1039/C6FO01503D PMID: 28229156

Zhou, Y., Yan, M., Pan, R., Wang, Z., Tao, X., Li, C., Xia, T., Liu, X., & Chang, Q. (2021). Radix Polygalae extract exerts antidepressant effects in behavioral despair mice and chronic restraint stress-induced rats probably by promoting autophagy and inhibiting neuroinflammation. *Journal of Ethnopharmacology*, *265*, 113317. DOI: 10.1016/j.jep.2020.113317 PMID: 32861821

Zhou, Y., Zhang, H., Zheng, B., Ye, L., Zhu, S., Johnson, N. R., Wang, Z., Wei, X., Chen, D., Cao, G., Fu, X., Li, X., Xu, H.-Z., & Xiao, J. (2016). Retinoic acid induced-autophagic flux inhibits ER-stress dependent apoptosis and prevents disruption of blood-spinal cord barrier after spinal cord injury. *International Journal of Biological Sciences*, *12*(1), 87–99. DOI: 10.7150/ijbs.13229 PMID: 26722220

Zhou, Y., Zheng, B., Ye, L., Zhang, H., Zhu, S., Zheng, X., Xia, Q., He, Z., Wang, Q., Xiao, J., & Xu, H. (2016). RETRACTED ARTICLE: Retinoic Acid Prevents Disruption of Blood-Spinal Cord Barrier by Inducing Autophagic Flux after Spinal Cord Injury. *Neurochemical Research*, *41*(4), 813–825. DOI: 10.1007/s11064-015-1756-1 PMID: 26582233

Zhou, Z. D., Yi, L. X., Wang, D. Q., Lim, T. M., & Tan, E. K. (2023). Role of dopamine in the pathophysiology of Parkinson's disease. *Translational Neurodegeneration*, *12*(1), 44. DOI: 10.1186/s40035-023-00378-6 PMID: 37718439

Zhou, Z., Doggett, T., Sene, A., Apte, R., & Ferguson, T. (2015). Autophagy supports survival and phototransduction protein levels in rod photoreceptors. *Cell Death and Differentiation*, *22*(3), 488–498. DOI: 10.1038/cdd.2014.229 PMID: 25571975

Zhu, L.-H., Bi, W., Qi, R., Wang, H., & Lu, D. (2011). Luteolin inhibits microglial inflammation and improves neuron survival against inflammation. *The International Journal of Neuroscience*, *121*(6), 329–336. DOI: 10.3109/00207454.2011.569040 PMID: 21631167

Zhu, Y., Fujimaki, M., Snape, L., Lopez, A., Fleming, A., & Rubinsztein, D. C. (2024). Loss of WIPI4 in neurodegeneration causes autophagy-independent ferroptosis. *Nature Cell Biology*, *26*(4), 542–551. DOI: 10.1038/s41556-024-01373-3 PMID: 38454050

Zielińska, A., Costa, B., Ferreira, M. V., Miguéis, D., Louros, J. M., Durazzo, A., & Souto, E. B. (2020). Nanotoxicology and nanosafety: Safety-by-design and testing at a glance. *International Journal of Environmental Research and Public Health*, *17*(13), 4657. DOI: 10.3390/ijerph17134657 PMID: 32605255

Zientara-Rytter, K., & Subramani, S. (2020). Mechanistic Insights into the Role of Atg11 in Selective Autophagy. *Journal of Molecular Biology*, *432*(1), 104–122. DOI: 10.1016/j.jmb.2019.06.017 PMID: 31238043

Zong, Y., Li, H., Liao, P., Chen, L., Pan, Y., Zheng, Y., Zhang, C., Liu, D., Zheng, M., & Gao, J. (2024). Mitochondrial dysfunction: Mechanisms and advances in therapy. *Signal Transduction and Targeted Therapy*, *9*(1), 124. DOI: 10.1038/s41392-024-01839-8 PMID: 38744846

Zou, S., & Kumar, U. (2018). Cannabinoid Receptors and the Endocannabinoid System: Signaling and Function in the Central Nervous System. *International Journal of Molecular Sciences*, *19*(3), 833. DOI: 10.3390/ijms19030833 PMID: 29533978

Zschocke, J., & Rein, T. (2011). Antidepressants encounter autophagy in neural cells. *Autophagy*, *7*(10), 1247–1248. DOI: 10.4161/auto.7.10.16520 PMID: 21642768

Zschocke, J., Zimmermann, N., Berning, B., Ganal, V., Holsboer, F., & Rein, T. (2011). Antidepressant drugs diversely affect autophagy pathways in astrocytes and neurons—Dissociation from cholesterol homeostasis. *Neuropsychopharmacology : Official Publication of the American College of Neuropsychopharmacology*, *36*(8), 1754–1768. DOI: 10.1038/npp.2011.57 PMID: 21508931

# About the Contributors

**Tahmineh Mokhtari**, Ph.D., is an neuroscientist with dual doctorates in Anatomical Sciences from Tehran University of Medical Sciences and neuroscience from thc University of Chinese Academy of Sciences. As a Visiting Scholar at the University of California, Davis, Tahmineh investigates neuroinflammatory pathways and novel therapeutic strategies for neurodegenerative disorders. Her interdisciplinary research integrates advanced methodologies, including in silico approaches, in vitro techniques, and in vivo studies. Tahmineh's work emphasizes elucidating disease mechanisms and translating discoveries into innovative treatments. She combines her expertise with computational proficiency to drive cutting-edge research in neuroscience and related health conditions.

**Kadir Uludag**, Ph.D., is a researcher in Applied Psychology, renowned for his contributions to various domains including schizophrenia research, machine learning, and educational psychology. With a strong academic background, he successfully completed his Ph.D. at the prestigious Chinese Academy of Sciences. Currently, Kadir is serving as a postdoctoral researcher at Capital Medical University Xuanwu Hospital, where he continues to make significant advancements in psychiatry.

***

**Amjad Islam** is a Ph.D. and an established international researcher. He earned his PhD (Education) from The University of Lahore, Pakistan, in February 2023. He was a researcher and data analyst at The University of Lahore for three years (2019–2022). He works as a reviewer and guest editor for several top-ranked international journals. He has taught subjects like Educational Research, Statistics in Education, Policy and Plan Implementation, Educational Philosophy, and Assessment and Evaluation. He has published several articles with leading publishers, such as Wiley, Emerald, Taylor & Francis, and Tech Science Press. His research interests include Positive Psychology, School Psychology, AI tools, Neuroscience, Inclusive education, Special education, STEM Education, Technology, and Education.

**Parvaneh Behi Shahreza**, PhD student, is a neuroscientic with a robust academic background. She earned her MSC in anatomical sciences from Kashsan university of medical sciences and worked on reproductive and cancer. Then she changed her academic field and started to work on neuroscience. During this time she was engaged with students to teach them anatomical sciences. Currently, Parvaneh is studing PhD at Tehran university of medical sciences. Her research primarily focuses on spinal cord injury, inflammasoms and autophagy that occures un this pathway.

**Shima Farajpour** is a medical student at Tehran University of Medical Sciences, driven by a deep curiosity for medical science and a strong dedication to research in this area. Currently pursuing her degree in medicine, she is actively building a foundation in the intricate mechanisms of human health and disease. She recognizes the critical role of research in advancing medical knowledge and improving patient outcomes, with a particular enthusiasm for exploring the frontiers of neuroscience. Shima is committed to contributing effectively to the field, seeking to uncover new insights that can ultimately enhance medical care.

**Farzaneh Majidi**, a medical student at the Tehran University of Medical Sciences, with great enthusiasm in the domain of Neuroscience. Currently attempting to gain knowledge and experience in this area and building the foundation for her research, focusing on the mechanisms involved in neurological and mental disorders. she aspires to contribute effectively to promoting knowledge in this field in the future.

**Shima Mohammadi** is a dedicated neuroscientist with a strong academic background in neurobiology. She earned her Ph.D. from Tehran University of Medical Sciences, where she built a solid foundation in neuroscience. Driven by her passion for academia, she joined Shahroud University of Medical Sciences, where she is currently serving as a faculty member. Her research primarily focuses on the pathogenesis of neurological disorders and the molecular pathways involved in these conditions, with a particular emphasis on circadian rhythm dysregulation. By investigating how disruptions in biological rhythms contribute to neuropsychiatric diseases, her work aims to deepen our understanding of their underlying mechanisms and potential therapeutic strategies.

**Pranav Kumar Prabhakar** is currently working as a Professor in the Department of Biotechnology, School of Engineering and Technology, Nagaland University, Meriema, Kohima, Nagaland, India.

**Reyed M. Reyed** is an Associate Professor of Microbial Biotechnology with extensive expertise in microbial genomics, computational enzymes, and environmental sustainability. He holds multiple academic credentials, including Dia.M.C, RN, PAND, PhD, MSc, and BSc in Microbial Biotechnology. His research spans a wide range of fields, from the nutritional human gut microbiome and food microbiology to water treatment and food safety. From 2008 to 2019, Dr. Reyed served as an expert in water treatment technology, applying his knowledge to critical challenges such as biofilm formation and marine biofouling in reverse osmosis systems. From 2020 to 2023, he led the Egyptian-Indian joint project on water treatment in collaboration with Indian leader Asifa Qureshi at the Environmental Biotechnology and Genomics Division (EBGD), CSIR-National Environmental Engineering Research Institute (CSIR-NEERI) in Nagpur, India, contributing to practical environmental solutions. In the field of food microbiology, Dr. Reyed has made notable contributions to food safety and has founded a winter school on the subject. He is currently the program coordinator at SRTA-CITY SCI EG. His interdisciplinary research includes advances in computational enzymes and bioinformatics, which complement his expertise in microbial biotechnology, as well as the integration of synthetic biology and systems biology in exploring gut health and environmental sustainability. Dr. Reyed has participated in the Diabetes Master Class (Dia.M.C) organized by Novartis Egypt and the Arabic Society for Diabetes and Metabolism in 2022, under the supervision of the Egyptian Medical Syndicates in Cairo, Egypt. His leadership extends beyond research; he serves as the Academic Editor of the International Journal of Environment and Climate Change and as the Editor-in-Chief of the Journal of Climate Change and Renewable Energy. Dr. Reyed has participated in major international conferences, including events in the UAE, such as the International Conference on Bacteria, Viruses, and Infectious Diseases (Dubai, 2018), and the Food Safety Exhibition in collaboration with Emirates Universities and ADNOC (Abu Dhabi, 2023). He will also serve on the Research Evaluation Committee at the 8th International Conference on Genetic Disorders in Dubai in May 2024. Dr. Reyed holds several certifications in Clinical Nutrition, including Code B N 0458 from Baheya Hospital for Breast Cancer Treatment in Giza, Egypt, in collaboration with Medix Foundation for Training and Medical Education. He is a Registered Clinical Nutritionist (Registration No. 44320/6068) in the Egyptian Syndicate of Scientific Professions and a Registered Physical and Health Nutritionist with the Egyptian Professional Syndicate Committee for Specialists in Cosmetic Professions and Health Services (900444). Additionally, he is a member of the General Syndicate of Egyptian Ambulance Workers (No. 2283) and the Egyptian Red Crescent (No. 1352). His commitment to environmental sustainability, public health, and microbial biotechnology is further reflected through his volunteer work in the UAE, where he holds multiple licenses, including General Volunteer No.

523273 (Abu Dhabi) and Sharjah Volunteer Center No. 61318. Reyed M. Reyed has participated in several workshops related to drug design. These include the 4th Computer-Assisted Drug Design Workshop (CADD-W) held at Kafr El-Sheikh University on November 6-7, 2021, and the InSilico Drug Design Course at Ain Shams Center of Genetic Engineering, Faculty of Agriculture, from September 28-30, 2021. Furthermore, he is scheduled to participate in the Drug Discovery, Delivery, and Therapeutics Workshop at New York University Abu Dhabi campus from May 6-8, 2024, in collaboration with the International Center for Genetic Engineering and Biotechnology (ICGEB), Italy. Reyed M. Reyed participated in several funded training programs and courses, notably in 2007 at Sandia National Laboratories, USA, focusing on biosafety, and at the International Centre for Genetic Engineering and Biotechnology (ICGEB) in Italy. He also contributed to a workshop on CRISPR/Cas9 Genome Editing. He is an active member of several professional societies, including the Arab Society for Diabetes and Metabolism, the Egyptian Society of Functional Medicine, and ESPGHAN (European Society for Pediatric Gastroenterology, Hepatology, and Nutrition). Reyed holds several certificates and has completed numerous courses. He earned a Certificate in Entrepreneurship through the Knowledge Project in cooperation with the Regional Office for Arab States of the United Nations Development Corporation and the Mohammed bin Rashid Al Maktoum Foundation for Knowledge, from Copenhagen Business School in March 2024. He completed a course on Understanding Obesity at the University of Edinburgh in May 2023, and a course on Nutrition and Lifestyle in Pregnancy at Ludwig-Maximilians-Universität München (LMU) in August 2022. Additionally, he completed two bioinformatics-related courses: Bacterial Genomes: From DNA to Protein Function Using Bioinformatics (Wellcome Connecting Science, January 2021) and Train the Trainer: Design Genomics and Bioinformatics Training (Wellcome Connecting Science, October 2021). Most recently, he took a course on Nutrition for Health and Sustainability with EIT Food in September 2023. Reyed also holds SCOPE Certification (the international standard of obesity management expertise) with 16 SCOPE points, having passed 20 modules. He was awarded the SCOPE Core Learning Path Scholarship for Healthcare Professionals in 2021. As an international arbitrator and reviewer, Reyed contributes his expertise to several prominent scientific journals, including the European Journal of Nutrition, B. SNF (Swedish Nutrition Foundation), Cureus Journal of Medical Science, Microbiology Research Journal International, Journal of Advances in Microbiology, Plant Cell Biotechnology and Molecular Biology, and Food and Health. Reyed has also been involved in various bioinformatics courses and workshops. These include the NCBI Virtual Workshop: An Introduction to NCBI Cloud Computing for Biologists in November 2021, and the Data Science Summer School of 2022 held at Hertie School Data Science Lab in Berlin, Germany. He attended the Virtual VeuPathDB

Bioinformatics Workshop in December 2021, and the Ensembl Workshop organized by Erasmus Postgraduate School of Molecular Medicine, Rotterdam, in April 2021. He further participated in the 2nd Ensembl Workshop (Virtual Ensembl REST API Course) from June 15-17, 2021, held by the European Bioinformatics Institute (EMBL-EBI) in the UK. In December 2020, Reyed contributed to the 3rd GED-PerMed Project: Big Data Meets Precision Medicine virtual workshop, organized by the DAAD German Egyptian Progress Partnership Program and Nile University, Egypt.

**Maryam Saadat** holds a master's degree in Anatomical Sciences from Semnan Medical University. She has a passion for neuroscience and neurological diseases, focusing her research on the cellular and molecular mechanisms underlying these diseases and exploring potential cures.

**Golnaz Shafiei** is an anatomist with a strong academic background. She earned her Ph.D. in Anatomical Sciences from Kerman University of Medical Sciences in Feb 2024, where she developed a solid foundation in anatomy and related fields. Following her passion for academia, she served as an assistant professor at Kashan University of Medical Sciences, where she closely engaged with students and honed her teaching skills. Currently, Golnaz,Ph.D., is actively teaching anatomy and conducting diverse research in the fields of reproductive science and neuroscience. She is dedicated to expanding her research horizons and aims to contribute to the advancement of science and education in her areas of expertise. Her research primarily focuses on unraveling the cellular and molecular mechanisms associated with neuropsychological diseases, with a particular emphasis on neuro-inflammatory signaling pathways in critical brain regions.

**Fatemeh Sheikhbahaei**, Ph.D., is an atomist. She studied in Kerman University and now she is an assistant professor at Keman University of Medical Science. Her research primarily focuses on unraveling the cellular and molecular mechanism of neuropsychological diseases.

**Şebnem Yıldırımcan Kadıçeşme** received her PhD from the Department of Medical Biology at Dokuz Eylul University Faculty of Medicine. In 2015, she started working at the Department of Medical Biology at Kafkas University Faculty of Medicine. He has studies particularly on cancer molecular biology, pharmacogenetics, cytogenetics, and autophagy. Current advances in cancer stem cells (2024), targeted therapies in non-small cell lung cancer (2023), pharmacogenetics and clinical applications (2023), and the role of autophagy in cancer (2024) are among the book chapters available.

# Index

## Symbols

3D Printing 101, 114, 115

## A

Alzheimer's Disease 12, 15, 16, 17, 21, 23, 27, 35, 49, 51, 54, 55, 57, 58, 60, 69, 70, 71, 72, 73, 74, 75, 82, 83, 85, 86, 87, 91, 94, 96, 100, 104, 110, 114, 115, 116, 120, 123, 129, 158, 161, 165, 188, 216
AMD 199, 200, 202, 203, 204, 205, 206, 208, 209, 210, 211, 212, 213, 214, 217, 221
Amyotrophic Lateral Sclerosis 13, 17, 20, 34, 37, 56, 60, 119
Antidepressant Agents 153
Antioxidants 22, 58, 63, 92, 97, 113, 144, 154, 190, 222, 244, 259, 271
Apoptosis 40, 43, 44, 53, 77, 78, 127, 128, 129, 130, 141, 142, 143, 152, 155, 156, 161, 164, 165, 168, 169, 171, 172, 173, 174, 175, 176, 177, 178, 179, 180, 181, 184, 185, 186, 187, 189, 190, 192, 193, 194, 195, 196, 197, 200, 205, 209, 211, 213, 215, 219, 233, 242, 271
Autophagy 1, 2, 5, 6, 7, 8, 9, 10, 11, 12, 13, 14, 15, 16, 17, 18, 19, 20, 21, 22, 23, 24, 25, 27, 28, 29, 30, 32, 33, 34, 35, 36, 37, 38, 39, 40, 41, 42, 43, 44, 45, 46, 47, 48, 49, 50, 51, 52, 53, 54, 55, 56, 57, 58, 59, 61, 62, 63, 64, 65, 66, 67, 68, 69, 70, 71, 72, 73, 74, 75, 76, 78, 79, 80, 81, 82, 83, 84, 85, 86, 87, 88, 89, 90, 91, 92, 93, 94, 95, 96, 97, 98, 99, 100, 101, 102, 103, 104, 105, 106, 110, 111, 112, 113, 114, 116, 118, 119, 121, 122, 123, 125, 126, 127, 128, 129, 130, 131, 132, 133, 134, 135, 136, 137, 138, 139, 140, 141, 142, 143, 144, 145, 146, 147, 148, 149, 150, 151, 152, 153, 154, 155, 156, 157, 158, 159, 160, 161, 162, 163, 164, 165, 166, 167, 168, 169, 170, 171, 172, 173, 174, 175, 176, 177, 178, 179, 180, 181, 182, 183, 184, 185, 186, 187, 188, 189, 190, 191, 192, 193, 194, 195, 196, 197, 199, 200, 201, 202, 203, 204, 205, 206, 207, 208, 209, 210, 211, 212, 213, 214, 215, 216, 217, 218, 219, 220, 221, 222, 223, 224, 225, 226, 227, 263, 269, 270, 271, 272, 273, 274, 275, 277, 278, 279
Autophagy Inducers 40, 48, 74

## B

Brain Organoids 111, 116

## D

Depression 13, 22, 39, 46, 61, 65, 70, 82, 86, 87, 91, 97, 102, 103, 104, 111, 112, 128, 129, 130, 134, 135, 136, 137, 138, 139, 140, 141, 142, 143, 144, 145, 146, 147, 148, 149, 150, 151, 152, 153, 154, 155, 156, 157, 159, 160, 161, 192, 229, 230, 231, 233, 234, 235, 236, 237, 238, 239, 240, 241, 242, 244, 245, 246, 247, 248, 249, 250, 252, 253, 254, 255, 256, 257, 258, 259, 260
Dysfunction 13, 16, 29, 34, 37, 38, 40, 42, 43, 50, 53, 58, 61, 62, 65, 66, 68, 70, 71, 73, 74, 76, 77, 78, 79, 80, 81, 83, 85, 103, 105, 106, 123, 126, 130, 153, 159, 161, 164, 168, 169, 171, 178, 181, 183, 188, 189, 200, 201, 203, 204, 205, 206, 208, 211, 213, 221, 230, 231, 232, 233, 235, 238, 239, 240, 245, 247, 256, 259, 265

## E

Epigenetic Regulation 23, 81, 84, 113

277, 278
Neurodegeneration 21, 22, 23, 24, 32, 33, 34, 35, 37, 40, 50, 51, 53, 55, 57, 68, 69, 70, 71, 93, 94, 98, 115, 116, 120, 128, 158, 168, 169, 171, 174, 184, 185, 186, 187, 201, 205, 210, 211, 216, 237, 247, 251
Neurodegenerative Diseases 1, 2, 3, 11, 14, 16, 21, 22, 33, 34, 35, 38, 40, 43, 45, 47, 49, 50, 51, 52, 56, 59, 71, 72, 82, 88, 92, 94, 95, 96, 100, 104, 105, 107, 108, 114, 115, 117, 118, 119, 122, 149, 164, 174, 201, 224
Neuroinflammation 20, 49, 61, 62, 63, 64, 65, 68, 69, 70, 71, 72, 73, 74, 76, 77, 78, 80, 83, 86, 87, 88, 89, 90, 91, 92, 93, 94, 95, 96, 97, 98, 99, 100, 101, 102, 103, 104, 105, 106, 107, 108, 111, 112, 117, 118, 119, 123, 131, 136, 138, 139, 142, 148, 149, 150, 153, 159, 162, 170, 174, 177, 178, 181, 188, 189, 190, 191, 195, 229, 230, 235, 236, 237, 238, 239, 240, 241, 242, 247, 248, 249, 250, 251, 253, 254, 255, 256, 257, 259, 260
Neuroprotection 56, 86, 87, 89, 91, 128, 129, 140, 157, 159, 164, 165, 169, 171, 172, 173, 174, 178, 181, 183, 187, 192, 193, 207
Neuropsychiatric Diseases 44, 47, 64, 155


# G


Glaucoma 199, 200, 203, 204, 205, 206, 207, 210, 211, 212, 213, 214, 216, 218, 219, 222, 225
Gut-Brain Axis 64, 92, 106, 121, 241, 259


# H


Huntington's Chorea 13
Huntington's Disease 13, 15, 16, 21, 22, 27, 37, 52, 53, 54, 55


# I


Inflammation 32, 40, 46, 62, 63, 64, 65, 67, 68, 69, 70, 71, 72, 74, 75, 76, 80, 81, 84, 85, 86, 89, 90, 91, 92, 93, 94, 95, 96, 102, 103, 104, 105, 107, 108, 118, 128, 130, 134, 136, 138, 139, 140, 141, 142, 146, 147, 149, 151, 155, 156, 157, 162, 164, 165, 168, 171, 172, 174, 179, 182, 183, 189, 192, 194, 195, 197, 199, 205, 213, 219, 230, 231, 232, 233, 235, 236, 238, 239, 240, 241, 242, 243, 244, 245, 247, 249, 250, 251, 252, 256, 257, 259, 261


# M


Macroautophagy 2, 3, 4, 5, 8, 24, 28, 29, 30, 31, 42, 43, 57, 59, 154, 165, 184, 200, 203, 204, 209, 220, 222, 226
Major Depressive Disorder 49, 73, 87, 89, 106, 125, 126, 142, 143, 145, 148, 154, 155, 236, 238, 246
Medical Education 270, 271, 273
Medical Literacy 263, 265, 266, 270, 273
Misfolded proteins 1, 31, 33, 35, 40, 43, 70, 71, 73, 94, 96, 102, 199, 201, 202, 203, 205, 208


# N


Nanotechnology 263, 264, 265, 266, 267, 268, 269, 270, 273, 274, 275, 276,


# P


Parkinson's Disease 12, 15, 18, 20, 22, 24, 36, 49, 50, 51, 53, 55, 57, 58, 69, 70, 71, 72, 73, 87, 91, 94, 96, 100, 107, 109, 111, 112, 113, 118, 119, 120, 139, 165, 224
PCOS 229, 230, 231, 232, 233, 234, 235, 237, 239, 240, 241, 242, 243, 244, 245, 246, 248, 249, 254, 255, 257, 258, 259
Polycystic Ovary Syndrome 229, 230, 245, 246, 247, 248, 249, 250, 251, 252, 253, 254, 255, 256, 257, 258, 259
Precision Medicine 122
Probiotics 64, 75, 92, 93, 105, 109, 113, 116, 120, 244, 246, 254

Protein Aggregates 1, 12, 14, 28, 29, 34, 37, 38, 42, 45, 66, 70, 89, 94, 99, 104, 171, 201, 207, 212, 222

## R

Retinal Degeneration 201, 208, 209, 211, 213, 220, 227
ROS 63, 69, 72, 73, 77, 78, 79, 80, 113, 120, 129, 133, 135, 137, 141, 155, 169, 171, 173, 176, 178, 179, 182, 183, 205, 224, 240, 241, 259

## S

Spinal Cord Injury 163, 164, 183, 184, 185, 186, 187, 188, 189, 190, 191, 192, 193, 194, 195, 196, 197

## T

Therapeutic Targeting 55, 210
Toxicity 42, 47, 49, 52, 111, 117, 263, 265, 270, 271, 272, 273, 275, 276, 277, 279

www.ingramcontent.com/pod-product-compliance
Lightning Source LLC
LaVergne TN
LVHW080013210926
848481LV00011B/3011
* 9 7 9 8 3 3 7 3 4 5 0 0 0 *